Willie,

    I hope this book brings together many of the interests central to your first 50 and your next 50 years.

Margaret.
9/10/98.

The Ascent of Science

The Ascent
of Science

Brian L. Silver

A Solomon Press Book

New York   Oxford
Oxford University Press
1998

Oxford University Press

Oxford   New York
Athens   Auckland   Bangkok   Bogotá   Bombay
Buenos Aires   Calcutta   Cape Town   Dar es Salaam   Delhi
Florence   Hong Kong   Istanbul   Karachi
Kuala Lumpur   Madras   Madrid   Melbourne
Mexico City   Nairobi   Paris   Singapore
Taipei   Tokyo   Toronto   Warsaw

and associated companies in

Berlin   Ibadan

Copyright © 1998 by Oxford University Press

Published by Oxford University Press, Inc.
198 Madison Avenue, New York, New York 10016

Oxford is a registered trademark of Oxford University Press

Library of Congress Cataloging-in-Publication Data
Silver, Brian L.
The ascent of science / Brian L. Silver.
p. cm. "A Solomon Press book."
Includes bibliographical references and index.
ISBN 0-19-511699-2
1. Science—History. 2. Science—Philosophy.
3. Thought and thinking—History. I. Title.
Q125.S5425 1998 303.48'3—dc21   97—15430

9 8 7 6 5 4 3 2 1
Printed in the United States of America
on acid-free paper

This book is for Sharon,
who knows the reasons that reason cannot know

Contents

Acknowledgments

It is a pleasure to acknowledge the help given me by the staff of the Oxford University Press in New York, especially the encouragement, friendliness, and constructive professionalism of Kirk Jensen and Helen Mules.

Please read before opening bottle.

The nineteenth-century Belgian mathematician and statistician Adolphe Quételet, in his *Treatise on Man and the Development of His Faculties, An Essay on Social Physics,* defined "l'homme moyen." I have never met an average man, but I frequently meet his far more charismatic brother: *l'homme moyen sensuel.* The French is usually puritanically translated into English as "the average man," thus, with Anglo-Saxon squeamishness, sidestepping the Gallic "sensual." We shall leave him with his hormones and call him "HMS." HMS remembers little or nothing of the math and science that he learned at school, he is suspicious of jargon, he is more streetwise than the average scientist, he is worried about the future of this planet, he may like a glass of single-malt whisky to finish off the day. Above all, he is curious. In about 50 percent of cases he is in fact she. It is primarily to such readers, to HMS, that this book is addressed.

Preface

That which we know is a little thing; that which we do not know is immense.

—Pierre-Simon de Laplace

Science, man's greatest intellectual adventure, has rocked his faith and engendered dreams of a material Utopia. At its most abstract, science shades into philosophy; at its most practical, it cures disease. It has eased our lives and threatened our existence. It aspires, but in some very basic ways fails, to understand the ant and the Creation, the infinitesimal atom and the mind-bludgeoning immensity of the cosmos. It has laid its hand on the shoulders of poets and politicians, philosophers and charlatans. Its beauty is often apparent only to the initiated, its perils are generally misunderstood, its importance has been both over- and underestimated, and its fallibility, and that of those who create it, is often glossed over or malevolently exaggerated.

The attempt to explain the physical universe has been characterized by perpetual conflict. Established theories have continually been modified or violently overthrown, and as in the history of art and music, innovations tend to be ridiculed only to become, in time, the new dogma. The struggle between old and new has rarely been dignified. Scientists come in many colors, of which the green of jealousy and the purple of rage are fashionable shades. The essence of scientific history has been conflict.

This book presents science as a series of ideas that changed the course not only of science itself but often of whole areas of human thought. Science, of course, has its practical benefits, but they will not be our primary concern. This is not a book about nonstick frying pans. We will be looking at ideas—admiring their beauty, occasionally standing awestruck before the towers of imagination, but always being prepared to doubt; always being aware not only of the ingenuity but also of the deep limitations, and the repeatedly demonstrable inertia, of the human mind.

Science, by its nature, is changeable. There is always some scientist, somewhere, who is disproving an explanation that another scientist has proposed. Usually these shifts of interpretation leave the fabric of society undisturbed. Occasionally, however, real revolutions tear down part of our system of established beliefs. Thus, in the seventeenth century, science presented us with a mechanical universe, a giant inexorable clock. Three centuries later, physics has cut some of the levers that bind cause to effect and has led us into a shadowy maze where we affect the universe by the act of observing it and are ignorant of the true meaning of our most basic concepts.

Some see the fragility of scientific theory as an indication of a basic inability of science to explain the universe. But scientific change is almost always accompanied by an increase in our ability to rationalize and predict the course of nature. Newton could explain far more than Aristotle, Einstein far more than Newton. Science frequently stumbles, but it gets up and carries on. The road is long. As we come to the end of this century, it is prudent to recall that at the end of the nineteenth century the general opinion amongst physicists was that nothing of any great importance remained to be done in physics. And then came radioactivity, X-rays, the discovery of the electron and the nucleus, a couple of hundred new fundamental particles, quantum mechanics and relativity, antimatter, dark matter, black holes, chaos, the Big Bang, and so on. Biology has been no less prolific. At the end of the twentieth century there are again voices proclaiming the imminent arrival of a theory of everything, a complete explanation of the Creation and workings of the cosmos. Maybe.

Science is not a harmless intellectual pastime. In the last two centuries we have moved from being simply observers of nature to being, in a modest but growing way, its controller. Concomitantly, we have occasionally disturbed the balance of nature in ways that we did not always understand. Science has to be watched. The layman can no longer afford to stand to one side, ignorant of the meaning of advances that will determine the kind of world that his children will inhabit—and the kind of children that he will have. Science has become part of the human race's way of conceiving of and manipulating its future. The manipulation of the future is not a question to be left to philosophers. The answers can affect the national budget, the health of your next child, and the long-term prospects for life on this planet.

This book is a report of the scientific campaign up to now. It is not a history of science but rather an account of the major battles, the frequently eccentric generals, and the ways in which science has deeply influenced man's picture of the world *and of himself*. It is not a final summing-up. We know that we are far from a real understanding of nature. We press on. Michelangelo's divine discontent gives us no rest. And although the history of science may be a trail littered with broken theories and discarded concepts, science is also a triumph of reason, luck, and above all imagination. There are few more successful, exciting, or strange journeys.

Introduction

This is a material world and I am a material girl.

—Madonna

The Argentine writer Jorge Luis Borges tells a tale of a man cursed not only with the ability to remember every leaf, every wave, every pattern of shadows that he has ever seen but also with the inability to realize that different leaves are united by the concept "leaf" or that the myriad shifting shapes of the waves are all examples of the one idea of "wave." This extreme fragmentation of experience is the antithesis of man's inherent need to find unifying themes in nature. The great scientists have often been those who saw such themes where others failed, who saw *wave* under the waves.

But if genius is, by definition, the possession of a tiny elite, how is the average man to satisfactorily comprehend that which took genius to reveal? In some notorious cases, such as quantum mechanics, it is a fact that many scientists can only form a blurred image of what the expert sees more or less clearly, and it is hardly encouraging that those experts are fighting to clarify some of the most basic concepts of science. Nevertheless, the vast majority of science is intelligible to the vast majority of high school graduates. The truth is that many of the great discoveries were made by someone who happened to have a torch to illuminate the darkness. The torch was the miracle: what was revealed was often simple—beautifully so. However, it also remains true that at the borders of our knowledge we run into real complexity, into questions that challenge our ability to define the nature of reality. Thus the very foundations of quantum mechanics, the central theory that we use to describe the physical universe, are the subject of deep controversy. We don't really understand quantum mechanics. We are like drivers who have learned to drive a Rolls-Royce without being told much about the workings of the internal combustion engine. The car works wonderfully, but we're not sure why.

It gets worse: we don't really know what "matter" is. A child thinks of matter as being something like clay. A particle physicist might tell you that matter is a "bunching up of a field," but if we are honest we have to admit that "matter" is little more than one of the concepts that allows us to deal with what our senses report of what we call the external world. As the seventeenth-century French scientist and philosopher Gassendi wrote: "Man establishes a system of signs, of names, which permit him to identify things perceived and to communicate with other men."

So before we set out, we should face the facts. Our capacity to predict the behavior of the physical world is often amazing; the theories which allow us to account for the behavior of matter, in terms of a few basic concepts, rank with the highest achievements of the human imagination. And we can be thankful that, in tackling practical problems, the theories of science work very well for a huge range of questions. But when it comes to fundamentals, we see through a glass darkly;

there is more faith involved in science than many scientists would be prepared to admit.

The Objectives

One purpose of this book is to set science in its social perspective. Scientific *ideas* have affected the relationship of man to society, his ideas of God, and his image of himself. Science has influenced the way people write poetry and the way they paint pictures. In the hands of bigots, it has provided a theoretical justification for the sterilization of some human beings and the enslavement of others. Science, *as a source of ideas,* is a major character in the human drama.

The other aim of this book is to explain for the layman the basic meaning of the great breakthroughs in science and to point out the chain of imagination that links the Creation with the movements of the planets and the formation of the chemical elements, that binds the evolution of man to the end of the universe. As far as the purely scientific content of the book is concerned, we can classify everything under one of three headings: matter, change, or field.

Matter

I am writing these words on a flower-fringed balcony overlooking the Mediterranean. The sun blazes. My snow-white cat, "Schwarz," stretches in the Sun. This morning I revel in being a material being in a material world.

My infinitesimal slice of the universe contains an apparent infinity of shapes, colors, textures, sounds, smells, and substances. Matter has endless expressions. Matter appears to be the source of everything that we know.

We are material beings. Only a puritan like Bernard Shaw could ask his audience to believe in a world populated by disembodied vortices. Shaw's projected scenario doesn't explain how our airy descendants will be capable of thinking. There is no evidence for thought existing independently of matter, living matter. If we want to begin to comprehend the universe, our starting point must be an understanding of matter—its forms, its organization, its movement, and its transformations.

As a counterweight to my paean of praise for matter, it is only fair to record that there have been extreme idealists who have questioned whether the material world of beer and hot dogs has any objective reality whatsoever. The eighteenth-century Irish philosopher Bishop George Berkeley, in his *Treatise Concerning the Principles of Human Knowledge,* suggests that the physical universe is nothing but a constant perception in God's mind. If you are a follower of Berkeley, I am willing to respect your belief, but I ask that in return you be prepared to admit that God's perception is infinitely worth studying—which is one of the objects of this book.

We will come to the conclusion that Beethoven and Scotland are constructed from a menagerie of subatomic particles stranger than the inhabitants of any medieval bestiary. We will meet forces that are pathetically weak and others that are hideously powerful. And we will meet the weak and the powerful human beings who have constructed, and quarreled over, these concepts.

Change

Matter is the flesh of the universe; chemical and nuclear change is its soul. As you read these words, you are kept alive by the microscopic chemical changes that characterize the living cell. Molecular change may not be the meaning of life, but life would cease without it. It would certainly stop without nuclear change. It is the apocalyptic fusion of nuclei in the Sun that releases the staggering amounts of energy that maintain life on this planet. Without the Sun's radiation the Earth would cool, the winds would die, the Earth would lay silent, barren, and dark. Now turn off the stars.

Without an appreciation of material change, we cannot understand the nature of the Creation, of life, of evolution, of our environment and the threats to that environment. And we will find that some changes appear to define the direction of time itself.

Fields

There appears to be more than matter in the universe. An apple falls to Earth, a nail is dragged toward a magnet. What we observe is motion. If you hold the apple or grasp the nail and attempt to stop that motion, then you can feel the action of what we call a force. Yet there is no *apparent* physical connection, no stretched spring, between Earth and the apple, or between the magnet and the nail.

We use a convention to describe what is happening. We say that there is a gravitational (or magnetic) *force* acting on the apple (or the nail), because the apple is in a gravitational *field,* and the nail is in a magnetic *field.* The word *field* suggests that there are invisible entities surrounding the Earth and the magnet, entities created by the Earth and the magnet but having a life of their own.

Fields have become an essential part of our description of the universe; one of the ways in which we talk of light is to call it an electromagnetic field. Some physicists talk of matter as being the "manifestation" of a field. Part of this book is about fields.

A Note on the Observer

There is someone who is living my life. And I know nothing about him.

—Luigi Pirandello, diary

Say to yourself: "I am reading this book." Who is "I"? Is it that part of you which you imagine to be outside the material universe? Not matter, not field? An ineffable something localized somewhere in your head, and quietly chatting with you? What is consciousness?

A tremendous amount of research has enabled us to map the microscopic structure of the brain. No one has ever found an "I" in there. If "I" exist, "I" have been extraordinarily elusive. Descartes, in some of his writings, assumed that the soul was not located in space at all, but elsewhere he localized it in the pineal gland, which sits in the brain. It fails to appear under the microscope. Our failure is understandable if "I" is a *function* of the brain, not an occupant. We would laugh at someone who took his car engine to pieces in order to find its "power."

So how do "I" fit into our scheme of things? "I" has, up to now, refused to be categorized as matter, change, or field. Yet "I" am part of the universe, "I" am the observer; without "me" there would be no science. Is there something besides matter, change, and fields? I doubt it. But in saying so I am relying on faith, not reason. This book has nothing to say about consciousness. Read other books, go to conferences. You will find that there are many theories. This is always a bad sign.

i

In which, among other things, we will look at the fallibility of science and of scientists, and ask if anyone can be trusted.

1

Newton Gets It Completely Wrong

Matter, in the everyday world, comes in three forms: solid, liquid, and gas. Gas is the simplest form of matter, the form that we understand best. We live in a gas, in fact, a mixture of gases—air. The way that scientists see gases is not the way that our eyes see them. In fact our eyes don't see air at all, but it is typical of science that it starts off with the somewhat thin evidence of the senses and then proceeds, by experiment and hypothesis, to build a model that explains the properties of whatever system we are interested in. One of the most successful models is that constructed to explain the behavior of gases. The key to the understanding of gases is to believe in the microscopic motion of the bits of matter of which a gas is composed—the molecules. Newton got this wrong, and we can profit from seeing both why he thought that he was right and why we think that he wasn't.

I am going to ask the reader to accept certain facts on faith. This is not a disreputable cop-out. We all do it every day. Scientists certainly do it—we have no choice. I am a physical chemist. The only way that I can intelligently apprehend modern genetics, for example, is by taking certain things for granted and using my reason to build on them. This blind faith in what I read in reviews or selected research papers is not to be likened to an unquestioning belief in flying saucers. No one has demonstrated the existence of flying saucers in a manner that would satisfy either a sloppy scientist or a critical layman. In this book I will, unless specifically stated, use only facts that have been accepted as such by the international scientific community. Where a fact is in reality only a hypothesis I shall say so, and I will tell you when I think that I'm on shaky ground. It is quite possible that some of my so-called facts will sooner or later turn out to be fiction—I am not the pope. For the moment all I ask is unqualified acceptance. This sounds like the spiel of a secondhand car salesman, but if you don't believe something in this book, you can go to standard textbooks. If you question their validity, you can take a specialized course of lectures. If doubt remains, follow your doubt; you may be the next Galileo.

Molecules

In this chapter, and elsewhere in this book, the word *molecule* will make many appearances. What is a molecule? For the moment think of it this way: Take a glass of water and divide its contents into four cups; you will still call the substance in each cup "water." Now take one of the four cups and divide the water in it into a thousand droplets. Each drop is still recognizably water. Take one of these droplets and divide it into a million droplets, each one visible only under a microscope. All tests, physical or chemical, will verify that each drop is a sample of water. Can we

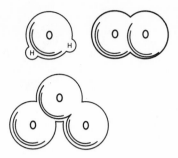

Figure 1.1. The approximate shapes of the molecules of water (H_2O), oxygen (O_2), and ozone (O_3).

go on with this game indefinitely? There have been those in the past who thought so, who believed that matter was a continuum, an infinitely divisible Jell-o. Today we know that the game stops somewhere. There comes a stage when the pieces of water, if split, divide into fragments that are very definitely *not* water. It's a little like splitting up a crowd of demonstrators; there comes a time when you get down to individuals who, if split up, are no longer people.

The smallest entity that still has the right to be called water is termed a *molecule* of water. Look at Figure 1.1. Most of you will recall that water is also known as H_2O, which is just a shorthand way of saying that each molecule of water consists of two hydrogen atoms and one oxygen atom. (The word *atom* also needs more explanation, and we will return to it later on.) The smallest part of a sample of oxygen gas that is still recognizably the oxygen that we breathe is the oxygen molecule, which consists of two oxygen atoms linked together, and which is therefore written as O_2. If you split an oxygen molecule in half, you will get two oxygen atoms whose physical, chemical, and biological properties are utterly different from those of the oxygen molecule. Ozone has the chemical formula O_3. Each ozone molecule has three oxygen atoms joined together (Figure 1.1).

It is not surprising that we can chop up a sample of water apparently almost endlessly and still have water. Molecules tend to be very small entities. A teaspoonful of water contains about 200,000,000,000,000,000,000,000 water molecules.[1] Textbooks like to suggest hypothetical counting games as an aid to the visualization of very large numbers. In accordance with this tradition, I can tell you that if the whole population of Earth set out to count the molecules in a teaspoon of water, each person counting at the rate of one molecule per second, it would take over a million years. But let's get back to gases.

The Invisible Sea

We live at the bottom of an invisible sea. Nature, working through gigantic astrophysical, geological, and biological processes, has generated for us a molecular atmosphere of gas within which we live and without which we cannot survive. Those few of us, such as divers or astronauts, who leave the gaseous sea have to take some of it with us, or we die. We need gases—and not just the air.

Man uses gases to weld steel, to fill laser tubes and neon lights, to anesthetize patients, to give sparkle to soft drinks, beer, and champagne, to fly airships, float

---

[1] It's easier to write this huge number as 2×10^{23}, where 10^{23} ("ten to the twenty-three") means 1 with 23 zeros after it, or 10 multiplied by itself 23 times. A thousand is 10^3; a million is 10^6.

wrecks, cook hamburgers, inflate tires, and commit genocide. Deep under the ground, the decomposition of primeval forests by heat and pressure has given us coal and inflammable natural gas, which is an important, if minor, source of energy. In the meantime, we pollute the atmosphere with gases that at best erode buildings and change the color of oil paintings, and at worst poison life and perilously disturb the delicate heat balance of this planet. Gases can be indispensable to man or catastrophically destructive. As Willy Loman might have said, attention must be paid to such substances.

If you walk into a chemical supply company, you can purchase samples of many different gases. Some are familiar enough: oxygen, nitrogen, carbon dioxide, carbon monoxide, helium, chlorine. These are substances that appear weekly in the daily press. The chemical properties of these gases are so different as to suggest that there is no important basic property shared by all gases. Consider two well-known gases.

Helium exists in the stars, in a few underground cavities, and in very small quantities in the air. It was first discovered in the Sun, by examining the nature of the Sun's light—hence its name, from the Greek for Sun, *Helios*. Helium is a complete wimp of a gas, almost incapable of interacting with any other substance, which is why it avoided detection for so long. Chlorine, on the other hand, attacks almost everything and is reminiscent of the mythical chemist who discovered a universal solvent but failed to find a bottle capable of holding it. It kills painfully, as British and French soldiers found out in 1917. It is pale green, one of the few visible gases.

In general, chemical experience suggests that each gas is unique, which is true, and has very little, if anything, in common with most other gases, which is not true. *That which is common to all gases is the way in which their molecules move.* That is the real subject of this chapter, and it has implications that reach far beyond the nature of gases.

The Violent Crowd

Look at Figure 1.2. It shows an imaginary cubical box constructed in the air of the room in which you are sitting. The way that the molecules in it behave has been the subject of controversy for at least a century. The secret of time may reside in this box.

Note how small the box is. The letters *nm* stand for nanometer, which is one thousand millionth part of a meter, which can be written 10^{-9} meter. If everyone in China had one of these boxes and, by order of the Party, lay the boxes end to end, touching each other, the total length of the line would be only about five meters.

Descartes thought that the universe was continuous, like jelly, but he didn't do

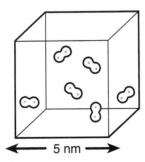

← 5 nm →   Figure 1.2. Molecules and empty space in the air around you.

any experiments to back up his guess. Look at the box. The six objects represent six molecules drawn to scale. Oxygen and nitrogen molecules look very much the same, so we don't differentiate between them in our picture. It is clear that *air is mainly empty space*. The average distance between molecules in the air around you is about ten times their largest dimension. It is as though the average distance between canaries in a cage was about 1 meter.

So that you don't forget the kind of numbers we are dealing with, bear in mind that a teaspoonful of the air in your house contains somewhat more than 10^{20} molecules (1 with twenty zeros after it). This is about 2000 times less than the number of water molecules in a teaspoonful of water. Water molecules in liquid water *are* crowded together; the living conditions in liquids are slumlike compared with those in gases.

Two vital facts are not revealed by the drawing:

- Molecules in a gas are in constant motion.
- The average speed of a molecule in a gas increases with temperature.

Be very clear what this motion is and is not. We are *not* talking about drafts, winds, hurricanes, or convection currents. Even in a tightly closed room in the absence of breathing creatures, moving machinery, or sources of light or heat, the gas molecules are always moving. In the cylinder of oxygen standing in the hospital storeroom, the oxygen molecules are endlessly in motion. And they are moving fast. Very fast.

An Aside: Why Molecular Motion?

There is no such thing as a motionless atom or molecule, and there is hardly an aspect of the nonliving or living physical world which can be completely understood without taking into account molecular motion. The concept of temperature and the nature of heat both have a deep connection with molecular motion, and molecular motion has a strange connection with the arrow of time and the death of the universe.

Back to the Madding Crowd

Professors have a weakness for analogies. So here's one: A gas, *any gas*, is similar to a crowd of flies. The analogy is dangerous, but we can learn from the dangers. First of all, flies can see; they don't normally bump into each other. Molecules are "blind"; in a gas they are continually blundering into each other. Every collision changes the speed and direction of both molecules involved, so that a molecule in a gas resembles a flying dodgem car, continually getting jolted. Another difference between flies and molecules is that the molecules in our box are presumed to fly in straight lines unless they hit something. Flies practice their aeronautical skills. An improved fly analogy is a crowd of straight-flying, blind, deaf flies, but this is still misleading. Flies get tired. They often relax, and in the end they die and lie on the floor with their legs up. Molecules don't do this; the molecules in an oxygen cylinder never stop moving—until the end of time, as they say at MGM. Again improving our analogy, we liken the molecules in a gas to a collection of straight-flying, blind, deaf, radarless, tireless, immortal flies. We're getting there, but the problem, as we will soon see, is that flies have a sense of smell and molecules don't. First, however, let's look at the speeds of molecules.

How fast do molecules move in a gas? That depends on two things, the *mass* of the molecule (less correctly, its *weight*) and the temperature of the gas. The question needs sharpening. Since any given molecule is continually changing direction and speed due to collisions, it is more informative to speak not of its "speed" but of its average speed. If you drive from home to office, your speed is hardly constant; it drops to zero at red lights and rises a little above the speed limit on the freeway. However, you can work out your average speed from only two observations: just divide the distance from home to office by the time for the complete journey. Average speeds for molecules in a gas can be both measured experimentally and calculated theoretically. Here are some approximate average speeds, in miles per hour, for molecules in some still gases at room temperature:

Hydrogen in a steel cylinder	3800
Helium in an airship	2800
Oxygen, on a dreamy day in St. Tropez	1030
Carbon dioxide in a soda-syphon refill	830

The masses of these molecules increase along the series: hydrogen, helium, oxygen, carbon dioxide. Here we see the mass dependence of the average speed: at the same temperature a heavier molecule travels slower *on the average* than a lighter molecule. The average speed of the molecules in a sample of gas is a constant quantity, the same today as it was 500 years ago, provided the temperature is the same. If the temperature goes up, the average speed goes up. The average speed of an oxygen molecule rises by about 7% in going from $0°C$ ($32°F$) to $40°C$ ($104°F$). Not a stupendous increase, but it has practical implications. We will see that it is because particles get faster as they get hotter that the Sun is able to radiate energy.

You might care to compare molecular speeds in gases to the speeds of more familiar objects, again in miles per hour:

Concorde in full flight	1450
Lamborghini, cruising	180
Ben Johnson, 1988 Olympics (assisted)	22.8
Carl Lewis, 1988 Olympics (unassisted)	22.6

Now consider the following problem, which is based on the fact that although solutions of alcohol, like beer, are liquid, the smell of alcohol can be easily detected. This is because in all liquids, molecules escape into the surrounding air. If they didn't, liquids would not evaporate. At room temperature the calculated average speed of an alcohol molecule is about 800 mph. Here's the problem:

A diligent, but worldly-wise, professor of physics returns from a late-night "seminar" and gingerly opens his front door. Why is it that his whisky-laden breath is not immediately detected by his loving, and equally worldly-wise, wife, who is waiting for him, 10 yards away, at the foot of the marble stairs? After all, molecules traveling at an average speed of 800 mph should be able to cover 10 yards in about one-fortieth of a second. In fact, if the air in the room were still, it would take months for the smell of alcohol to be detectable at that distance. Why? Why is the professor's secret safe until he plants a guilty kiss on his wife's skeptical cheek?

The answer lies in the astronomical number of collisions that a molecule makes as it blunders about in a gas. If you could follow the path of a single molecule of, say, oxygen in the air, you would find that in round numbers, it collided with other molecules about 6000 million times in one second. Molecules have a hectic social life, even by jet-set standards. An oxygen molecule in air changes direction at an incredible rate. It would take a very long time to fight its way across a room. The fact that you can quickly detect the smell of cooking all over the kitchen is due to the help given to the molecules by drafts or convection currents which move whole chunks of air rapidly from place to place. Similarly, if you put a drop of ink in a bath of motionless water, the color will spread out because of molecular motion, but it will spread very slowly. Drop the ink in a river, and it will travel downstream at a rate determined by the bulk flow of the water.

The difficulty that a molecule has in fighting its way across the room is analogous (here we go again) to a passenger trying to get to his platform during rush hour at Grand Central Station and being jostled by other passengers going in different directions. Again we can learn from the imperfect analogy. The passenger *knows* where he wants to go and directs his feet in that direction, when allowed to between pushes. In contrast, molecules in a static sample of gas have no predetermined destination: they are not heading anywhere, they just move. After every collision they travel in a direction determined by the details of the collision, like billiard balls, but in three dimensions, not two. Molecules have no will of their own. That is why the fly in our previous analogy had to have no sense of smell, since otherwise, even if blind and deaf, it might direct its flight toward food. The essence of molecular motion in gases is that it is *completely* determined by the last collision. We say that the motion is *random*.[2] The flight of the fly is not random.

The professor's wife knows that the alcohol molecules leaving her husband's lips, once they have slowed down from the initial push given by his lungs, travel between two well-separated points at less than a snail's pace. Calculation shows that in six months the average, point-to-point distance traveled between starting point and final destination, by an alcohol molecule in still air, is about 10 yards. The total zigzag path length traveled is about 35 million miles!

War would be farcical if projectiles traveled through the air in the way that molecules do. Billy the Kid would have looked even stupider than he was if his bullets had performed a random dance after leaving his gun. However, a bullet weighs about 10^{23} times as much as an oxygen or nitrogen molecule and is far too heavy to be diverted by such puny objects. Linebackers aren't usually stopped in their tracks by gnats. They are, however, slowed down by air resistance, which is basically the cumulative effect a myriad of collisions between a moving macroscopic object and the molecules that it encounters. The same effect works if you try sprinting through a cornfield.

---

[2]There is a medieval riddle that illustrates the dangers of assuming that a process is genuinely random. The problem was published in a book by Claude-Gaspar Bachet de Meziriac (1581–1638). A boat, crewed by 15 Turks and 15 Christians, runs into a storm. The boat can be prevented from sinking only if half the crew are thrown into the sea. A Christian suggests that they form a circle and throw in every ninth man until there are 15 left. This is done, and, lo and behold, all the Turks are thrown in and all the Christians survive. The secret is that the organizer of the fiendish scheme knew his Renaissance puzzle books. He arranged the circle like this: CCCCTTTTTCCTCCCTCTTCCTTTCTTCCT and started counting from the first Christian. The Turks had assumed that the process was random. In Turkey they tell it differently.

A Pause for Doubt

We have converted the air from a structureless, thin, invisible soup into a horde of frenetic particles—molecules—forever on the move, battering into each other at tremendous speeds and incredible frequency. The average speed of the molecules in the sample remains *constant* as long as the temperature remains constant. What we have found is genuine perpetual motion. Stubborn, deluded inventors are still seen sitting confidently in the corridors of patent offices, carrying the plans of ingenious contraptions that once set in motion are supposed to run forever, in the absence of any subsequent input of energy. They don't run and they won't run; only nature has taken out a patent on perpetual motion. At least that's the story that I've been telling. *But is it true?* I have given you the basic assumptions that go into the scientist's model of a gas. Could there be an entirely different and more correct model?

History shows that even the greatest minds can construct theories that work but that have one minor flaw—they are completely erroneous.

Newton Gets It Wrong

Thomas Willis (1621–1675), a professor of natural philosophy at Oxford, discovered the crippling disease myasthenia gravis.[3] He was an iatrochemist, a physician who believed in the chemical basis of the body's workings, so it was natural for him to be aware of his famous contemporary, the chemist Robert Boyle. One of Willis's brighter students was Robert Hooke, and when Boyle was looking for an assistant, Willis recommended Hooke. Thus it was that in 1655, the twenty-year-old Hooke came to work with the twenty-eight-year-old Boyle.

The conjunction of Boyle and Hooke was heaven-sent. Hooke knew how to construct air pumps of the type that had been invented in 1650 by Otto von Guericke of Magdeburg. Hooke built a pump for Boyle, who, by creating a moderate vacuum in a glass vessel, showed that air was necessary for life (mice died), for combustion (fires went out), and for the passage of sound. (Von Guericke had already shown that light, but not sound, traveled through a vacuum.) In a famous experiment, Boyle trapped a volume of air inside a glass tube and varied the pressure on it. You can do the same by putting your finger over the nozzle of a bicycle pump and pushing the handle in. As the handle enters farther into the barrel of the pump, the volume of the trapped air decreases and you must push harder to keep the pump handle where it is. It is as though you are striving to compress a spring. Boyle's *qualitative* observation—that when the pressure on the air sample increases, the volume decreases—is not very surprising. But Boyle did something that was still a rarity in the seventeenth century; he made *quantitative* measurements. He recorded the pressure and the volume of the air during the experiments. He found that when he doubled the pressure on a sample of air, the volume halved. When the pressure went up three times, the volume went down to a third of the original volume. All his experiments were done at about the same temperature, that of his laboratory. He concluded that for a sample of gas kept at constant temperature, the pressure of the sample multiplied by its volume was unchangeable. As one got bigger, the other got smaller. We write this result as: pressure times volume = a constant, and call it *Boyle's law.*

---

[3]Willis was in charge of the dissection of the corpse of Anne Green, hung for infanticide in 1650. The "corpse" came back to life on the operating table.

Boyle's law, published in 1662, is one of the first examples of a physical law. Any theory about the nature of gases has to lead to Boyle's law; if it doesn't, it must be wrong. (In fact Boyle's law it is not an exact law, but it is not bad under normal circumstances.)[4]

Boyle sought an explanation for the "spring" of the air and proposed that air consisted of "many slender and flexible hairs," like a sheep's fleece, "each of which may indeed, like a little spring be easily bent or rolled up." The problem with this idea is that as a theory it is useless. It *qualitatively* explains Boyle's observations, but you can't derive Boyle's law, or any other property of a gas, from Boyle's theory.

The Curse of the Fat Sumo

What happened next was an excellent example of a completely incorrect theory agreeing perfectly with experimental results. This has happened more than once in the history of science, and the present case contains a warning that applies to any kind of explanation, scientific or not. Thus the theory that sumo wrestling tends to makes you fat is beautifully confirmed by the fact that almost all sumo wrestlers are enormously bulky. In the absence of any other experimental data, the theory looks good. The curse of the fat sumo fell on Newton. He succeeded in elegantly explaining Boyle's law. Incorrectly.

Newton proposed that the particles of the air (we would call them molecules), were *motionless* in space and were held apart by repulsive forces between them, so that any attempt to reduce the volume of a sample of gas was analogous to the compression of springs. He assumed that the repulsive force was inversely proportional to the distance between the particles: double the distance and the force is halved. He showed that, on the basis of this assumption, a collection of *static* particles in a box would behave exactly as Boyle had found. *His model led straight to Boyle's law.* Probably the greatest scientist ever, Newton managed to get the right answer from a model that was wrong in every possible way. Experiment shows that molecules in a gas are not motionless and, contrary to Newton's second assumption, that molecules almost always *attract* each other unless they are actually touching. Nevertheless, if all that we knew about the behavior of gases was Boyle's law, there would be no grounds for challenging Newton's theory. It gives the right answer. Unfortunately, Newton's model does not predict or explain any other property of gases. He had been struck by the curse of the fat sumo. Beware.

We believe that we have a better model than Newton's. It explains far more facts than his does, as we will see in Chapter 16, and the final test of a model is whether it explains the experimental facts. Nevertheless, we again ask: Why should we believe it to be true? Are molecules really moving continually, even in an undisturbed, closed container? Maybe there is a better theory waiting around the corner. Maybe we will also find that new facts about gases will be completely inconsistent with our present ideas.

Science is not carved on tablets of marble. Theories have arisen, have worked, have been challenged, and have been superseded. So why should you believe *anything* that I have written up to now? How skeptical should you be about the pronouncements of scientists?

---

[4]You can see one reason why it can't possibly be true always, since if you compress a gas enough you will force the molecules to touch each other. At this stage you can increase the pressure by a factor of two but the volume will be almost unchanged; it will not be halved, as the law claims.

2

I Believe

Read my lips . . .
—George H. W. Bush

The book that you are holding purports to give a description of the physical world. *Why do you believe what I tell you?* Because I am a professor of physical chemistry and have canvased your vote?

On an overcast autumn afternoon in 1976, I wandered down a nondescript side street off Charing Cross Road in London and strayed into a dingy secondhand bookshop that could have been the film set for a whimsical tale of a kindly old bookseller. There was no apparent order, either in the physical stacking of the books or in their subject grouping. The poor lighting and the autumnal colors of old books combined to give a mistaken impression of dirt. There was an affable-looking grayhaired lady shifting books from one pile to another. Against one wall, surrounded by shelving, stood a glass-fronted mahogany cupboard containing four or five rows of leather-bound books. A half hour after I had come in, I walked out into the dusk, clutching a parcel. I was carrying a block of gold. I had paid 100 pounds for a 1688 edition of *Les Principes de la Philosophie* by René Descartes.

When I reached home I searched through the opening pages like a schoolboy looking for the dirty words in a dictionary. There it was on page 4: Je pense, donc je suis. I think, therefore I am. Starting from the most famous sentence in the history of philosophy, Descartes attempted to build an all-embracing account of the nature of man and the universe.

The question that, above all others, worried Descartes was: Why should I believe anything—*what can I know for a certainty?* We still lack a final answer. It is a question that could be asked by HMS every time a scientist claims that something has been proved—like perpetual molecular motion in gases.

We are going to look at three very different attempts, not to find certainty but at least to avoid error in our ongoing attempt to understand the universe. Before we start, I would point out that this chapter is not a treatise on epistemology, and it is a very long way indeed from a summary of any individual philosopher's lifework. It is a signpost, as is much of this book.

Descartes (1596–1650)

Part of what we claim to know, we derive from the evidence collected by our ears, eyes, nose, and sense of touch, but by far and away our major sources of information are the written, visual, or word-of-mouth accounts of what others have seen, heard, smelled, or touched.

Now, you can't see, hear, or smell the molecules in the air around you, so why should you believe it when I tell you that molecules in a gas are in constant motion, or even that molecules exist? You have never directly experienced a molecule with your senses, nor are you likely to. As late as the beginning of the twentieth century, there was at least one prominent physicist who refused to accept the known experimental facts as being evidence for the existence of atoms. Today there are instruments that apparently produce pictures of atoms. But ask yourself a tough question: Can my physical senses be trusted? In the end, however complex the scientific equipment that we use, the final observer is the human brain. Can it be trusted? Descartes wrote "Pourquoi on peut douter de la verité de choses sensibles." Roughly and clumsily: "Why one can doubt the existence of things revealed by the senses."

Descartes's questioning of the senses was not out of place in an era in which it had become quite clear, despite the contrary evidence of the eye, that the Earth went around the Sun. He was not the first to suspect the senses; the roots of his doubt went back to Plato. There are still many today who maintain that all we can know of the "external" world are the images that we carry in our minds and the relationships that our minds create between those images. There is no guarantee that what goes on in our heads corresponds even roughly to what goes on in the external world—if such a world exists.

Although Descartes was not the first to doubt the validity of observation, he was one of the most ruthless in his conclusions. He wanted absolute certainty; he wanted, in a world of apparent illusion, to penetrate the sensory mists and reach the undeniable truths behind man's perception of his world. He used the method of doubt: if something could be doubted on any grounds, then it must be mercilessly stripped away from the body of accepted knowledge: "I was convinced that I must once and for all seriously undertake to rid myself of the opinions which I had formerly accepted, and commence to build anew from the foundation, if I wanted to establish any firm and permanent structures in the sciences." Notice the aim. It is *science*, not philosophy, although unlike most present-day scientists, Descartes would have seen them as inseparable.

Since the senses are suspect, we must reject them, at least initially, as a source of knowledge. This is, to put it mildly, a drastic step. The senses are our source of information. What can we know without them? Descartes answered: "Cogito ergo sum." This is usually translated as "I think, therefore I am," but a philosophical acquaintance of mine prefers "I experience, therefore I exist." Take your pick.[1]

Three Latin words formed the one tiny rock to which Descartes clung in the sea of his doubt. But it was hardly enough, this safe little subjective island. He wanted to find a way back to the world because his ultimate ambition was to construct a truthful account of how the physical universe worked. The fact that "I exist" is important, especially to me, but it hardly provides a basis for understanding the movements of the planets or the mysteries of magnetism. If Descartes was to explain nature, he had to prove more than his own existence. To do so he cheated, not in his terms but unquestionably in ours: he "proved" the existence of God:

> From this one idea, that the idea of God is found in me, or that I exist possessing this idea, I conclude so clearly that God exists, and that my existence

---

[1]Saint Augustine (354–430) had related thoughts, as seen in the passage in *Confessions* (Book 11, Chapter 14), which begins: "Everyone who observes himself doubting observes a truth, and about that which he observes he is certain; therefore he is certain about a truth."

depends entirely on Him in every moment, that I am sure that nothing could be known by the human mind more evidently or more certainly. And it seems to me that I now see before me a road which will lead from the contemplation of the true God (in whom all the treasures of science and wisdom are hidden) to the knowledge of other things.

Notice Descartes's attempt at consistency; since he knows nothing but his own existence, it is to his thoughts that he turns to look for God, not to an as yet unproved outside source. This was a slap in the face for the Church and the Holy Scriptures—it never even occurred to Descartes to appeal to their authority.

Having, in his own estimation, proved that both he and an all-knowing God existed, Descartes proceeded to the next stage: to establish what he deemed to be a foolproof criteria for determining in which cases our senses *could* be trusted. This is the keystone of his attack on the problem of knowledge. After all, the $64,000 question in epistemology is What can I be sure of? If Descartes had an infallible truth detector, then he could begin to investigate the universe, accepting only those messages that were validated by his wonderful gadget.

We are bombarded by a continuous stream of sensory impressions, but in Descartes's view not all of these impressions have equal status—many of them may be suspect, perhaps even sent to us by Satan. Nevertheless, there are certain facts about the external world that Descartes labeled "clear and evident." Thus, it would be "clear and evident" that iron is harder than butter; one could hardly doubt that. Now Descartes realized that being sure of a fact was not proof of its truth. So he cheated again. He appealed to the God that he had invented: *God would not deceive us.* If a fact appears to us to be "clear and evident," it does so because God, knowing the fact to be true, allows *us* also to see it as clearly and evidently true.

This is not an argument that a modern scientist or philosopher, or even HMS, would accept. Notice that there is absolutely no question of Descartes's invoking reason here.[2] The "proof" of God's existence and the subsequent assumption of his honesty are firmly within the fiefdom of faith, not the realm of reason. Worse is to follow. If God guarantees our unerring judgment, ensuring that *we* know what is true, then how can we err at all? Descartes admitted that he could err, and he gave the problem a lot of attention. He solved it by invoking free will, also a gift from God. At this point the reader may well be raising his eyebrows. For if I suppose something to be true, how do I know that it is because God has made it "clear and evident" to me, or because I am using my free will and have made a mistake? There are more problems with Descartes's philosophy, but they are not for us.

In the end, Descartes's philosophy rests on God. He lived in a society bathed in religious belief, and to most of his contemporaries the existence and omnipotence of God was self-evident. Descartes's acceptance of God was his only major concession to the prevailing framework of knowledge established by the Church and the Greeks. In a world in which these were the unassailable authorities in all matters, Descartes brushed them aside and proclaimed that all you needed to reach the truth was your own (God-given) intuition. For Descartes, Almighty God guaranteed

---

[2] As Suzanne says to her betrothed, Figaro, in Beaumarchais' *The Marriage of Figaro,* "Giving a reason for being right amounts to admitting I could be wrong. Are you my humble servitor or aren't you?"

the correctness of his inner feeling that something was clearly and evidently true. But although he gave the Almighty an indispensable place in his philosophy, it is limited to the roles of Creator and guarantor; he is rarely appealed to otherwise. After he made the universe, it developed according to the laws of nature. He is not the mover of the planets, only their builder. The universe proceeds on its way mechanically. This belief lies at the core of deism, a doctrine we will meet again in the eighteenth century and that still speaks to many scientists.[3]

Descartes was inevitably accused of fostering atheism, a strange accusation in the light of the essential role he gave to God as the guarantor of truth. Descartes actually denied to atheists the capacity to really know the truth about anything, as the following extract clearly and evidently shows:

> That an atheist can clearly know that the three angles of a triangle are equal to two right angles, I do not deny. I merely say that this knowledge of his is not true science, because no knowledge which can be rendered doubtful should, it seems, be called science. Since he is supposed to be an atheist, he cannot be certain that he is not deceived, even in those things that seem most evident to him. . . . He will never be safe from [doubt] unless he first acknowledges God.

Despite this, Descartes's works were put on the banned list of the Catholic Church in 1663. He had not even given Rome's authority the courtesy of a challenge; he had just ignored it. God existed because Descartes said he did, not because of Holy Writ. The Church had already faced a similar sidestepping of its authority when Martin Luther had opened up a personal path to God, a path that made the Papacy and its hierarchy irrelevant, just as Descartes's path to philosophical truth asked no help from centuries of philosophy and religious dogma. Descartes's younger contemporary Baruch Spinoza also believed that "truth manifests itself." He was excommunicated by the Amsterdam Jewish community in 1656.[4]

Descartes is classified by history as a philosopher, but the bulk of his writings are on what we would call science. Most of his scientific writings, as distinct from his mathematical work, are almost meaningless. He theorized about nature instead of observing it. He rejected the idea of atoms and built a strange physical world consisting of vortices in a continuous fluid. This hypothesis had no basis in experiment. Descartes's faulty system of science was taught in most European universities for about a century, until Newton and Galileo supplanted him in the eighteenth century.

Descartes's reputation as a scientist was short-lived. During the Enlightenment, his science was savaged by Newton and Newton's followers, but his stature as a mathematician and philosopher is now unquestioned. His *Discourse on the Method* heralded the age of the Enlightenment (see Chapter 6). In Europe, more than perhaps anyone else, he laid the foundations for modern philosophy by un-

---

[3]The universe was not purely mechanical. Descartes believed that man had a soul, although animals didn't, but for him the soul was not the *causal* difference between a dead and a live body. We don't die because our souls leave us. A dead body is a broken clock; the soul deserts it *because* it is dead.

[4]Descartes was not the stuff from which martyrs are made. On hearing of Galileo's indictment by the Inquisition for his belief in a heliocentric solar system, Descartes chose not to publish his *Treatise on the Universe*, in which he too supports the same system.

dermining every other system of philosophy going back to Aristotle. In his own terms, he was completely honest about the problem of knowledge. In the twentieth century, Sartre singled out *Cogito ergo sum* as the point of departure for existentialism, as "a truth which is simple, easily attained and within the reach of everybody; it consists in one's immediate sense of one's self." Voltaire, in his *Lettres Philosophiques*, puts his finger on an essential point: "Descartes gave sight to blind men; they saw the errors of the ancients and his own. . . . He destroyed the absurd chimeras with which young minds had been filled for two thousand years; he taught the men of his time to reason. . . . He was estimable even in his mistakes." For scientists, Descartes's refusal to be brainwashed by precedent is a lesson that they, hopefully, keep permanently in mind. Does he bolster your confidence in the hypothesis of perpetual molecular motion in gases? In no way: he does not prove that molecules in a gas do what I say they do. But he does do something more useful. He tells you that you don't have to accept what I say just because I am classified as an "expert."

To remind us that intellectual idols can be human, I quote from John Aubrey's biographical sketch of Descartes: "He was too wise a man to encumber himselfe with a Wife; but as he was a man, he had the desires and appetites of a man; he therefore kept a good conditioned hansome woman that he liked and by whom he had some children (I thinke 2 or 3)."

Francis Bacon (1561–1626)

In complete contrast to Descartes's skepticism, Bacon trusted the senses *completely*. Observation was everything, it revealed the truth, and no supernatural guarantees were needed. Nature, for Bacon, was "an open book" that could not possibly be misread by an unprejudiced mind. As Leonardo da Vinci had said a century before: "Experience never errs." Bacon would have found absurd Descartes's conclusion that the knowledge acquired by the senses of an atheist was somehow less reliable than that acquired by a believer. The suggestion that the Devil might deliberately be deceiving us has no place in Bacon's scheme of things. He was a straightforward, no-nonsense thinker, although there are those who would rephrase this description to read "a naive, unsophisticated dilettante."

HMS in general believes what he sees, and scientists generally put their faith in experiment. For this we have to thank Bacon, a few Greeks and Romans, and a succession of European thinkers, mostly living in the sixtenth and seventeenth centuries, some of them in an intellectual atmosphere that modern man has recreated only in twentieth-century dictatorships. The dominance of dogma in Bacon's day, and in the previous few hundred years, was almost absolute. Most scholars in fourteenth-century Oxford accepted almost anything written by Aristotle or the fathers of the Church, whether or not it was confirmed by their senses. At the trial of Galileo the defendant found that the evidence of his eyes, as aided by his telescope, was confronted not by conflicting observations but by centuries-old dogma. For his inquisitors, Holy Writ was the final arbiter. Galileo claimed to have observed four previously undetected moons of Jupiter. The Church ruled that they could not exist. Francesco Sizi, an almost forgotten contemporary of Galileo, explained why. There were only seven heavenly bodies (seven has a strong place in mystic lore), and each had an astrological significance. The proposed moons had no astrological significance, therefore they could not influence man, therefore they could not exist. Cesare Cremonini, an Aristotelian colleague of Galileo at Padua, refused to look through Galileo's telescope. Why would anyone want to see what no one but Galileo had seen? "And anyway, peering through those spectacles gives me a headache." The professor of mathematics at the Collegio Romano declared that if he were allowed to first build the supposed four satellites of Jupiter into some glasses, then he too would see what Galileo saw. Bacon would have asked to look through the telescope, and if what he saw clashed with the tenets of the Church or the ancients, then too bad for them. His single-minded insistence on the primacy of observation was no less a rejection of authority than Descartes's. He believed what he saw, not what he read.

Bacon was not a practicing scientist, but he affirmed that knowledge could only be built on the observation of nature. In parallel with this "commonsense" view of nature, he saw the purpose of science as "the relief of man's estate," the betterment of our material environment and health—antibiotics rather than relativity. This concept of the role of science was to profoundly influence the seventeenth-century attitude toward science in England and the philosophy of the Enlightenment in eighteenth-century France.

The trouble with Bacon was that he distrusted theory. We agree with his belief that the basis of science is observation, but fact gathering is not enough. Without theory, science is merely picking up shells on the beach. Science is not a Baconian flea market of unrelated data, but Bacon, with English down-to-earthness, was suspicious of hypothesizing intellectuals: "The intellect, left to itself, ought always to be suspected." And again: "For the wit and mind of Man . . . if it work upon itself, as the spider worketh his web, then it is endless and brings forth indeed cobwebs of learning, admirable for the fineness of thread and work, but of no substance and profit." Like Descartes's vortices. Bacon was particularly contemptuous of Greek philosophy, which he considered to be "puerile, or rather talkative than generative . . . fruitful in controversies, but barren in effect."[5] Bacon's allergy to contemporary theory is understandable; for an intelligent man with a skeptical mind, the largely theoretical framework of learning built up by the Church and the disciples of Aristotle was too much to swallow. Unfortunately, his contempt for empty hypothesizing blinded him to the work of Copernicus and, possibly because of his meager knowledge of mathematics, he also ignored the monumental advances of Kepler and Galileo.

Despite his avoidance of theory, Bacon's works are not devoid of the kind of questions that troubled other philosophers. Thus he revolted against the concept of final causes (the belief that things were as they were for a purpose) and championed the idea of efficient causes (things were as they were because they were the effects of a preceding cause). And he used a very significant phrase which qualifies, in a most important way, his lauding of observation. He says that the mind has "a power of its own." He was aware that a mind that receives sense impressions without interpreting them and putting them into some kind of workable framework is incapable of grasping reality. For Bacon, reality, or rather our knowledge of it, contained an input from the mind.

---

[5] He was less blunt than Hobbes in his ranking of the Aristotelian tradition: "When men write whole volumes of such stuff, are they not mad, or intend to make others so?"

Bacon grew up in the court circles of Queen Elizabeth I, and his ambitions were stupendous—politically as well as scientifically. The Lord Chancellor of England proclaimed famously, "I have taken all knowledge to be my province," by which he meant that he was going to define the directions and organize the categories of existent knowledge, an undertaking he thought he had completed in *Instauratio Magna* (1620).

There was a marvelous self-confidence about the way Bacon wrote. Despite his avowal to avoid the arrogance of the Aristotelians in his approach to nature, his implied humility in the presence of nature sometimes smells of a prosperous nineteenth-century Yorkshire mill owner avowing, "I'm a humble man." He was put on trial for corruption in 1621. For the last five years of his life, he was barred from holding public office and banished from the court of James I. He died of a fever that developed from a chill following an excursion into the snow: "It came to my lord's thoughts, why flesh might not be preserved in snow, as in salt. . . . They alighted out of the coach, and went into a poore woman's howse at the bottome of Highgate hill, and bought a hen, and made the woman exenterate it, and then stuffed the bodie with snow, and my lord did help doe it himselfe."[6] Perhaps the only experiment he ever did killed him. He died in debt, having failed in his attempt to create a system of philosophy to replace that of Aristotle.

Above all, Bacon left a legacy of belief in the evidence of the senses and in experiment, and a vision of the role of science as *the* means of improving the material condition of man. Bacon was not a scientist; neither can he be ranked among the great philosophers. Nevertheless, his sponsoring of *inductive reasoning*, arriving at general conclusions from the accumulation of facts, was enormously influential. He expounds the inductive method in his *Novum Organum* (1620), a title that could be translated as the *New Instrument*, meaning a new tool for arriving at the truth. Collect enough cases and you can generalize—that was his message. But first *observe*. It is natural that he admired Machiavelli for describing men as they are, not as the moralists hoped they would be.

Bacon's belief in the primacy, and reliability, of observation, and the magnificent prose that he used to convey his ideas indirectly affected the whole subsequent history of science. Although, like Descartes, he contributed little directly to knowledge, he helped to shape the intellectual climate of the Enlightenment and the following centuries. The Age of Reason, that great blossoming of free thought, acknowledged three prophets: Newton, Locke, and Bacon. Bacon's bold and prophetic vision of a world enriched by the practical consequences of science underlies nearly all applied research. Ironically, that vision was not based at all on the inductive reasoning that he so espoused, since science had done almost nothing in his day to improve the lot of man.

When the Dutch scientist Christiaan Huygens (1629–1695), wrote to the politician Colbert to explain the aims of the newly established Académie des Sciences, in Paris, he explained, "La principale occupation de cette assemblée et la plus utile, doit être, a mon avis, de travailler a l'histoire naturelle a peu pres suivant le

---

[6]The account comes from Anthony Powell's entertaining edition of John Aubrey's *Brief Lives*. Bacon's great contemporary, Descartes, also died as a consequence of cold, in this case the raw Swedish winters.

dessein de Verulamus."[7] Bacon had been created Lord Verulam in 1618. A century and a half after Bacon's death, when Emmanuel Kant wrote his greatest work, *The Critique of Pure Reason,* he dedicated it to Bacon. In 1847, William Whewell, the Oxford scientist and philosopher, wrote, "If we must select some one philosopher as the hero of the revolution in scientific method, beyond all doubt Francis Bacon must occupy the place of honour."

The search for facts, the sacrosanct standing of observation, the implicit belief in reason—these were not the invention of one man. There is something to be said for the "time is ripe" school of history, the supposition that it is the total social, political, and scientific environment that makes inevitable the changing direction of man's thought. Nevertheless, every change has its standard-bearers, those whose flags are seen above the battle. Bacon was such a one. Video-clip summaries of world philosophy tend to oppose Bacon to Descartes, labeling one an empiricist and the other a rationalist. The classification has its justifications, but in both cases it needs qualification, and it is sometimes forgotten that they both, despite their different paths, saw reason and science as the hope of a better future for mankind. This belief is firmly linked to Bacon's name, but it is also stated very clearly in Descartes's *Discours de la méthode.*

Does Bacon's observational approach at least help us to believe that molecules exist? He asks us to observe and, by employing inductive reasoning, use our observations to come to conclusions. This doesn't work too well with molecules since they are not observable, except in favorable cases in highly specialized laboratories. Nevertheless, if you are prepared to believe in the collective honesty of the international scientific community, you can accept their observations as standing in for yours.

In fact, molecules were part of the scientist's explanation of nature long before we could observe them. Their existence was *deduced* from the behavior of matter. The basic, experimentally determined laws governing the behavior of gases were shown to be consistent with the simple nineteenth-century model of a gas described in the previous chapter. The model was based not on the observation of molecules but on the supposition of molecules. Up to now there is no aspect of the behavior of gases that is not consistent with the model. Does this *prove* that the model is correct? Asking a much weightier question: What does "proof" mean? Neither Bacon nor Descartes helps us here. One trusted his senses; the other trusted his God. Neither helps me to dismiss the suggestion that there could be another model for gases that is just as successful, or even more so, than the present one. How do I prove that a given model is the correct one? Where does certainty lie in this world?

Karl Popper: The Certainty of Uncertainty

In so far as a scientific statement speaks about reality, it must be falsifiable:
and in so far as it is not falsifiable, it does not speak about reality.

Karl Popper, *The Logic of Scientific Discovery*

---

[7]"The principle occupation of this assembly, and the most useful, should be, in my opinion, to work on natural history in a manner similar to that of Verulam."

Despite the fact that knowledge is based on observation (at least that is the credo of HMS), the fact is that the Oscar winners in the history of science have almost always been creators of *theory*: Newton and universal gravity, Darwin and evolution, Maxwell and electromagnetic fields, Einstein and relativity. It is theory that allows us to comprehend a multitude of facts in the light of a handful of concepts. Theory is the outcome of man's apparently inherent drive to order the universe. Theory explains, theory predicts, theory suggests new directions for thought and experiment. It is, in the end, because of theory that we can plan the paths of spacecraft, construct compact disc players, synthesize new pharmaceuticals, and reveal the structure and mode of action of the AIDS virus.

There are some extremely successful theories, but we should not be misled by their successes. *Science is not a means of obtaining absolute truth.* There is an enormous amount of highly persuasive data supporting evolution; most scientists believe in the theory, but it has not been *proved*.[8] Facts may be regarded as indisputable; theories are not.

What criteria, if any, are there for proving that a theory is true? Is there a cast-iron method of proving the truth of our model of a gas? Or of any other scientific theory? Most scientists will tell you that a theory is true as long as it works. Thus it was once "true" that the Earth was flat, in the sense that for early man the assumption of the Earth's flatness led to no incorrect conclusions. Ships disappearing over the horizon might be assumed to be too far away to be seen properly. The flat-Earth theory was demolished (except in the eyes of a few eccentrics) by a flood of facts that were consistent only with a roughly spherical Earth—and we can now photograph the globe from space. At a higher level, Newton's system of mechanics was true for a couple of centuries. In the twentieth century it was shown to be lacking, to be an approximation to a new "truth," the theory of relativity. However, for almost all practical purposes, Newton still suffices; we don't need relativistic mechanics to compute the path of a Scud missile. In this limited sense, Newton's theory is still true.

Theories come and theories go, because science is, at the deepest level, skeptical. The motto of the theoretician could be "It's right until one fact proves it's wrong." Is this a fatal weakness? My few antiscience acquaintances harp on the historically high death rate for scientific theories—another one bites the dust. They have an ally in the philosopher Paul Feyerabend, who saw science as merely a fashionable ideology and a very faulty way of understanding the world. And, he warned, ideologies, like fairy tales, contain "wicked lies." Scientists, he accuses, are all too ready to abandon their beliefs: "We can change science and make it agree with our wishes. We can turn science from a stern and demanding mistress into an attractive and yielding courtesan who tries to anticipate every wish of her lover." I never realized that I was in such an erotic profession. Feyerabend exaggerated, and he knocked down large numbers of straw dolls. Not unnaturally, scientists see the flexibility of theory in a very different, and totally unerotic, light.

Scientists rarely claim infallibility; the ugly fact that destroys the beautiful theory is just one more challenge (although if the defunct theory is one's own, it is

---

[8]There are ongoing battles in the United States about whether high schools should teach creationism. In Israel, in 1992, the ultra-orthodox community was deeply offended by an advertisement campaign run by the Pepsi-Cola Company, which portrayed man's evolution from the apes—and his final enlightened realization that he should buy Pepsi-Cola.

useful to have a large glass of whisky handy). The *strength* of science is that it admits to its limitations, and its errors. Having said which, there are theories, like evolution, that appear to be unshakable. Surely there must be a criterion for the correctness of a theory? I do not believe there is, and in this I am, as are probably the great majority of my scientific colleagues, a disciple of Sir Karl Popper (1902–1995).

Popper is a philosopher whose best-known work has been in the area that we are presently discussing—the validity of scientific theory. Popper takes as a basic property of any scientific theory the possibility of its being *disproved*. In other words, that which cannot be disproved is not a theory. Take the statement that a newly minted coin falling from a small height onto a hard, level, smooth surface can end up only as heads, tails, or on its edge. This is not a theory; it is a statement of fact. It certainly will never be disproved. On the other hand, the medieval belief that garlic destroys the power of magnets could be termed a hypothesis, a miniature theory. It *can* be disproved, and it was disproved by the Italian Porta who, in 1589, wrote: "Not only breathing and belching upon the lodestone after eating garlic, did not stop its Virtues, but when it was anoynted over the juice of Garlick, it did perform its office as well as if it had never been touched with it." End of hypothesis. That, according to Popper, is the nature of theories—they can be disproved. But Popper goes further:

Theories can *only* be disproved, they can never be *proved*.

In other words, science is a lie detector. Newton provides a magnificent example. His mechanics accounted for a myriad of observations on this Earth and in the heavens. Over the years, every new consistent fact and calculation hammered home the absolute correctness of Newton's theories. They accounted for the motion of the moon, the trajectory of artillery shells, the movements of molecules in a gas, the flight of airplanes. It took a brave man to doubt a theory that *never* gave the wrong answer. There were a few who questioned Newton's assumption of an absolute time and space "out there," independent of man—but anyone doubting the validity of the laws of motion was not taken seriously by the scientific community. The Newtonian pudding had been proved by repeatedly eating it. And then, after two centuries, strange observations began to undermine the infallible laws, observations that seemed to deny the validity of our (Newtonian) understanding of time and space. At the end of the nineteenth century, all the armies of previously determined facts could not save the Newtonian universe because for a growing number of problems Newton gave catastrophically incorrect answers. Observation clashed with theory, and, as it always must, observation won. Humpty-Dumpty could not be put together again. Bacon would have smiled. Popper understood: no one ever proved Newton to be right, but in the end his theory was *disproved*, or, more correctly, shown to be an approximation to a more general theory. This classic example of scientific fallibility can be taken as an illustration of Popper's statement that the multiplication of positive cases does not increase the probability of a thesis being correct. This proposition goes against our intuitions, and there are those who disagree with Popper on this point. It is interesting to compare Hume: "A wise man proportions his belief to the evidence."

Popper has underlined what many scientists implicitly accept: science demands absolute open-mindedness, and those who claim infallibility for science are betraying it. We must always prepare to be wrong. No theory is final, and Popper believes this to be true in the social and political spheres as well. In Locke's words, "broad daylight" may be unattainable, but we may reach, in his beautiful phrase, "the twilight of probability." Even Popper's theory may be disproved!

On Having One's Feet on the Ground

There have been philosophers, like Hume, who would weaken, or destroy, the claim of science to understand reality by questioning the very existence of cause and effect. Berkeley went further, denying the existence of matter in the common-sense meaning of the word. Ludwig Wittgenstein stated that nothing that was unobserved could ever be validly inferred from what we can observe, a belief which, if taken seriously, could stop much of theoretical science dead in its tracks. Why have we ignored these skeptical voices and singled out only three representatives of world philosophy to present their case for scientific truth? In fact, why bother with philosophy at all?

The attitude of the majority of scientists to philosophy is: "Science is one thing, philosophy is another. Science has proved its basic validity, not only by rationalizing the workings of Nature but by predicting observations and entities which subsequently materialized. I have no need for epistemology." As Leibniz said, "Although the whole of this life were said to be nothing but a dream, and the visible world nothing but a phantasm, I should call this dream or phantasm real enough, if, using reason well, we were never deceived by it." Most of us would agree, so what place has philosophy anyway? Why is Leibniz not enough?

One reason is that twentieth-century findings on the behavior of the physical world have plunged us into a universe where "using reason well" often completely fails to provide an explanation of what we observe. It is our good luck that these dilemmas, which force us to face complex and fascinating philosophical problems, are not relevant to most fields of science. Thus a set of accepted concepts are enough to account for such questions as the stability of bridges, the flight of birds, and the chemical properties of matter. It is when we start to dig beneath the surface and ask what determines the behavior of the elementary particles of matter, and of light, that we start to lose contact with such primitive concepts as matter and causality and begin to suspect that the skeptical philosophers will have the last laugh, that we will never find a rational explanation for our perceptions. Today some scientists are asking questions that echo the age-old concern of the philosophers with the nature of knowledge. This is the main concern of Descartes, Bacon, and Popper. They were chosen partly because they are what I would call, in the broadest meaning, commonsense philosophers. They all, in the end, accept that our perceptions give us an image which is, if not a duplicate, then a good approximation to the external world, and that that world operates according to laws that, even if they are man-made and temporary, can be used to predict and systematize our perceptions. This is what most of us, including scientists, believe.

Scientists, and HMS, generally *behave* as though they sympathized with Bacon and Popper. Most of us behave as though we believe in the scientist's working concept of cause and effect, and that our senses tell us something real about a real world, even though modern physics reveals a cosmos that is ultimately enigmatic, a "dream and a phantasm." For truth, as we shall see throughout this book, plays very hard to get.

Truth in Practice

Truth. three sorts therof Natural, Mathematical and Moral.

—Bishop Berkeley, notebooks

We have seen three classic approaches to scientific truth, none entirely satisfying. Scientists should have a vested interest in "truth," despite Popper! In searching for that truth in the laboratory, the practicing scientist often leans toward Bacon, even though she probably does so unconsciously.

Bacon used inductive reasoning, which roughly means that if you have enough evidence it provides the basis for a safe conclusion. Thus if I drop fifty brandy glasses from 30,000 feet onto a concrete parade ground and they all break, Bacon would be amazed at my stupidity, but he would confidently state that the next glass that I drop will certainly break and that I can formulate a general law as to the breakability of brandy glasses dropped from 30,000 feet. Most of us would take this law, obtained by inductive reasoning, to be inviolable. Is inductive logic our answer?

Inductive reasoning is used by all of us, every day, and it is widely used in science. But it is defective because the next case may always be the one that doesn't fit in. This is the weakness of inductive reasoning, a weakness that Hume focused on. The method, said Hume, depends on the axiom "Instances of which we have had no experience, must resemble those of which we have had experience, and that the course of nature continues always the same." But this cannot be proved logically.

We must be aware that if we have used inductive reasoning to construct a theory, then the first case that doesn't fit could bring that theory to its knees. This is exactly the case that Popper focuses on. Popper sees science as advancing by a process of falsification. To idealize his thesis: We work within an accepted framework, say Newtonian mechanics, and then, when something doesn't fit in, we realize that Newtonian mechanics is wrong or incomplete and we look for a new framework. In the best of all possible worlds, this is how things should happen, but what usually happens in the laboratory is altogether different.

When I find a fact that doesn't fit in with a well-established theory, I instinctively look for experimental error. This is certainly a judicious first move. In the meantime, I slander the new fact, I say it is "puzzling," "a controversial finding," "anomalous," "unverified"—and I may be right. If the awkward observation does eventually turn out to be experimentally spotless, I have a choice: I can call the existing theory in doubt or I can document the "anomaly" and carry on as before. This happened with the "anomalous" movement of the planet Mercury, which was observed in the nineteenth century but did not overthrow the Newtonian mechanics that failed to solve the problem of its motion. In the twentieth century Einstein showed that it could be explained by the theory of relativity, a more general theory than Newton's.

Thomas Kuhn, particularly in his *Structure of Scientific Revolutions* (1962), has stressed what he sees as the inordinate inertia of the scientific community and its exaggerated dependence on paradigms—established and accepted frameworks of thought. Kuhn claims that paradigms are not, as Popper suggests, given up when disproved, but only when they are replaced by a new theory. There is some truth in this, but it does nothing to invalidate Popper's central thesis.

Induction, Bacon's method, gives us only a probabilistic definition of truth. As we saw, those who claim that this probability is increased as the number of confirming cases grows, are challenged by Popper. Is Popper too pessimistic about the possibility of proof? Is there a better way than Bacon's to arrive at truth?

Deduction: A Surer Path?

We go back to Aristotle, who first clearly defined inductive and deductive logic, and try the latter approach. The idea in its classic form is to start off with two irreproachable statements and then, by the rules of logic, to arrive at an irreproachable conclusion. This is the famous *syllogism*. For example:

All elephants have a long memory and a long trunk.
Dumbo is an elephant.
Therefore Dumbo has a long memory and a long trunk.

If the first two statements are true, then the last statement *must* be true, because the logical step that took me there is valid. The syllogism was a cornerstone of logic until the beginning of this century. It has come under criticism, partly because the final statement is already contained within the premises—you are not proving anything that you didn't really know.

The essence of a syllogism is a comparison of its conclusion against observed fact, so that it is the fact that can destroy the theory. Thus, starting off with a basic assumption (theory), one can work out its various consequences and wait until one of them is at variance with fact. At this point you have disproved the theory; unfortunately, in all the successful cases you have *not* proved the theory, you have merely given yourself growing confidence that the theory is correct.

This method is sometimes called the hypothetico-deductive method: you have a hypothesis, you use deductive logic to examine its consequences, and you compare them with the observed facts. For example:

All homosexuals are evil. (Hypothesis)
Benjamin Britten was a homosexual. (True)
Therefore Benjamin Britten was evil. (False)

From all accounts he wasn't; therefore, the hypothesis is wrong. It sounds very much like Popper: "there is no more rational procedure than the method of trial and error—of conjecture and refutation; of boldly proposing theories; of trying our best to show that these are erroneous; and of accepting them tentatively if our critical efforts are successful."

The method is clearly related to the inductive method: in practice, after a few observations, one attempts to set up a theory and then to deduce consequences from the theory. The inductive method places little or no initial emphasis on a hypothesis, while the *hypothetico-deductive method* brings in a hypothesis at a very early stage of the game. It is a fancy name for the approach we have taken toward the credibility of moving molecules. No one had observed molecules when the model was first proposed, but a hypothesis was constructed and it has weathered the storm; it agrees with all the facts known at the time or revealed since. The test is always the observed facts.

Where does "the scientific method" fit into all this? There isn't *a* scientific method. There are many—and none of them is foolproof. A given scientist may use quite different approaches to different problems. She might start one investigation by collecting large amounts of data, hoping to detect a pattern; she might start another by considering a well-known anomaly and come up with a hypothesis. And there is no guarantee that the roots of the hypothesis can be traced directly to the data; they may originate in an association of ideas, an analogy drawn with another problem, a chance remark.

One point is incontestable: the "truth" of science must always remain open to critical scrutiny and will sometimes have the status of a beauty queen: looks good today, but next year she'll be dethroned. That is because the real test of a scientific theory is not whether it is "true." The real test is whether it works.

The Fragility and Strength of Science

What does all this mean for this book? Will the "truths" of present day science be here today and chuckled at tomorrow? Some of them, yes. The current theories concerning the origin of life can't all be right. And what about the question that prompted this chapter? Are we really sure that molecules move in the fashion that we have described? I would bet my life on it.

The kinetic theory of gases is based primarily on the assumption that a macroscopic sample of a gas is composed of a huge number of molecules in continual motion, their average speed being constant at constant temperature. Our faith in this assumption, and in the theory as a whole, is based on the fact that the theory allows us to mathematically derive conclusions about the expected behavior of a gas, and that these conclusions have been *verified experimentally* time and again for all kinds of gases and all kinds of properties. In the spirit of Popper, we can say that we have constructed a theory that could, in principle, be shattered by one set of data that didn't fit in. Until that happens, we accept the theory. I believe in our model for a gas, and I am asking you to believe in it, because no one has disproved the model. The analogy of the senseless flies is not perfect; real molecules experience forces from each other, and it is primarily because these forces are often not accurately known that the exact calculation of certain properties is not possible. The nineteenth-century theory is too simple, but it still, for many purposes, suffices to deal with the behavior of most gases, provided they are not at too high a pressure or too low a temperature. I will go out on a very short and sturdy limb and predict that, in its general principles, the kinetic theory of gases will never be shown to be wrong.

I would risk betting my salary on 75% of this book being undisturbed for the next decade or so. But I might lose. I had an uncle who bet obsessively on dog races. Many years ago, there was a race at the White City stadium in London in which one of the six greyhounds was the track recordholder for the distance and another was the holder of the world record. It would have been madness to bet on any other dog. The starter's gun rang out, the traps opened, and the two superdogs, who were in adjacent traps, leaped forward, bumped into each other, and staggered. My uncle, who claimed to have a "system," bet on another dog, which won. In racing and in science there are no "sure things," only odds-on favorites. My uncle died bankrupt.

In the end we are forced to say of theories, and of dogs, "I believe," rather than "I can prove." Science goes to the courthouse backed only by circumstantial evidence. That's the best we can do. But, as Popper points out, in *Conjectures and Refutations* (1962), it's more than can be said for most human activities: "The history of science, like the history of all human ideas, is a history of irresponsible dreams, of obstinacy, and of error. But science is one of the very few human activities—perhaps the only one—in which the errors are systematically criticized and fairly often, in time, corrected." This philosophy fits badly with those who want eternal verities, and the faithful often contrast the Rock of Ages with the shifting seascape of science.

Faith

I believe in the Almighty, but I protest.

—Isaac Bashevis Singer

Personally, I find that the faith-versus-science controversy has the smell of past centuries. I am happy to follow Bacon and agree to a separation of forces: "We do not presume, by the contemplation of nature to attain to the mysteries of God. . . . only let men beware . . . that they do not unwisely mingle or confound these learnings together." For Bacon, reason and common sense were irrelevant to religion: "The more absurd and incredible any divine mystery is, the greater honor we do to God in believing it; and so much the more noble the victory of faith." Galileo likewise sidestepped the question: "Both the Holy Scriptures and nature originate in the Divine Word. . . . [T]wo truths can never contradict one another." Later, the famous seventeenth-century chemist Robert Boyle was to declare that "there is no inconsistence between a man's being an industrious virtuoso (*an amateur scientist*), and a good Christian." Many thinkers have explicitly stated that reason is a tool that can only be applied to nature, that God and the spiritual world lie outside its range. But there is always a suspicion that, in societies in which the religious authorities could make life hard for heretics, the statement that reason was irrelevant to religion was a ruse to keep those authorities out of the natural philosopher's realm. It is very possible that this ploy was used by Bacon. It was certainly not used in this way by the Franciscan William of Ockham (c. 1285–1349), who had so little fear of papal authority that he accepted excommunication rather than retract what were regarded as heretical statements. He very firmly placed the natural world within the realm of reason, and God outside it, although he was too much of a believer to deny God the credit for the Creation or the right to interfere with the workings of the world, if he so wished. In this he was very similar to Newton.

If the Creator does not appear as a real player in this book, it is because I know nothing about him or her. As Georg Lichtenberg (1742–1799) asked: "After all, is our idea of God anything more than personified incomprehensibility?" In a letter to Voltaire, Diderot dismisses the idea of God as not philosophically necessary and suggests that it had in all likelihood been thought up by an enemy of the human race, since it had caused so much conflict. Voltaire saw things slightly differently: "The world embarrasses me, I cannot conceive of so exquisite a clock without a maker." Personally, I am tempted to believe that if God did once exist, she created the Universe but died in childbirth.

One thing that nearly all of us believe in is the reality of an external world. In that supposed real world there are two elements that all scientists, and most laymen, have taken as fundamental: matter and motion. Thomas Hobbes (1588–1679), the great English proponent of materialism, believed that there was nothing *but* matter in motion. His books, and Descartes's, were read by the young Isaac Newton. And seven years after Hobbes's death it was Newton, the jealous and sometimes dishonest genius, who was to publish the laws of force that were to explain the movement of molecules and the paths of the planets, finally establishing the authority of science in Western culture.

ii

In which Reason nearly triumphs and is then maligned

Civilization has periodically renewed itself after long periods of stagnation. Athens in the sixth century B.C., the emergence of Islam in the sixth century A.D., the twelfth-century renaissance in Europe, and the "official" Renaissance were all wakenings to new dawns. Such a time was seventeenth-century Europe, when the blossoming of science redirected world history and forever changed the way in which man saw himself and his universe. The scientific revolution had many heroes, but towering above them all were Galileo and Newton. It is their achievements that we look at in this section, but not only their scientific achievements; for Newton, albeit unknowingly and unwillingly, was to be the man whose reputation was to underpin the rationalism of the coming Age of Reason—the Enlightenment of the eighteenth century. In the next few chapters we will see what Newton did, and how a handful of simple physical laws united Earth and Heaven, liberated men's minds, and subtly eroded faith. But we will also see Newton's enemies massing on the horizon.

3

Thomas Aquinas versus Neil Armstrong

Projectilia perseverant in motibus . . .

—Isaac Newton, *Principia*

On 4 October 1957 the first man-made satellite to go into orbit was launched from the Soviet Union (R. I. P.)[1] Sputnik's speed and trajectory were calculated using laws published 280 years previously. Sir Isaac Newton's laws of motion are still generally sufficient, even in the age of relativity and chaos, to account for the movement of most bodies in the everyday world. Their near infallibility in that world has prompted me to curse Sir Isaac on three occasions: in 1944 (London, Hitler's V2), in 1973 (the Golan Heights, Assad's artillery), and 1991 (Haifa, Sadaam's Scuds). Waiting for the impacts, I spared a warm thought for the sixteenth-century Italian Tartaglia, who is said to have abandoned research on the dynamics of moving bodies because he was fearful of possible military applications.

Chandrasekhar, a Nobel laureate in physics, puts Newton among the greatest three intellects who have ever lived, in any field. He didn't state who the others were. Newton's work on the motion of bodies was a turning point in intellectual, not only scientific, history. Both he and Galileo realized the nature of the connection between movement and force, a connection that will be our concern in this chapter. It was above all the laws of motion that gave validity to the concept of an ordered universe, a cosmos understandable by human reason.

Newton's marriage of force and motion did not at all conform with the Aristotelian concepts that had been basically unquestioned for two millennia, but even today Aristotle's totally incorrect picture of motion seems more natural to most people. Common sense can be a dangerous guide.

We start by looking at the nature of force.

Force

We are pushed into the light by the force exerted on us by our mother's wombs, and for the rest of our lives we generate, exploit, and are subjected to force. Physical force dominates mankind's earliest memories: the weight of material objects, the gentle pressure of encircling arms, the battering force of the gale.

If we could see forces instead of objects, we would be dazzled by an incredibly

---

[1]On the same evening I was singing to my guitar at a rather wild party in Paris. After the first song there were cries of "Sputnik! Sputnik!" I am still unsure if this was the "highest" acclamation or a hint that it would be better for everyone if I were sent into orbit.

dynamic infrastructure, a close-knit, shifting jungle, underlying the physical world: the strange forces holding together the atomic nucleus, the electric forces that stabilize atoms and shape molecules, the gravitational force that binds the planets to the Sun and holds the galaxies together.

All of which leaves us without a definition of what force is. We need more than the tension in our biceps as a guide to the nature of force. We are not helped overmuch by everyday language; common usage is a frequent enemy of understanding. Of course, science has no monopoly on any word, and no harm comes from talking of the force of de Gaulle's personality. More harmful, for HMS's mental health, are those who debase the language by purveying nonexistent yet profitable commodities such as cosmic force. And what is one to make of Tolstoy, in *War and Peace*? "Man's free will differs from every other force in that man is directly conscious of it; but in the eyes of reason it in no way differs from any other force." Oh yes it does.

So what is the scientist's force?

What Is This Thing . . . ?

When it comes to understanding the basic components of nature, I seek solace in that school of philosophy that defines things by their behavior and regards the question of what things *are* as meaningless. Many of us are tempted to concur with Noam Chomsky, who suspects that the brain is an inadequate instrument with which to arrive at the complete truth about the universe. The Big Questions may be beyond the capabilities of the human computer, just as dogs will never understand jokes. We understand a very great deal about what forces *do* but are far from finalizing the discussion about what forces *are*. Maybe we never will. Newton very specifically refused to commit himself as to what gravitational force was, but he nevertheless accounted for the movements of Earth and Moon and deduced the masses of the Earth and the Sun by knowing only what gravity *does*. We have discovered forces that Newton never knew, but basically we still only define force by what it does. So don't look down on dogs.

What Force Does

We are going to see what force does—any force. Galileo and Newton will be our guides. Since we will often be using gravity as an example of a force, I will assume that:

- You are aware of the attractive force of gravity exerted by the Earth (otherwise you'd fall off).
- You are prepared to believe that all the heavenly bodies exert a pull on any material body (otherwise Neil Armstrong would have fallen off the Moon).

We start with Newton's first law of motion because of the very deep connection between force and motion. The law appeared in Latin in 1687 in *Philosophiae Naturalis Principia Mathematica*—arguably the greatest scientific publication in history.

The First Law

Paradoxically, the first law tells us what happens to the motion of a body in the *absence* of force. It says that *nothing* happens:

Every body continues in its state of rest, or of uniform motion in a straight line, unless it is acted upon by a force.

If the first law seems less than a showstopper, you should recall that Steven Hawking (who, like Newton, holds the position of Lucasian Professor of Mathematics at Cambridge) has referred to Newton as "a colossus without parallel in the history of science."[2] A major component of Newton's unparalleled achievement was his formulation and use of the laws of motion.

Historians will tell you that Newton's first law owed much to Galileo and that both had been peeking at Descartes and his contemporary Pierre Gassendi. They have a point, but we are not here to portion out credit. We are here to show that the first law makes sense, even though Aristotle would have dismissed it out of hand.

It helps to rephrase the law, splitting it into two statements that both stress the *resistance to change* of a body, whether it is moving or not:

(i) **A body at rest will remain at rest unless acted on by a force.** This sounds like common sense; rocks rarely walk without encouragement.

(ii) **A body moving in a straight line will continue to do so *at constant speed* provided that no force acts on it.**

This appears to deny daily experience. Aristotle and Spinoza would have contradicted Newton had they been around. Descartes, over forty years before Newton, had proposed two laws resembling (i) and (ii), so he was a believer, but Saint Thomas Aquinas would definitely have shaken his head. And he had common sense, popular opinion, the sages of antiquity—and God—on his side.

Neil Armstrong, Thomas Aquinas, and the Hand of God

Aquinas lived in a world where carts stuck in the mud and ships became becalmed. It was obvious that if you stopped pushing (or pulling) a cart, a log, or a galleon, it would quickly stop moving—the evidence of one's eyes could not be denied. Thus, in apparent contradiction to Newton's first law, it was generally believed that a moving body will *not* continue moving at constant speed in the absence of a force. A modern Aquinas would smile patronizingly and point out that if your car continued moving indefinitely after you switched off the engine, you would save a tidy sum in running expenses.

But Aquinas had a problem: the Sun rose every day. He believed that it circled the Earth; it returned day after day, just as the Moon returned night after night. How could the heavenly bodies remain in constant motion without a force acting on them? Ships didn't. On level ground, carts didn't, and rocks certainly didn't. In Aquinas's world the only forces permitted were those that involved the direct contact of one body with another, a hand pushing a broom, the wind pressing on a sail. What hand was pushing the Sun?

At the beginning of the *Summa theologiae*, the bulky summing up of the relation of God to man that Aquinas produced in the thirteenth century, he states that God's existence can be proved by reason. And he presented five proofs. So when Aquinas

---

[2]He was, perhaps unknowingly, echoing a phrase in a letter of recommendation written for Newton by his teacher at Trinity College Cambridge: "unparalleled genius."

asked himself whose hand was pushing the Sun and the Moon, he knew the answer—the hand of God.

Neil Armstrong knew better. At about 5:00 P.M. (Houston time) on 17 July 1969, the *Apollo 11* spacecraft was a little under 30,000 miles from the Moon on its return journey to Earth. At this point in space the gravitational tug of the Moon on the *Apollo* was roughly canceled out by the equal and opposite pull of the Earth. The craft was thus traveling through space effectively unacted upon by any force, its rockets were turned off, and it was subjected to negligible *net* outside (gravitational) force. This was a golden opportunity to test Newton's first law. Here was a body moving in a straight line in the absence of any significant force. Newton, if asked, would have said that *Apollo* would travel onward at constant speed. Aquinas would have said that it would stop, unless God was pushing it.

The speed was almost constant: at 4:30 P.M. it was 2946.8 mph; at 6:00 P.M. it was 2933.9 mph, a difference of about 0.44%. From ground station data on subsequent journeys into far space, it is very clear that once spacecraft are well away from heavenly bodies, and have switched off their rockets, they maintain an effectively constant speed—*in the absence of any significant external force.* The first law really works.

The mistake of those who think like Aquinas is that they don't recognize that *resistance is a force.* This is what they leave out when they describe the forces acting on a body. When I stop pushing a cart, it slows down because there is *still a force acting on it*, the combined forces of friction and air resistance. If they were absent, the cart would go on rolling forever. The cart is not disobeying the first law.

Initially it is difficult for most people to believe in the first law; they can't recall seeing a body moving steadily in the absence of a force acting on it. There are always forces of resistance that will slow down a moving body that has no means of propulsion. But nevertheless if, as in the case of our spacecraft, a body is moving and there is no net force acting on it, it just keeps on moving. Newton realized that all material bodies have this strange built-in property, which we call *inertia*, a stubborn resistance to changes in motion. If they are at rest they "like" to stay so; if they are moving at constant speed they "prefer" to continue doing so. We don't know why. We believe it because Galileo suspected it, and Newton showed that he could correctly calculate the paths of the planets, only if he assumed it. It is a correct theory because it works. The huge advantage that Newton has over Aquinas is that Newton's laws allow the *quantitative* calculation of the paths of almost any moving body on Earth or in the heavens. If that word *almost* worries you, we'll clear it up later.

The first law comes with Newton's name attached to it, but Galileo really got there before him. End-of-twentieth-century man has a laboratory in space denied to Galileo and Newton, but Galileo had physical intuition—and he reasoned. He watched spheres rolling on plane surfaces and noted that if they rolled *down* an inclined plane they accelerated; if they rolled *up* a plane, they slowed down. He guessed that this was an effect of gravity tugging at the spheres. A sphere set rolling on a horizontal plane was neither slowed down nor speeded up by gravity, and the smoother the plane, the further the sphere rolled. Galileo realized that, even in the absence of a propelling hand, a sphere would roll on forever *if there was no resistance.* In real life, at least down here on Earth, there always is resistance, and it slows down the rolling sphere on the horizontal plane.[3]

I occasionally look through my copy of Descartes's *Principes de philosophie*, partly for the sensory pleasure that I get from the feel, color, and smell of the yellowed seventeenth-century pages.[4] The text contains the French equivalent of the phrase "state of motion." This appears to be the first time this phrase appears in print. It allows us to express the first law very concisely:

A body maintains its state of motion unless acted upon by a force.

Gravity is not the only force that science recognizes. We will meet others, but the laws of motion apply to all of them.

By the way, if you look in the right places, you will find simple examples of the first law. Here are three.

The Mass Turnpike and the First Law

To test the first law on Earth it would be nice to have force-free conditions, but these are not obtainable on Earth—there is always a source of resistance. It is, however, possible to cheat and attain a condition in which two or more forces cancel each other in such a way that there is zero *net* force acting *in the direction of motion* of a body moving in a straight line. According to Newton, such a body should just keep on moving at constant speed along that line. An example of such a body is Carl Lewis, during the last 80 or so meters of a 100-meter sprint, during which his speed is more or less constant. The first law applies: Lewis travels at *constant* speed, because his legs are exerting exactly the same force as the air and track resistance to his motion—the *net* force acting on him is zero. The same principle holds, in less exciting circumstances, when you cruise steadily along the Massachusetts Turnpike. And it applies to a parachutist falling steadily earthward, pulled down by gravity but resisted by the equal and opposite force of air resistance. Zero *net* force implies *constant* speed, not zero speed.

I am not suggesting that the preceeding examples prove the first law. In no case did we measure the forces and show that they add up to zero. But reflect on the first law the next time you watch the end of a sprint race, or cruise along Route 66. And try to dismiss common sense!

Aristotle Gets It Wrong Again

It took man centuries to grasp the first law. The commonly accepted, and incorrect, reasoning up to the Renaissance was, of course, due to Aristotle. He argued thus: motion is possible only if the force, **F**, acting on a body is greater than the resistance, **R**. If the propelling force is equal to the resistance, that is, **F=R**, there is no net force acting on the body and so, according to Aristotle, it will not move. This sounded like common sense, but it was wrong, as a parachutist could have told him. It means that a parachutist, at the point at which the air resistance, acting on him and his parachute, became equal to the pull of gravity, he would stop in midair. This spectacular party trick seems to be an extremely rare phenomenon. Aristotle came to the wrong conclusion. Zero force does *not* mean zero speed; in

---

[3]Galileo believed that if the experiment was performed on a grand scale, the ball would go on rolling around the center of the Earth forever if there were no resistance.

[4]1688 was one year after Newton's *Principia*. Who first bought my copy? I doubt if it was Newton; he was alive but he must have read the Latin version some years before. I would like to think that it was John Locke, Newton's friend.

fact, it means zero *change* of speed. His theory works in one special case, when a body is at rest, because there is no force acting on it. Then zero force really does go with zero speed.

Weightlifters regularly demonstrate the validity of the first law for bodies at rest. They exert exactly the same force upward on the *motionless* bar as the downward force due to the weight of the iron that they hold above their heads. Zero net force on the bar, zero change of motion.

Principia, Book II

Newton grasped that there were negligible frictional forces acting on the planets and that this was the reason they kept moving—there was effectively zero net force acting on them *in the direction in which they were moving.* God was not pushing them, and nothing was hampering their progress through the heavens. They had been given a push, long ago, when the solar system was formed, and since then the force acting on them, in the direction of their flight, was negligible. Zero force—zero change of motion.

Newton saw that if there was a resisting medium in space, say air, it would continually slow down the planets, just as your car slows down when the engine is switched off. The car does not stop immediately; it exhibits the property of inertia, the tendency of a moving body to keep moving. If the engine is turned off, the only force acting is the resisting force, and it is this that slows down the car. Newton was aware, from the observations of astronomers, that the planets were not slowing down, at least not detectably. He reasoned that there was effectively no interplanetary matter and that space was a vacuum. He was very nearly right; there is a tiny concentration of molecules in space, but not enough to significantly affect the motion of the planets, not in your lifetime.

Newton's conclusion that space was empty clashed with Descartes's belief in a universe filled with matter. But Descartes was a philosopher; Newton was a scientist.

On Earth, resistance to motion is a practical problem, and Newton attempted to deal with it in the least known of the three parts of the *Principia*, the second book, which is largely devoted to motion in a resisting medium.

Inertia

The first law of motion has sometimes been called the principle of inertia, the term *inertia* implying in general the unwillingness of a body to change its state of motion. We accept inertia as a property of mass because we are brainwashed by the everyday world into accepting certain facts as "common sense." Heavy bodies, like railway wagons, seem to almost consciously resist our efforts to get them moving if they are stationary, or to stop them if they are in motion (facts that should help you believe in the first law.) But being used to something doesn't explain it, and it may come as a surprise to HMS that we have no explanation of inertia. In 1883 the German scientist Mach (the speed-of-sound Mach), who was an original, iconoclastic physicist, proposed that inertia, at least for bodies in this part of the universe, was a property caused by their interaction with *all* of the other bodies in the universe. Albert Einstein called this a "bold idea," but neither Mach nor Einstein ever explained how inertia arose from interaction with the heavenly bodies, and Einstein eventually abandoned the trail. Kant was also puzzled by inertia, which he saw as a *force* resisting changes of motion.

Inertia remains an unsolved problem, another example of a scientific concept that we have become used to, that has had extensive practical use, but that defies our understanding. Dogs and jokes again? Perhaps, but inertia cannot be avoided and it will pop up again when we come to Newton's second law.

It is not only movement in a straight line that persists in the absence of force. A spinning top would keep on spinning forever if there were no air resistance and no friction between top and floor. *Rotational* motion also demonstrates inertia. In certain types of machine a flywheel is used to keep the machine turning over smoothly. The heavy metal wheel keeps things going even when the source of power is not acting; once it is spinning around rapidly it "wants" to keep going. There is research on the possibility of using massive flywheels as a source of motive power for vehicles that travel short distances in limited areas, such as delivery trucks. If Mach was right, they will be driven by the stars.

The Straight and Narrow Path

Two of our protagonists, Aquinas and Newton, were victims of the excesses that often accompany hero worship.

Saint Thomas Aquinas was careless enough to die in a monastery. The "Angelic Doctor" was promptly decapitated by awestruck monks, his body boiled and the bones preserved as holy relics. When the widely admired Saint Elizabeth of Hungary was lying in state in 1231, fervent fans took cuttings of her nails, her hair, and (tremble Madonna) her nipples. Newton was spared this kind of ghoulish attention (it wasn't done in George II's England), but he could well have been in danger a couple of centuries earlier. Dr. Samuel Johnson was of the opinion that Newton would have been worshiped as a God had he lived in antiquity. This was acceptable flattery, but a stranger phenomenon arising from the adoration of Isaac was the attempt to apply Newton's famously successful laws of motion to politics, medicine, and human behavior. Thus, Nicholas Robinson published "A New Theory of Physick and Diseases, founded on the principles of the Newtonian Philosophy." And there was the painful poem of Newton's friend Desaguliers, entitled: "Newtonian System of the World, the Best Model of Government" (1728). Two lines are enough to illustrate its quality:

> Newton the unparallel'd whose Name,
> No time will wear out the Book of Fame.

Newton should have sued, but he was dead. An intellect of an entirely different caliber, David Hume, also came under the spell of the laws of motion and transferred the first law to the workings of the mind: "The imagination, when set into any train of thinking, is apt to continue, even when the object fails it, and like a galley put in motion by the oars, carries on its course without any new impulse."

The best-known effort to force Newton's first law into a farcically unsuitable mold was that of the mathematically minded theologian John Craig, who in his *Theologiae Christianae Principia Mathematica* (1699) (flattery or plagiarism?) proposed his own first law, which could be called the principle of the inertia of pleasure: "Every man attempts to prolong pleasure in his mind, to increase it, or to persevere in a pleasurable state." For a devout Christian this law describes a suspiciously hedonistic trajectory through life; Craig speaks like an undercover agent for Hugh Heffner. A more theologically correct version might read: "Every man continues in a state of spiritual rest or constant devotion to the straight and narrow path unless acted upon by a force of evil."

The Real World

The first law of motion is not enough to deal with the complexity of the real world, the world of cars, stars, people, and other moving objects. The law says that nothing happens to the motion of a body in the *absence* of a net force, but it doesn't say *quantitatively* what happens in the *presence* of a net force. To proceed we need to go to the second law of motion.

4

The Second Law

Good god what a fund of knowledge there is in that book.... [D]oes he eat
& drink & sleep, is he like other men?"

—The marquis de L'Hôpital, on Newton's *Principia*.

The last thing Newton would have wished to be known as was a revolutionary. He
was not to know, nor could he have guessed, that his second law of motion would
help to change the intellectual atmosphere of Europe.

The start of the Indianapolis 500, the circulation of your blood, the paths of the
planets—anytime that a body moving in a straight line *changes* its velocity or direction, the second law is at work. Anytime that a body at rest *starts* moving, we
need the law to predict its behavior. Anytime that a *body in motion* is acted on by a
force, the second law tells us, quantitatively, how that motion changes. Unlike the
first law, it was almost entirely Newton's own achievement; he owed little to
Galileo.

The Stubbornness of Matter

Newton's second law of motion as it appears in the *Principia* can be expressed in
English as follows:

**The acceleration of a body is proportional to the force acting on it and occurs in the
direction of that force.**[1]

Three of the basic conceptual components of the universe are contained in the
second law: mass, force, and motion. They are related by the equation $\mathbf{a} = \mathbf{F}/m$,
which we can write in words as:

Acceleration = Force divided by Mass

To seek out the meaning of this relationship, we will need to define the three concepts involved. *Acceleration* will not be a problem. In discussing *force* we have already adopted a typically British empiricist approach and settled for a description
of what force does rather than what force is. The nature of *mass* is more problematic.[2] Before you persist in knowing what mass and force *are*, listen to that talented

---

[1] If you want to make your own translation, the original Latin goes: *Mutationem motus
proportionalem esse ui motrici impresse, & fieri secundum lineam rectam qua vis illa imprimitur.*
[2] Not only is the concept elusive; we are not even sure where the name comes from. There are
those who point to the Latin *massa*, which means a piece of dough or clay. This could be derived
from the Greek maza for barley cake, a word occurring in the *Agamemnon* of Aeschylus. I favor
the Hebrew *matza*, the unleavened bread eaten at Passover. Anyone who has eaten a few squares
of this bland comestible has a good intuitive feeling for mass.

Danish goal keeper and physicist Niels Bohr: "It is wrong to think that the task of physics is to find out how nature *is*. Physics concerns what we can *say* about Nature."

Bearing which in mind, we now go to the second law, which describes what happens to a body when a force acts on it. I am not talking about a hammer descending on a Dresden shepherdess. We will limit ourselves to bodies that do not shatter or permanently distort.

Acceleration: The Fast Lane

Car companies like to boast about how few seconds it takes for their latest macho model to go "from zero to sixty." If you are rash enough to keep your eyes on the speedometer for the first six seconds of blastoff you might see the needle climb from 0 to 60 mph. This is exactly what we mean by acceleration: an increase in speed. (We can define negative acceleration as a decrease in speed.) There is no guarantee that the needle on the speedometer of an accelerating car will climb smoothly, such that every second that passes gives an increase of precisely 10 mph in the speed. If this were to occur in our case, then after 1 second the speed would be 10 mph; after 2 seconds, 20 mph; after 3 seconds, 30 mph, and so on. If the car were to behave in this way, we would say that it is *accelerating uniformly* at 10 mph per second. *Uniform motion*, on the other hand, is what occurs when a body travels at a constant speed, thereby covering equal distances in equal increments of time. Anyone who drives a car knows what constant speed is and also knows how acceleration feels. Anyone who has sat in an airplane that is taking off knows that acceleration—"picking up speed"—is connected with force. You can feel the back of your seat pressing against you. What the second law says is that if there is a *net* force acting on a body it will accelerate and the acceleration can be calculated from the equation. If we knew the net force acting on a racing car (the sum of the force exerted by the engine and the frictional force opposing its motion), we would only have to know the total mass of car and driver to work out the acceleration. The equation says that for a given mass, m, the acceleration, $\mathbf{a}$, is proportional to the force, $\mathbf{F}$.[3] The bigger the force, the faster the speed increases. Neglecting air resistance, a Boeing 747 that takes off on all four engines will accelerate off the runway at twice the rate it would if it used two engines. If it reaches 200 mph in a minute using four engines, it will only reach only 100 mph in a minute if it uses two engines.

Both Galileo and Newton realized that when a net force acts on a body it *changes* the velocity of the body. Thus, if a football rests motionless on the ground and I kick it, its velocity *changes* from zero to some finite value. As the ball flies through the air, it is opposed by another force, air resistance. This is the only force now acting on the ball, and it must change the ball's velocity. It does: the ball slows down. If another player kicks it, its motion will again be changed by the force exerted by the player's boot. It could be said that the very fact that the ball changes velocity *by definition* means that a force has acted on it.

Notice that the first law is really superfluous; it is a special case of the second law. The first law says that the motion of a body is undisturbed in the absence of a

---

[3]The reason that $\mathbf{a}$ and $\mathbf{F}$ are written in bold type is that they are both *vectors*: they have both size and direction. The velocity of the wind, or of a car, is a vector. Mass is not a vector; it has a size but not a direction. Weight, which is a force, is a vector.

net force acting on it. But this is exactly what the second law says when you put **F** = 0. You find **a** = 0, which means that the body neither speeds up nor slows down, which in turn means that its state of motion is unchanged, whether it is stationary or moving.

The second law tells us how to *quantitatively* relate the mass and acceleration of a body with the force acting on it. It reveals one of the deepest secrets of the physical universe:

The resistance to changes in motion (to acceleration) increases as the mass increases.

That's why the Hell's Angel on his Harley Davidson motorbike has a good chance of leaving you and your Porsche standing when the lights change. You have far more mass and therefore far more inertia than he has.

Apples Again

It was probably Galileo who first measured the acceleration of bodies as they fell to Earth. They accelerate because there is a force acting on them, the force of gravity. This *acceleration due to gravity* is commonly called "g," and, although it varies slightly depending on the exact location at which it is measured, we can, for most purposes, take an average value of 9.8 meters per second, per second. This means that after falling freely for 1 second a body will attain a velocity of 9.8 meters per second, that is about 22 mph, the speed of an Olympic sprinter. After 10 seconds its speed, *in the absence of air resistance*, will reach about 220 mph.

On the Moon the force of gravity is about one-sixth of that on Earth; that means that the *force* on a given mass, m, is one-sixth of what it is on Earth. It follows from the second law that the *acceleration* of a falling body must be one-sixth of that on Earth, that is, one-sixth of g, namely, 9.8/6 = 1.63 meters per second, per second. After falling moonward for one second an astronaut would have a speed of only 1.63 meters per second, or less than 4 mph. Remember how gently Armstrong and Aldrin floated downward after taking a hop? And how high they could jump without great effort?

Notice that if we drop two bodies, one with a mass of, say, five times the other, they will fall with the *same* acceleration—provided we do it in a vacuum to avoid air resistance. This is what Galileo claimed, and it is consistent with the second law. The force **F** acting on the two bodies is just their weight, **W**. The force on the more massive body (its weight) is five times the force on the lighter body, but its mass is also five times bigger so that if we write $a = W/m$ for the lighter body, we have to write $a = 5W/5m$ for the heavier body. **a** is thus the same, for these two and any two falling bodies.

The Units of Force

The unit of length is the meter, and of mass the kilogram. The scientific unit of force is the *newton*. The symbol for the newton is N. It is useful to have a sense of the magnitude of the newton, and since weight is a familiar force, for which we have a feel, we use it by way of illustration. (If you want to convince a colleague that weight is a force, stand on his stomach.) As has been frequently pointed out, the weight of a smallish apple is, pleasingly, about 1 newton, or 1 N. That is the downward force that the apple exerts on your hand and the force that you exert upward on the apple if you want to hold it steady. Newton said that force equals mass times acceleration. My mass is 80 kilograms. If I fall from a roof, my acceleration is 9.8 meters per second, per second. This gives 784 newtons for the force that I exert on the Earth—my weight. Newton probably weighed about 700 newtons.

An Understandably Giant Step

Anyone who has watched the blurred shots of astronauts walking on the Moon can sense that it took little effort for the heavily loaded explorers to leap into the airless. This rubs in the fact that mass and weight are very different animals. If we define the mass of a body as "the amount of substance" that it contains, then it is clear that Neil Armstrong's mass remained effectively constant during his journey to the moon. Plus or minus small percentages, the number and type of atoms in his body were unchanged. In other words, his *mass* was not changed significantly by his journey, but his *weight* was. What changed in going to the Moon was the force with which he was attracted to the ground that he was standing on. The force of gravity on the surface of the moon is only about 16% of what it is on Earth. It is the force of gravity on a body that we call its weight, and that force varies with the planet or other piece of matter that the body is close to. *Weight is a force.* In contrast, *mass is an amount of material.*[4] Armstrong's weight on the Moon would be about 30 pounds weight, or about 136 N—by which is meant that the force pulling his body toward the Moon was equal to the force exerted by the Earth on a mass of 30 pounds. That's why he jumped about so easily. He probably weighed about 180 pounds weight (about 818 N) in the Space Center at Houston. Don't confuse mass and weight. Seen from the point of view of a space traveler, weight is a thoroughly local property, depending on where you are in the universe.

Now consider Phobos, one of the two potato-shaped moons of Mars. Its dimensions are roughly 12 by 13 by 17 miles. The force of gravity at its surface is about one-thousandth of what it is on Earth. Neil Armstrong would weigh less than 3 ounces on Phobos, about the same as a Hershey bar on Earth. The weakness of gravity is apparent. The fact that galaxies are held together by the force of gravity is a reflection of their immense masses. It is only in cosmology that gravity reveals awesome power.

The Third Law

If you dive off a cliff, you will accelerate as you fall because there is a net force acting on you—your weight. When you reach the ground, the same force is acting on you, yet you do not move. This is not a contradiction of the second law. You are stationary because there is no *net* force acting on you. The explanation is that the Earth is pressing upward on you, and that this force is exactly equal to your weight. This is an example of Newton's third law. Newton saw forces as acting *between* two bodies, not as one body exerting a force on the other. He spoke of *action* and *reaction,* which he declared to be equal and opposite.

In the case of someone standing on the Earth, this equality is obvious: what is perhaps less obvious is that while you were falling off the cliff the pull of the Earth on you (which you can call the action) is equal and opposite to the gravitational

---

[4]I know that this is an unsatisfactory definition, but it is understandable and useful. The real truth is that we don't really know what mass is.

pull that you exert on the Earth (the reaction). Actually it doesn't matter which you call what. There are two equal forces, as though there were a spring stretched between you and the Earth, pulling both of you together with the same force. But there's something wrong here. I'm moving because the Earth pulls me; why isn't the Earth moving if I pull it?

It is! The force between you and the Earth is given by your weight, say 600 newtons. This is also, according to Newton's third law, the force that you are exerting on the Earth. We can therefore find the acceleration, **a**, of the Earth by putting the right numbers into $\mathbf{a} = \mathbf{F}/m$, remembering that m is the mass of the Earth, that is 6 x 10^{24} kilograms. Putting $\mathbf{F} = 600$ newtons we find that $\mathbf{a} = 10^{-22}$ meters per second, per second. If it took you 10 seconds to fall to Earth, the Earth's velocity at the end of those 10 seconds, because of the pull that you are exerting, would be 10^{-21} meters per second. In familiar units: 2.25×10^{-21} mph. To give you an idea of how slow this is, if you traveled at this speed, it would take you about a billion years to travel one-thousandth of an inch. Little wonder that the Earth doesn't appear to move for you. The difficulty of fighting inertia is very clear here; it takes a mighty push, or pull, to get the Earth to detectably change its motion.

The Limits of Newton's Laws

... had Newton not been a mere beginner at gravity he might have asked how the apple got up there in the first place. And so might have discerned an ampler physics.

—Les Murray

1. Before Newton, there was no way of predicting or understanding the way in which bodies moved under the action of forces. For the world of everyday objects and for the known heavenly bodies, Newton gave an answer, except in a tiny number of curiosities, such as the exact path of the planet Mercury. However, although the path of an artificial planet revolving about the Sun will be described reasonably well by Newtonian mechanics, that mechanics can predict neither the existence of the new planet nor the path in which man puts it. Newton himself realized that the fact that nearly all the planets revolve about the Sun in the same plane was not a consequence of his laws. There is no reason they couldn't all be moving in different planes. He assumed that the plane in which they move was chosen by God. There are other, more prosaic explanations.

2. In the twentieth century, Newtonian mechanics was shown to be entirely inadequate to deal with the behavior of atoms. The answers that Newton couldn't provide were supplied by quantum mechanics.

3. Cosmology, the very area where Newton had had his greatest triumphs, was to demonstrate, in one of the most spectacular experiments ever performed, that there is a complete schism between the way Newton saw time and space and the way Einstein's theory of relativity described them.

We come to these matters later on. Meanwhile, don't be tempted to downgrade Newton. He remains a giant, and it is worth remembering that the experimental facts that led to twentieth-century science became available only long after Newton's death.

5

Predicting Catastrophe

In other words, I am buried in my work, like Newton (who incidentally died a
virgin) in his famous discovery of the law of gravity (I think that was it).

—Eugène Delacroix, letter to George Sand

Circular No. 5636 of the International Astronomical Union, published on 15 October 1992, stated that the Swift-Tuttle comet (named in 1862) might collide with the Earth on 14 August 2126. The comet is large enough to threaten the continuation of civilized life on this planet. The second sighting of what could be the most important object in human history was by an amateur Japanese astronomer. The announcement triggered a rash of measurements, and a more accurate prediction of the comet's path showed both that it is the comet that was seen in China in 69 B.C., and that we will be safe in 2126 but may be in danger on 26 August 3044. More measurements will enable the accurate prediction of the nearest approach (hopefully not zero) to the Earth in 3044. We know that this can be done because astronomers have accurately predicted the return of comets before.

The Norman Conquest of England in 1066 was commemorated by a wall hanging that can still be seen in the town of Bayeux in Normandy. One of the panels shows an object in the sky that can be identified as Halley's comet, so called after the English astronomer who observed its return in 1682. (The Englishman Thomas Harriot, an early visitor to America, observed the return of 1607.) Halley predicted that the comet would return in 1758. His calculations were improved on by others, who gave the expected date as March 1759. The comet appeared in April, not bad agreement for an age without computers. How is it that we can make accurate predictions of this kind? What is their basis?

Newton's Revelation

At the age of fifty, Isaac Newton, already renowned throughout Europe as the greatest natural philosopher since antiquity, began an intense study of the books of Daniel and Revelation. His purpose was to predict the date on which the world would end. Now, although Newton had no doubt that there was an all-powerful Creator who determined the overall path of history, he didn't believe that the Almighty spent his days and nights chaperoning the universe, a permanent backseat driver for the material world. Once the heavenly bodies had been set in motion, that motion was controlled by laws. And the laws were universal. They controlled the path of the Moon and the path of an arrow, the trails of the comets and the surge of the tides. Newton was too religious to deny God the right and ability to manifest his power by interfering with the workings of the world, but in general, although God might be responsible for every sparrow that fell from the tree, he did not usually decide *how* the sparrow fell; that was determined by the laws of motion—and the law of universal gravitation.

Gravity is one of the handful of forces that mold the universe. It is almost certainly the force that will overwhelmingly determine the end of the present universe. Newton revealed just how ubiquitous it is.

I Shot an Arrow . . .

Until the space age, all man-made missiles fell back to Earth. The Greeks had only the most evasive explanation of this fact. The general attitude was that things fell because that was their nature. In this vein, Aristotle declared that the natural place for all heavy things was on the ground—it was obvious. The moon presented a difficulty: what kept it up there? Aristotle decided that it was made of ether, which solved everything. However, arrows and apples fell to Earth because the Earth was their natural home, and they moved faster as they approached the ground because of their happiness in returning home. Galileo, doubting whether jubilation was within the range of emotions normally expressed by the average apple, contented himself with determining *how* bodies fell, not *why*. He measured the position of falling bodies as a function of time, rather than creating anthropomorphic reasons for their behavior. He accepted that the Earth attracted all material bodies, but he ventured no explanation of why this should be so, or indeed of why the Moon did not fall.

Newton, born in the year that Galileo died, contemplated the second-most famous apple in history and realized, like Eve, that perhaps *any* two bodies could attract each other, not only Earth and apple. In a superb gesture, he reached for the stars and extended Galileo's earthbound ideas to any two material bodies in the whole universe.

Universal Gravity

Newton said three fundamental things about gravity:

1. It is a *universal* force. Any two bodies attract each other. Gravity operates between you and the coffeepot, between you and the farthest galaxy. The pull of the Earth on a falling object is just the most familiar example of a phenomenon that is a built-in property of all matter.

2. A simple mathematical expression allows the calculation of the strength of the gravitational force of attraction, **F**, between any two bodies. The expression is known as Newton's law of universal gravitation:

$$\mathbf{F} = \frac{G \times m_1 \times m_2}{\mathbf{R}^2} \tag{1}$$

In this equation m_1 and m_2 represent the masses of the two mutually attracting bodies and $\mathbf{R}$ the distance between them. The dependence on $1/\mathbf{R}^2$, is the reason we refer to the equation as an "inverse square" law. Regarding G: many of the equations that govern the behavior of the universe contain a small number of *constants,* numbers that Nature hands out to us with no explanation as to their magnitude. We can only determine the values of such constants by experiment, and this has been done very accurately for the so-called gravitational constant G.

3. Newton refused to explain what gravity was. "I have not been able to discover

the causes of those properties . . . and I frame no hypothesis [*non fingo hypothesis*]. . . . [I]t is enough that gravity does really exist, and act according to the laws which we have explained."

Don't judge Newton too quickly on this last issue. Science is very good at using shadowy concepts to make astonishingly accurate predictions. It is a mark of Newton's sophistication as a scientist that, living when he did, he realized the limitations of his postulate of universal gravitation: "To tell us that every Species of Things is endow'd with an occult specifick Quality [*he meant like gravity*] by which it acts and produces manifest Effects, is to tell nothing." Nevertheless, Newton expressed the belief that the cause of gravity is not "incapable of being discovered and made manifest." It was his refusal to be dogmatic, his belief in the ultimate triumph of observation and reason, that was to be one of Newton's greatest contributions to the Enlightenment. His scientific honesty comes like a glass of ice-cold beer after a route march in the desert. It is a pity that he was sometimes less than honest in dealing with his scientific rivals.

It is worth pausing to consider Newton's famous rejection of hypothesis. On the face of it, he appears in this respect to be a disciple of Bacon, but it is very obvious from his work that he did hypothesize. His statement *non fingo hypothesis* can perhaps be understood, not literally, but as a protest against the placing of hypothesis above empirical laws. In this context it is interesting that Alexandre Koyré has suggested that *fingo* actually means "feign." What was *observed* had to be believed; what was hypothesized could be found wanting. This point was central to Newton's philosophy, as stated very clearly in the *Principia*.

The idea of gravitational attraction between the bodies in the solar system had also been suggested by the remarkably productive physicist Robert Hooke but not developed. Newton realized that gravity was the key to the motion of the heavenly bodies. He wrote down the equations by which it is possible to account for the paths and speeds of the planets. But he had to have facts against which to check that his calculations made sense.

The Basic Facts

For Newton's theories of planetary motion to have any credibility, it was necessary that his calculations of the paths of the planets agree with their observed paths. The most accurate data on the positions of the stars and the movements of the planets were those of Johannes Kepler (1571–1630). Kepler, a Lutheran born in southern Germany, was a meticulous data gatherer and a mystic. His great good fortune was to have inherited the notebooks of the Danish astronomer Tycho Brahe (1546–1601), "without whose observation books everything that I have brought into the clearest light would have remained in darkness."

Brahe was the pleasure-loving, star-intoxicated eldest son of a Danish nobleman. In 1566, on a legendary occasion during his student days at the University of Rostock, he fought a duel over the question of who was the better mathematician, he or his opponent. The bout took place in near darkness, and a slice of Tycho's nose was carved off. He replaced it with an artifact made of silver and gold.

The king of Denmark gave Brahe the island of Hven on which to construct an observatory. In Uraniborg (the Heavenly Castle) Brahe spent twenty years gazing at the stars. All in all, over his lifetime he catalogued the positions of a thousand stars, thus supplanting the far less accurate tables of the second-century Alexandrian

Figure 5.1. If the time for a planet to travel from a to b is equal to that from c to d then, according to Kepler, the (shaded) areas swept out by the line joining the planet to the Sun are equal.

astronomer Ptolemy. When Brahe lay on his deathbed and handed his records to Kepler, it was on the understanding that Kepler would use the data to prove that the Earth was at the center of the solar system, in contradiction to Copernicus.

For some quarter of a century, driven by his Pythagorean belief that number ruled the universe, Kepler pored obsessively over Brahe's data, and out of the huge mass of numbers he conjured three simple and striking laws (Figure 5.1):

1. Planets travel in elliptical paths about the Sun. Aristotle and Copernicus had said circles.
2. In equal intervals of time, a line from the planet to the Sun sweeps out equal areas. This implies that the speed of the planet is not constant, contradicting Aristotle.
3. For any planet, the square of its time of revolution around the Sun, divided by the cube of its mean distance from the sun, is the same as for any other planet.[1]

Kepler was ecstatic: "I yield freely to the sacred frenzy; I dare frankly to confess that I have stolen the golden vessels of the Egyptians to build a tabernacle for my God far from the bounds of Egypt. If you pardon me, I shall rejoice; if you reproach me, I shall endure. The die is cast, and I am writing the book, to be read either now or in posterity, it matters not. It can wait a century for a reader, as God himself has waited six thousand years for a witness." In any event, he had to wait seventy years before Newton showed that his laws had a rational explanation.

Kepler did not know *why* the planets took the paths that they did; his laws were empirical laws, extracted from Brahe's data by trial and error. Notice that his third law suggested that the planets, in some sense, collectively formed a separate unit from the stars. The concept of the solar system, implicit in Copernicus's work, was finally crystallizing.

With the power of genius, Newton devised the law of universal gravitation and, combining it with the laws of motion, constructed an all-embracing theory that accounted for Kepler's laws and for the fall of apples. It is not consistent with our present purpose to follow Newton's analysis of planetary motion, but the magnitude of Newton's breakthrough was clear to his contemporaries, and it should be clear to us. It is one of the very greatest examples of the application of mathematics to facts derived from observation of the natural world. This was the hypothetico-deductive method in all its glory. Starting with a hypothesis (the laws of motion and gravitation) based on meager facts, Newton had deduced the consequences, and shown that they were compatible with Kepler's laws, laws based on observation of Nature. By showing that the planets followed paths that were based on the same laws of motion and gravitation that applied to the movement of bodies on Earth, Newton

---

[1]The students of the Japanese astronomer Asada Goryu (1734-1799) claimed that he independently discovered Kepler's third law.

demolished Aristotle's belief that celestial and terrestrial bodies were subject to different laws. The cosmos had been united.

When he wrote his masterpiece, Newton named it with concise accuracy: *Philosophiae Naturalis Principia Mathematica* (Mathematical Principles of Natural Philosophy). The title page of the first edition is inscribed "Julii 5. 1686," but it appeared in the following year.[2]

In 1686, Robert Hooke spoke to the astronomer Edmond Halley, who was actively involved in preparing Newton's manuscript for publication, and asked for his contribution to Newton's work to be mentioned. Newton deeply resented Hooke's claims to priority, dismissing him as a man of "strange unsociable temper" and impulsively threatening to withdraw Book 3 of the *Principia* rather than have Hooke's name appear. Fortunately, he relented, but this was not the only occasion when the two men clashed, and it cannot be said that either emerged with great credit from their quarrels. Scientists can be fiercely competitive. In our day we have witnessed an unedifying priority wrangle over the question of which laboratory first identified the AIDS virus.

Newton's use of the laws of motion, together with the law of universal gravitation, to explain the motions of the heavenly bodies is legendary. But he did not close a chapter of scientific history; he opened one. After 300 years, gravity still occupies the attention of theoretical physicists; gravity is very peculiar.

The Strangeness of Gravity, Part I

Newton, as we have seen, did not lack intellectual enemies, and the strangeness of gravity gave them fuel. How could two bodies be pulled toward each other without being in contact in some way? Force-at-a-distance: no strings, no springs, how does it work? It smelled of necromancy. Leibniz affirmed that "gravity must be a scholastic occult quality, or the effect of a miracle." Perhaps he was getting in a dig at Newton's obsession with alchemy and magic. Newton, Leibniz declared, "had a very odd opinion concerning the work of God." In one of his letters Leibniz complained, "In the time of Mr. Boyle and other excellent men, who flourished in England under Charles II (in the early part of the reign) no body would have ventured to publish such chimerical notions. . . . But it is men's misfortune to grow, at last, out of conceit with reason itself, and to be weary of light."

The chemist Robert Boyle was puzzled by the idea that a body could, because of gravity, move according to certain laws. How, he asked, could matter "obey" laws, as though it had a will of its own? How does a piece of inert matter *know* that it should obey laws? Bishop Berkeley also had deep doubts about Newtonian physics:

> *Force, gravity, attraction,* and terms of this sort are useful for reasonings and reckonings about motion and bodies in motion, but not for understanding the simple nature of motion itself. . . . As for attraction, it was certainly introduced by Newton, not as a true, physical quality, but only as a mathematical hypothesis . . . to be of service to reckoning and mathematical demonstration is one thing, to set forth the nature of things is another.

---

[2]The title page also carries the inscription: "Imprimatur. S. Pepys, Reg. Soc. Præses." The famous diarist was the president of the Royal Society.

Berkeley, who was often astonishingly in tune with modern thinking on the nature of science, also had doubts about the deterministic philosophy of cause and effect, implicit in the *Principia*. Things fell to Earth all right, but why conjure up a hypothetical attraction to explain it? All that was happening was that every time we release something it falls. Therefore, all we can say is that these two events always appear together, which doesn't mean that one has a causal connection with the other.[3]

This was not the last time that science met philosophy.

The Strangeness of Gravity, Part II

Suppose that you got into an elevator on the hundredth floor and had the unsettling experience of hearing all the cables snap. The elevator plunges to Earth. Free fall. I expect that you will spend your last moments doing experiments. Take out your car keys, hold them in front of you and release them. They stay where they are, hanging in the air opposite your belt buckle, because they fall together with you. Galileo said so: all objects fall at the same speed, and since the air in the elevator is falling with you, there is no problem of air resistance. Now suppose that you had been asleep when the elevator started to fall and had woken up in midflight. The elevator has no windows. You would not know that you were falling. In fact, as far as you are concerned, the elevator is not moving. Yet your keys don't fall. Dr. Moriarty has apparently finally thrown the antigravity switch; gravity doesn't exist anymore *for you,* and this remains true as long as the elevator falls freely, accelerating as it goes. *There is no detectable gravitational force acting on you or the keys.* Furthermore, there is no experiment that you can do, *inside the elevator,* that will tell you whether you are accelerating and the force of gravity still really exists "out there," or that you are stationary and the force of gravity has vanished. We conclude that, for an observer in a freely falling elevator, acceleration appears to be able to cancel out the force of gravity. But a force can be canceled only by another exactly equivalent force acting in the opposite direction, as in the case of two balanced tug-of-war teams. It looks as if *acceleration induced by gravity is completely equivalent to the force of gravity.*

That acceleration can mimic the effect of force can be verified by any airline passenger. If, against safety regulations, you leave your table down and place a glass on it, the glass will slip toward you as the plane accelerates on takeoff. This does not happen if the plane is cruising at constant speed. During acceleration a force appears to be acting on the glass. Note that to an outside observer, looking through the window of your plane, the glass appears to be displaying the property of *inertia*; it tries to stay motionless with respect to the runway, as it was just before takeoff. For the motionless external observer there is no force acting on the glass. For you there is. We have a first hint here that the observed behavior of a system can depend drastically on the motion of the observer.

The strange equivalence of gravity and its associated acceleration could, in principle, have been noticed since Galileo's time, but it took the genius of Einstein to

---

[3] As Thomas Reid (1710–1796), the Scottish originator of what came to be called the "commonsense school of philosophy," pointed out, day invariably followed night, but they were not linked by cause and effect; days do not cause nights.

realize its importance. To most of us the equivalence would appear to be a mean-ingless fact, a simple consequence of the way we have constructed our experiment. For Einstein it was the trigger that led him to construct the general theory of relativ-ity, one of the keys to the large-scale structure of the cosmos.

Black Holes and Dirty Harry

When we come to discuss cosmology, we will need the concept of escape velocity. As you get farther from the Earth, you feel its gravitational pull lessening. That is a fact and it is reflected in equation (1); the bigger **R**, the smaller **F**. Increase R a thou-sandfold, and the force drops by a factor of a million.[4] The drop-off in gravitational force with distance suggests that if a projectile, a bullet say, is fired vertically up-ward with enough initial speed it will not eventually be conquered by gravity and fall back to Earth, but will escape, speeding outwards forever, perhaps to circle the Sun.

Travel into deep space is obviously possible only if we can overcome gravity. The "easy" way to do this is to fire an object into the sky at high velocity and also provide it with its own rockets. In this way, if it is tempted to slow down too much and eventually stop and turn around, it can give itself a boost by turning on its rockets. Alternatively, the initial blast can be prolonged so that it is still operating high above the Earth where the pull of gravity is weak. The hard way to escape the Earth's gravity is to fire a simple missile, having no propelling source of its own, say an arrow or bullet, into the deep blue yonder. What would the initial velocity of such an arrow have to be to escape the Earth forever?

Using equation (1) it can be shown (as they so annoyingly say in the textbooks) that the minimum velocity needed for a body to escape permanently from any planet is given by:

$$\text{Escape velocity} = V_E = \sqrt{\frac{2 \times G \times M}{R}}$$

In this expression G is the constant in equation (1), M is the mass, and **R** is the ra-dius of the (assumed) spherical planet from which the projectile is hoping to es-cape. The expression does not take into account the slowing effect of air resistance, but for reasonably aerodynamic bodies—say, rockets—the expression gives a very good rough answer.

The important point to notice is that nowhere in the expression does the mass of the *escaping body* appear. This means that the escape velocity is the same for all missiles, a shotgun pellet or an artillery shell.

We have the data needed to work out the escape velocity from the Earth: M, the mass of the Earth = 6×10^{24} kg, **R** = 6.37×10^6 m, and the numerical value of G is 6.67×10^{-11}, from which we find that V_E = 11.2 km/sec. In more friendly units this is a little over 25,000 mph, or 7 miles per second. This, then, is the minimum ve-

---

[4]The drop-off in gravitational pull at greater heights means that the Earth is pulling on your feet more than on your head. This effect is too tiny to have any serious consequences, but if you were approaching a neutron star or a black hole (Chapter 34), both of which have stupendous gravitational fields, you would start stretching and eventually be pulled to pieces!

locity with which an object would have to leave the ground to ensure that it would never return. The muzzle velocity of Dirty Harry's (Clint Eastwood's) Smith and Wesson 29, using 0.45 Magnum ammo, is only about 1600 mph.

On the Moon, V_E is about 5400 mph. This is still too large for Armstrong and Aldrin's lunar module to escape from the Moon under their own power, which is why they did it in two stages, first joining the circling space module. On Phobos, one of Mars's two tiny moons, V_E is about 90 mph. Ivanisevic's first service against Agassi in the 1992 Wimbledon final was timed at 116 mph. If directed vertically upward from Phobos it would never return—though it is unlikely that this moon will ever be included in the Grand Slam circuit.

We will return to escape velocities when we consider black holes.

Jumping on the Bandwagon, Part II

Just as the laws of motion were invoked to explain everything from politics to morality, so the law of universal gravitation was dragged into politics, biology, and chemistry. The Enlightenment scientist Maupertuis, who in 1732 published the first French text to accurately explain Newton's theory of gravitation, hazarded that the law of gravity might have a parallel in the functioning of living organisms.[5] David Hume spoke of the association of ideas in Newtonian terms. There were those who likened a king and his courtiers and advisers to the Sun and its planets, the latter held to the former by an attractive force. These attempts to incorporate major scientific breakthroughs into other fields now appear ludicrous, but the same phenomenon was to recur with Darwin and Einstein.

---

[5]It was typical of the unpleasant side of Voltaire's character that he dismissed Maupertuis's role and falsely claimed that he was the first Frenchman to explain Newton to his compatriots. And this after Maupertuis had helped Voltaire to write the chapters on Newton in the latter's *Lettres philosophiques*.

6

From Newton to De Sade: The Partial Triumph of Reason

Reason must be our last judge and guide in everything.

—John Locke, *An Essay Concerning Human Understanding*

Glory to God who created the worlds and Newton!

—The abbé Jacques Delille

The Enlightenment

Something encouraging happened in seventeenth-century Europe. It became more fashionable to shout, "That's not reasonable!" than to whisper, "That's heretical!" Some call the period that culminated in the 1760s the Age of Reason, some speak of the Enlightenment. The French, who were at the heart of the Enlightenment, called it "le siècle des lumières." It was a time of intense intellectual excitement, a time when men began to feel that they had the ability to understand and control the world and that the means by which to achieve those aims were not revelation and dogma but intelligence, reason, and science. Not since the Greeks had reason and science had such a good press. Newton's role was central.

Reason, for the philosophers of the Enlightenment, was not only a means of reaching truth but also a principle to be defended. Consistently, the Enlightenment was a time when tolerance was increasingly advocated and the influence of the established Churches continued to decline. Belief grew in the rule of rational, not arbitrary, law, both within and between nations.[1] The men and women of the Enlightenment emphasized that morals should not be divorced from politics and economics. They believed passionately in education. Above all they took as axiomatic the primacy of the individual, his religious and political freedom, his right to believe or disbelieve. They of course rejected the concept of original sin. Many of them saw pre-Christian Europe, and the civilization of Greece and Rome, as a legitimate source of morality. The Enlightenment was to irrevocably alter the mental mold of Western man; it began to seem possible that Bacon (quoting a Roman source), was right: man is the architect of his fortune.[2]

---

[1]Michael Polyani has controversially suggested that the glorification of a rational secular society, associated with the Enlightenment, led to the totalitarianism and nihilism that in turn are the direct roots, he believes, of the Communist and Nazi movements that have so scarred our century. Others have held similar views, but the topic, though provocative, is not appropriate to this book.
[2]To be politically correct you can write *people* wherever I have written *man* or *men* in this paragraph.

The roots of the Enlightenment were many and ran deep. We focus on a small number of outstanding themes, which are of particular significance here because they are inextricably intertwined with the scientific revolution of the seventeenth century. We will glance at skepticism, at reason, and at deism. Above all, we will consider the role of Newton.

Newton and the *Principia* did not create the Enlightenment, although he has his champions, as have Bacon and *The Advancement of Knowledge,* Descartes and the *Discourse on the Method,* Locke and his *Essay Concerning Human Understanding* or Montaigne and his *Essays.* They all played a part, as did others, but if I had to choose one hero I would choose Newton. More specifically, I would focus on the *Principia* and, if forced to absurdly and perilously narrow my field of fire, I would target the second law of motion. Newton was very definitely not an Enlightenment man, but the *Principia,* that towering monument to reason, was a perpetual presence hovering behind the Enlightenment.

Whatever portion is finally allotted to Newton, there is no question that science played a star role in the Enlightenment. In his influential *Origins of Modern Science* (1949), Herbert Butterfield writes of the scientific revolution of the sixteenth and seventeenth centuries, that "it outshines everything since the rise of Christianity and reduces the Renaissance and Reformation to the rank of mere episodes, mere internal displacements, within the system of medieval Christendom. . . . [I]t changed the character of men's habitual mental operations . . . and the very texture of human life itself." He was carried away, but understandably.

The Roots: Skepticism

"We know nothing, not even whether we know or do not know, or what it is to know or not know, or whether anything exists or not." So said Metrodorus of Chios, in the fourth century B.C. Skepticism in this extreme form can question our ability to know anything. It can question the validity of reason itself.

A doubter of the highest order was the third-century B.C. Greek thinker Pyrrho of Elis, whose negative reflections were collated by his follower, Sextus Empiricus. Pyrrho was one of a long line of philosophers, going through Plato and Descartes, who emphasized the doubtful validity of the information provided by our senses. When Empiricus's Greek text was translated into Latin by Henry Estienne, it fell into sixteenth-century Europe like rain on budding wheat. The seedlings of skepticism were there waiting, and none was more thirsty than the essayist, Michel de Montaigne (1533–1592) and his distant relative Francesco Sanchez. They both veered toward a path that Descartes was to follow later, a path that led to the thoroughly skeptical conclusion that the only sure knowledge was of oneself. Sanchez states clearly that nothing can be known, all that can be done is careful empirical research, with the hope that patterns will emerge which will help us know how to act. Montaigne's *Apology for Raymond Sebond* (1575) (which, incidentally, says very little about Sebond) says a great deal about man's *inability* to know. Montaigne makes himself absolutely clear: "I determine nothing; I do not comprehend things; I suspend judgment; I examine." His motto, *"Que scay je?" (What can I know?*) was Descartes's departure point in his search for knowledge.

Skepticism had a powerful revival in sixteenth century Europe, eroding dogma, fertilizing the ground for the seeds of the scientific revolution. In the Western world, that skeptical, questioning spirit was to become a permanent part of our culture. It began to be accepted that people have a right to question all things without being burned alive or broken on the wheel by pious clerics. There were to be no more Galileos.

Perhaps the most representative skeptic of the period proceeding the Enlightenment was the son of a Huguenot pastor and the writer of *Historical and Critical Dictionary*, a book that had a strong effect on many of the Enlightenment thinkers, including Voltaire, who said of the author that "he taught doubt." Pierre Bayle (1647–1706) attacked superstition, prejudice, and dogma, and made quiet fun of the improbabilities of the Bible. He held that comets are natural, not supernatural, phenomena, and that the bones of a dog are as efficacious as those of a saint if only the worshiper believes in them. He was a skeptic, but not a super-skeptic. The portly eighteenth-century philosopher David Hume was.

Hume challenged man's ability to ever explain the world. He set out to use Newton's scientific method to build a moral philosophy modeled on Newton's natural philosophy. Early on he dismissed religion, but the skeptic in him took over completely, and eventually he was prepared to state: "I am ready to reject all belief and reasoning." He even proceeded to challenge the validity of the concept of cause and effect.

Metrodorus and Hume went a doubt too far for the philosophers and scientists of the Enlightenment, whose skepticism was more moderate, but still sharp enough to challenge the scientific and religious establishment. They were empirical in outlook: the world existed, it ran in conformation to certain laws, these laws could be discovered, even if (and here their skepticism also revealed itself) there were limits on our understanding of the true nature of the physical world.

One thing was certain: all our knowledge of the natural world comes, as Bacon knew, from our senses. This is the attitude that we find, modified by the character of each, in Newton and in Locke.

The Enlightenment atmosphere of skepticism and empiricism was highly congenial to the advance of science, especially since, to this antimetaphysical mixture was added an ancient and highly metaphysical component—the belief in the existence of order in the cosmos. The Pythagorean ideal of natural order, a universe governed by immutable laws that were immune even to supernatural interference, is one of the great pillars of the Enlightenment mind.

The ground cleared by skepticism in the sixteenth century was not immediately occupied by the armies of reason.

The Roots: Reason

I speak of reason, not of rationalism. They are far from synonymous. For the professional philosopher, "rationalism" is a technical term, including not only the belief in the *use* of reason to arrive at truth but also the proposition that we can arrive at important truths about the world by the use of reason *in the absence of experience*. This philosophy, represented in different ways and degrees by Descartes, Spinoza, and Leibniz, is not what I mean when I talk of reason. Reason, for us, refers to the use of logic, to that rational discourse which has, with frequent and woeful exceptions, been accepted in the West as the normal tool for settling intellectual differences.

The Greeks, as usual, had laid the foundations. Plato said that we must "strive to reach ultimate realities without any aid from the senses," which makes him a rationalist as well as an advocate of reason. Aristotle was less of a philosopher than his mentor, but almost single-handedly he created the formal theory of logic; for him man was "a rational animal." Plato's teacher, Socrates, died for his belief in the right to logically examine and question all things. He was not the last. Zeno tested the power of reason by constructing brain-teasing paradoxes, and Euclid wrote the earliest of logically built texts, his *Elements,* which was a series of geometric proofs, each carefully constructed on the preceding proof, the whole structure being supported by a set of axioms.

The apparent certitude of mathematics, in a sea of doubt, struck deep into the minds of several seventeenth-century thinkers, notably Thomas Hobbes and René Descartes. Is there a more beautiful story of the stepwise seduction of one mind by another (across two thousand years) than that of Hobbes opening Euclid in the middle and, not convinced by the proof before him, going back to the preceding proof on which it rested and continuing to work his way backward theorem by theorem to the opening axioms? Hobbes saw Euclid's step-by-step construction of a system from (presumably) unassailable axioms as the perfect model for the scientific method.

There was no lack of reasoning men before the eighteenth century, so why do we single out the Enlightenment as the Age of Reason? Why did Voltaire declare the eighteenth century to be "that century when reason was perfected"? Because the standing of reason and logic in the millennium before the Enlightenment was not comparable to that of irrational faith. Logic was an established subject in the medieval university, but chiefly as an accessory to law and theology, not as a tool for examining the workings of the natural world. In theology, reason was certainly not a universally welcome guest. At the Council of Trent in 1551, the papal legates criticized those within the Church who use the path of reason instead of confirming "their opinions with the holy Scripture, Traditions of the Apostles, sacred and approved Councils, and by the Constitutions and Authorities of the holy Fathers." Anselm, the eleventh-century Archbishop of Canterbury had put it more concisely: "I believe in order that I may know, I do not know in order to believe."

It is true that Aquinas applied reason to the question of God's existence but would never have said so if he had doubted what the result would be. The implicit view of the Catholic Church was that reason could be used to explain and reinforce the Scriptures but not to challenge either them or the Church's description of the cosmos.

The notoriety of Galileo's clash with the ecclesiastical authorities is often used to support the thesis that the rise of science in the seventeenth century was opposed single-handedly by the Catholic Church. This is quite untrue; if anything it was the *Protestant* churches on the Continent that took the lead in questioning the fruits of rational science. Luther may have questioned the workings of the Papacy, but he regarded "how?" and "why?" as "dangerous and infectious questions." It was the Lutherans who perhaps put up the first serious opposition to the Copernican suggestion of a heliocentric universe, which they saw as a direct attack on the literal truth of the Scriptures. Luther said of Copernicus, "The fool wants to turn the whole art of astronomy upside down." In 1549, his friend Melanchthon said of the heliocentric hypothesis: "Now it is want of honesty and decency to assert such notions publicly, and the example is pernicious." The straightlaced Protestant John Calvin also blasted Copernicus, but reason was not exactly the strong suit of

Calvin, whose God arbitrarily and irrevocably sorted out souls for Heaven or hell, before birth. There was a limited number of tickets to Heaven—and the stadium was not that big.

The Catholic Church was not so quick on the draw. It took the Vatican about 80 years to wake up to Copernicus and attempt to ban his teachings. Like Copernicus, Galileo learned that using reason in discussing nature was not the way to win friends and influence people. However, it would be a mistake to highlight the rational aspect of the struggle between Galileo and the Church. The pope did not challenge the most rational of Galileo's works—his mechanics; he challenged his *observations* on the solar system and his conclusion that the Sun, not the Earth, was at its center.[3]

In the religion of Reformation and Catholic Europe, the bounds of reason were firmly prescribed by faith, and in some countries the torturer was there to ensure that this cardinal principle was not forgotten. Protestant England breathed slightly freer air. The seventeenth-century chemist Robert Boyle could risk saying of other creeds: "Why a man should be hanged because it has not yet pleased God to give him his spirit, I confess I am yet to understand." In the sixteenth and early seventeenth century, England produced a number of influential, reasoning, antiauthoritarians with scientific leanings. Thus Bacon had trumpeted the virtues of observation as against accepted wisdom, and Thomas Hobbes had attempted to build a sound basis for knowledge by using the kind of mathematical reasoning that had so enthralled him in Euclid. The scattered beginnings of dogma-free science could be detected. More and more thinkers were prepared to use their common sense, and their senses, rather than appeal to the so-called wisdom of the ages.

Why, then, is sixteenth-century England not included in the Age of Reason? Primarily because reason had failed to show any earth-shattering popular successes. Science and scientists were not held in particularly high regard. Bacon's prediction of a science-based Utopia had led nowhere in practice. Hobbes was (unjustifiably) under suspicion as an atheist.[4] The popularly perceived achievements of Galileo were confined to the wonders revealed by his telescope, and this is reflected in the way that writers and poets referred to him. His highly rational work in mechanics was barely taught in the universities. Physics was in its childhood, alchemy was a cookbook heavily laced with mysticism, and biology had hardly advanced from the inaccurate catalogue of birds and beasts left by Aristotle. As John Donne cuttingly commented,

> We see in authors, too stiffe to recant,
> A hundred controversies of an ant.

He was, understandably, far from alone in his low estimation of science. Observation was too often undirected and weakly linked to theory—if at all. Reason had few triumphs to flaunt.

Expanding our view to take in the spirit of the times in general, it is clear that the

---

[3]It is a curiosity of intellectual history that, in the eighteenth century, when the freethinker Rousseau, in his *Discours,* blasted science, it was a Jesuit publication that came to its defense.
[4]Hobbes nowhere says that there is no God; he merely says that God is completely incomprehensible. On the other hand, the term *atheist* tended to be used against anyone who strayed outside the conventional beliefs of the time.

first decades of the seventeenth century were an era of profound uncertainty for thoughtful men. Skepticism flourished, religion had splintered into many cults, reason was weak. Donne, not knowing that a new dawn was near, spoke famously for the times when he lamented, "I was born in the last Age of the World."

And then came Newton. He was not alone. As an unwitting heralder of the Enlightenment, he was preceded by Bacon and followed by Locke. As natural philosophers, his English contemporaries Robert Boyle and Robert Hooke were princely figures, as was the Dutchman Huygens. But, among those whom we associate with the scientific roots of the Enlightenment, Newton was king.

The Roots: Newton, the Catalyst

> We are all his disciples now.
>
> —Voltaire

Newton was not only a genius, he was popularly and almost universally perceived as such, and his achievements colored the intellectual life of Europe. He moved reason into the limelight, he catalyzed, although certainly not single-handedly, the Enlightenment. One has only to list a smattering of the adulatory comments about him by his contemporaries and to skimpily trace his effect on the thought of later generations to realize that if one wanted to elect a champion of the Age of Reason, Newton was the man of the hour. Rousseau was one of the very few prominent thinkers who took almost no interest in the scientist or his work, but the comparatively undereducated Rousseau was the Age of Reason's odd man out.

Newton's influence stretched from Russia to America. Jefferson kept a portrait of Newton on the wall of his study and considered that Newton, Bacon, and Locke were the greatest men in history. As late as 1851 Schopenhauer wrote: "The almost idolatrous veneration in which *Newton* is held, especially in England, passes belief. Only a little while ago *The Times* called him 'the greatest of human beings,' and in 1815 (according to a report in the *Examiner*) one of Newton's teeth was sold for 750 pounds sterling to a Lord who had it mounted on a ring."

The early spread of Newton's influence in Europe was mainly due to the French. It was Voltaire who was perhaps primarily responsible for initiating the French love affair with English philosophy and science. Voltaire visited England in 1726 and became an Anglophile, as reflected in his *Lettres sur les Anglais* (1734). It was chiefly through his eyes that the French began to see England as more than a nation of uncouth louts. As he wrote, "It is inadvertently affirmed in the Christian countries of Europe that the English are Fools and Madmen."

In 1738 Voltaire wrote an account of Newton's work: *Élémens de la philosophie de Neuton* (*sic*). Earlier, aided by the scientist Maupertuis, he had given a simpler outline in his *Lettres philosophiques* (1734), where he lauded Newton's empirical investigations and his tolerance in matters of philosophy and religion:[5] "a man like M. Newton (we scarcely find one like him in ten centuries) is truly the great man, and those politicians and conquerors (whom no century has been without) are generally nothing but celebrated villains."

Voltaire saw the philosophical implications of his hero's work. He realized that

---

[5]Voltaire was also tolerant. He wrote of the Jews: "In short we find them not only ignorant and barbarous people. . . . Still we ought not burn them."

Newton had shifted reason onto a new plane and that his rejection of dogma unsupported by facts was a model for an attack on the Church, Voltaire's favorite enemy. An antiauthoritarian, pro-scientific, skeptical attitude was to characterize much of the anticlerical writings of the Enlightenment.

One of Voltaire's lovers shared his enthusiasm for Newton and the new science. Gabrielle-Émilie Le Tonnelier de Breteuil, marquise du Châtelet, was an unusual woman for her time. Voltaire used to perform physics experiments in her salon. She admired Newton because she sensed the significance of his work, both mathematical and physical. The only French translation of the *Principia* is that of the redoubtable Madame du Châtelet. Possibly she got help with the math from the mathematician Alexis-Claude Clairaut, another Newton fan—and another of her lovers. Her admiration stopped short of that of a compatriot of hers who suggested that the calendar be restarted with Newton's birthdate as the first day of the new era.

In Germany, Lessing compared Newton to Homer. An Italian, Francesco Algarotti, published (1727), the patronizingly titled, *Il Newtonianismo per le dame,* which was translated by Elizabeth Carter and appeared in English, as *Sir Isaac Newton's Philosophy Explain'd for the Use of Ladies* (1739).[6]

Back in England, David Hume proclaimed Newton to be the greatest and rarest genius that ever arose. The astronomer Halley wrote, "Nearer the Gods no mortal may approach," and Pope's endlessly quoted verse ran, "Nature and Nature's laws lay hid in night, Then God said 'Let Newton be!' and all was light." This may be doggerel, but it was polished compared with some of the verse about Newton that appeared in the early eighteenth century.

Well before Newton died in 1727, it was seemingly bad form *not* to mention him in a poem. Thus, one William Somerville, at the end of a poem devoted to "The Chase," dutifully tacked on some lines on Newton. For the first half of the eighteenth century, bookshops were never lacking in third-rate philosophical texts or excruciating books of verses looking to Newton and science for an intellectual crutch.

To grasp the stature of Newton in the eyes of the eighteenth century, it is sufficient to examine the plan of the cenotaph designed, but never built for him, by the visionary French architect Boullée. Supported by a huge cyprus-covered socle was an immense hollow sphere pierced by holes that let in sunlight simulating the positions of the stars and planets. At the bottom of the sphere would be the austere tomb of Newton. Visitors, enclosed by the upwardly sloping walls, would be drawn, as though by gravity, to the immediate surroundings of the tomb, where they would turn their eyes toward the limitless heavens.

The source of this extraordinary near worship was without doubt the fabulous success of the laws of motion combined with the law of universal gravitation. The Babylonians, the Chinese, the Maya, and the Arabs had all been meticulous observers of the heavens. Ptolemy, Brahe, Kepler, and Copernicus had, in different ways, documented and systematized the movement of the heavenly bodies, and Galileo had discovered new moons. But no one, since the beginning of history, had *rationalized* the movement, and it was the law of universal gravitation and the sec-

---

[6]Elizabeth Carter (1717–1806) was a 'bluestocking', one of those women who defied the prejudice of the age and engaged in pursuits considered more suitable to men. Apart from Italian, she translated from the French and Greek and wrote poetry.

ond law of motion that were the keys. As Voltaire wrote in his notebook: "Before Kepler, all men were blind, Kepler had one eye, and Newton had two eyes." By a huge stroke of genius, Newton had welded his and Galileo's ideas on forces and had written down mathematical equations that described the paths of the planets. He had unified the heavens in one great scheme, extending the rule of physical law to the whole of the solar system, and by implication to the whole universe. It was rightly seen as an intellectual triumph of unequaled splendor. Man, using observation and *reason,* had stretched out to touch and understand the cosmos. It was not Copernicus who destroyed the Aristotelian universe. Copernicus was an Aristotelian who accepted most of the master's teachings. And Galileo was a great pioneer who struck at the basis of medieval physics; but Newton's achievement was incomparable, and was acknowledged as such.

Voltaire commented that Newton had very few readers, "because it requires great knowledge and sense to understand him. Everybody however talks about him." It mattered little that few of those who praised or chatted about the *Principia* could have understood its contents. He was a cultural idol, rather like the even more popularly obscure Einstein. In 1705 he became the first scientist to be knighted, and his status was confirmed in 1727 by his burial in Westminster Abbey alongside the kings of England, a ceremony that made a strong impression on Voltaire.

Newton and Newtonianism became part of European culture to an extent previously exceeded only in the case of Aristotle. It is true that Galileo's astronomical observations had been widely lauded, but his telescope tended to get more praise than he did. (Kepler compared it to a magic wand, a cupid's arrow, "more precious than any scepter!") But it was because of Newton that the men of the Enlightenment believed in a *rational* explanation for the workings of the cosmos, and they extended that belief to a conviction that man himself was rationally explicable, perhaps only in terms of matter, force, and motion, like the heavens.

It was Newton's spirit that, together with Descartes's mechanical doctrines, fueled the extreme materialism of Julien Offroy de La Mettrie as expressed in his *L'Homme machine* (1747), where it is proposed that man himself is nothing but matter and motion. The Marquis de Sade quoted de La Mettrie more often than any other writer, and he accepted completely La Mettrie's materialistic view of the universe, including man. One can sense de Sade in La Mettrie's incorporation of sexual ecstasy into philosophy, and in his (convenient) belief that he had to live a life of sensual pleasure in order to achieve his philosophical aims.

The success of Newton lurked behind the theories of Rousseau's pet bugbear, the fanatically antireligious Paul, baron d'Holbach, whose *Système de la nature* (1770), coauthored with Diderot, propounded the view that matter in motion was all that existed. On a different plane, it was the apparent complete success of the mechanical description of the universe and *the implied invincibility of reason* that were to partially underlie the flavor of much of the political and social thought of the Enlightenment. The leaders of the Enlightenment both respected science and used it as a weapon against superstition and religion. Many of them saw in the scientific method the only way to attain knowledge.

Newton neither foresaw nor intended any of this. He was not the John the Baptist of the Enlightenment, and he would not have been at home with its ideals. The heralder of the Age of Reason left over 2 million written words devoted to religion and alchemical thinking. It was not as a *propagandist* for reason that he was worshipped. It was his immense personal standing and his scientific triumphs that elevated the prestige of reason and the empirical approach. And although Newton philosophized, he built no system of philosophy. His friend John Locke did, and it was his and Newton's empiricism that infused the Age of Reason.

Locke's *Essay Concerning Human Understanding* (1690) is a key document of the Enlightenment. It has a claim to be the bible of empiricism, and as such is very relevant to the history of science. For Locke, experience was the only source of knowledge. Locke was, as a student at Oxford, fascinated by science, especially in its experimental aspects. The interest remained with him most of his life. In the *Essay* he writes, obviously referring mainly to Newton, "The commonwealth of learning is not at this time without master-builders whose mighty designs, in advancing the sciences, will leave lasting monuments to the admiration of posterity."[7] Locke saw the scientific method as a less than perfect path to true knowledge, but he nevertheless stated, in *Some Thoughts Concerning Education* (1693), that Newton had shown "how far mathematicks, applied to some parts of nature, may, upon principles that matter of fact justifie, carry us in the knowledge of some . . . particular provinces of the incomprehensible universe." Locke's empiricism, a typically British "commonsense" philosophy, and his belief that the basis of knowledge is experience followed by rational thought melded effortlessly with seventeenth-century science.

Half a century after Locke's great text appeared, Immanuel Kant, a young theological student, was spending a large part of his time studying Newton. In 1756, before he was twenty, Kant published *Monadologia physica*, contrasting Newton's methods of thought favorably with those of his rival, Leibniz, who was much more abstract in his approach to knowledge. The young philosopher believed that the only road to scientific knowledge was via the mechanistic approach of Newton.

Locke, Bacon, and Newton had a deep, and acknowledged, influence on the central document of the Enlightenment in France, the huge *Encyclopédie,* edited by Denis Diderot, which was an attempt to summarize human knowledge.[8] This antireligious and empiricist text was the most typical printed expression of the Age of Reason. The mathematician and philosopher Jean d'Alembert wrote a *Discours préliminaire* to advertise the *Encyclopédie*. His hero worship is typified by his opinion that "Newton, whose way had been prepared by Huygens, appeared at last, and gave philosophy a form which apparently it is to keep." No less.

---

[7]Newton's relationship with Locke ran aground in 1693, when Newton was suffering from what appears to be a mental breakdown. A notorious letter which he wrote to Locke starts, (with modernized spelling): "Sir, Being of the opinion that you endeavored to embroil me with women and by other means, I was so much affected with it as, that when one told me you were sickly and would not live I answered t'were better if you were dead. I ask you to forgive me this uncharitableness." It was at this time that Newton broke off his connections with Samuel Pepys.
[8]The three Englishmen were all Protestants. Bacon's mother was fiercely Protestant, Newton was an anti-Trinitarian who grew up in Cromwell's England, and Locke's parents were Anglican with strong Puritan sympathies. Robert Merton has attributed the blossoming of science in seventeenth-century England to the fact that the Puritan virtues of methodical persistent action, empirical utilitarianism, and antitraditionalism were all congenial to the same values in science. Incidentally, Boyle, the pioneer chemist, and William Harvey, the discoverer of the circulation of the blood, were both Anglicans. On the Continent, Kepler was a Lutheran and Huygens and Leewenhoek both belonged to the Dutch Reformed Church.

The Diminution of God

Read Voltaire! Read D'Holbach! Read the Encyclopedia!

—Gustave Flaubert, *Madame Bovary*

It is understandable that the skepticism and reason that underlay the scientific revolution fed the Enlightenment, but how did religion come to give qualified support to the Age of Reason? The answer lies partially in the nature of deism.

Newton, echoing Descartes in this respect, had dispensed with the need for a celestial driver; once the Creator had set the planets in motion, they moved according to his laws and not by his active intervention. God could intervene if it was necessary, but in general it wasn't. This semimechanistic view of the universe struck a responsive chord with the deists.

Deism was first clearly defined in 1624 in Lord Herbert of Cherbury's *De Veritate*. Deism disposed to a lesser or greater degree, depending on its proponents, with the mystic elements of Christianity: miracles, the Virgin Birth, the Trinity—in fact, most of what is called revelation. Deism suited those who gave reason a very high priority but were not prepared to take the plunge into atheism. Herbert's writings became progressively more acceptable toward the end of the seventeenth century. Newton himself subscribed to the Arian heresy that denied the Trinity. (It is, after all, not *reasonable* to believe that the Son is also the Father.) Many of Newton's contemporaries found their credulity strained by mysteries and miracles, and this doubt was expressed in numerous publications, especially in England. Newton's triumphs eased the path for such texts.

John Toland's *Christianity Not Mysterious* (1696), published when the author was twenty-six, argued the case for reason in religion, as against divine revelation: "A Man may give his verbal Assent to he knows not what, out of Fear, Superstition, Indifference, Interest, and the like feeble and unfair Motives." His book, which triggered a series of deistical texts, was condemned to be burned by the common hangman, but Toland pressed the same line in *Tetradymus* (1720), where, in the new scientific spirit, he attempted to give natural explanations of a number of miracles in the Old Testament. He had allies, including Karl Bahrdt (1741–1792), who explained that the thunder that was heard when Moses was on Mount Sinai was caused by the patriarch's experiments with explosives!

Deism probably reached its peak in the writings of Matthew Tindal, who, at the age of seventy-three, published a bombshell entitled *Christianity as Old as the Creation* (1730). This text savages the Bible and claims that the true God is Newton's God, the creator of the awe-inspiring universe that moves in accordance with his eternal laws.

All deists believed that there was a creator, but most of them, especially the French deists, denied that God took a part in the day-to-day running of the physical universe. Many of the deists also believed that God made no *moral* demands on man. He was manifested in the ordered working of the universe, which operated according to his universal laws. Thus compatibility was forged between religion and the Enlightenment, whose leaders generally wished to believe both that God existed and that his universe fell completely within the realm of reason.[9] This di-

---

[9] The shift away from the traditional doctrine of God as ever-present puppet master had an echo

chotomy fits in beautifully with Newton's universe, created by God but running according to changeless physical laws. The *Principia* was seen to justify deism, and many deists saw natural philosophy (science), rather than the writings of the Church, as the sure way to religion.

Deism was effectively confined to a handful of European countries. Where science was widely practiced and respected, deism flourished. In scientifically backward Italy, in the Iberian Peninsula, the Hapsburg empire, and eastern Europe, deism was almost nonexistent. Benjamin Franklin, who spent much time in France, declared himself to be a deist, although deism did not grip the American colonies. A notable deist, although he arrived a bit late for the Enlightenment, was the English freethinker Tom Paine (1737–1809). Science, he said, proved the existence of God. It was he who wrote the book that gave its name to an era—*The Age of Reason,* a book he described as a "march through Christianity with an axe."[10]

Voltaire was a deist. His contempt for organized religion was made clear in statements, books, and letters and in his courageous defense of the victims of religious wrath, as in the case of the young Chevalier de La Barre, who insulted a religious procession and damaged a crucifix. His tongue was wrenched out and he was beheaded, in spite of Voltaire's protests. Voltaire's deism revealed itself particularly in his response to human misfortune and natural disaster. He was profoundly affected, and his deism strengthened, by the huge earthquake that left Lisbon in ruins on the morning of 1 November 1755. The sheer intensity of the shock, the consequent fire, the huge tidal waves, and the great loss of innocent life made a deep impression on Europe. Voltaire included the incident in his novel *Candide,* but his first written response was to write a poem on the disaster, mocking Leibniz's claim that this was the best of all possible worlds, and questioning God's involvement with the world. Such sentiments sat very happily with deism, whose adherents saw God in retreat from the natural world.[11]

The deists never had an official creed, but deism was the unofficial religious affiliation of most of the leaders of the French Enlightenment, and the influence of Newton should not be underestimated in this connection.

Newton's Legacy

Physics and mathematics are now on the throne.

—Edward Gibbon

The Newtonian breakthrough cast a long shadow. The hundred years between the death of the chemist Robert Boyle in 1691 and the French Revolution saw a remarkable peak in the popularity of science, coinciding with a discernible decrease of interest in religion, particularly in France. The masterly chronicler of the times, Tocqueville, noted the "universal discredit" into which religion had fallen toward

---

in the concept of natural law, of which the Dutch lawyer Hugo Grotius (1583–1645) was a great proponent. Natural law was a product of man's nature, not of Christianity or God's will, and Grotius regarded it as so fundamental that not even God could change it.

[10]Although William Blake pointed out, "The Perversions of Christ's words & acts are attack'd by Paine & also the perversions of the Bible. Paine has not attacked Christianity."

[11]But somewhat less happily with Pope, "And, spite of pride, in erring reason's spite, / One truth is clear, whatever is, is right."

the end of the eighteenth century. When Napoleon asked Laplace why God did not appear in his great *Traité de mécanique céleste,* the author quotably replied, "Sire, I have no need of that hypothesis."

The scientists of the seventeenth-century scientific revolution were on the whole not Enlightenment men. They generally aligned themselves with moderate religious and political views, eschewing the growing disrespect for tradition and authority that were a feature of metropolitan society in their day. But when, in 1767 the Baron von Grimm wrote "Enlightenment is spreading on all sides," the image of science had never been brighter.

The Ascent of Science

The Enlightenment is inseparable from the ascent of science in the seventeenth century. The prophets of a new age nearly all saw Newton as one of its main builders, and science as one of its essential building blocks.

By the middle of the eighteenth century, science had attracted a rapidly expanding audience. At the beginning of the century there were only four scientific journals, of which the heavyweights were those produced by the Royal Society in London and the Academie Royal des Sciences of Paris. By 1800, excluding medical journals, about seventy-five journals were appearing regularly. Scientific societies sprang up all over provincial France and were imitated in England, most notably in the "Lunar Society" of Birmingham (which was attended by James Watt of steam engine fame, Joseph Priestley the chemist, and Charles Darwin's grandfather Erasmus Darwin). The Literary and Philosophical Society of Manchester was founded in 1785. In America the efforts of the tireless Benjamin Franklin led to the formation of the American Philosophical Society in Philadelphia in 1744, and John Adams had a strong hand in the establishment of the American Academy of Arts and Sciences in Boston in 1780. But the influence of science went far beyond the growth of a new class of scientific amateurs.

During most of the eighteenth century, due in great measure to Newton, the highest quality of man was seen as his ability to use his reason. Which is not to say that religion died, or that reason ruled, but the civilizing effect of reason, its appeal to the moderate side of man, to his tolerance, flowered in Europe. Literature was not unaffected.

The writing of an age provides insight into its spirit. It would be misleading to suggest that the Age of Reason bred reasonable writers, and yet many of the greatest English and French writers of the age were characterized by qualities that appealed as much, if not more, to the intellect of the reader as to his passions.[12] In general, the authors flourishing during the earlier years of the century display an intrinsic respect for order and reason, and although it cannot be said that many had serious scientific interests, or that they were comfortable with the scientific outlook, it is certainly arguable that science, usually meaning Newton's science, affected the way that men of letters wrote.

In England, elegance largely displaced the vigor of Restoration prose and the blood and sweat of Elizabethan writing. Burns and Blake were not enamored of sci-

---

[12]Gibbon's *Decline and Fall of the Roman Empire* is a wonderful example of the Enlightenment's balanced rationalism and clear-sighted view of man's estate. And his style is finger-lickin' good.

ence and reason (or elegance), but Pope, Johnson, Addison, Steele, and Defoe, with their wit and their urbane satire, spoke to the thinking as much as to the feeling man, even if they had doubts about the benefits that science, especially in the hands of the Royal Society, was supposed to bring to mankind.

Alexander Pope, the outstanding writer of his age, was subtly shaped by the rise of science. His *Essay on Man,* very popular in Enlightenment Europe in translation, was typical in this respect. In it Pope says that we will "Laugh where we may, be candid where we can, / But vindicate the ways of God to man." He then goes on to speak very much of man and very little of God. A deist would have felt reasonably comfortable with Pope. God was the Creator, and there his role ended. Voltaire was ecstatic: "the finest, most useful, sublimest didactic poem ever written." Pope, although he was far from averse to making fun of the new science, was a great admirer of Newton's genius. Pope's thoughtful humanism spoke to the English and French HMS of the time, with its clarity, its balance of optimistic reason and reasoning doubt. Today's blind antirationalists contrast crudely with Pope's belief that man's passions need not be the enemy of his reason but could coexist constructively with it. The much quoted lines:

> Know then thyself, presume not God to scan;
> The proper study of mankind is Man.

are followed by a very much unquoted line that reveals the tenor of the times and Pope's underlying respect for science:

> Go Wond'rous creature, mount where Science guides.[13]

Other English writers were similarly supportive. As early as 1668, well before Newton's reputation was made, the poet John Dryden had enthused, "Is it not evident in these last hundred years (when the study of philosophy has been the business of all the Virtuosi in Christendom), that almost a new Nature has been revealed to us," and "more noble secrets in optics, medicine, anatomy, astronomy, discovered than in all those credulous and doting ages from Aristotle to us?"

Not everyone was so enthusiatic. Jonathan Swift's praise was selective. The king of Brobdingnag declared that scientists "do more essential Service to Country, than the whole Race of Politicians put together," but Swift ridiculed what he saw as useless science. In *Gulliver's Travels* (1726) he poked fun at the apparently pointless experiments reported to the Royal Society. In the fictional Grand Academy of Lagado there were plans for extracting the sunshine in cucumbers and building houses from the roof downward. Swift prophesied the downfall of Newton's cosmology, but it is significant that science, and the activities of the Royal Society, had a high enough profile to warrant his attention. Where he and other early eighteenth-century English writers, such as Addison, were critical of scientific matters, their wit was rarely directed at the major natural philosophers, but rather at the "virtuosi," the dabblers in science.

---

[13] As Pope lay dying, he was asked by a friend if he wished to see a priest. Typically of the age, he replied, "I do not think it essential but it will be very right and I thank you for putting me in mind of it."

In eighteenth-century France the glorification of reason exceeded in intensity anything in the rest of Europe. Within a period of a few years there were at least sixty printings of Pope's *Essay on Man,* as well as enormously influential translations of the books of John Locke. A plethora of books, essays, pamphlets, and meetings advocated deism or atheism. Translations were made of all those works, from the Greeks to Hobbes, considered to be atheistic or anticlerical enough to serve the cause of rationalism, although much of the output was anti-Church rather than pro-reason. The power of the Catholic Church in France was still such that French intellectuals often had to choose between dangerous defiance of the authorities or humiliating compromise. Thus, in 1749, when the great naturalist Comte Buffon published the first volume of his *Histoire naturelle,* a theological committee at the University of Paris made it clear that he would run into difficulties unless he modified the text, especially where it dealt with the creation of the Earth. Buffon backed down on all "that might be contrary to the narration of Moses." As to his description of the birth of the planets, Buffon explained that they were "pure philosophical speculation." He got away with it, but one can understand the Church's nervousness; Buffon had presented a picture of the Earth, and its life-forms, that reached far into the past, one that took into account the possibility of the gradual emergence of new species. This hardly tied in with Genesis.

The French took much from the English. Voltaire vigorously advertised the *Principia,* Locke's *Essay Concerning Human Understanding,* and Bacon's ideas. Grimm declared, "Without the English, reason and philosophy would still be in the most despicable infancy in France."

In France, the demand for religious toleration, the atmosphere of skepticism, the esteem accorded to the non-Christian classical past and to science—in short, the Enlightenment—found vigorous expression in the writings of the philosophes, an informal group of writers, thinkers, scientists, mathematicians, and others. They had (thank God!) no manifesto. They saw both church and state as dogmatic, dishonest organizations and searched for eternal values not in religious, national, or cultural roots but in reason and personal experience. Their faith in human reason, and specifically in science, was almost unlimited. With reason all could be explained, without need for revelation or mysticism. The attitude of the philosophes to Newton was a major influence on the French Enlightenment.

The philosophes were not in general knowledgeable enough to understand the full meaning of Newton's work, although a few were mathematicians or naturalists. In France the officially accepted physics had been that of Descartes. Perhaps partly because of chauvinism, the vast superiority of Newtonian mechanics took some time to be accepted, but once it was, it became all the rage in France.

Newton's conquest of France was overwhelming. It was not so much the mechanics in itself that fired the French but rather a conglomeration of ideas that included the superiority of the scientific method and the concept of an overall order in the universe, an order that was in principle subject to inquiry and ultimately understandable. To the philosophes, the scientific method seemed to be applicable to almost any field of human interest. Above all, *if the cosmos could be shown to be explicable by simple universal laws, then perhaps there really was an ordered universe, and perhaps the affairs of men could be ordered by reason.* No comment.

Voltaire was not alone in his hope that the scientific method could be applied to history, and Condillac suggested that the whole of philosophy be rebuilt using the sciences as a model. D'Holbach was convinced that science was the salvation of mankind.

Baron d'Holbach, a German millionaire, ran a salon in Paris that was a notorious meeting place for scientists, atheists, deists, and antiestablishment types in general. The English chemist Joseph Priestley reported that all the philosophical persons to whom he was introduced there were "unbelievers in Christianity, and even professed atheists." D'Holbach was one of the many at that time who believed, like Priestley, that science would bring about a better life for mankind, and that the Church was standing in the way, preventing the spread of scientific teaching: "It is as a citizen that I attack religion, because it seems to me harmful to the happiness of the state, hostile to the mind of man, and contrary to sound morality." But caution was the rule: when he published his *Système de la nature* in 1770, he did so under the name of a man who had been dead for ten years. The book, which is an enormously long and unorganized defense of materialism and atheism, denied the existence of the soul and of God, and singles out necessity as the driving force behind behavior. Predictably, it was condemned and ordered to be burned by the Parlement of Paris. The author was to be identified, arrested, and punished. It is said that the ten people who knew that d'Holbach was the author kept silent for twenty years, until he died. A taste of d'Holbach's viewpoint is contained in the following passage, which carries definite Newtonian undertones:

> Man is the work of Nature; he exists in Nature; he is submitted to her laws. He cannot deliver himself from them, nor can he step beyond them, even in thought. Instead therefore, of seeking Nature, let him learn her laws, contemplate her forces, observe the immutable rules by which she acts; let him apply these discoveries to his own felicity, and submit in silence to her mandates, which nothing can alter; let him cheerfully consent to ignore causes hidden from him by an impenetrable veil; let him without murmuring yield to the decrees of universal necessity which can never be brought within his comprehension, nor ever emancipate him from those laws that are imposed upon him by his essence.

Reading these lines, and those of scores of other eighteenth-century writers, one could be excused for hoping that the Age of Reason was indeed just over the horizon. It was a time in which Condorcet believed in the "infinite perfectibility" of man and of science, and could write of a day, coming soon, when, "the sun will shine only on free men." The political Freedom Train has been delayed, but Condorcet's optimism was a product of the times, times in which the standing of science and scientists was higher than it had ever been.

The Scientist as Man-about-Town

Everybody has begun to play at being the geometer and the physicist.

—Voltaire, *Correspondence, IV.*

Science had become a matter of popular concern. Newton's fierce (and often underhanded) campaign against the great German philosopher and mathematician Leibniz—fought over the right to be considered the originator of the calculus—was a talking point in the court of Charles II, although very few of that licentious assembly could have known what the calculus was about.[14] The same TV talk-show level of discussion was to befall relativity in the twentieth century. From the middle of the eighteenth century, scientists were more than welcome in the fashionable society of Europe. In Paris, public lectures on science attracted enthusiastic audiences. Bewigged gentlemen and fan-fluttering ladies exchanged chitchat about physics and chemistry. The English author Oliver Goldsmith wrote, "I have seen as bright a circle of beauty at the chemical lectures of Rouelle as gracing the court of Versailles." The effects of static electricity were a popular party piece; men-about-town linked hands and jumped about as a weak electric current passed through them. Because of Newton, the orrery, a moving model of the solar system, became popular in the eighteenth century.

The prestige of science is reflected in the large number of aristocrats, and even heads of state, who took enough interest to personally carry out experiments, or have them carried out. Science penetrated to the bedrooms of kings. The next time you are in the Louvre, look at the painting of Madame de Pompadour by Maurice Quentin de La Tour. One of books on the table next to which she sits is entitled *Encyclopédie.* There is no doubt whatsoever that this is a volume of the enormous and notorious encyclopedia edited by Denis Diderot: *Encyclopédie: ou dictionnaire raisonné des sciences, des arts et des métiers.*

Diderot was in many ways the most human character among the philosophes.[15] He appears to waver between atheism and deism, and although no scientist, he shared the belief of many of the *philosophes* that scientific knowledge is the only reliable knowledge that we have. The encyclopedia, the first volume of which came out in 1751, specifically declared its debt to Bacon, Newton, and Locke. It contained numerous articles on science and hundreds on technology—of which d'Holbach contributed over 400. Rousseau provided an article on music. A glance at the encyclopedia will disillusion anyone who believes that technology was practically nonexistent before the Industrial Revolution. It was the unfulfilled dream of the philosophes to unite science and technology, but science had little to offer technology at that time. The editor also failed in his wish to supersede religion with science, but the great text helped to change the intellectual flavor of Europe, edging it toward reason and science and away from unquestioning faith. As the editors wrote, "Everything must be brought to light boldly, with no exceptions and unsparingly." In a letter to Voltaire, Diderot, who was an atheist, wrote, "Science has done more for mankind than divine or sufficient grace." And this came from a man who was practically scientifically illiterate and who distrusted mathematics.

The encyclopedia was far too subversive for the religious authorities. The Jesuits

---

[14]In the course of this prolonged argument, each implied that the other was a deist. The fact that Newton's mechanics had shown that the planets could revolve without the help of God was one of the standard examples that deists would pull out to prove their case. But although Newton was a Dissenter, he was certainly not a deist, and he had stated that if necessary God could make minor adjustments to the paths of the planets if they strayed. Leibniz declared that "Newton's God is a clumsy watchmaker." God was a better workman than this and had no need whatsoever to interfere with the mechanism of his universe. In this sense it was Leibniz's universe that should have been the model for the deists, but in fact it was Newton who was almost always invoked in this context.

[15]His letters to his mistress, Sophie Volland, provide a fascinating insight into his times.

and the Church were enraged. Diderot was interrogated by the civil authorities.[16] When seven volumes had appeared, the Council of State banned the encyclopedia: "The advantages to be derived from a work of this kind, in respect to progress in the arts and sciences, can never compensate for the irreparable damage that results from it in regard to morality and religion." It was said that the subsequent volumes, 8 to 17, were finally published because of La Pompadour's influence on King Louis. Perhaps that is why, in her portrait, she has volume 9 next to her. (I wonder if Louis was aware of the poem in which Diderot recommends strangling kings with the bowels of priests?)

It is of significance that it is the *Encyclopédie,* predominantly a text on science and technology, that is often taken to be the most representative publication of the Enlightenment. The presence in the encyclopedia of a large number of articles on medicine is understandable in the light of the Enlightenment's obsession with health and the concept of science as the means for curing man's ills, both literally and metaphorically. The greatest teacher of medicine at the time, the famous Hermann Boerhaave, professor at Leyden in the Netherlands, regarded himself as a disciple of Newton, on whose work he lectured and whose methods he attempted to employ in his work. In 1715 he published *Oratorio de comparando certo in physicis,* in which the Newtonian theory of gravitation is given the place of honor. Boerhaave insists that we must always be guided by experiment and observation, aided by mathematics and the power of reason. His influence was remarkable; he and his students and colleagues helped to spread the Newtonian message throughout western Europe. The Leyden school of medicine also paid homage in their lectures to the empirical principles of Bacon and Locke, the other two members of the British trio. One of Boerhaave's students was Julien de La Mettrie, the extreme materialist whom we met earlier. As we noted, de Sade was a follower of de La Mettrie, and so we can trace a direct line of thought from Newton's mechanical universe to de Sade's hedonistic, materialistic picture of man. It would be hard to imagine two more dissimilar characters, and yet the shadow of Newton hovers behind the Frenchman's mechanistic philosophy. He specifically mentions Newton in part 3 of *Juliette,* where he accuses the physicist of sidestepping the problem of free will in a deterministic universe—a problem to which we will return.

Another book on La Pompadour's table appears to me to be Montesquieu's masterpiece, *De l'Esprit des lois* (1750). This rambling, highly controversial, and highly influential treatise on the origins of law, now regarded as a seminal text in sociology, was reprinted twenty-two times in eighteen months. There was no doubt as to Montesquieu's inclinations: "Reason is the most perfect, most noble, and most exquisite of all our faculties." And again: "I have not drawn my principles from my prejudices but from the nature of things." It could have been Newton speaking, and it is clear that Montesquieu felt himself to be part of the scientific tradition. He was hardly suitable bedside reading for the king's mistress. His book attracted the wrath of the Jesuits, who saw through its feeble lip service to Christianity. But they had no hold over La Pompadour. "Madame Whore," as some called her, was a well-read woman who read what she liked and whose favorite acquaintances included mathematicians, natural philosophers, and assorted rationalists.

---

[16]Diderot fought an incredible uphill battle to get the encyclopedia published. On top of official opposition, he was unaware for some time that his publisher was censoring some of the articles without his knowledge.

Science was fashionable, and not only among the aristocracy. The middle class had also been infected. In France an increasing number of schools abandoned the almost total emphasis on the humanities and religious studies that had typified education in most European countries for so long. The atheist Swiss millionaire and philosophe Claude-Adrien Helvétius, in his popular book *De l'esprit* (1758), advocated a complete overhaul of the educational system in France, and incidentally hinted that a drastic change in education could only be made if there was a political revolution first. Theology and history were both branded as weapons of the establishment. The essential step was to remove education from the control of the Church. Education was to be provided to all, irrespective of sex or age. Dead languages were to be removed from the syllabus and replaced by the study of science and technical subjects.

Political and religious dictatorships burn books. *De l'esprit* was burned—and the sciences began to be widely taught. For the first time, many schools established teaching laboratories.

When the Old Regime fell and when the young Republic had to face the threat of war, scientists willingly helped in setting up a war industry. At that time, Lavoisier and the great mathematicians Laplace and Lagrange were among the members of a committee appointed to revise weights and measures. They produced the metric system, officially adopted on 25 November 1792.

As has happened after other revolutions, political control of the French Revolution passed out of the hands of the moderate leaders. The Reign of Terror was driven by ambitious fanatics who had little patience with intellectuals. The revolutionary (and would-be scientist) Jean-Paul Marat led a mini revolt against mathematically based science, probably partly motivated by his own academic limitations. The epithet "elitist" (as applied to the monumental mathematical work of Laplace, Lagrange, Fourier, and others) and the banner: "science for the people" are unpleasantly reminiscent of the populist invective used in more recent times by the scientific lackeys of dictatorial regimes. The Academy of Sciences was abolished. But in 1793 Marat was murdered in his bath by Charlotte Corday, and the political movement that he had led—the Jacobins—was destroyed. French science survived a politically motivated crisis, the Reign of Terror passed, but the Age of Reason was over.

7

From Rousseau to Blake: The Revolt against Reason

I am the eye with which the Universe
Beholds itself and knows itself divine.

—Percy Bysshe Shelley, "Hymn to Apollo"

When he was at school at Eton, Shelley spent much of his leisure in the study of "the occult sciences, natural philosophy and chemistry," and his pocket money was spent on "books relative to these pursuits, on chemical apparatus and materials." Shelley was fascinated by the voltaic pile of Volta and had dreams of a world transformed by electricity. At Oxford University the contents of his room included "crucibles, . . . an electric machine, an air pump, the galvanic trough, a solar microscope, and large glass jars and receivers." This was the door to nature chosen by the archetypal Romantic poet who was later to approach the natural world almost solely through the gates of his imagination: "Earth and Ocean seem, To sleep in one another's arms, and dream."

Shelley had, in his youth, fallen under the influence of one of the Lunar Society, Dr. James Lind, and had read such antireligious writers as Lucretius and Condorcet. It was the publication of his *Necessity of Atheism* that resulted in his being expelled from Oxford in 1811. That year, this son of a moneyed baronet wrote to his friend Elizabeth Hitchener that "I am no aristocrat, nor 'crat' at all, but vehemently long for the time when men may dare to live in accordance with Nature and Reason—in consequence, with Virtue, to which I firmly believe that Religion and its establishments, Polity and its establishments, are the formidable though destructible barriers." Like all the Romantics, he had been shaped by the French Revolution—but also by science.

In Shelley's long poem *Prometheus Unbound*, Heaven is asked, "Heaven, hast thou secrets?" to which the author supplies the Newtonian reply, "Man unveils me; I have none." The young poet saw the natural world through scientific glasses. On his honeymoon with his sixteen-year old bride, Shelley, himself only nineteen, began to translate the works of the French naturalist Buffon. Reading the quotation at the beginning of this section, it is therefore natural to assume that "the eye with which the Universe beholds itself" is Science. A reading of the whole poem makes it completely clear that the eye is definitely not science. It is poetry.

Prometheus Unbound was written in Italy in 1819; the men of the Enlightenment had died long ago, and times were changing. By the 1820s the cause of rationalism was in retreat. The Industrial Revolution was approaching, and the love affair between science, reason, and thinking men was cooling. In his introduction to the poem, Shelley, the once youthful reformer, says that its aim is "simply to familiarize the highly refined imagination of the more select class of poetical reader with beautiful idealisms of moral excellence." Oh dear.

Shelley was not the first intellectual to become disillusioned by science and the spirit of the Enlightenment and to move toward the views expressed by Rousseau as far back as 1749. Rousseau, in his *Discourse on the Arts and Sciences*, had seen a threat in the development of science, technology, and European culture in general. For him they had all led man away from the naturally good state of the noble savage and brought about his moral corruption.

The Enlightenment emphasis on reason was a provocation to those who feared that imagination would be smothered by mechanics. When the backlash came, history followed its predictable course: the heroes of the Old Regime became the villains of the new, and those who revolted against what they saw as the rule of reason usually saw Newtonianism and science as their archenemies. William Blake, who associated rationalism with the rise of the machine-based industry that he loathed, was deeply offended by Bacon, Newton, and Locke. "Bacon's Philosophy," wrote Blake, "has Ruin'd England," and he identified Locke's philosophy and Newton's mechanics with the hated machine:

> I turn my eyes to the Schools & Universities of Europe
> And there behold the Loom of Locke, whose Woof rages dire,
> Wash'd by the Water-wheels of Newton: black the cloth
> In heavy wreathes folds over every Nation: cruel Works
> Of many Wheels I view, wheel without wheel, with cogs tyrannic
> Moving by compulsion each other, not as those in Eden, which,
> Wheel within Wheel, in freedom revolve in harmony and peace.

Newton as culprit appears again in the poet Cowper's remark to Blake: "You retain health and yet are as mad as any of us all—over us all—mad as a refugee from unbelief—from Bacon, Newton, and Locke." (the Trinity again). Later, the poet Coleridge was to write that "the souls of five hundred Sir Isaac Newtons go to the making up of a Shakespeare or a Milton." Even in this century Newton still remains a target for those attacking rationalism and the mechanical universe:

> Here we all are,
> Back in the moderate Aristotelian city
> Of darning, and the Eight-Fifteen, where Euclid's geometry
> And Newton's mechanics would account for our experience

> W. H. Auden, "For the Time Being"

When the tables were turned it was in Germany that the revolt against Newton and reason sprung up most virulently.

Goethe versus Newton

> Anyone who attempts to determine the extent of the Kingdom of God with the yardstick of reason faces utter misguidance.
>
> —Al-Qazwini

The Enlightenment took on different colors in different countries. It was above all a movement that grew up in Britain and France. Rationalism failed to grow healthy roots in eighteenth-century Germany, where the antirationalists were busy as early as the 1770s. It has been remarked that Germany passed straight to Romantic irrationalism without stopping at rationalism. Rationalism in Germany, in contrast to France and England, hardly touched the middle class, or indeed the rest of the population, making a limited mark among a handful of writers and intellectuals, such as Lessing. When the middle class did wake up, it was to the music of Romanticism. By a perverse twist of history, it was Newton who was partly to blame. An important turning point came when Goethe read Newton and became interested in the nature of color.

Goethe considered himself to be primarily a natural scientist. He viewed Newton's use of a prism to split light into a spectrum of colors as being symbolic of an attempt to "split the unity of Nature." Newton had been "putting Nature on the rack." Notice the vague use of evocative, imaginative, but meaningless generalities. Goethe magisterially proclaimed, "A great mathematician was possessed with an entirely false notion on the physical origin of colors; yet owing to his great authority as a geometer, the mistakes which he committed as an experimentalist long became sanctioned in the eyes of a world ever fettered in prejudices." Goethe set out to construct an alternative theory of color.

It took Goethe nearly twenty years to produce his *Theory of Colors,* his longest book, and the one that he thought the most of. It contains very little science. Colors, in sharp contrast to Newton's treatment, are regarded primarily as objects of sensory enjoyment. It appealed more to painters than scientists, and indeed it was eventually annotated by the antiintellectual English painter Turner, one of whose paintings was verbosely entitled *Light and Colour (Goethe's Theory): the Morning after the Deluge, Moses Writing the Book of Genesis.*

Colors were just a beginning, Goethe had a greater venture: to topple the spiritually empty edifice of Newtonianism. Why he turned against Newton is a matter for speculation.

The Sturm und Drang (Storm and Stress) movement, with which Goethe was associated, lasted less than a decade and was a middle-class, antirationalist movement, a reaction to the perceived materialism of the French philosophes.[1] This materialism was seen as an inevitable consequence of the mechanical, Newtonian universe. To the French and English Enlightenment the universe was potentially explicable using reason and science; to the Sturm und Drang movement it was an enigma hidden behind a curtain of mystery, to be reached as much through the imagination as through reason, and, from man's point of view, basically meaningless. A new buzzword emerged: *Naturphilosophie.* The German intellectuals abandoned the rational approach to knowledge and substituted intuition and metaphysics of the obscure variety.

Naturphilosophie was a flag around which many German writers and philosophers rallied. Heinrich Heine's comment on one of them (Schelling) says much about the whole movement: "he does not feel himself at home in the cold heights of

---

[1] An exception to the rejection of the philosophes was the welcome given to their writings, and those of Locke, by German Protestants. The Thirty Years' War had left them with a thirst for religious toleration.

logic, he is happy to slip over into the flowery valley of symbolism." Madame de Staël was wittier and bitchier, avowing that German Romantic philosophy "spreads over all things the darkness which preceded Creation but not the light which followed." *Naturphilosophie* has been used to cover a variety of philosophical stances. Its adherents empathized with Rousseau's emphasis on man's instincts and sensibilities, and his desire for a culture that would include the virtues that he saw in ancient civilizations and in primitive societies. *Naturphilosophie* shared Spinoza's doctrine that God reveals and fulfils himself in the material world. Goethe said, "The world is the living visible Garment of God." But that world could not be understood by Newtonian mechanics. As Schiller put it, in *Maria Stuart,* "reason finally leads men astray."

Observation, the supreme Baconian virtue and the foundation of science, was demoted. Artistic genius was now the crowning human virtue: "The genius does not observe. He *sees*, he feels," said Johann Lavater the Swiss founder of the antirational religious and literary movement known as *Physiognomics.*

The period of the Sturm und Drang movement in Germany has been labeled "Romantic," but Goethe himself was not a romantic. Like the philosophes, he was suspicious of mysticism and was not religious; unlike them, he was violently opposed to the mathematical basis that Newton had constructed beneath the physical universe. Likewise, his compatriot Schelling, although he conceded that reason had its place, rejected the idea of a rational universe.[2] The predominant power in the universe was evil, and man himself was driven as much by his irrational natural impulses as by his reason.

For Goethe and his compatriots, man's greatest gift was that he could say "I." They may be right, but this doesn't excuse their irrational anti-Newtonianism, of which Hegel was one of the saddest examples. Hegel completely misunderstood the *Principia*, and repeatedly misrepresented it in his lectures in Berlin.[3] But he was the Herr doctor professor, and not one of his audience ever had the nerve to point out his errors. The man who wrote that "Reason is the Sovereign of the World" was prepared to sacrifice reason if it suited his prejudices.

The antipathy to Newton in Germany was bolstered at the turn of the century by Schlegel's translation of Shakespeare's plays, which set off a strong Romantic revival: Shakespeare was hardly a typical Enlightenment man. The German Romantics believed, or wanted to believe, despite clear evidence to the contrary in Newton's writings, that he had confused mathematics with reality, a reality that was not susceptible to scientific analysis.

The Infection Spreads

Naturphilosophie was picked up enthusiastically on a visit to Germany by Wordsworth's great friend, the Germanophile poet Samuel Taylor Coleridge. Coleridge was one of those who held that higher reason, which is basically intuition,

---

[2]Antirationalism, often combined with mystic nationalism, has a distinguished history in Germany. Perhaps the greatest German writer of the twentieth century, Thomas Mann, also pokes quiet fun at the rationalist intellectual, as represented by Settembrini in *The Magic Mountain* .
[3]Hegel claimed (incorrectly) that Newton had not proved Kepler's laws and that he had "flooded mechanics with a monstrous metaphysics." See *Hegel's Philosophy of Nature*, trans. A. V. Miller (Oxford: Oxford University Press, 1970).

has the power to uncover truths that were beyond the reach of the senses and of the "lower" reason, which of course included science.

Naturphilosophie, and its explicit anti-Newtonianism, appealed strongly to a number of English authors. Thomas Carlyle enthusiastically embraced its assumption that science is an acceptable human activity but not adequate to explain reality. In his almost unreadable *Sartor Resartus*, he contrasts the superficiality of scientific reason (*Verstand*) to the "higher" reason (*Vernunft*): "To the idea of vulgar Logic what is man? An omnivorous Biped that wears Breeches. To the eye of Pure Reason what is he? A Soul, a Spirit, and divine Apparition. . . . Deep-hidden is he under that strange Garment." The garment, as for Goethe, was matter. It is clear that the "higher" reason was not really reason at all. But then reason was no longer fashionable.

Naturphilosophie, at first sight paradoxically, had an important role in the scientific motivation of some of the greatest scientists of the nineteenth century. Coleridge was a friend of the prominent English chemist Humphry Davy, with whom he discussed philosophy. The Danish physicist Hans Christian Oersted was a friend of Schelling, and both he and Ampère in France were deeply affected by the spirit of *Naturphilosophie*. It was the all-inclusiveness of its vision of the world and its emphasis on the existence of universal underlying causes that appealed to the scientific impulse to find such causes for the behavior of the physical world. To this day, the same impulse drives those who are attempting to find a theory that explains all the forces of nature.

One can still find those who see Goethe's holistic view of nature as being an antidote to the divisiveness that they claim to see in modern science. This claim is not too intelligible in the light of the fact that modern universities are packed with interdisciplinary (holistic?) research projects—and not because of Goethe but because scientists realized long ago that nature is a single entity. It doesn't need the higher reason to see that.

The Individual Mind

It is fair to say that by 1820 the highest quality in man was considered to be no longer his reason but his ability to understand himself and his place in nature. Reading much of the literature of the time, one can detect the shift in the balance between thinking and feeling. The individual increasingly became the hero of poems—not the individual of Donne's poems, who could address God in times of trial, but a post-Enlightenment hero who had to face life and mortality alone. English poetry became subjective. The secondary title of the central poem of the Romantic era, Wordsworth's great *Prelude*, was *A Poem on the Growth of an Individual Mind*.

Wordsworth was not unusual for the times, in that his attitude toward science, like Shelley's, suffered sea changes. The young man who wrote, "Science with joy saw Superstition fly" later wrote, in the preface to the second edition of his *Lyrical Ballads*, that *if* men of science ever did anything that really mattered, then poets would be willing to write about it. Perhaps under the influence of his friend the great Irish mathematician William Rowan Hamilton, Wordsworth condescended to write his lines honoring Newton:

> And from my pillow, looking forth by light
> Of moon or favouring stars, I could behold

> The antechapel where the statue stood
> Of Newton with his prism and silent face,
> The marble index of a mind for ever
> Voyaging through strange seas of Thought, alone.

Wordsworth as a young man had had a brief love affair with a French girl, and with the French Revolution: "Bliss was it in that dawn to be alive." But the Reign of Terror brought disillusionment, and he turned inward. The introverted themes of Romantic poetry contrast sharply with the naive optimism of the philosophes. With Wordsworth, mortality is often just under the surface, as it was with Keats, another child of his time, who believed, because of the Enlightenment, that we are material beings in a material universe and that we must accept that fate. We are mortal, but with no divine shoulder to lean on, and we will never understand the deepest truths, which, contrary to all the protestations of the Enlightenment, neither reason nor science can reach. Keats had the tragic sense of life. He is recognizably a Romantic; there was no Enlightenment Utopia waiting for him.

Reason in the Shadows

What Reason weaves, by Passion is undone.

—Alexander Pope

At the end of the eighteenth century, rationality (*Verstand,* the "lower" reason) was being downgraded all over Europe. The Romantics ruled the intellectual roost. At a dinner given in 1817 by the second-rate English painter Benjamin Haydon, and attended by Wordsworth, Keats, and other writers, it is said that the toast was "Newton's health, and confusion to mathematics."

Was the Enlightenment dream of a Utopia ushered in by science killed by the Industrial Revolution, which was born between 1780 and 1820? Is it simplistic to suppose that many saw the social ravages of industrialization and felt that science, mistakenly identified with technology, had deluded man into a false belief in the Garden of Eden? One who definitely saw things this way was William Blake's friend, the painter Samuel Palmer, who looked at the expanding and ugly urban industrial landscape and cried out, "If so-called science bids us give up our faith, surely we have a right to ask for something better in return!" And the claim of the French Revolution that it would put scientific principles into practice rang hollow in the ghastly reality of the guillotine. The culmination of the Enlightenment, the great intellectual revolt of rational secularism, had ended in the slaughter of the Reign of Terror.

The status of science was visibly eroded during these years. Thomas Carlyle dismissed the scientist's claim to have revealed nature's secrets: "Scientists have been nowhere but where we also are; have seen some handbreaths deeper than we see, into the Deep that is infinite, without bottom as without shore." Typically of the times, the poets Shelley, Coleridge, and Keats (who was a pharmacist by training) had all been attracted to science in their youth and dabbled with scientific experiments. (Coleridge attended the lectures of the great chemist Sir Humphry Davy in order, so he said, to improve his stock of metaphors.) But the nature that appeared in their mature poetry was much closer to the unspoiled nature of Rousseau than the matter-in-motion of the Royal Society.

Was it by chance that the great flowering of Romantic poetry in England occurred in the years when the status of science declined? Wordsworth, in the year of
the French Revolution, wrote,

> Enough of Science and Art;
> Close up those barren leaves;
> Come forth, and bring with you a heart
> That watches and receives.

Those barren leaves doubtless included the pages of the *Principia*.

Galileo and Newton, Boyle and Lavoisier, Hooke and Huygens, had begun the
demystification of the natural world. But, as the eighteenth century progressed,
many thought that things precious had been lost in that great adventure, not least
the mystery of existence. De Quincey would have said that imagination had died;
Blake branded science the tree of death.

The scientific revolution had had little patience with what was seen as unclear
thinking. When Boyle wrote of atoms in *The Sceptical Chymist*, he cited the mystic
Pythagoras, but in the end he stressed that it was the prosaic evidence of the senses
and the cold operation of reason that were the final arbiters. The day of the balance
and the ruler, of the microscope and the telescope, had arrived. In Voltaire's Europe
there could be no return to the days when the sixteenth-century alchemist and
physician Paracelsus saw God as an alchemist and the cosmos as a chemical distil-
lation. After the seventeenth century, no self-respecting natural philosopher would
dream of saying, as did Paracelsus, "Man is a Sun and a Moon and a Heaven filled
with stars." Nature had been kidnapped by the professors.

The models to which the Enlightenment had looked had been cast in the stern
mold of the Roman Republicans. *Enthusiasm* was a dirty word in the vocabulary of
Locke and Voltaire: it smacked of religious excess. The Scottish philosopher Adam
Smith wrote, in his classic book, *An Inquiry into the Nature and Causes of the
Wealth of Nations* (1776): "Science is the great antidote to the poison of enthusiasm
and superstition." But the imagination, and the supposed intuitive ability of man
to understand the universe, could not be ignored for long.

There arose a new skepticism that, in contrast to that of the sixteenth- and seventeenth-century Europe, was directed not at the fossilized teaching of the Church
and the ancients but at the claims of science and reason and their now hollow-
sounding promises of progress and justice. If science appeared to many to literally
have taken the magic out of life, it had, by way of recompense, promised to rescue
man from the daily round of work and disease and deliver a "glorious and paradisiacal" world, as prophesied by Joseph Priestley. By the turn of the century it was
clear that it hadn't.

The poet William Cowper saw practical science "building factories with blood,"
and the Industrial Revolution's image was immortalized in Blake's condemning
phrase: "those dark Satanic mills" although he was possibly referring symbolically
to Newton's mechanics rather than to real machinery. Scientists will point out,
rightly, that to discover iron is not to make swords. The discoveries of science in
the sixteenth and seventeenth centuries were in the main not directly connected
with the technology that blackened the Midlands of England. The creation of machines for weaving was the work of inventive engineers, not scientists. But few in

the late eighteenth and early nineteenth century saw it that way. For them the natural philosophers had brought misery.

When the demolition team came to remove reason from its pedestal and replace it with imagination, poets were in the forefront. Wordsworth had once hoped that the "impassioned countenance of the sciences" would shine on the soul of the poet, but in the end it seemed to him that the mechanical, rational cosmos of Newton had substituted a "universe of death" for the true universe, "which moves with light and life instinct, actual divine, and true." He had pinpointed the loneliness of the mechanical universe.

The Enlightenment concluded that we are responsible for our own destinies, but few were prepared to be alone in the godless universe of d'Holbach, that "dead" universe which was the apparently inescapable end point of the natural philosopher's probings. Few would accept Diderot's compensation for losing the afterworld: "What posterity is for the philosopher, the other world is for the religious man." For HMS this offers cold comfort, and even in the Age of Reason, humanity clung to things irrational, steering clear of the depressing conclusion that all is matter and mechanics. Thus, in the eighteenth century we find apparent devotees of reason displaying intense interest in phrenology and in such shady characters as Mesmer (whence mesmerism), an admirer of Paracelsus and the proposer of "animal magnetism."[4] Although a committee, set up at the request of Louis XVI and including Lavoisier and Benjamin Franklin, returned a negative report on the medical benefits of Mesmer's "cures," he still attracted attention and patients. They *wanted* to believe in him. Even more successful was Count Cagliostro (real name Guiseppe Balsamo), who, at the height of the prestige of the philosophes, told fortunes by gazing into water and speaking to the dead. Communication with the dead has a more obvious appeal than the silence of the galaxies. In the face of the cold comfort offered by reason, the appeal of immortality is irresistible. In this respect the Enlightenment threw little light into the nineteenth century. In 1874, William Crookes, a notable British physicist who dabbled in spiritualism, studied the evidence for the existence of Katie King, the spirit attendant on the medium Florence Cook. Crookes came to the conclusion that Katie was genuine. He was subsequently, but one hopes not consequently, knighted and elected to the presidency of both the Royal Society and the British Association. As late as the early 1900s, another prominent British physicist, Sir Oliver Lodge, was completely taken in by the tricks of spiritualist mediums. Had he approached the phenomena with the eye and mind of a scientist, he would have detected the frauds, but he wanted to believe—so he did. The same need provides a good income for TV evangelists and psychics.

Although reason trembled as the eighteenth century progressed, it did not fall. But if it was far too late in man's intellectual history to bury reason, at least it could be rebuked for its arrogance and put in its place, which in a way is what Romanticism did. In fairness to the philosophes, they had given what they generally called "passion," or "the passions," a place of central, sometimes dominant, importance in their scheme of things. Diderot wrote to his mistress Sophie Volland, "I forgive everything that is inspired by passion," and it would be a mistake to interpret that

---

[4]Mozart's operetta *Bastien and Bastienne* was dedicated to his fellow Austrian Mesmer, and at the end of Act 1 of *Cosi fan tutte* the two heroes are mesmerized.

as lover's language, as an inspection of Diderot's writings will show. Indeed many of the philosophes' statements sound as though they come from the mouths of the Romantics. The philosophes had a more balanced attitude to "imagination" and science than the Romantics. They rarely denied the place of the arts, nor the absolutely essential place of the imagination in the creative process.

It is a mistake, made by Blake and the Romantics, to see the philosophes and the other heroes of the Enlightenment as extremists in a war between reason and imagination. Much of their attractiveness comes from their recognition that, in general, the excesses of both reason and passion are equally to be avoided.

Romanticism and *Naturphilosophie* were themselves doomed after 1830, ironically from the very expansion of industrial power that they so loathed. A prosperous upper middle class grew up in England and Germany, its wealth usually based on industry. Science was now seen, in the guise of technology, to be producing welcome results in the real world, while the philosophers, especially the German variety, merely turned out obscure and impotent tomes. The middle-class became increasingly materialistic and politically conscious, and in the nineteenth century science would regain its prestige, sometimes for the wrong reasons.

The Enlightenment: A Different Perspective

In recognizing the long-range liberating role of the Enlightenment, we should realize its limitations. It was not a mass movement. At its peak it involved only a few tens of thousands of people, mostly rich or middle-class. The Enlightenment and Newton meant little or nothing to the average man. As Kant said, he was living in an Age of Enlightenment, but the age was unenlightened. The "radical" leaders of the Enlightenment believed in universal education, where "universal" almost always meant the professional and merchant classes, not the proletariat. Only Condorcet, of the Enlightenment's French luminaries, believed that women should have equal status to men, although Diderot blamed men's historical dominance of women for their "inferiority."[5] D'Holbach declared that women's weaker organs prevented them from attaining genius. Rousseau, despite a nod toward equal educational opportunities for women, still proclaimed that women had no originality and recommended that they concentrate on pleasing men. The editors of the *Encyclopédie* had no qualms about including illustrations of factories employing child labor.

It would be difficult to prove the thesis that the Enlightenment seriously damaged religion at the time. The philosophes and a significant number of English intellectuals may have been deist or atheist, but the general population knew little of these things. In fact, many of the clergy were among those who bought the *Encyclopédie*, and some of the philosophes saw reason as, in principle, compatible with theology. In England in particular, the clergy saw science as a legitimate interest that offered no threat to belief.

The "radicals" of the Enlightenment left a permanent mark on Western culture and on man's concept of his role, rights, and potential, but they were not revolu-

---

[5]Diderot's attitude toward women might repay study. He sympathized with Clarissa, the hard-done-by heroine of Richardson's sentimental novel, and he had a distinct, and documented, weakness for the nubile breasts of Greuze's insidiously erotic peasant girls.

tionaries in the image of Lenin or Che Guevara. They were writers and talkers, not guerillas, although some of them, like Voltaire, Rousseau, and Condorcet, suffered or risked exile or imprisonment for their views. Some of them might have preferred a secular state, but few were democrats, and there is little evidence to show that they foresaw or desired the bloody revolution that was to come. They never created a party or a popular movement. All the "big names" except Condorcet died by the 1780s, so that the movement died socially before the Revolution. Marie-Jean-Antoine Nicolas de Caritat, marquis de Condorcet, perhaps the wisest of the philosophes, died in 1794 in a prison cell, persecuted by those who had perverted the principles of the Enlightenment. Perhaps, in his last days, he recalled Diderot's resigned lament, "The philosopher may have to wait. He may find that he can only speak forcefully from the depth of his tomb."

Nevertheless . . .

The Enlightenment had its triumphs. We no longer, in the West, officially burn heretics or rack and then strangle servant girls who bewitch children, as happened in Switzerland as late as 1782. We can still echo Condorcet's cry for "reason, toleration and humanity."

Perhaps the best epitaph for the Age of Reason was written by Tom Paine when he claimed that he was "contending for the rights of the living and against their being willed away, and controlled, and contracted for, by the manuscript-assumed authority of the dead."

In retrospect, the Enlightenment's near-total belief in reason and science as a universal and almost imminent panacea now appears as naive optimism. And yet the seventeenth-century scientific revolution and the eighteenth-century Enlightenment liberated us, giving us the confidence to use our reason and the confidence to doubt it. The Enlightenment gave us the right to question and the right to doubt, even the existence of God. It gave us the courage to challenge the wisdom of the ancients, no matter how sonorous the prose in which it was couched, nor how established the hierarchy of its priests. Kant, who saw the Enlightenment as a revolt against superstition, suggested for it a quote from Horace, *Sapere aude! Dare to know!* We dare.

There was, however, a price to pay. The Enlightenment gave man a lonely dignity, a home in a mechanistic cosmos in which God had a doubtful status. For those who accept this fate, it may be, as the materialist Hobbes said, that "there is no such thing as perpetual tranquility of mind while we live here." Three centuries later he was echoed by Camus, who, accepting the absurdity of the cosmos, advises us to pull back on our bowstrings and face the enemy: "There can be no contentment but in proceeding."

As for Romanticism, thankfully it still speaks to us, a reminder of the individual mind. In book 5 of the *Prelude*, Wordsworth tells of a dreamer who meets an Arab fleeing from a tidal wave that will swallow up the world. The Arab carries a shell and a stone, and explains that both are in reality books: the stone is Euclid's *Elements of Geometry*. The shell is "something of more worth": "An Ode, in passion uttered."

When we say that Newton's laws can explain the movements of the planets, we don't ask of what the planets were made. That kind of knowledge doesn't appear in his equations. All that we need to know about the planets in order to use Newton's equations are their mass and the forces acting on them. This is a classic example of the scientist looking at an immensely intricate system, the universe, and singling out a specific aspect, with the hope, fulfilled in Newton's case, that an explanation can be found *for that aspect*. Newton used one of the classic variations of the scientific method—he made a *model*. We did the same thing in Chapter 1.

The box of molecules in Figure 1.2, does not contain real molecules. Real molecules are complex entities containing subatomic particles called protons, neutrons, and electrons. We completely ignored the internal structure of atoms and molecules, and constructed an extremely simple model. The reason was that we were concentrating on a simple property and asking a simple question: How do they move? For that limited purpose, it was sufficient to take structureless blobs of matter, knowing that they were not accurate representations of reality. The complexity of a model is not required to exceed the needs demanded of it.

Science very often employs models. Sometimes it is because we can do no better, we just haven't got the information. If the model works, we feel that it is justified. Often we know that the model will be insufficient to form a basis for the explanation of all the physical properties of a system. If we had attempted to explain the interaction of light with our molecules-in-a-box, the simple model that we used would have been useless. We would have to construct another model, one that would be much nearer to reality.

Throw a stone into the sea. Newton's laws will tell you how it moves through the air. Switch on a bedside lamp. Newton would be mystified. He would be incapable of using his laws to explain what is happening. Nearly a century was to pass, after Newton's death in 1727, before a new force took its place in the description of the cosmos. Electromagnetism was the key to the nature of light, and the intimate structure, and properties, of matter.

8 | Lodestone, Amber, and Lightning

Lodestone, amber, and lightning were nature's hints that there were strange forces in Heaven and Earth. Lodestone, a naturally occurring magnetic oxide of iron, attracted nails and, if suspended, aligned itself roughly along a north-south line. If rubbed with suitable material, a piece of amber attracted pieces of thread or paper. These weird properties still intrigue children—and adults. The compass needle fascinated Einstein as a child. How did the needle know which way to point? Even as late as the seventeenth century, many believed that magnets had souls. The pull of a magnet was likened to the lure of a siren for a sailor. Katherine Philips (1631–1664), a minor English poet, saw the turning and trembling of a needle in response to a lodestone as a manifestation of a kind of love.

Over the centuries, out of simple experiments and observations, two ill-defined concepts emerged: electricity and magnetism. They appeared to be quite independent phenomena; you didn't get a shock from a magnet, and an electrically charged rod didn't behave like a compass needle.

How Many More Forces Are There?

We don't know. No one would have thought up magnetism and electricity if their effects had never been detected; they are not a consequence of Newton's laws. So it's worth keeping an open mind regarding the existence of as-yet-undiscovered forces.

As far as we know at present, only four forces shape the universe. Two of the four, the so-called *weak* and *strong* forces, are of very short range and operate only within the nucleus of atoms, the tiny core at the center of each atom (see Chapter 12). The other two, long-range, forces that control Creation are gravity and *electromagnetic* force.

Why *electromagnetic*? Why combine two phenomena that seem so very different? You can't run a laptop computer on a magnet. And try cutting a magnet in half to isolate the north pole and the south pole. It doesn't work, you just get two smaller magnets, each complete with its own north and south poles—no matter how far you carry the process. It is impossible to obtain a single, isolated north or south pole of a magnet.[1] This is in contrast to the independent existence of two kinds of electricity: positive and negative electric charge. Electricity and magnetism appear to be very different from each other. We will shortly show that they are madly in

---

[1]In 1931 the outstanding theoretician Paul Dirac predicted that monopoles should exist. Other theoreticians have predicted that they should have been formed in great numbers around about the time of the Big Bang. It has been claimed that the results of one experiment show the effect of a magnetic monopole passing through the apparatus. This has not been confirmed. If you want a Nobel Prize the hard way, you know what to look for.

love; one (almost) cannot live without the other, but it took a long time to realize that they are two faces of one coin.

Immobile Charge

Static charge, the kind that makes some artificial fabrics stick to your skin, has been known for centuries. If you rub two pieces of amber, or plastic, with a piece of suitable material, they repel each other. Since they are presumably charged with the same kind of electricity, you have shown that like charges repel each other. You can observe the same phenomenon if you rub two glass rods on a cooperative cat. However, if you hold a rubbed glass rod next to a piece of rubbed amber, they will attract each other. You have shown that there are two kinds of electricity and that unlike charges attract each other. The names *positive* and *negative,* given by Benjamin Franklin, are purely conventional; there is nothing either positive or negative in the algebraic sense about electric charge. It is convenient, however, to use the conventional labels because positive and negative charges appear to cancel each other, just as +3 cancels -3. If equal amounts of negative and positive charge are combined, the resultant net *macroscopicically measurable* charge is zero.

Until the beginning of the nineteenth century, electric charge was usually generated with a variety of "electrical machines," all based on the friction of a material such as leather or silk on a rotating disk or cylinder, often made of glass, occasionally of sulfur. This is a mechanical version of dragging a comb through your hair. The charged body carried a static electric charge, which it could lose if connected directly or indirectly to the ground. The discharge of charged bodies took a fraction of a second during which a momentary flow of electricity might be induced to flow along a wire or a thread, but that was that. No one knew how to produce a constant, long-lasting flow of charge, and thus experiments on electricity were almost entirely confined to systems having a static, not moving, charge.

To do any kind of quantitative experiments on electricity it is advisable to have a unit of electric charge, just as there is a unit for mass and for length. One familiar unit of electricity is the amp (the ampere). The amp is a unit of *flow* analogous to the unit *cubic meters per second* that we use for the flow of liquids. The unit of electric *charge* is named after the Frenchman Coulomb. The coulomb is defined as the amount of electric charge that flows past a given point in a wire in 1 second if the current is 1 amp. Most household lighting is run on a 5-amp current, which means that if you took a specific point on a lamp filament, then in 1 second 5 coulombs of electric charge would pass by. A current of about 200 amps (200 coulomb per second) through the human body will kill. Actually households run on *alternating* current. The current flowing in the wiring changes its direction about 60 times a second, so you would have to measure the total charge passing by a certain point per second, irrespective of the direction in which it was flowing. Most heating appliances use 15-amp current. A total of about 3000 to 5000 coulombs passes through the heating coil when you boil an electric kettleful of water.

Several attempts were made to find the law that holds for the force between electric charges. Most of those who searched for this law assumed (correctly) that they would find an inverse square law, by analogy with Newton's law for gravity. As the mathematician and physicist Henri Poincaré (1854–1912) once remarked, "It is often said that experiments should be made without preconceived ideas. That is impossible."

In the 1760s both Daniel Bernouilli and Joseph Priestley looked, unsuccessfully, for an inverse square law. The man who first succeeded was Henry Cavendish, who experimentally demonstrated the law, sometime in the 1760s. However, Cavendish's results only came to public notice in 1879 when James Clerk Maxwell published them. Thus, for over a century, priority was accorded to the French military engineer Charles Augustine Coulomb, who published his work in 1785 and 1787.

If you've forgotten what the gravitational law looks like here it is again:

$$F = G\,[m_1 \times m_2]/R^2$$

Electrically charged bodies obey a similar law. The force between two charges is given by:

$$F = K\,[q_1 \times q_2]/R^2$$

in which the places of the two masses are taken by q_1 and q_2, the magnitudes in coulomb of the two charges, and R is the distance between the charges. Increasing the distance between two charged bodies by a factor of 5, reduces the electrostatic force between them by a factor of 25, whether the force is repulsive or attractive.

In contrast to the value of G, the value of K in the equation for the electrical force depends strongly on the medium in which the two charges are situated. For example, the force between two charges in water is only about one-eighteenth of that between the same charges in air, at the same distance. The gravitational force between two bodies is effectively independent of the nature of the material between them.

Although machines for producing static charge by friction were common in the eighteenth century, there was, until the beginning of the nineteenth century, no reliable way of artificially producing a reasonably strong magnet. Here is Michael Faraday's description of how to make a magnet:

> Taking a bar of iron, place one end ". . . pointing about $24^1/2$ degrees west of north, and downwards, so as to make an angle of $72^1/2$ degrees with the horizon. When the bar is held in this position, with one end on a large kitchen poker in the same position, and the other end struck three or four times with a heavy hammer, it will become a good magnet."
>
> *Chemical Manipulation* (1827)

Before the nineteenth century, the highest fields attainable from magnets were no more than a few times stronger than that of the earth. Today there are commercially available scientific instruments that produce magnetic fields well over 350,000 times as strong as that of the Earth.

Mobile Charge

The concept of *current* electricity was probably first grasped by Otto von Guericke (1602–1686), who showed that if he connected a charged ball of sulfur to a nearby body, by a thread, the body displayed the same electrical properties as the ball. Charge had moved along the thread.

The next ingenious amateur to significantly advance the cause of current electricity was the charismatic Benjamin Franklin, friend of the philosophes. He stated that charge could be neither created nor destroyed. In other words, he formulated a *conservation* law: charge is conserved. Discharging a body does not mean that the charge is destroyed; it just goes somewhere else or is apparently canceled by an equivalent influx of charge of the opposite sign. Franklin guessed that electricity came in discrete particles and that lightning consisted of moving electric charge. The 1752 experiment in which he flew a kite in a thunderstorm in order to demythologize the thunderbolts of the gods, has become part of the mythology of science. Electricity flowed along a wet thread, from kite to ground.

The nineteenth-century chemist Joseph Priestley declared that the discovery that lightning is an electrical phenomenon was perhaps the greatest since the time of Sir Isaac Newton. Less well publicized is an earlier experiment conceived of by Franklin in 1750 but not carried out by him. He suggested constructing a bizarre apparatus consisting of a man-sized box, on the floor of which was an insulating pad, on which the rash experimenter stands, and through the door of which rises a long, pointed rod, extending 20 to 30 feet in the air and terminating in a point. When lightning strikes the rod, the suicidal experimenter can, using an insulating wax handle, hold a grounded wire near the rod and see sparks jump across the gap. Rather you than me. The experiment, designed to show that lightning was electricity, was first actually performed by Dalibard, in France in 1752. It should have come with a government health warning. Georg Richter (1711–1753) a rash Estonian, constructed a variation of Franklin's box in St. Petersburg. He was electrocuted in his box during a storm. Someone commented that they envied him his glorious death in the cause of science. (Someone else remarked that it was a shocking tragedy.) In the same year, Franklin received a medal from the Royal Society. I wonder if he spared a thought at the ceremony for poor reckless Richter.

A Note on Conservation

Franklin's suggestion that charge was conserved is the first example of a *conservation law*. Such laws play a central role in science. Conservation laws are general; for example, in *any* process in nature the total charge of the universe remains unchanged. Charge cannot be created or destroyed by mortal man. We shall be meeting more conservation laws, each saying something very fundamental about the nature of nature.

Things Start to Move

Lightning being a somewhat inconvenient source of electric current, it was fortunate that Franklin's friend, Count (to be) Alessandro Volta (1745–1827) found a way of producing steady currents of electricity in the laboratory. This discovery, which was the key to the practical use of electricity in the early nineteenth century, owed a lot to Volta's compatriot, Luigi Galvani.

In 1791, Galvani had observed that a frog's legs jerked when he passed a momentary current through them. He got similar responses by passing short-lived currents through corpses. His findings galvanized Europe. Galvani's source of current was a charged body, but subsequently, and fatefully, he found that he got the same effect by simply touching the legs of frogs with two connected rods made of different metals. Galvani's interpretation was that living matter contained, or could generate, "animal electricity." It can, but this had little to do with his experiments. Today electroencephalography (EEG) and electrocardiography (ECG) are routine techniques based on the body's own electricity, and of course there are electric eels. But the response of the frogs was to an external, not internal, source of electricity. Galvani didn't realize it, but it was the pair of metals that were generating the current of electricity. Volta became interested in Galvani's twitching corpses.

Alessandro Volta was a self-confident type. His friend Georg Christof Lichtenberg (1742–1799) said of him that he "understood a lot about the electricity of women."[2] Volta sneered at Galvani's suggestion of animal electricity: "Physicians are ignorant of the known laws of electricity." He had little reason to snipe. Volta had once put a coin on his tongue, a coin of a different metal underneath his tongue, and connected them with a wire. He experienced a salty taste which he, like Galvani, (wrongly) interpreted as electricity generated by living tissue. When he heard of Galvani's work, Volta proceeded to experiment with whole frogs and rapidly came to the conclusion that a current flowed when two different, and physically connected, metals were applied to *any* moist body. He constructed piles of alternating disks of copper and zinc separated by wet, absorbent paper or even damp wood. This was the first man-made battery.[3] Since a battery is just a means of pumping electric charge, placing batteries head to tail in a row (in the jargon, "in series") means that their total pushing power is the sum of that of all the pumps. By using many pairs of copper and zinc plates, Volta obtained strong currents that lasted for a considerable time. The "voltaic pile" was the first convenient source of current electricity. Volta invented the phrases "electromotive force" and "electric current," and commenced to persecute the animal world. He showed that an electric current could induce movement in a variety of unfortunate animals and insects: "It is very amusing to make a headless grasshopper sing." Volta, who had a great sense of humor, conveyed his findings to the Royal Society in 1800. The announcement of the construction of a source of steady electric current created a sensation. Napoleon contacted Volta, created him count, coined a medal in his honor, and provided him with financial assistance for many years.

Volta's pile had the disadvantage of all chemical batteries: the chemicals get used up. Nevertheless, the pile provided, for the first time, a controlled, continuous source of electric current, opening the way to a number of almost immediate practical applications such as the telegraph and electroplating, and also to a whole new range of experiments. The concept of voltage was born. Voltage (from Volta, of course) is the electrical equivalent of pressure, measuring the force with which a generator or battery can push electrons. Typical voltages are 1 to 2 volts for simple batteries, 6–12 volts for most car batteries, 120–240 volts for domestic electricity, and about 1000 volts for the shock delivered by an electric eel. ECG involves voltages in the single-figure millivolt range, compared with 20 to 30 millivolts for EEG.

The voltaic pile remained the source of electricity for commercial use, such as telegraphy, until well into the 1800s. No equivalent technical advance was made in magnetism until 1820.

---

[2]Maybe Lichtenberg was just envious; he had a desperately unhappy marriage.
[3]Electric eels have a similarly structured electric organ, but not of course based on metal disks.

Magnetic Personalities

Magnetism's uncanny manifestations provoked a rich folklore from the time that the lodestone was discovered, possibly by the Chinese in the third millennium B.C. One of the earliest rational references to magnetism in the West was in the Roman Lucretius's poem *De rerum natura* (see Chapter 14). It is typical of the down-to-earth atomist that he refused to countenance mystical explanations of magnetism.

The earliest known treatise on experimental physics, *Epistola Petri Peregrini de Maricourt ad Sygerum de Foucaucourt Miletem de Magnete* (A letter of Peter the Pilgrim of Maricourt to Sygerus of Foucaucourt, Soldier; on the Magnet), contains an account of the author's experiments with lodestones. Peter completed his manuscript "in the year of our Lord 1269, eighth day of August" while he was serving in the army of Charles I of Anjou, at the siege of Lucera in Italy.[4] He was the first to record that magnets had two "poles," one seeming to look for the north, the other the south. Like poles repelled each other, but north attracted south. Peter's book was taken to heart by William Gilbert, physician to Queen Elizabeth I and James I, who, despite his conviction that magnets had souls, was an experimentalist in the Baconian mold.

It was Gilbert, in his book *De magnete,* who first suggested that the Earth itself was a giant magnet, thus explaining the behavior of the compass needle. The magnetic north pole, where the magnet points vertically downward, was discovered in Canada in 1831. Gilbert used the phrase "magnetic coition" rather than magnetic attraction, "because magnetic movements do not result from attraction of one body alone, but from coming together of two bodies harmoniously (not the drawing of one by the other)." He seems not to have realized that there can be coition when only one body attracts the other. Gilbert thought that the attraction between magnets could provide an explanation of the Moon's path about the Earth, an idea that was accepted by Kepler. They were both wrong.

In 1778 a blacksmith's daughter named Ami Lyon worked as a provocatively dressed hostess in the Temple of Health, an establishment in the Strand. Billed as "Vestina the Rosy Goddess of Health," she was better known in later years as Emma, Lady Hamilton, Lord Horatio Nelson's mistress. The activities of the Temple of Health were centered on a large "Celestial" bed constructed by a Dr. Graham and equipped with 1500 pounds of "artificial and compound magnets" that, the mercenary doctor claimed, revivified slackening sexual vigor. The bedroom reeked of expensive Oriental perfumes, and it usually cost a couple twenty pounds a night to use the bed. Some gentlemen were charged fifty pounds a night, but it is not recorded what extra services were provided for this sum. Considering the ancient belief that if one put a piece of lodestone under the pillow of an unfaithful wife it would make her confess her sins, it is a wonder that English fashionable society flocked in such large numbers to Graham's establishment. Graham died a poor man. The age-old mystique surrounding magnetism could not compensate for a flaccid succession of disappointed clients.

Dr. Graham's French equivalent was Friedrich Anton Mesmer (1734–1815), a

---

[4]Dame Kathleen Lonsdale, who lectured to me in crystallography, wrote a scientific text while in prison, but to write a scientific text in the midst of a military campaign appears to be a feat unique to Peter and to the Greek physician Dioscorides, who served as a physician in Nero's armies (see Chapter 20).

doctor who took many of his ideas from the Jesuit priest Maximillian Hell. Mesmer exposed his patients to magnets while they sat by the side of a large tub containing a murky solution of chemicals. The committee appointed to examine his methods and claims included Benjamin Franklin. Their conclusion, that "the imaginaytion does everything, the magnetism nothing," implicitly acknowledges the power of hypnosis, which was almost certainly responsible for some of Mesmer's genuine successes. He appears to have initiated the medical use of hypnosis in western Europe.[5]

Unification

All physical experiences are due to one force.

—Immanuel Kant

On 21 July 1820, the Danish physicist Hans Christian Oersted (he was a friend of H. C. Anderson) published four pages in Latin entitled: "Experimenta circa effectum conflictus electria in acum magneticum" (Experiments on the effects of an electrical conflict on a magnetic needle). This paper made Oersted's name. He had shown that the passage of an electric current through a wire influenced the orientation of a compass needle placed near the wire. It was known, of course, that a compass feels a force from a magnet, so Oersted concluded that the electric current must have been acting like a magnet. For the first time, magnetism and electricity were shown to be related, and Oersted used the term *electromagnetism* in referring to the phenomenon he had discovered. *He had pointed the way to the unification of two of nature's basic forces.*

Oersted made his observation while lecturing to a class of students, and it has often been stated that the discovery was accidental, apropos of which two comments are in order. As Pasteur remarked, "In the field of observation, chance favors the prepared," and second, Oersted had been looking for such an effect for years.[6]

Oersted's paper was rapidly translated into several European languages, Latin no longer being readily understandable by the whole scientific community. It had an almost immediate technological effect. William Sturgeon, who as a professional soldier in the British Royal Artillery performed experiments on electricity in his spare time, wrapped some copper wire around an iron horseshoe and passed a current through the coils. He had made an electromagnet. Across the Atlantic, also in 1820, Joseph Henry did likewise but, being American, used more coils of wire. His electromagnet could raise a ton of iron.[7]

---

[5]Despite the frauds perpetrated by charismatic imposters, there is good evidence that magnetic fields can affect living organisms. It is well established that certain iron-containing bacteria are affected by magnetic fields, which is not really surprising. It has been claimed that there is a good correlation between magnetic storms and suicides, but this remains to be verified.

[6]It is a curiosity that the scientist Johann Ritter, who dabbled in astrology, had previously written a letter to Oersted prophesying that the next great discovery in the field of electricity would be in "1819$^{2}/_{3}$ or 1820."

[7]Henry, like Michael Faraday, came from a working-class background and was an apprentice—to a watchmaker at the age of thirteen. He also, like Faraday, is reputed to have had his interest in science sparked by a book, which he apparently found under some floorboards. He tore up his wife's petticoat to provide insulation for the copper wire that he wrapped around his magnets.

Oersted's paper was presented in Paris by Dominique Arago. Among those in the audience was a professor of mathematics at the École Polytechnique: André-Marie Ampère. Ampère's personal life reads like a soap opera. He was born in 1775 to a father who was guillotined when Ampère was eighteen. Ampère's wife died four years after the marriage, his second wife (and her mother) made his life unbearable, and they divorced. Ampère's son fell under the spell of a woman of a certain age, the formidable, charming, but not over compliant hostess, Madame Julie Récamier. (You can judge his taste by looking at the portrait that David painted of her.) He spent about 20 years doing nothing constructive, unless you include intermittently gazing at the lady's décolleté. Ampère's daughter married a mentally unstable alcoholic officer. Ampère said that he had only two happy years in his life. His tombstone reads, *Tandem felix,* Happy at last.[8] He took up science after finding, in his father's library, a eulogy of Descartes, which convinced him of the nobility of a life in science. Just another of those chance encounters that so often determine our fates.[9]

Within a week of hearing of Oersted's work, Ampère, announced that he had observed an attractive force between two parallel wires through which a current flowed in the same direction. Currents flowing in opposite directions produced repulsion. The force depends on the strength of the current, and we still use the force between two current-carrying wires as the fundamental basis for the definition of the unit of electric current—the ampere. Arago, recalling the school experiment in which iron filings on a piece of paper are formed into a pattern by a magnet beneath the paper, passed a current through a wire piercing a piece of paper on which iron filings were sprinkled. The filings lined up to give concentric circles about the wire. A current indeed acted like a magnet, and the force between two current-carrying wires can be interpreted as a magnetic force that is exerted on a *moving* charge in one wire by the current in the other—and vice versa.[10]

The current in a wire, or any magnetic influence, will not exert a force on a *static* charge. Static charges and static magnetic fields are completely blind to each other,[11] which is one reason that the connection between magnetism and electricity took so long to be discovered.

Since Oersted had shown that currents act like magnets, Ampère decided that he could explain Oersted's and his own results if "one admits that a magnet is only a collection of electric currents." His detailed explanation of the origin of the currents in a magnet was wrong. Nevertheless, between them, Oersted and Ampère

---

[8]Would he have been cheered up could he have foreseen the daily stream of customers asking for "a 15-amp plug, please"? Prominent scientists run the risk of being immortalized in this way. Among the physical units that we use are a volt, an ohm, a coulomb, a farad, and a Faraday, a maxwell, a newton, a dalton, a curie, a Kelvin, a becquerel, and an einstein.

[9]Who can say what chains of scientific influence were initiated by a certain priest in the Jesuit College of la Flèche, who on 6 June 1611 read out a poem composed to mark Galileo's discovery of Jupiter's moons? One of the listening pupils was the fifteen-year old René Descartes.

[10]The effect of a magnet on a moving electric charge is used in all television cathode tubes. A stream of electrons (negatively charged particles) pass between the poles of a small magnet, the strength of which can be varied. The force on the electrons, and therefore the deflection from their straight path, depends on the strength of the magnet. This is the mechanism by which the electron beam is made to sweep out the 650 lines that cross the screen.

[11]You might ask why an electric current running through a wire, which is, after all, a flow of electric charge, does not exert an *electric* force on an external charge. The answer is that the wire is electrically neutral. The atoms of which it is made have positively charged cores (nuclei), and their total charge equals the total negative charge on all the electrons in the wire.

had shown that electricity and magnetism were two faces of the same animal. The next step was taken in England.

Working-class Hero

I love a smith's shop and anything related to smithery. My father was a smith.

—Michael Faraday, journal, 2 August 1841

Michael Faraday is one of the fairy tales of science—the working-class apprentice who rose to be courted by royalty. Handicapped by his lack of a formal education, he would never have been a theoretical physicist, since his knowledge of mathematics was poor. He couldn't follow the math in Ampère's papers. What Faraday did have was an intuitive feeling for the workings of nature, which he translated into visual rather than mathematical terms. And he had an extraordinary talent for choosing the right experiment. The blacksmith's son had marvelous hands, and the natural world could not resist him. He was a scientific Casanova.

Faraday's life was science. He avoided honors, turning down the presidency of the Royal Society and a knighthood. The portrait of the man who never attended university and whose school education was effectively limited to the "three Rs," hung on the wall of Einstein's study. At the age of thirteen, he went to work for a French émigré bookbinder and printer, in Blandford Street, off Baker Street in London, beginning as a paperboy. The turning point in his life came when he was rebinding a volume of the *Encyclopaedia Britannica* that included an article on electricity. The author's suggestion that electricity had the nature of a nonmaterial vibration caught his attention. Another book that fell into his hands, and which awakened his interest in chemistry, was Jane Marcet's *Conversations in Chemistry,* an early example of popularization. At the age of nineteen, he began to make up for his educational shortcomings by attending public lectures, at first at the City Philosophical Institute, where there was a voltaic pile. In 1812 a customer at the bookbinders, a certain Mr. Dance, was so taken by the curiosity and intelligence of the young apprentice that he gave him tickets to attend four of the series of lectures being given by the renowned chemist Humphry Davy at the Royal Institution. Davy, like Faraday, never went to university, and probably never wrote down a mathematical equation in his life, which would have made it easier for Faraday to appreciate his lectures. Faraday bound his lecture notes and sent them to Davy with a request for a position. The notebook still exists. Davy, who had been temporarily blinded by an explosion, accepted him as an amanuensis. Fate played its next card when an assistant at the institution was sacked for fighting and Faraday was given his position. Davy had just been knighted and it had gone to his head, but more so to his wife's, who treated Faraday as though he were her personal servant. Under these not too promising conditions, one of the greatest experimentalists in the history of science began to experiment, with the talent that genius often has for asking the essential questions and finding simple experiments to answer them.

Faraday was a giant who pointed theoretical physics in a new direction, and whose experiments were to lead directly to technologies which were to change the face of civilization.

In 1800, Nicholson and Carlisle constructed the first voltaic pile in England, and the availability of steady current began to pay off immediately in terms of new discoveries. They found that a current decomposed water, an observation probably made in the same year by the German Johann Ritter. This was the first electrochemical reaction. In the same year Humphry Davy began a series of experiments on the effect of electric current on chemicals, which led to the discovery, among other things, of the metals sodium, potassium, and calcium. In the 1830s Faraday showed that when an electric current was passed through a solution containing salts and there was decomposition to give new chemical species, the amount of decomposition was directly proportional to the amount of current passing. If a certain quantity of oxygen gas is released from water by the passage of a current for five minutes then twice as much gas will be produced in ten minutes. The implication was that electric charge was an integral part of matter and somehow involved with the way that atoms were bound together. After consulting on nomenclature with William Whewell, a distinguished Oxford professor who coined the word *scientist* in 1834, Faraday adopted the words *electrode, cathode, anode, ion, electrolyte,* and *electrolyze.*

Today Faraday's experiments on electrolysis are taken to be direct evidence for the existence of a discrete "packet" of electric charge, the electron. We can almost see the current going in, electron by electron, each electron joining onto a molecule or ion in the solution and changing its chemical form. Why didn't Faraday realize that electricity came in discrete packets? Why didn't he discover the electron? Because for him, passing electricity into a solution was like pouring in a continuous fluid with no structure. It resembled the passing of time: you could select as much or as little time as you liked, but there was no question of time coming in discrete packets. Consistently, Faraday did not believe in atoms.

Faraday's electrochemical experiments were important, but it was three entirely different findings of his that were to turn the course of history:

1. He showed (Figure 8.1a), on 21 October 1821, that when a current was passed through a rod that was free to move, and one end of the rod was in the vicinity of a magnetic pole, the rod circled around the magnet. Electrical energy could be continuously converted into mechanical energy. This was the origin of the *electric motor.*

2. He showed (Figure 8.1b), on 17 October 1831, that when a magnet *moved* through a coil of wire it induced an electric current in the wire, but only during the time that it was moving and the strength of the magnetic field near the wire was changing. The mechanical energy of motion could be turned into electrical energy. The *dynamo* was born, and Faraday subsequently built the first primitive electric generator based on this principle.

3. In the same year, he showed that a current could be induced in a circuit by *changing* the current in an adjacent circuit. The *transformer* and the *induction coil* were born. At about the same time Joseph Henry, in America, discovered the same effect.

Notice that what Faraday had shown, among other things, was that:

A *changing* magnetic field created an electric current (2), and
a *changing* electric current created another electric current (3).

These two facts, and Oersted's discovery that a steady current could produce a

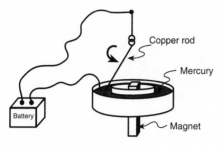

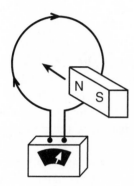

Figure 8.1a. Faraday's demonstration that a magnetic field produces a force on a moving electric charge. The copper rod revolves when a current flows.

Figure 8.1b. A changing magnetic field exerts a force on an electric charge. When the magnet moves, a current of electrons flows.

magnetic field, were to form the experimental basis for the complete melding of electricity and magnetism in one theory.

Every power station and every electric motor in the world uses the technical consequences of Faraday's discoveries. But they had more than technological consequences. Faraday invented the idea of *electric and magnetic fields*. If he had done nothing else, he would have changed science. That is the reason that he appeared on Einstein's wall.

The Power of Space

Magnets and charged bodies act through space to exert forces at a distance, on other, appropriate bodies. Faraday pondered on the problem of action at a distance, and there grew in his mind the idea that, surrounding a magnet or a charged body, there was an invisible, immaterial "sea," an entity that exists in space, rather like the waves that spread out from a stone thrown into a pond.

If a piece of wood is placed in the ripples generated in water by a vibrating body, and *we were unable to see the waves,* we might say that the vibrating body acts at a distance on the piece of wood. Faraday would say that, despite appearances, there is a field of waves which exists over the *whole* surface of the water, not just where the wood bobs up and down. Despite the fact that it could never have come into being without the vibrating body, the field of waves is an entity in its own right; it will persist for some time even if the body stops vibrating. It is capable of moving bodies submerged in it, acting as an intermediate in transferring movement from the vibrating body. For water this is an easily acceptable idea because the waves are of course visible. For magnets or electrical charges it was a completely new concept. Faraday did not envisage a field of waves but rather a static field that could be mapped out in terms of *lines of force.* Thus a charged body can be considered to be

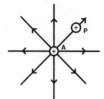 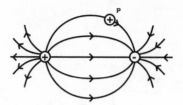

Figure 8.2. Visualizing an electric field. The arrows surrounding the charge A indicate the direction of the force felt by a positive test charge, P, placed at several positions in the plane of the page. Imagine an infinite number of similar arrows above and below the page. A pair of opposite charges produce the pattern shown on the right.

surrounded by an electric field, which could be mapped out by drawing, at every point in space, a tiny arrow whose direction would be in the direction of the force exerted by the field on an electrically charged body put at that point (Figure 8.2). In the case of a magnet these lines could be visualized by examining the pattern made by iron filings sprinkled on a piece of paper lying on a magnet. In their entirety these lines define a magnetic field. Faraday admitted that he did not know what a line of force or a field really was, any more than Newton knew what gravity was, but breakthroughs in science have often ridden into town on the backs of enigmatic new concepts. The simplest way to picture the fields created by magnets or static charges is to define them by what they do. Notice how, as in the case of force, we have once again fallen back on an operational definition of a basic scientific concept. You are what you do.

Faraday's concept of fields was the product of a brain that worked with visual images. Faraday *saw* fields. His intuition filled the universe with silent, ever-present seas. When he lectured on the magnet, he would hold one up with one hand, knock on it with the other, and declare, "Not only is the force here, but it is also here, and here, and here," moving the other hand through the air around the magnet. Today the concept of fields has a major role in theoretical physics, one of the reasons being that fields are very convenient in quantitatively, and qualitatively, explaining countless phenomena in the material world. This is possible only because mathematical equations have been constructed which give the form and point-by-point strength of fields. Faraday could not do this, but he did not have to wait long before someone did. Before we make that step, we return to findings 2 and 3, and look at them from the field point of view.

Finding 2 (Figure 8.1b) can be interpreted as a changing magnetic *field* producing an electric *field*. It is this electric field that drives the electrons around the wire coil in which they sit. Likewise, finding 3 shows that the changing current in the first coil (the so-called primary) creates a magnetic *field* which, as it changes, creates—in accordance with finding 2—an electric field at the second coil (the secondary), which drives the current in that coil.

These interpretations are at the very root of the unification of electricity and magnetism. They show that these two forces are merely the different faces shown by one force under different conditions:

A change in a magnetic field produces an electric field.
A change in an electric field produces a magnetic field.

Faraday's fields took into account the possibility of curved lines of force. Thus the iron filings near a current-carrying wire were arranged in circles going around the wire, and a compass needle placed near such a wire would point at a tangent to these circles. When Ampère found that two current-carrying wires repelled or attracted each other he assumed, in analogy to gravitational force, that the force was a simple "straight line" interaction between two currents. Faraday realized that the force arose from the *tangential* magnetic field due to the current in one wire, acting on the current in the other wire (Figure 8.3). Ampère couldn't break away from Newtonian pictures.

Initially, most contemporary physicists saw Faraday's fields as a convenient fiction rather than having any reality. The notable exception was William Thomson, better known to history as Lord Kelvin. Thomson attempted to formalize Faraday's ideas in mathematical terms and in doing so he became convinced that there might be a connection between electromagnetism and light. He suggested as much to Faraday, knowing that it would be wise to leave any experiments to the master. Faraday did not disappoint. He passed a beam of light through a piece of glass sitting in a magnetic field and found an effect on the light. There was a connection between light and *magnetism.* The Faraday effect is still used by scientists interested in the detailed structure of certain types of molecule.

Probably driven by a sense of the unity of nature, Faraday wrote prophetically that he had "the very strongest conviction that Light, Magnetism and Electricity must be connected." But Kelvin failed to progress with his attempt to construct a formal theory of fields, a theory that would unite light with electricity and magnetism. And as Faraday moved into an old age clouded by illness, it looked as if his dream of unification would not be realized in his time.

Faraday's intensive workload led to a breakdown in his health and clear signs of mental strain. In 1840, at the age of forty-nine (the age at which Newton had a mental breakdown), Faraday wrote some anxious and revealing notes: "this is to de-

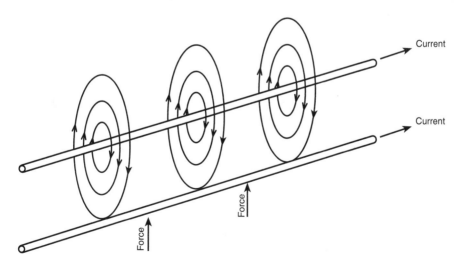

Figure 8.3. The current-carrying wires feel a force perpendicular to both the direction of the current in them and the direction of the magnetic field acting on the wire. Only the magnetic field due to one wire has been indicated.

clare in the present instance, when I say I am not able to bear much talking, it means really, and without mistake, or equivocation, or oblique meaning, or implication, or subterfuge, or omission, that I am not able, being at present rather weak in the head and able to work no more." There was a three-year gap in which he did nothing, not even reading science. In his final years Faraday lived in a house provided for him by Queen Victoria, his health probably affected by mercury poisoning. This was a then unrealized health hazard that undoubtedly affected many nineteenth-century scientists who worked in the laboratory, where spilled mercury often accumulated between and beneath floorboards and generated a permanent atmosphere of mercury vapor.

Faraday carried on experimenting until he was over 70. He died in 1867, the year that *Das Kapital* was published, and he was buried in Highgate cemetery, in North London, where Marx was to join him later. He lived long enough to see his prophecy of a unified field theory for electricity and magnetism fulfilled, by a theoretical genius.

Dafty Sees the Light

It was left to the Scottish mathematical physicist James Clerk Maxwell (1831–1879), who was known as "Dafty" at school, to sum up, in the form of equations, everything that was known about electromagnetism—the way in which electric and magnetic fields behaved. What Maxwell did was to express mathematically the form of electric and magnetic fields produced by charges or electric currents, whether the currents were steady or fluctuating. Thus if you wave a charged body about, the electric field it produces will also fluctuate at any given point in space. But we have seen that a changing electric field creates a magnetic field, and this too will change strength and direction as you wave the charged body, in its turn create a changing electric field, and so on. Maxwell's equations describe quantitatively the changing strength of these fields at all points in space. His book *Electricity and Magnetism* appeared in 1873, almost 200 years after the *Principia* and 100 years after Lagrange's *Mécanique analytique*. Maxwell's equations are as basic as the laws of motion, and together they form the foundations of classical physics—the framework within which the physical universe was explained until the end of the nineteenth century.

I will not write down Maxwell's equations, which add almost no new *physical* idea to what had been found before him. What they did do was put the ideas of Oersted, Ampère, and Faraday into mathematical form:

The *first* law gives the shape and strength of the electric field if we know the distribution in space of the electric charges that are producing it.

The *second* law says, among other things, that all magnets have two poles of equal strength; you can't have a tremendously strong pole at one end and a feeble one at the other.

The *third* law gives the strength and direction of the electric field at any given point, caused by a *changing* magnetic field. It can be used, for example, to find the electric force acting on the electrons in a coil of wire when a magnet moves relative to the wire.

The *fourth* law gives the form of a magnetic field caused by the *movement* of electric charge or the change in an electric field.

Sometimes when you have a set of equations and you sit down with a pencil and paper, you find that they contain more than you thought they did. When Maxwell started to play with the equations, his excitement mounted. In 1862 he wrote that "we can scarcely avoid the inference that *light consists in the transverse undulations of the same medium which is the cause of electric and magnetic phenomena*" (Maxwell's italics). In 1865 he wrote to his cousin, "I have also a paper afloat, which, till I am convinced to the contrary, I hold to be great guns." We saw earlier that the movement of an electric charge induces a changing magnetic field, which sets off a train of changing electric and magnetic fields, each causing the changes in the other. When Maxwell analyzed the form and time dependence of the electric and magnetic fields produced by a fluctuating electric current, he realized that his equations showed that the changing current would result in the formation of electromagnetic *waves* traveling through space at a speed equal to the known speed of light, and with a frequency equal to that at which the current fluctuated. This was one of the major breakthroughs in science: the proof that *light behaves like an electromagnetic wave,* that wherever you are in a ray of light you can detect fluctuating electric and magnetic fields.[12]

The significance of Maxwell's results was little appreciated at the time, but Einstein and Infeld, in a book published in 1938, wrote that Maxwell's equations were "the most important event in physics since Newton's time, not only because of their wealth of content but also because they form a pattern for a new type of law." This was a law that dealt only in fields, not in matter at all. Newton had expressed Galileo's (and his own) concepts in the language of mathematics; Maxwell had done the same for Faraday.

Maxwell's mathematics put his papers out of the reach of many contemporary scientists, including of course Faraday, who wrote to Maxwell and asked if, when a mathematician has arrived at his conclusions, "may they not be expressed in common language as fully, clearly, and definitely as in mathematical formulae? If so, would it not be a great boon to such as I to express them so?"[13] But Maxwell had little choice. His laws involved time and three spatial dimensions—and the formalism of differential calculus to describe them.

Meanwhile, back in Germany, Hermann von Helmholtz (1821–1894), the director of the Physikalische Technische Reichsanstalt, set out to verify Maxwell's equations experimentally. The student that he chose to actually do the work was a bright lad called Heinrich Hertz. In Karlsruhe, in 1886, Hertz showed experimentally that an oscillating electric current produced waves that spread through space. Their wavelength was typical of what we would now call radio waves. Moreover, he showed that the waves traveled at the speed of light. The scientific world acknowledged his achievement and finally fully realized the magnitude of what Maxwell had done. It was completely clear that light was an electromagnetic wave

---

[12]In Chapter 28 we discuss why I write "behaves like" and not "is."

[13]Faraday had an ally in Goethe, who was also allergic to mathematicians. He compared them to Frenchmen: "Talk to them and they will translate it into their own tongue where it at once becomes something altogether different."

and that other such waves, of almost any frequency, could be produced, or maybe existed in nature, a hope that was spectacularly realized with the discovery of X-rays, microwaves, and γ-rays, which took their place beside the already known infared (IR) and ultraviolet (UV) radiation.

It was also clear that Faraday's idea of fields had to be taken seriously, even if neither he nor Maxwell knew what fields were.

The Microscopic Origins of Electromagnetism

Later on we are going to find that:

1. The material cosmos is built of three kinds of long-lived particles, *all* of which are permanently magnetic.
2. Two of these three particles are permanently electrically charged.

If these facts are true, why aren't we giving off sparks in all directions, and why aren't there magnets everywhere?

Electricity

Matter is built up from very small entities called atoms. There are many different kinds of atoms, but they all have the same basic structure. All atoms have a massive central core, the nucleus, which is *positively* charged. This charge originates in one of the two types of particle that compose the nucleus—the *protons*. The other type of particle in the nucleus, the neutron, is electrically neutral, or, if you prefer, has no electric charge. Different types of atom have different numbers of protons in their nuclei, but the total electrical charge on a normal atom is zero. This is because the nucleus is surrounded by a cloud of *negatively* charged particles—the *electrons*. The electric charge on an electron is *exactly* equal in magnitude to that on a proton but is negative. In normal circumstances the great majority of atoms have in their immediate neighborhood as many electrons as they have protons in the nucleus, which is why the *net* charge on the atom vanishes. It is the attraction between unlike charges that holds the atom together, and it is also electrostatic force that is responsible for the stability of molecules, holding the constituent atoms together. The oxygen and nitrogen molecules-in-the-box are neutral: each molecule has as many electrons as it has protons. There is no electrical force between the molecules until they come really close to each other, almost justifying the assumption in our model that there were no forces acting on the molecules between collisions.

Incidentally, when you rub amber on a cat, a tiny number of electrons are transferred to the amber; when you rub glass on your sleeve, a minute number of electrons go from the glass to your sleeve. We are not always very good at predicting whether electrons will leave substances when we rub them on other substances. Sometimes the simplest phenomena are really complex.

If you could shrink yourself to the size of an electron and wander about inside an atom, you would feel electric forces on you from all directions. But we are so big compared with atoms that all we normally detect is the net result, well outside the atom, of their collection of oppositely charged particles, which is usually zero. Some atoms and molecules have a net charge because their number of electrons is either more or less than their number of protons. In other words, the number of negative and positive charges in the atom is unequal. Such *ions* are always found in nature in conjunction with effectively equal numbers of oppositely charge ions. The classic example is common salt, sodium chloride, which consists of orderly arrays of equal numbers of positively charged sodium ions and negatively charged chlorine ions. Once again, if we are well away from the surface of such a crystal, we cannot detect a net electric field, although we have sophisticated instrumentation capable of mapping the electric charge on the surface of macroscopically neutral matter. All electric charge in the everyday universe originates in protons and electrons, and we have good reason to suppose that the total negative charge in the universe is equal to the total positive charge. The universe appears to contain equal or very nearly equal numbers of positive and negative charges. It is tempting to see the Creator separating the void into positive and negative matter in equal quantities. This numerical balance effectively holds for all bodies of macroscopic size and is negligibly affected by charging a body. This is because it takes only tiny amounts of charge to produce very strong electric fields.[14]

Attempting to seriously disturb the equality of positive and negative charge in a macroscopic body is not easy. Trying to add more charge to an already charged body is a process that meets with a rapidly growing resistance because the charges repel any incoming charge and also repel each other. The tendency of like charges to run away from each other is the reason that you ground a piece of electrical equipment. Thus any static charge on a negatively charged body will prefer to escape to the Earth, where the electrons can put more distance between themselves.

The discovery of the electron and proton at least allowed a visual image of electricity to be formed: electric current was not a fluid, it was the flow of charged particles. In principle, either positively or negatively charged particles can flow, but electric currents through solid matter are almost always carried by electrons, the common example being the flow of current through metals. Ions, which are charged atoms or molecules, are normally too large to move rapidly through solid matter. In liquids or gases, both positive and negative ions can move easily under the influence of electric force.

The basic question remains: What is charge? Is it something painted onto elementary particles? Is it an intrinsic property of certain kinds of matter, or can we scrape it off? For almost every practical purpose that you could encounter, the simple early-twentieth-century picture of intrinsically charged particles suffices. Later, when we consider quantum mechanics, we will go deeper.

Magnetism

All the components of matter are magnetic: the proton, the neutron, and the electron are all tiny magnets. The electron is a far stronger magnet than either the proton or the neutron, and it effectively accounts for all the magnetism normally encountered by HMS from bar magnets. The magnetic fields in medical imaging instrumentation are produced by electric *currents* flowing through coils—Oersted's experiment.

The electron thus has two quite different magnetic strings to its bow. It is magnetic in itself, but because it is also electrically charged it can also give rise to mag-

---

[14]Take two sugar cubes, and from one of them take *one* in every billion electrons and transfer them to the second lump. Hold the lumps one *kilometer* apart. The force of attraction between them is about 1000 newton, equivalent to roughly the fighting weight of Mohammed Ali.

netism when it moves. Electrons in some atoms can be regarded as moving around the nucleus, thus creating the microscopic equivalent of an electric current, which produces a magnetic field, just as Oersted did macroscopically. The electron has a property called *spin*, as has the proton, so that one visually attractive way of explaining the *intrinsic* magnetism of both particles is to imagine them to be charged spheres that are rotating about their own axes, and to suppose that the rotating charge produces tiny electric currents, which in turn give rise to magnetism. This is a nice simple picture, but it is not very satisfactory. One problem is that the neutron, which has no net charge, is also a magnet.[15] There are other more serious problems with the picture.

We can understand how nature generally succeeds in hiding from man what is a colossal quantity of electric force. Most of the material in the cosmos contains a mixture of equal amounts of positive and negative charge. But how does it manage to hide all that magnetism, if *every* particle in the cosmos is a magnet?

If you take two identical bar magnets and place them head to tail, with the north pole of each magnet next to the south pole of the other, you will find that their joint ability to attract paper clips at a distance has been reduced tremendously from that of a single magnet. This is the principle behind the explanation of the nonmagnetic behavior of the majority of matter. In most of the molecules in the universe there is an even number of electrons—each one not only having a charge but also acting like a magnet. The quantum theory shows that in the vast majority of such molecules the electronic magnets "pair up" two by two, pointing in opposite directions, north to south, thus canceling each other out completely as far as the outside observer is concerned. Some atoms have an odd number of electrons, so that when all the electronic couples are dancing the last waltz, one lonely guy is left over, his magnet uncanceled. Such atoms are automatically magnetic, but it often happens that they too fail to exhibit this magnetism because when they are present in solids, adjacent atoms arrange their magnets head to tail. Molecules that have an odd number of electrons—*free radicals*, as they are called—are strongly suspected of being implicated in the induction of cancer. Tobacco smoke contains such free radicals.

Nuclei are built up of protons and neutrons, both types of particle being magnetic. It is not surprising that many, but not all, nuclei are magnetic. This magnetism is extremely weak compared with that of electrons. The nucleus which is the strongest magnet is that of hydrogen, which is a single proton. It is about 700 times weaker as a magnet than an electron. Despite the feeble magnetism of nuclei, it has been used to great effect in two applications. Nuclear magnetic resonance (NMR) is a technique that allows the internal structure of molecules to be examined in great detail. It relies on the fact that there are magnetic fields inside molecules, and, if not, then such fields can be induced. The interaction of magnetic nuclei with these fields can be detected, and since the fields depend on the internal structure of the molecule, much about this structure can be deduced from the behavior of the nuclear magnets. Many atoms have magnetic nuclei, and chemists and biochemists in particular have studied thousands of molecules by this method. The techniques were first used in the 1950s with small molecules, but it is now possible to examine much larger molecules, in particular, small proteins and other biologically signifi-

---

[15]We will see in Chapter 31 that the proton and the neutron have an internal structure, both being made of charged particles.

cant entities. In effect, the magnetic nuclei act as informers, revealing the nature of their close environment within the molecule. The same principles are used in one of the most important modern methods of medical diagnosis, magnetic resonance imaging (MRI). This technique normally makes use of the magnetism of the proton. The proton is the nucleus of the hydrogen atom, so that all the hydrogen atoms in your body line up something like (but not exactly like) little compass needles, when you are placed in the magnet of the MR Imager. If an electromagnetic wave is now passed into you (radio waves don't harm you), the magnetic field of the radiation can act on these magnetic needles, flicking them around. These movements, which are influenced by the chemical surroundings of the proton, can be monitored. It is possible to "see" different organs and tissues, by detecting the different rates at which the protons flick round in different environments.

The survival of an animal depends on its ability to collect sense impressions, process them, and react. For this purpose, evolution has provided a nervous system in which electric impulses rush between the brain and other parts of the body, including the sense organs. The origins of animal electricity are not mysterious; there are comparatively well-characterized chemical and physical processes within cells and their membranes, which are capable of producing the kind of voltages detected experimentally.[16]

The consequences of commercial electricity have been monumental, as you can appreciate by looking around your house or waking up to find that your car battery has run down. Probably the most significant *social* consequence of the availability of controllable electric current is in the field of communications; radio, television, telephones, Internet, and fax machines rely on electricity, although there is a rapidly growing use of fiber-optic links to carry messages as light signals which are converted to electrical impulses at the receiver. Many a Ph.D. in sociology has been, and will be, based on the communications revolution of the twentieth century.

The transmission of electricity either through power lines or on a smaller scale usually makes use of copper wires. Passage of an electric current through such a wire always results in the heating of the wire. Sometimes this is exactly what we want, as in heating coils, but more often it is just a waste of energy.

An important scientific discovery, made at the beginning of this century, has recently borne fruit in technological advances, which may overcome the problem of heat loss and thus have enormous economic repercussions. It is a good example, not merely of the practical use of basic science but as a double example of the primacy of observation over theory.

Superconductivity

The passage of electrons through metal has some similarities to the path of a molecule in a gas. When a wire is connected to a source of voltage, the electrons feel an electric field that exerts a force on them and, according to Newton, results in their

---

[16]As with magnetism, the con men have hyped electricity as a cure for all manner of ills. The following quotation from the 1913 volume of *Scientific American* comes from an article entitled "Electrified Chickens": "Chickens living in electrified flats reach, in five weeks, the normal weight of chickens three months old. . . . Chickens so weak that they could not stand up, and who in the ordinary course would infallibly have died, have been put in the electrified flats and became healthy and strong."

acceleration. As they move through the wire, they collide with the static metal ions and experience a resisting force that grows as the speed of the electrons increases, just as air resistance grows with the speed of a moving body. When the resisting force and the force on the electron due to the voltage are equal, the electron settles down to a steady speed (remember your laws of motion!). This happens very quickly.

The collisions of the moving electrons with the other components of the wire result in an increase in the vibrational energy of the atoms of the wire. In straight language, they wobble more violently. This is reflected in a rise in the measured temperature of the wire; recall the connection between temperature and motion in the case of the gas in Chapter 1. That's why the filament of an electric lightbulb or a heating coil heats up and glows. The slowing down of electrons by collisions is the cause of what we technically call the *resistance* of the wire, and it is clear that this resistance, resulting as it does in a heating of the wire, is a source of energy loss, through loss of heat, in the transmission of electrical energy.

As in the case of a gas (remember the alcohol molecules and the professor in Chapter 1?), the speed of the electrons between collisions far exceeds their average speed along the wire. In the average household wire, which is about 1 millimeter in diameter and carries a current of 5 amps, the average *net* speed of the electron along the wire is only about 2 meters in an hour. This sounds ridiculous in view of the fact that electric telegraph messages cross the Atlantic in a matter of seconds. But think of it this way: the transatlantic cable is a tube of electrons that move forward *as a whole* if there is a voltage difference between the ends of the wire. Variations in the voltage at one end cause changes in the overall voltage difference, which in turn cause almost immediate variations in the speed at which the *whole collection* of electrons move forward; the current varies. That's how messages are sent. Note that the current varies almost simultaneously at all points along the wire. It is not the electron that leaves London that carries the message to New York; it is the electron that is already in New York.

Thanks to a discovery made in 1911 by the Dutch physicist Kamerlingh Onnes, we can dispense with resistance. He found that as he cooled mercury the resistance of the metal dropped until suddenly, at about -269° C, the resistance falls to zero, where it remains, provided the temperature is not allowed to rise. This was totally unexpected and certainly totally unpredicted. Any scientist would have told you that it was impossible; electrons cannot move through material without bumping into things. Observation won out over common sense, not for the last time in science, and a slowly growing number of substances, not all metals, were found to display zero resistance at very low temperatures. It was not until 1956 that Bardeen, Cooper, and Schrieffer found a Nobel Prize winning explanation—known as the BCS theory, for the phenomenon of *superconductivity*. This theory relies on quantum mechanics.

Zero resistance means that once an electron starts moving along a superconducting wire it should never stop—until it reaches the end of the wire. There is no loss of energy—because there is no resistance. It is possible to take a loop of superconducting wire, induce a current in it, and leave it. The current will flow for years, provided, of course, that you keep the coil cooled. The energy-saving possibilities of superconductivity are staggering when one takes into account the enormous energy loss involved in transmitting electricity along high-power lines.

So why don't we use superconducting transmission lines? Because there is one drawback to the miracle of superconductivity: it only sets in at low temperatures; no known substance is superconducting at normal temperatures. It is possible to cool a number of metals, and other substances, down to superconductivity by using liquid helium, a liquid that boils at 4.2 K, or in more familiar terms -269° C. Unfortunately, liquid helium costs about the same per liter as whisky, which precludes its use to cool thousands of miles of transmission lines. It is, however, feasible to use liquid helium for relatively small installations, and its main use is in the production of strong magnetic fields such as those used in MRI, and in the magnets used in particle accelerators (see Chapter 31). In these types of apparatus the magnetic field is produced by an electromagnet, the size of the field depending on the strength of the electric current in the coils. The current in electromagnets was once limited by heating problems caused by resistance, but these have been enormously reduced by the introduction of superconducting coils, cooled with liquid helium. This allows higher currents and higher magnetic fields.

The possibility of using superconductivity on a large scale is dependent on finding materials that are superconducting at temperatures well above that of liquid helium. The search for such materials was regarded by many theorists as being a waste of time. The official theories of superconductivity suggested that so-called high-temperature superconductivity was impossible. Observation won out again.

In the 1980s a series of materials were found that exhibit superconductivity at temperatures much higher than that of liquid helium, temperatures that are attainable by cooling with liquid nitrogen, which boils at -196° C. Liquid nitrogen costs only about the same as milk, which is certainly an advantage but still does not make long-distance superconducting transmission lines a realistic possibility.

The discovery of high-temperature superconductors stimulated an explosion of interest, primarily because of the obvious commercial possibilities but also because the finding contradicted theory. In the meantime, the search goes on for new superconducting compounds in the hope that at least one of them will work at room temperature. Anyone finding such material could become a multibillionaire.

Electro-magneto-gravitational Force?

Einstein said that the history of physical science contains two couples of equal magnitude: Galileo and Newton, and Faraday and Clerk Maxwell. The recognition of the nature of electromagnetic force was a turning point in the history of science.

As far as the coarse structure of the universe is concerned, gravity and electromagnetism are the only forces that matter. Gravity dominates cosmology, and electromagnetism dominates the structure of atoms and molecules. Gravity is the weaker of the two. The enormous quantitative difference between the forces is illustrated by the fact that, at a given distance, the gravitational attraction between a proton and an electron is about 10^{39} weaker than the electromagnetic attraction between them.

The amalgamation of electricity and magnetism, in spite of their apparently distinct nature, was in accord with many scientists' intuitive, but rationally unjustified, feeling that simplicity is right, and more specifically that all forces are aspects of a single force. Einstein was driven by the desire to amalgamate electromagnetism with gravity in a unified field theory, but he never succeeded, which doesn't mean that it can't be done. Or that it can be done. The discovery of another two, very short-range, forces that operate within the nucleus did not made life easier for the unified field fans, but in the middle of this century one of those forces was shown to be related to electromagnetism. The total unification of nature's forces has not yet been achieved, but many physicists have an intuitive feeling that it will be sooner or later. This is an example of unrationalized belief driving rational research. Is this how science works?

9

Belief and Action

Those who carry out scientific research are the recipients of a culture
developed by previous generations.

—Thomas S. Kuhn, *The Structure of Scientific Revolutions*

Do our beliefs affect our work? To what extent do metaphysical reflections, religious faith, or scientific dogma influence the *science*, as distinct from the weekend thoughts, of scientists? It would be nice to think that nothing would influence a scientist's search but the objective facts. Reality often says otherwise. Scientists are part of the society in which they live, and it is natural to look for the part their environment plays in their professional lives, and the effect it has on their motivations and the way in which they see nature. At one time that environment included philosophy.

In Pythagoras, science and philosophy were intermingled. So it was with the alchemists; they saw more than the eye beheld. Substances, besides their observable characteristics, had other properties that were only divinable with the inner eye. Gold was a metal but it was, in a very real way, the symbol of regenerate man, shining with purity of spirit, resisting every temptation, and immune to evil. The basest metal, lead, represented man the sinner, made ugly by his sinning, susceptible to temptation and prone to evil.

Although science and philosophy drifted apart during the sixteenth and seventeenth centuries, they have never stopped interacting, and there have been several periods in the last three centuries when science has deeply affected social and philosophical thinking. We have already seen an outstanding example of science, in the shape of Newton and determinism, infecting philosophy.

The influence of man's beliefs on his scientific activities has been less direct. The religious beliefs of Newton and Boyle strongly influenced the way they thought of man and the universe, but as far as their science was concerned, philosophy and mysticism were confined to their alchemical pursuits. Although they saw their science as the means to reveal God's creation, nevertheless, their physics is physics. What is true is that their *motivation* sprang partly, and perhaps predominantly, from irrational sources. What can be loosely termed irrational motivation is far more common than the nonscientist might suspect, and it is well documented that philosophical considerations that bordered on mysticism were never very far from the minds of several of the leading scientific figures of the early nineteenth century. Can we discern an overt influence on their scientific *results*? Can you distinguish an airplane built by an atheist from one built by a believer? (Stronger wings?) How can the results of science, based as they are on observation, depend on the convictions of the observer? The decline in religious faith over the centuries has not blunted the edge of this question, because in considering the effect of belief on the scientist, we have to include not only conventional religious faith but also, and far more important in modern times, the belief in existing scientific dogma. For while there is perhaps a separation between the professional life of most scientists and their formal religion, there are other kinds of faith besides the belief in a supreme being.

Galileo initially refused to accept the elliptical orbits that Kepler proposed for the planets. For the ancients, circles were perfect forms, and Galileo, in this instance, was trapped inside a medieval mind-set. The abstract mystical principle overcame the concrete observable fact. The problem is general. We tend to see what we expect to see, and those expectations can be governed by accepted scientific theory, by philosophical convictions, or by religious beliefs. Preconditioning of the mind affects scientists no less than other men, although the present-day scientist is, by the nature of things, more susceptible in his professional work to the compulsions that arise from existing scientific theory than to the promptings of philosophy or religion.

Part of the scientist's system of faith is existing science. In practice I *believe* in much of existing science, even if faith in a theory occasionally turns out to be unjustified. The medieval shoemaker *knew* that the Saint of Shoemakers was protecting him; a century ago, my scientific predecessors *knew* that Newton would never be proved wrong—faith in both cases, but that faith could affect action, in scientists as well as shoemakers. An undergraduate student who, on the basis of his classroom lectures, has reason to believe that his laboratory results, when plotted, will give a straight line, is capable of drawing such a line through a series of experimental points that resemble a silhouette of Mae West.

Science is not carried on in a sociological or philosophical vacuum. If it is, then how do we account for the complete failure of whole civilizations, such as the Chinese and Islam, to contribute significantly to rational scientific theory (as distinct from mathematics or technology) until they were exposed to the West?

The crudest fashion in which ideology can affect science occurs in political or theological dictatorships. Marx generally made no appearance in Soviet scientific papers, but his influence was felt indirectly in the official frown aroused at one time by quantum mechanics, and the quashing of Darwinian evolution in the old Soviet Union.

Chairman Mao makes appearances in Chinese scientific papers published in the 1970s. The author of one chemical paper, for example, explains that following Mao's precept that "the strong will overcome the weak," he used concentrated, not dilute, acid. The absurd political kowtowing was obviously tagged on after performing the experiments. One can find a quotation to go with anything—as I realized when I was trapped for five days at the same ship's table as a pair of Latin scholars.

In the West, in our time, except in the midst of sick episodes like McCarthyism, a scientist's proclaimed religious and philosophical beliefs are, one would like to think, genuinely his own. How do they affect his science?

In the twentieth century scientists are not, as a group, *professionally* influenced by religion and philosophy. Some are prepared to discuss the wider relevance of science, or to peep through the chinks through which some of them see God or oblivion. But few would admit that a single experiment or theory bears the traces of her philosophical or religious convictions, which is as it should be. An ecologist may have entered his field because of a mystic belief in the sanctity of life, but his professional papers will usually read much the same as those of a less idealistic colleague. A believing *experimental* physicist today keeps his religion or philosophy for one set of journals and his experiments or theories for another. Theory is another matter. There are physicists and biologists whose attitude toward evolution and cosmology is demonstrably affected by beliefs connected with the "meaning" of human life.

If "faith" is defined broadly enough to include the unquestioning acceptance of most of scientific lore, it has been claimed that faith does reach into all laboratories. The best-known proponent of this thesis is Thomas Kuhn (1922–1996). Kuhn has postulated that, at any given time in its history, each branch of science builds for itself a commonly accepted collection of concepts and theories, an official way, as it were, of looking at the world. Scientists work within the borders of this "paradigm," this faith. There may be more than one paradigm in the same discipline. Thus in the sixteenth century you could look at the mechanics of bodies through the eyes of Aristotle or the eyes of Galileo. Whereas Galileo saw a body being attracted to the Earth by a force, the followers of Aristotle saw it "coming home" to where it belonged. They each started off with a different preconception of how nature works. Kuhn would say that they had chosen different paradigms, but he points out that both of them have used the same observed facts to support their paradigm. Science changes, according to Kuhn, not by the falsification method of Popper, not by attempting to disprove what it believes. On the contrary, science is an activity that occurs primarily within the borders of the paradigm; experiments are designed to reveal facts that support, rather than challenge, the paradigm. In fact, anything that doesn't fit in is regarded as anomalous and is either treated with suspicion or ignored for as long as possible. Kuhn's picture of the scientist is even more unflattering than this; he suggests that if a fact doesn't fit in with the paradigm it is simply ignored or not even "seen." He gives the example of sunspots, which, before Copernicus, were not recorded by Western astronomers because the heavens, including the Sun, were supposed to be unchanging, while sunspots in fact appear to move across the face of the Sun. Almost immediately after the publication of Copernicus's book, sunspots began to be recorded. Kuhn notes that in China, where there was no doctrine of immutability of the heavens, sunspots had been observed for centuries.

Kuhn further rejects Popper's claim that when a paradigm is falsified it is abandoned. According to Kuhn, abandonment of an old paradigm occurs only when a new one is available. In other words, Popper says that when the raft is uninhabitable we jump into the sea, while Kuhn says we jump only when another raft is available.

Popper, in reply, concedes that much science is not carried out with the object of falsifying theories, and he sees such science as second-class. He insists that science as a whole jumps forward by the process of falsification. There is something in what both Popper and Kuhn have to say, but working scientists would surely wish to modify both points of view. Working within the paradigm is not necessarily second-class science. The discovery of the structure and role of DNA falsified no major theories and smashed no paradigms. There were a number of previous suggestions as to the structure, but they were hardly in the class of a paradigm, a complete system of thought, like Newtonian mechanics or Darwinian evolution. Is Popper saying that Watson and Crick were given the Nobel Prize for second-class science? As for Kuhn, I feel that he has been telling the story of science with very carefully chosen examples of "scientific blindness." For every story like that about the sunspots one can find another in which a completely unexpected observation has not been ignored or downplayed. The fact that a strange observation is treated with caution is just as it should be, provided it is not rejected out of hand because of the existing paradigm: first skepticism, then experiment. In fact Kuhn himself, in attacking Popper's falsification criterion, stresses the fact that an "anomalous" observation may be mistaken and should not therefore be taken as the basis for the disproof of a theory.

Although Kuhn's emphasis on paradigms has been criticized (including by himself in his later work) he can strike uncomfortably close. There is no doubt that young scientists are conditioned by their education and the prestige of established theory (and established scientists), so that they are likely to interpret what they see in terms of existing concepts and theories. They will usually try to rationalize the unexpected in terms of the current paradigm. This may not always be such a bad thing to do. After all, if the anomalous path of one planet had cast such grave doubt on Newtonian mechanics that it collapsed soon after the *Principia* was published, there is no doubt whatsoever that the progress of science would have been seriously impeded. The tendency of a theory to push forward regardless of troubling inconsistencies is stressed in particular by Imre Lakatos (1922–1974), as a model for the actual historical development of science.

The history of science should dispel any idea that HMS has that science is an activity carried on by completely unprejudiced searchers-after-knowledge, floating free of established dogma. That is the Saturday Morning Post, white-toothed, smiling face of science. An equally mistaken view of science is typified by the antimasculine polemics of Sarah Harding in her book *The Science Question in Feminism*. Here we read of oppressive masculine concepts that have, in her opinion, resulted in science being primarily concerned with the exploitation of nature. Science has to be emancipated, to be imbued with new values—feminist values.

A scientist's beliefs and actions may be as strange as any other man's, but what drives a scientist is almost irrelevant to those of his discoveries that stand the test of time. Faraday's motives have nothing to do with the *validity* of his work. Nevertheless, HMS may be surprised by the extent to which many of the greatest scientists have been driven by irrational forces. Newton had alchemy and religion. Einstein clung to a stubborn belief that God would not build a universe in which cause and effect seemed to vanish, as implied by the quantum theory, a belief that was enough to prevent an astonishingly creative man from accepting quantum mechanics. In the present context it is not relevant whether or not he was justified; what is significant is that a philosophical belief caused him to stand outside the community of physicists, for three decades.

Oersted, Ampère, Faraday, and Maxwell all had deep religious or philosophical convictions. For all of them, the concept of the unity of nature, derived immediately from *Naturphilosophie* and ultimately from Pythagoras, had a direct or indirect influence.

Oersted, Ampère, Faraday, and Maxwell

Oersted (1777–1851) was a propagator of Kant's philosophy from his student days.

His doctorate thesis was based on a defense of a lesser-known book of Kant: *Metaphysische Anfangsgründe der Naturwissenschaft* (The Metaphysical Foundations of Knowledge) (1786). Kant had a serious interest in science, and the particular point in his book that impressed itself on Oersted was the argument that we experience only force, and that there are only two basic forces. This belief, for it was no more than that, in the unity of nature's forces, was to guide Oersted for the rest of his scientific life.

The revolt against the Enlightenment certainly affected Oersted. In particular, he was a friend of and read the works of the philosopher Friedrich Schelling, a practitioner of *Naturphilosophie*. Schelling wrote much on the philosophical aspects of science. He was fascinated by the "animal electricity" of Galvani. Oersted attended lectures on *Naturphilosophie* and was again particularly struck by the concept of a single unifying force accounting for all natural phenomena. If Oersted had followed Schelling blindly, he would never have discovered anything. Schelling didn't believe in observation: "Coming after the ruination of philosophy in the hands of Bacon and of physics in the hands of Newton and Boyle, a higher perception of nature begins with *Naturphilosophie*; it forms a new organ of intuition for understanding nature. . . . [T]he concept of empirical science is a mongrel notion." Oersted did believe in experimentation; he said of Schelling, "He wants to give us a complete philosophical system of physics, but without any knowledge of nature except from textbooks." But on two things he was firmly in Schelling's camp: his sympathy for the general principles of *Naturphilosophie*, and Schelling's belief that a program should be initiated to establish the nature of the single unifying force and demonstrate its ubiquity. Oersted took this task upon himself. About ten years before his historic discovery that an electric current produced magnetic effects, he announced his belief that light, heat, chemical affinity, electricity, and magnetism were all forms of "one primordial power." Any doubt as to the influence of Kant and *Naturphilosophie* on Oersted's scientific impulses should be removed by an excerpt from Oersted's own account, written in the third person, of his crucial experiment, published in 1830, (italics mine):

> Electromagnetism itself was discovered in the year 1820, by Professor Hans Christian Oersted, of the University of Copenhagen. Throughout his *literary* career, he adhered to the opinion, that the magnetical effects are produced by the same powers as the electrical. He was not so much led to this, by the reasons commonly alleged for this opinion, as by the philosophical principle, that all phenomena are produced by the same original power.

This was not retrospective cosmetics, Oersted also made attempts to fit chemistry into the same philosophical scheme.

Oersted accepted a basic axiom of *Naturphilosophie*, which was that, because man had been created in the image of God, human reason was a reflection of divine reason. But God had also created nature so that human reason could, by itself, construe the laws of nature. This triple linkage between human and divine reason and the laws of nature was one of Leibniz's beliefs. It also brings to mind the more direct linkage sensed by the fifteenth century physician and mystic Paracelsus, who held that man contained the whole Creation in himself and could therefore reach a far deeper knowledge of nature than could be attained by reason, through the association between an object and its corresponding representation in man. Oersted attempted to weld philosophy to the natural world. One of his university colleagues wrote that "Oersted was searching for this connection between two great forces of nature. . . . [T]he thought of discovering this still mysterious connection constantly filled his mind." By linking magnetism and electricity, he showed that the concept of the unification of natural forces was not merely a Pythagorean dream. He instigated the experimental quest after the unity of forces, which carries on until this day. His last, unfinished, paper was entitled "The Soul in Nature."

Ampère was a devout Catholic who stumbled upon that bible of secular rationalism, the *Encyclopédie,* and was tempted sorely. As a mathematician, he must have appreciated the logical, rational tone of many of the articles, and the great text remained an influence for much of his life, challenging his faith but basically leaving it untouched. He, too, was an early disciple of Kant in France. He adopted Kant's division of our knowledge of the world into two types: *phenomena,* which are our direct sensations (what goes on in your head) and *noumena* (their objective causes). However, like most scientists he believed in the real existence of the external world, a belief that he felt was challenged by Kant. Ampère convinced himself not only that matter was a reality but also that God and the soul existed. Ampère believed that we can, by the rational operation of the mind, reveal the nature of the *noumena.* His reasoning was that the interactions between the *phenomena,* which occur in our minds, must be related to the interactions between the unobservables, the *noumena.* He agreed with Kant that *noumena* could not of themselves reveal their true nature, and he held that all hypotheses as to that nature stand or fall on their success in leading to true predictions—which is a very modern approach to science, and reminiscent of Popper.

Naturphilosophie, and Kant, reached England in part through the poet Samuel Taylor Coleridge (1772–1834). Coleridge's writings contain ideas that are almost indistinguishable from those of Schelling. Coleridge was a close friend of Faraday's first mentor, Humphry Davy. The poet had scientific interests, and Davy wrote (rather awful) verse. Did the flow of ideas reach Faraday through Davy, and influence his attitude to nature? There are differences of opinion on this point. Like Oersted and Ampère, Faraday was deeply concerned with the nature of forces, but he lacked the broad education of his Continental colleagues and shared the British aversion to mystic philosophies. Faraday was a religious man, belonging to a small Protestant sect originating in Scotland—the Sandemanians, who were fundamentalists, but not of the tight-lipped kind. They regarded the accumulation of wealth as unscriptural and believed in the power of love (sacred, not profane) and the separation of church and state. There is piquancy in the poorly educated Faraday reaching the peak of the scientific establishment, in view of the fact that the Sandemanians elected their bishops with no regard whatsoever to the candidate's education. Professor John Tyndall, the materialist scientist, said of Faraday: "He drinks from a fount on Sunday which refreshes his soul for a week."[1] But although Faraday was not a man with formal philosophical inclinations, he would have sympathized with that trend in *Naturphilosophie,* borrowed from Spinoza, which identified God with the material world. This would have been in accord with the

---

[1]Marie-Antoinette read pornographic books during long services, but it is not recorded if this refreshed her for the week.

Sandemanian belief that God was continually involved with his universe and that the proof of his existence lay in the contemplation of nature. It was his universe, a universe crossed by many forces, all of which were the faces of one force.

In his scientific writings, Faraday, unlike his Continental colleagues, was reticent about the beliefs, religious or philosophical, that sustained him, but his continual efforts to show the unity of natural forces are in complete accord with the ideas that Coleridge brought back from Germany in 1799. Apart from his work in electromagnetism, he attempted to find a connection between gravity and magnetism, and his work on electrochemistry, in which he sees a connection between the forces of chemical affinity and electricity, fits into the general scheme.[2]

Maxwell was also deeply religious, his belief in God so strong that he wrote, "If we had seen Him in the flesh we should not have known Him any better, perhaps not so well." It is very doubtful that he was directly influenced by the German school of philosophy, but there is at least one influence that can be traced back to *Naturphilosophie.*

Considering his intellect and his mastery of electromagnetism, the question could be asked: Why didn't he believe in a unit of electric charge? In other words, why didn't he foresee the electron? It is obvious to us that electricity comes in discrete particles, but we believe in atoms and molecules, and the electron is "real" for us. In Faraday's case we have conjectured that *Naturphilosophie* was the root of the trouble. The *Naturphilosophen* rejected the idea of particles in nature. It is almost certain that Coleridge's friendship with Humphry Davy accounts for the latter's disbelief in atoms, which he passed on to Faraday. Maxwell admired Faraday greatly, and it is interesting to speculate that he too inherited the antiparticulate view of nature. When he dealt with electrolysis in his treatise, he felt himself forced to refer to a "molecule of electricity," but made excuses, terming the phrase "gross" and implying that it is merely a convenience, useful for calculations but having no reality.

Naturphilosophie is not completely dead; once in a very long while Goethe is invoked in criticizing science—which is a pity. Goethe's Faust is one of the great metaphors for man's condition. To seek inspiration from his mystical pseudoscience is like taking spiritual sustenance from Shakespeare's laundry list.

Scientists are the product of their environment, but they are not unique in this respect. There is a generally applicable sermon in the English poet Ted Hughes's admonition: "The progress of any writer is marked by those moments when he manages to outwit his own inner police system."[3]

---

[2]This research may have been stimulated by the German Ritter, who, influenced by Kant, and before there was any experimental evidence, attempted to unify the two phenomena.

[3]For an amusing comment on Hughes's remark, see the poem "A Policeman's Lot" by Wendy Cope in "Making Cocoa for Kingsley Amis," Faber &Faber, London, 1986.

iv

The structure of atoms has been hinted at in the previous section, where we saw that atoms consist of a massive, positively charged, central body—the nucleus, surrounded by small negatively charged particles, the electrons. We cannot see these particles and, even until the early years of this century, there were those, including several notable scientists, who refused to believe in the existence of atoms. In this section we will go deeper into the structure of the atom and of the many forms of matter that nature and man have made.

The story really takes off at the very end of the nineteenth century. The basic principles of atomic and molecular structure were completed by the middle of the twentieth century. We will look at the way in which the structure of the atom was revealed, but not at the explanation for that structure. This is a little like noting the paths of the planets without knowing about Newtonian mechanics. In fact, that mechanics does not help us to understand the way in which atoms are built or the manner in which they behave. For that we need quantum mechanics, which we will tackle later. Newtonian mechanics can describe the paths of the gas molecules in our opening chapter, but it says nothing about the internal structure of those molecules, or how they react chemically, and why they very often don't react.

In Chapter 12, we are going to make a tourist's bus ride around the material world—on your left, metals; on your right, halogens. The study of the behavior of even a small class of molecules can fill a lifetime, but, as the man said, I haven't got that many changes of socks; so we are going to take a panoramic view of the basis of existence.

Modern material science and chemistry are built on the achievements of those who took matter out of the hands of the alchemist and into the scientific laboratory. We start this part by acknowledging their pioneering contributions.

10

The Demise of Alchemy

... a foul and pestilent congregation of vapours!

—Hamlet

"I have immortal longings in me." Cleopatra, waiting to take the asp from Iras and preparing to meet eternity, declares that she is "fire and air; my other elements I gave to baser life." Most of the spectators in Shakespeare's audience knew what she was talking about.

Since antiquity the four elements: earth, water, air, and fire—had been part of the accepted system of beliefs of Western man. Earth was lowest, and water covered the Earth. The fact that in some places the Earth rose above the waters was just another piece of evidence for the constant war between the elements. Above the waters were air and fire in that order. These, like Cleopatra, were the Elements that strove to rise upward, leaving behind the base elements earth and water, which remained below in their natural habitat.

The workings of the natural world were an almost complete mystery to sixteenth-century man, but the elements explained a lot. The four elements *controlled* the nature of men and matter rather than serving as building blocks. It was the proportions that mattered. Marc Antony says of Julius Caesar, "This was the noblest Roman of them all. . . . [T]he elements were so mixed in him that Nature might stand up and say to all the World, This was a man!"

Medicine, from the second century, had taken its language from the same source. It was then that the great physician Galen wrote the idea into his immensely influential texts. The four *humors* that governed man's health and sickness corresponded one-to-one to the four elements. Earth and melancholy were both dry and cold; water and phlegm, moist and cold; air and blood were moist and hot; and fire and choler, dry and hot. Most people had an excess of one of the four humors. Caesar had, according to Antony, an ideal balance, but in most men the balance was never static. Tamberlaine, in Marlowe's play, famously restates the commonly accepted notion of perpetual struggle between the elements: "Nature that framed us of four elements, Warring within our breasts for regiment."

The elements were, conveniently, so vague in their properties as to allow an explanation of almost anything. They could even change into each other. Donne says so: "Ayre condensed becomes water, a more solid body. And Ayre rarified becomes fire, a body more disputable and in-apparent." In explaining human disease, if everything else failed, one could always appeal to the influence of the stars on the elements, which accounts for the fact that a knowledge of astrology was still needed to qualify in medicine in Paris as late as the second half of the seventeenth century. A Paris physician who, in 1607, suggested that one way to fight the plague in Paris was to keep the streets and gutters clean, was completely ignored; but in

1658, one of Louis XIV's surgeons advised that the king should be bled in the first and last quarters of the Moon, "because at this time the Humours have retired to the center of the body."

It was into this world that Robert Boyle was born in 1627. Robert was the fourteenth of fifteen children of Roger Boyle, earl of Cork. It was said of Robert Boyle that, like Newton, he "hath never been hurt by Cupid." He remained a bachelor all his life, living with his sister, the remarkable Lady Ranelagh, a woman whose high, self-imposed level of education was a distinct rarity in a society that regarded education for women as an unquestionable waste of time. In her house he set up a well-equipped laboratory in which he carried out scientific experiments, helped by assistants. He died in 1691, exactly one week after his sister. They are buried together in St. Martin's-in-the-Field in London.

Boyle, like Newton, was a deeply religious man. Between them they did perhaps more than anyone else to build the mechanistic basis of physics and chemistry that was to finally dispense with the mysticism that pervaded medieval physics and especially alchemy. Typically of the transitional nature of the age, neither of them lost their belief in alchemy. In Newton's case this belief included the occult aspects of alchemy, but Boyle was more interested in the ancient attempt to create gold. In 1689 he persuaded the government to repeal a law, dating from Henry IV, forbidding the "multiplying of gold."

Boyle's scientific impulse developed early, inspired particularly by his reading of Latin authors who "conjured up that unsatisfied greed for knowledge that is yet as greedy as when it first was raised." He was intrigued by "the new paradoxes of the great star-gazer Galileo," who died when Boyle was in Florence, where "did he sometimes scruple . . . to visit the famousest bordellos . . . (but) retained an unblemished chastity." When the time came that he wished to start research into chemistry he decided to leave Ireland, "a barbarous country where chemical spirits were so misunderstood and chemical instruments so unprocurable that it was hard to have any Hermetic thoughts in it." *Hermetic* referred to Hermes Trismegistus, Hermes the thrice-great, the mythical creator of alchemy.

Both Boyle and Newton lived at a time when science was felt by many to be not too compatible with religion. Boyle felt the need to refer to this specifically, as witnessed by *The Excellency of Theology Compared with Natural Philosophy* and the explicitly titled *Some Considerations about the Reconcilableness of Reason and Religion*. In 1690, the year before he died, Boyle published *The Christian Virtuoso: Shewing That by Being Addicted to Experimental Philosophy a Man Is Rather Assisted Than Indisposed to Be a Good Christian*. (A "virtuoso" was an amateur scientist.)

Newton and Boyle interacted. Newton studied Boyle's work and offered an (incorrect) explanation of Boyle's law (see Chapter 1), and Boyle, aware of Newton's work, declared God to be a watchmaker. He frequently used the great cathedral clock at Strasbourg as an analogy for the workings of nature. "I consider the frame of the world, already made, as a great and, if I may so speak, pregnant *woman* with twins in her womb. . . . Such an engine as comprises or consists of several lesser engines." Three centuries before, Bishop Nicole d'Oresme had coined the phrase "the clockwork universe" and had of course seen God as the supreme clockmaker. For Boyle, as for Newton, the mechanical universe strengthened his belief in a Great Designer. Boyle spoke often of God's "concurrence" with the operation of the universe, but miracles were not ruled out. Both Newton and Boyle were affected by Descartes—Newton perhaps mostly by Descartes's mathematics and Boyle more by Descartes's attempt to build a mechanistic, logical theory for the material world.

Sometime in the 1640s a group of scholars and others interested in "natural philosophy" began to meet regularly in Oxford to discuss what was then known as experimental philosophy, but which we would call science. Boyle was one of the members of this "Philosophicall Clubbe," also known as the Invisible College, which was patronized by his sister, Lady Ranelagh. At one time it split into two groups, one in London and one in Oxford, the latter group meeting in Boyle's rooms in the university. In 1662 the informal club was given a charter by Charles II and became the Royal Society of London for Improving Natural Knowledge. Five years earlier the moneyed Henri-Louis Habert de Montmor had initiated a series of meetings in his home in Paris, where scientific subjects were discussed. The constitution of the so-called Montmor Academy stated, in the spirit of Bacon, that its aim "shall not be the vain exercise of the mind on useless subtleties" but "the improvement of the conveniences of life."

The original members of the Royal Society, which met on Wednesdays, were more lovers of science than scientists. It is a sign of the growing standing of science that, in addition to scientists such as Robert Hooke, Boyle, and others, there were, among the members of the society, figures such as the architect Christopher Wren, the poet Dryden, the diarist John Evelyn, and fourteen peers of the realm, including the duke of Buckingham. The motto of the new society was *Nullius in Verba*, which can be roughly translated as "Don't trust anything that anyone says," another very Baconian sentiment.

Around about 1626, Sir Francis Bacon published *New Atlantis*, which, though fictional, was a supplement to his scientific writings. In it he foresaw the establishment of scientific societies and research institutes modeled on his "Salomon's House." It is not surprising that the Royal Society saw Bacon as its spiritual forefather, especially in view of his utilitarian view of science. The new society quickly provided a focus for the collection of scientific and technological reports and the encouragement of research. It contributed in an indispensable way to the internationalization of science. Thus scientists in Europe were invited to publish their works in *The Philosophical Transactions of the Royal Society,* and one of the first members to be elected after the society was established was the prodigiously talented Dutch scientist Christiaan Huygens, who corresponded with Descartes, Newton, and Boyle. The pioneer Dutch microscopist Antoni van Leeuwenhoek was also elected to the society, in 1680. Scientists in England and the Netherlands were corresponding about new experiments while their respective navies were bombarding each other.[1]

Science, by the time of Newton, had become a supranational secular church. Latin, although it was not completely replaced by French as the secular lingua franca until toward the end of the eighteenth century, was ceasing to be universally understood, and Huygens, for example, used six languages in his correspondence. (The "unsociable" Hooke was rumored to have doubted the veracity of any docu-

---

[1]This scientific crossing of the trenches was echoed in 1806 when the English chemist Humphry Davy was awarded a prize of 3000 francs by L'Institut Francais although England and France were at war. Both Davy and the institute came under criticism in their respective countries.

ment written in French.) Today the international language of science is English, and the scientific community covers the entire globe. The Invisible College still lives.

The activities of the Royal Society in its early days attracted a good measure of ridicule. Samuel Pepys's diary entry for 1 February 1664 reveals that he was at Whitehall that day and heard the king laughing at the society "for spending time only in weighing ayre and doing nothing else since then." "Virtuosi" became an accepted target for wits about town. Samuel Butler wrote a mediocre play called *The Elephant on the Moon* (1676), in which some of the virtuosi are characterized as men "who greedily pursue / Things wonderful, instead of true." Thomas Shadwell, a minor playwright, satirizes the virtuosi in a play of that name. But these were early days. After Newton the attitude toward science changed.

Among the trunk of books that Sir Walter Raleigh took to sea with him was Robert Boyle's book *The Sceptical Chymist* (1661), which marked the first stirrings of modern chemistry. Written in the form of dialogues, it was, in the author's estimation, "a book written by a Gentleman, and wherein only Gentlemen are introduced as speakers."

The Sceptical Chymist contains the first clear statements of the concept of a chemical element (although not a modern definition) and of the difference between a mixture of elements and a chemical compound. Boyle realized that in a compound such as water, composed of oxygen and hydrogen, the atoms are more than mixed together; they cling to each other in small groups. Water is not the same as a mechanical mixture of hydrogen and oxygen gas, as you can prove by throwing a burning match into a sample of each.

Boyle was the first to thoroughly discredit the idea of the four elements, and he dismissed the three principles (salt, sulfur, and mercury) of the famous sixteenth-century physician and alchemist Paracelsus. His concept of *corpuscles* is an atomic theory in that it assumes matter to be composed of small, indivisible particles, and it makes a significant advance on the theories of Democritus and Descartes by postulating that there were a large number of *different* types of particles, each type associated with a different elementary substance. This accords with modern theory.

Boyle accepted that he lacked a systematic method to identify elements: "I have not yet, either in Aristotle or any other writer, met any genuine and sufficient diagnostic for the discriminating and limiting of species."

Boyle stands between alchemy and chemistry. His great contribution was to realize that matter is composed of a range of *elements*, each of which, in its pure form, is a collection of identical *corpuscles* or atoms. The atoms of each element were different from the atoms of any other element. It is a reminder of the nature of the intellectual world of the seventeenth century that Boyle was worried that his revival of atomism would arouse suspicions of atheistic tendencies because atomism was associated not only with Democritus but also with the atheistic materialist, Lucretius.

Boyle built on the ideas of others; but he drew together diverse and unclear ideas, added to them, and expressed them in a way that was to influence the course of chemistry. Sometimes dubbed the Father of Chemistry, he was responsible for placing the emphasis on careful experimentation and the abjuration of Scholastic or occult explanations. In the manner of a modern scientist, he chronicled the details of his experiments, even the unsuccessful ones. He brought the concept of the chemical element into the limelight. He was among those who, for good or bad, began the process of separating philosophy from science. But he could not make that break completely, any more than Newton could, because he was too much a child of his time. He was not a great chemist, but he opened the door to chemistry.

The Father of Chemistry?

If you ask a French scientist to name the father of chemistry, he will chauvinistically, but justifiably, answer "Lavoisier." If you were to ask an end-of-the-eighteenth-century French merchant who Lavoisier was, he would probably spit. Lavoisier picked an unfortunate profession for someone living at the time of the guillotine; he was a tax collector, and he became a millionaire because of it. It did not help him, in the eyes of the Revolution, that he was also born into wealth and that his father had bought him an aristocratic title in 1772, when Lavoisier was twenty-nine.

Louis XVI funneled huge amounts of money into the state's coffers by taxing salt. Everyone was obliged by law to buy a minimum designated quantity of salt and was taxed on the basis of that quantity. Tax collectors like Lavoisier did well, but salt smuggling was common, so Lavoisier suggested that a wall be built around Paris in order to control the passage of carts into and out of the city. The wall was literally to be the death of him. Designed by the famous architect Ledoux, it was 10 feet tall, 18 miles in circumference, and punctuated by 54 custom posts. It cost a vast amount of money and was vastly unpopular. The duc de Nivernois suggested that Lavoisier should be hanged. Forty of the gates were sacked in the Revolution.

In 1780, Jean-Paul Marat, the anti-Newtonian revolutionary-to-be, published a letter claiming that he had succeeded in making visible the "secret component" of fire. Lavoisier scoffed, but Marat was to have his day when the Revolution came. The fiery, self-proclaimed "Friend of the People" accused Lavoisier of imprisoning the city and restricting its air supply: "Would to heaven that he had been strung up on a lamp-post." Lavoisier, who had supported the Revolution when it started, was arrested. He was tried on 8 May 1794 and executed on the same day, in the great square now known as the Place de la Concorde. His body was thrown into an unmarked grave.

Lavoisier's contribution to the understanding of matter arose from his insistence that it was not sufficient to report the qualitative results of chemical reactions; *quantitative measurements* had to be made. This sounds trivial today, but it does so primarily because of Lavoisier. He was a quantitative experimentalist, which was just what chemistry needed so badly. The alchemists often used ill-defined materials, and there is no evidence that they attached any fundamental importance to the weights of the materials they used in their experiments. It was rather the *nature* of the products that was their main concern. It was not surprising that it had proved impossible to bring any kind of order into chemistry. Boyle's concept of elementary corpuscles, what we would call the atoms of the elements, broke through the mist, but real progress was impossible without quantitative studies.

In Lavoisier's time, chemistry was dominated by the phlogiston theory, which stated that when something was burned it gave off an invisible substance called phlogiston. If this were true, it implied that burning a substance would result in a decrease in its weight. The fact that burning often resulted in an *increase* in weight worried some of the proponents of the theory, but they solved the paradox by giving phlogiston a negative weight, so that when it escaped from a body that body did indeed increase in weight. One can regard this "explanation" either as a demonstration of creative thinking—the ability to break away from preconceived ideas—or as a disgraceful subterfuge. I have a weakness for the former view. Strange ideas can be true. After all, how many people believed in antimatter before it was found? And didn't Kepler propose the "strange" but correct idea that the attraction of the Moon caused the tides, only to be dismissed by Galileo for lending "his ear to occult properties and such-like fancies"? Phlogiston proved to be a myth, and it was Lavoisier who killed it.

By way of illustration of his way of thinking, consider Lavoisier's investigation of the burning of tin. He weighed a piece of tin, placed it in a large vessel containing air, sealed the vessel, *weighed it*, and heated it. There was an obvious change in the appearance of the tin, which turned to a powder. An eighteenth-century scientist would have said that combustion had occurred. We would say that it had been *oxidized*, that it had combined with the oxygen in the air. After the sealed vessel had been allowed to cool down, he weighed it again. *There was no change in weight.* Matter had been conserved in a chemical reaction. The conservation of matter (in chemical reactions) became a fundamental principle in science. When Lavoisier opened the flask, there was an inrush of air. He reasoned that, contrary to the supposition that the tin had given off something when it burned, it had taken something from the air. Consistently, the burned tin was heavier than the original metal. At about this time he was visited by the Englishman Joseph Priestley, who had also performed experiments involving combustion but had interpreted them in terms of the phlogiston theory. Lavoisier saw the true meaning of his and Priestley's work: they had both discovered the gas oxygen (the name given to it by Lavoisier), and burning was merely the combination of a substance with oxygen, not the loss of a hypothetical substance with negative weight. Phlogiston was an unnecessary concept, holding up the progress of chemical science. Priestley was not convinced, but Lavoisier had set modern chemistry on its path.

Another Conservation Law

The conservation of mass is a reflection of the fact that in ordinary life we can neither create mass nor destroy it, just as charge is inviolable. In the twentieth century Einstein showed that mass and energy were in fact interconvertible, but this is almost irrelevant to everyday life, except in the conversion of mass to energy in nuclear reactors and nuclear weapons.

Chemistry Takes Off

Lavoisier proceeded to examine a large number of substances, using his balance and his logic. He burned two small diamonds and showed that they were converted into carbon dioxide, the same gas produced by the complete combustion of charcoal. This demonstrated that diamond is a form of carbon. He burned alcohol, showing that the reaction produced water and carbon dioxide, and deduced from his measurements that alcohol contained carbon, hydrogen, and oxygen. This kind of experiment opened the door to the elucidation of the atomic composition of complex substances. He, like Priestley, showed that oxygen was essential to animal respiration: "One fifth only of the volume of atmospheric air is respirable." He realized that combustion of a number of organic materials, such as sugar, produced the same gas as respiration, namely, carbon dioxide. The analogy that he saw between combustion and respiration was of seminal importance in the development of physiology.

Lavoisier grasped the importance of Boyle's concept of elements. When he found a material that could not be decomposed by any means available, he assumed that it was an element. Inevitably, considering the state of chemistry in his day, he made mistakes. He mistakenly thought that silica (silicon dioxide) was an element because he found no method to break it up. Lavoisier listed thirty-two elements, some of which we now know to be compounds, and two of which, light and heat, are not material. Strangely, he did not accept the idea of the atom. There was no direct evidence for the existence of atoms in his day, and as late as the beginning of the twentieth century the eminent physical chemist Wilhelm Ostwald felt that the atom was an unnecessary concept.

Lavoisier laid out the principles of chemistry, as he saw them, in the text that should be regarded as the true beginning of modern chemistry: *Traité éleméntaire de chimie*. There he stated explicitly the principle that matter is conserved in chemical reactions. The book was published in 1789. On 14 July of the same year the Bastille fell.

On 14 July 1791 a group of incautious, radical-minded Englishmen gathered in the Royal Hotel in Birmingham to celebrate the second anniversary of the fall of the Bastille. A hostile mob stoned the hotel and then marched to Joseph Priestley's house. The scientist was not only a supporter of the American Revolution against the king of England, and the French Revolution against king and God, but moreover was hardly an orthodox Christian. He had published a book entitled *History of the Corruptions of Christianity* (1782), in which he had rejected miracles, original sin, and the Trinity, and, although a Dissenting minister, had declared that everything was matter, including the soul. The standard early-nineteenth-century mob in England was anti many things: Bonaparte, the French, the pope, and Dissenters. They burned down Priestley's home, containing his laboratory and library. What hurt him most was the loss of his scientific writings, "manuscripts which have been the result of the laborious study of many years, and which I shall never be able to recompose; and this has been done to one who never did, or imagined, you any harm." The riots persisted for three days as the eternal, mindless mob howled for the blood of "philosophers." Many fearful citizens put up notices on their houses proclaiming "No philosophers here." Priestley emigrated to America three years later and settled in Pennsylvania, where his lectures on Christianity led to the formation of a Unitarian Society. He died in 1804, firm in his belief in phlogiston.

The burning of Priestley's house is symbolic of the dwindling of the Enlightenment dream of endless progress through the operation of reason and man's humanity to man. Priestley lived with this dream: "Nature . . . will be more at our command; men will make their situation in this world abundantly more easy and comfortable . . . and will daily grow more happy. . . . Thus, whatever was the beginning of this world, the end will be glorious and paradisiacal beyond what our imaginations can now conceive." Or will the crowd decide that there will be no philosophers here?

Another Father of Chemistry

The man who finally unambiguously established the idea of the chemical element in modern science was John Dalton (1766–1844), the son of a Quaker weaver. He was put in charge of the village school at Kendall, in the Lake District, at the age of twelve. The appointment was not an outstanding success.

Dalton was perhaps the first full-time professional scientist. Far from enjoying the inherited wealth of a Lavoisier or a Boyle, he earned money by teaching and lecturing.

Dalton was not averse to speculation, but observation came first. By a series of simple experiments, he set out to show that certain substances could not be broken up to give simpler components, and he began the accurate determination of the relative weights of the atoms of different elements.

It is still impossible to directly weigh a single atom, but there are ways of getting the *relative* weights of different atoms, if certain assumptions are made. Thus, suppose you tell me that you have an unknown number of place settings consisting of only a knife and fork for each person. If you now separate all the knives from the forks and give me the two piles, then by weighing them I can tell you the *relative* weights of a single fork to a single knife—without knowing how many knives and forks I have, or what their absolute weights are. The only assumption I have made is that the two piles contain equal numbers. In 1803 Dalton published a memoir that contains the following sentences: "An enquiry into the relative weights of the ultimate particles in bodies is a subject, as far as I know, entirely new; I have lately been prosecuting this enquiry with remarkable success. The principle cannot be entered upon in this paper; but I shall just subjoin the results, as far as they appear to be ascertained by my experiments." Using simple assumptions and arguments, Dalton arrived at the first-ever list of relative atomic weights, which he gave at the end of his paper. No one took much notice at first, but gradually it dawned on the scientific community that Dalton's data opened the door to the experimental determination of the composition of molecules, in terms of atoms. The first step in determining the nature of any molecule is to find out which types of atoms it contains and how many there are of each type. Without relative atomic weights this was impossible.

Lavoisier had not believed in atoms; Dalton did. After Dalton, it was accepted by almost all scientists that atoms existed, that they had fixed weights, and that they were indivisible.

Dalton's life illustrates the growing respectability of science and its continuing international flavor. In 1816, one year after the defeat of Napoleon at Waterloo, the French Académie des Sciences elected Dalton to be a corresponding member. In 1833, after lobbying by a group including Charles Babbage, the pioneer of the computer, the British government awarded Dalton an annual pension of £150. When he was buried in 1844, about 40,000 people filed past his coffin. After his death, one of his eyes was dissected, in accordance with his wishes. Dalton was color-blind, a defect which he supposed to be caused by a blue coloration of the fluid in his eyes. The result of the dissection did not support his hypothesis.

A succession of Italian and French scientists—Avogadro, Cannizzaro, Gay-Lussac, and others—built on Dalton's ideas and corrected his errors, but the real breakthrough was Dalton's, whom many refer to as the father of modern chemistry. After Boyle and Lavoisier, he is the third to be so dubbed, but there is little point in championing one or the other: chemistry evolved, it had its heroes, but it never had a Newton.

The work of Faraday, Lavoisier, and Dalton underlay the science-based transformation of everyday life that blossomed in the nineteenth century.

11

The Nineteenth Century

The Origin of Species, published in 1859, inaugurated an intellectual
revolution such as the world had not known since Luther nailed his thesis to
the door of All Saints Church at Wittenberg.

—E. R. Pease, secretary of the Fabian Society

Science Brings Home the Bacon

In 1834, during a discussion in the Chambre des Députés, the distinguished French
physicist Arago was asked to justify his request for government support of the sciences. His colleagues demanded to know what practical benefits had accrued from
the discoveries of science? Arago could only think of one example: the lightning
conductor.

By 1900, HMS could have provided a far more impressive catalogue. By then,
the practical applications of basic science were everywhere to be seen, and the Baconian conviction that man could determine his own destiny through science was
renewed. But not everyone joined in the applause.

It was inevitable that theory, in the shape of Darwinism, and practice, in the
form of science-based technology, were seen as threats to belief. At its best, science
ignored faith; at its worst, it challenged Holy Writ. The materialism engendered by
technology had also tainted the social sciences, where economics and the law of
supply and demand were replacing more philosophical notions, such as the social
contract, that had occupied the thinkers of previous generations.

Blind Admiration

One of the books that I, as a ten-year old, found in my father's wildly heterogeneous
library was Winwood Reade's *Martyrdom of Man*, from which I copied out some
frequently quoted lines:

> When we have ascertained, by means of science, the methods of Nature's op-
> eration, we shall be able to take her place to perform them for our-
> selves . . . men will master the forces of Nature; they will become themselves
> architects of systems, manufacturers of worlds. Man will then be perfect; he
> will be a creator; he will therefore be what the vulgar worship as God.

That prediction of a scientific Utopia was penned in 1872. The socialist Beatrice
Webb summed up, in 1926, her understanding of the prevalence of such sentiments:

> It is hard to understand the naive belief of the most original and vigorous
> minds of the 'seventies and 'eighties that it was by science, and by science
> alone, that all human misery would be ultimately swept away. This almost

fanatical faith was perhaps partly due to hero-worship. For who will deny that the men of science were the leading British intellectuals of that period?

Naive or not, *The Martyrdom of Man* went through many editions, and its exaggerated views of the potential of science and the evils of religion had a wide appeal. Reade's success owed a great deal to Darwin, but just as much to the aura bestowed on science by the successes of the scientists of the nineteenth century. Few of his admiring readers fully understood the meaning of the great scientific breakthroughs, but those who did, or thought they did, formed an influential and voluble minority. His audience was by no means universally sympathetic to his atheism, but it was very aware of why he thought that science had superseded the blood of the Lamb. Never had so many been both spiritually influenced and materially benefited by science. Paradoxically, the impetus behind many people's respect for science was the Industrial Revolution, a revolution that had very little to do with science.

The Iron God

The nineteenth century opened as the Industrial Revolution was gathering momentum, and by the end of the century the western European nations dominated the world. That domination depended heavily—militarily, commercially, and psychologically—on the technological gap between Europe and the rest of the world. This gap was initially opened up by what can be termed the *first* Industrial Revolution, initiated in late eighteenth-century England.

The first Industrial Revolution, which was based on the power of steam-driven machines, owed almost nothing to science. It was created not in laboratories but by practical men, men who had learned their engineering in factories and mines. The counsel for Richard Arkwright, the inventor of the spinning frame, said it all in court:

> It is well known that the most useful discoveries that have been made in every branch of art and manufactures have not been made by speculative philosophers in their closets, but by ingenious mechanics, conversant in the practises in use in their time, and practically acquainted with the subject-matter of their discoveries.

So much for those, Romantic poets included, who had laid the blame for the social disasters of the Industrial Revolution at the doorstep of science. Established science was not a social force. The Royal Society had strayed far from the days when Hooke and Boyle and their like had created it, having become a social club for scientific nonentities and aging aristocrats. As Richard Steele wrote in the *Tatler*, as early as 1709, "When I meet a young fellow that is a humble admirer of science, but more dull than the rest of the company, I conclude him to be a FRS."

British engineering impressed "the natives" on four continents. Half the world's ships were built in the dockyards of Tyneside. One of the factors that raised the level of engineering was Charles Babbage's attempt to build a mechanical computer.[1] In

---

[1]The remarkable Ada Lovelace, Byron's only legitimate child, translated and annotated lecture notes taken in Italian during Babbage's visit to Italy, where his work had aroused great interest. A computer language called ADA has been developed, named in her honor.

1822 the British government granted him £1500 toward the costs of construction. The degree of accuracy needed in the machining of the parts pushed machine tool technology to new heights. The era's monument was the Great Exhibition of 1851, in which Great Britain was presented as what it had come to be called: the workshop of the world.[2] That workshop was not the child of the *Principia*, but the turn of science would come, and it would be Galvani, Volta, and above all Faraday, Lavoisier, and Dalton who were to lay the foundation for what could be termed science-based civilization, a phrase that I use to describe the specifically material infrastructure of life.

The Magic Fluid

In the 1880s a second industrial revolution took place. This time it was heavily based on science: it was the practical fruit of Volta's and Faraday's experiments on electricity, and the realization of Dr. Samuel Johnson's dream, a century before, that "electricity, the great discovery of the present age," be made to serve man.

Initially, the production of current electricity was dependent on cumbersome batteries and was largely limited to scientific laboratories. Napoleon backed the assemblage of a massive galvanic pile for the École Polytechnique in 1813. Despite the inconvenience of huge arrays of batteries that had to be regularly tended and topped up with chemicals, it was such arrays that provided the source of electricity for the newly invented telegraph system. In 1837 the first electric telegraph line was set up, between Euston and Camden in London. It depended on Ampère's 1820 discovery (see Chapter 8) that a magnetic needle suspended near a wire was deflected by the passing of a current. Ampère had in fact suggested the use of this phenomenon as a basis for sending messages. The opening of the first line created a sensation; posters declared, "The Electric Fluid travels at the rate of 280,000 Miles per Second." The few miles of cable that were laid were the forerunner of the first cable laid across the Atlantic, in the 1860s. In 1845, after several failures, Samuel Morse opened a line between Washington and Baltimore using his code, which was to become accepted throughout Europe in 1846. In Germany in 1847, Siemens and Halske founded a company that constructed a telegraph network in Russia and subsequently ran a telegraph cable from London, through Berlin to Calcutta. The telephone had a harder birth, as evident from these sentences, taken from the report of the committee set up by the *Telegraph Company* to examine Alexander Graham Bell's proposal:

> Technically, we do not see that this device will ever be capable of sending recognizable speech over a distance of several miles. Messrs. Hubbard and Bell want to install one of their "Telephone" devices in virtually every home and business establishment in the city. This idea is idiotic on the face of it. Furthermore, why would any person want to use this ungainly and impractical device when he can send a messenger to the local telegraph office.

---

[2]Not everyone was impressed by the goods that the workshop produced. At the Great Exhibition, William Morris "stood aghast at the appalling ugliness of the objects exhibited, the heaviness, tastelessness, and rococo banality of the entire display." But confident English capitalism had little time for aesthetes.

Indeed. In the 1840s the increased availability of electric current facilitated the establishment of an electroplating industry. Up to the 1880s, the main source of electric current for this industry, and in general, was the Daniell cell, a glass container containing two solutions separated by a porcelain partition. An alternative battery, the lead-acid accumulator, invented in 1881, can still be found in cars today but is obviously not a convenient source of electricity for long-term, maintenance-free, commercial or domestic use. It is not surprising that gas remained the main source of domestic lighting into the early twentiethth century. But a totally new way of generating electric current was to almost completely replace chemical cells. Basic science was again poised to initiate a torrent of technical innovation.

Albert, Victoria's prince consort, appears to have sought out and enjoyed the company of scientists and writers. He visited Michael Faraday's laboratory, in the days when the scientist's experiments with electricity looked like high school demonstrations. Faraday went through his party pieces—rubbing rods on cloth, charging metal spheres. Albert asked, "But of what use is it, Mr. Faraday?" The reply was, "Of what use is a baby, sire?"[3]

One of Faraday's simple experiments had shown that electricity could be generated by the relative motion of a coil of wire and a magnet. This particular baby was to burgeon into a giant. In 1882 the Edison Electric Illuminating Company began operating the world's first electricity-generating plant based on dynamos. Electric current could be continuously and conveniently manufactured and sent anywhere. Faraday's research had borne practical fruit in the dynamo and the electric motor. In the 1880s, Albert Einstein's father had a modest factory in Munich, making dynamos and other electrical equipment. His son was to be fascinated by Faraday's theoretical concept of a field, rather than by the field's practical uses.

Electricity spread everywhere: in industry, the home, and the electrification of transport.[4] In 1860, Sir Joseph Swan passed a current through a carbon filament inside an evacuated tube. He had created a 25-watt lightbulb. It was not long-lived enough to be commercially viable, but 20 years later electrical lighting began to illuminate Europe and America. The practical benefits of basic research in physics had never been more evident. Technology, for the first time in history, was becoming dependent on basic science.

The Wireless

Maxwell's purely theoretical breakthrough, the discovery and characterization of the electromagnetic field, opened the way to Hertz's radio waves and to the discovery and utilization of a spectacular range of electromagnetic waves. X-rays, discovered in 1895 by Roentgen, evoked lyrical panegyrics such as that of Silvanus P. Thompsons's: "November the eighth 1895, will be ever memorable in the history of Science. On that day a light which, so far as human observation goes, never was on land or on sea, was first observed." In 1901 Roentgen was awarded the first Nobel

---

[3]Benjamin Franklin is said to have made a similar remark when he was asked about the significance of the first balloon ascent, which he witnessed in 1783.
[4]Electricity provided a fruitful fishing ground for frauds such as the "Electric Infant," reported in the *English Mechanic* (1869): "This interesting but inconvenient infant was, it is stated, so endowed with electricity that nobody could enter the room where it was without receiving constant electric shock."

Prize in physics. The new rays excited prurient fancies; the popular press was replete with cartoons of solid bourgeois gentlemen peering at young ladies through what were supposed to be X-ray glasses. But people were also well aware that medicine had acquired a powerful diagnostic tool. Sadly, the pioneers were unaware of the dark side of radiation, and many of them eventually died of overexposure to the new rays.

In the next century, ultraviolet, infrared, microwave, and γ-radiation were to become familiar and to reveal their uses and their dangers.

The huge technological consequences of Faraday's experiments, and Maxwell's mathematics, remain the archetypal example of the growth of useful technology from basic science. Politicians who have to make decisions on the funding of basic research should learn a simple lesson from history: basic research very often pays off. Another convincing example arises from man's efforts to understand the nature of matter.

The Manipulation of Matter

The advance of chemical theory, sparked primarily by Lavoisier and Dalton, had by midcentury brought chemistry to the stage where the synthesis of novel molecules, and the identification and characterization of naturally occurring molecules, had become standard laboratory practice. Chemistry was no longer cookery. The first half of the nineteenth century saw the growth of chemistry from a largely empirical activity, carried on primarily in small private laboratories, to a major scientific discipline that was to change the materials, colors, foods, and pharmaceuticals of the world.

In 1826, Justus von Liebig founded a chemical laboratory in Giessen, Germany, and from 1842 onward he used it as a training ground for young chemists, most of whom went into industry. German chemistry advanced rapidly and, before long, industrial companies began to establish their own research laboratories.

New materials were developed. The availability of electricity allowed the commercial development of an electrolytic process for producing aluminum from its ore. In 1884, Compte Louis-Marie-Hilaire de Chardonnet obtained a patent for rayon. The Paris police opened a chemical laboratory not only for forensic purposes but also to check on the contamination of food by chemicals. The synthesis, at the age of seventeen, of the aniline dye Tyrian purple by the English chemist William Perkin, who patented his work in 1856, opened the way for brighter fabrics. Perkin also achieved the first laboratory synthesis of an aminoacid (glycine) and the first synthesis of a perfume (coumarin). Such feats would have been utterly beyond the capability of chemists working half a century before. They were now possible as a consequence of the developing understanding of the way in which atoms were linked together to form molecules.

The commercial use of aniline dyes was developed primarily by the German chemical industry, just as most of the commercial applications of electricity were pioneered in Germany and America. These were far from being the last cases of British scientific breakthroughs being brought to the marketplace by other countries. Maybe it can be put down to what Lord Cherwell saw as the intellectual snobbery concerning technology, which took a long time to die out in England. In Germany things were different. Max Maria von Weber, the son of the composer Carl

Maria von Weber, became an internationally influential expert on railway construction. A contemporary biographer said that this choice of technology as a career was "typical of a phenomenon decisive in our culture: the transition from the sphere of fancy and thought to that of applied activity, realistic creativity." When war broke out between the two countries, the khaki dye used for British army uniforms in 1914 came from a factory in Stuttgart.

The march of physics and chemistry was joined by biology. It was in the nineteenth century that the intrusion of the scientific method into biology initiated the construction of a rational base for medicine.

The Science of Life

In the field of biology the three most significant advances in the nineteenth century, apart from the great Darwinian watershed, were the development of the cell theory of life, the discovery of the electrical nature of nervous activity, and the germ theory of disease. Of these, the germ theory was undoubtedly the one having the major immediate consequences for HMS. In France, Pasteur was a national hero. The image of the child cured of rabies impressed itself deeply on the public.[5]

Perkin's aniline dyes led to a major advance in biological science when the German scientist Paul Ehrlich realized that if the dyes could color the animal fibers wool and silk, they might be able to stain the components of the living cell. He presented his thesis "Contributions to the Theory and Practice of Histological Staining," to Leipzig University in 1878. It was by the use of dyes that chromosomes were first seen under the microscope. They were named by Erhlich: "colored bodies," (i.e. *chromosomes*).

In 1876 the German scientist Robert Koch, for the first time, identified a germ connected with a specific disease, anthrax. Medicine was becoming both more scientific and more effective than it had ever been. The fact is, as Lewis Thomas pointed out in 1977, that before the advent of science, medicine was "an unrelievedly deplorable story. . . . It is astounding that the profession survived so long, and got away with so much with so little outcry."

"Scientific Tests Have Shown That . . ."

From the time of Newton until well after the middle of the nineteenth century, a growing number of humanists saw natural science as a subject warranting serious attention. This was partly because science was considered to be a part of philosophy and also because most science was readily intelligible to an averagely educated man. In the seventeenth century Locke was so interested in the *Principia* that Newton took the trouble to explain to him personally the elliptical paths of the planets about the Sun, and, as we have seen, this interest in science blossomed among intellectuals during the Enlightenment. In the nineteenth century Shelley, Coleridge, Keats, Tennyson, and others dabbled in science, or read and discussed it. Hegel considered that the central position of the arts in civilization was no

---

[5]In the late 1950s I worked in the Institut Pasteur in Paris. There was a small collection of memorabilia and of Pasteur's scientific instruments in a building across the road from the laboratory in which I worked. Whenever I was there I felt as though I was visiting a shrine.

longer justified and that they should be replaced by science, which for him embodied the spirit of reason, "the Sovereign of the World," "the substance of the universe"—provided, of course, that Newton was kept out of things. Émile Zola was quite explicit about the influence of science: "My aim has been a scientific aim, above all," and, "We have experimental chemistry and physics. . . . We shall have the experimental novel." The Goncourts proclaimed even more strongly: "The Novel has taken up the studies and duties of science." Scientific terms crept into the vocabulary of writers and of HMS. August Strindberg's wife, Frida, avowed that the nerves of the somber dramatist were so sensitive to "electricity" in the atmosphere that storms communicated themselves to them as though they were "wires." As early as 1816, Mary Wollstonecraft Godwin, Shelley's second wife-to-be, had been influenced by Galvani and Volta. At eighteen, impressed by newspaper accounts of the effect of electricity on corpses, she wrote *Frankenstein*. It was electricity that brought the monster to life. Byron was doubtless more interested in the supposed beneficial effect of "galvanism" on virility, a myth that was commercially exploited by enterprising quacks.

The acknowledgment of science as an intellectual force of wide significance was not new, but in the nineteenth century, for the first time in history, science was reaching out into practical, everyday life, building the prestige of the scientist as a benefactor and an authority to whom to turn for advice.

The Cadbury company in England was one of the first to use the newly acquired popular prestige of the scientist to promote its products. It produced an advertisement showing a bearded, bald-domed, authoritative figure in a lab coat holding a tube of liquid. A distilling flask, a microscope, and other apparatus are also depicted. On one of the notebooks open on his table one can read the words "Absolutely pure Cocoa. No chemicals." All that is missing is the comparison with "other leading brands." The standing of science had been restored following its decline at the turn of the century. But James Clerk Maxwell laid his finger on the danger of uncritical kowtowing to the new god: "Such . . . is the respect paid to science that the most absurd opinions may become current, provided they are expressed in language, the sound of which recalls some well-known scientific phrase." His warning still reads well.

A popular audience for science, and for technology, blossomed in Europe and America. Botanical and zoological gardens became popular. In England the prince consort, Albert, gave his active backing to the building of a large complex of scientific and natural history museums in the Knightsbridge area of London. Crowds came to see the skeletons of the dinosaurs. In 1851 Foucault suspended a long pendulum beneath the dome of the Pantheon in Paris. Thousands flocked to watch as the plane in which it swung apparently rotated as the Earth turned. In the 1890s the children of the bourgeoisie began to receive educational science kits as presents, and their parents meddled with simple chemicals and magnets. Science was becoming part of popular culture, although this was predominantly a middle-class phenomenon.

Science Comes Out of the Lab

Looking back from 1900, we see a century in which science and technology increasingly established themselves as major forces in the shaping of nations. In the seventeenth century Newton had given science a previously unequaled status, but science had done very little to affect the life of HMS. Newton's mechanics and Boyle's chemistry had no conspicuous practical consequences. Harvey's discovery of the circulation of the blood did not significantly change medical practice, and Gilbert's sixteenth-century investigations into magnetism had no technological follow-up. Technology up until about the middle of the eighteenth century was rarely the outcome of scientific advances. The art of the watchmaker and the skill of the machine-tool manufacturer needed to support that art, did infinitely more to improve industrial technology than the laws of motion. It was only around the middle of the nineteenth century that science began to feed technology, a process that gathered momentum as the century drew to a close.

It was the seemingly endless benefits inherent in science that produced the profound shift, during the nineteenth century, in the public attitude toward scientific education and science as a profession.

The Birth of the Professional Scientist

In early seventeenth-century Britain, nearly 60% of scientists had been educated at a well-known public school, and two-thirds of them had graduated from either Oxford or Cambridge. By the end of the eighteenth century, only about 20% had come from public schools or were graduates of Oxbridge, and fewer than 20% came from the upper classes. Science was no longer the pastime of gentlemen; it had shifted primarily into the hands of the middle class. The chemists Dalton and Priestley, and many others like them, taught not in universities but in Dissenters academies where science was given a respected place in the syllabus. The roll call of Quaker, Unitarian, and generally dissenting scientists is impressive. Science, for the Dissenters, was a route to the understanding of the workings of God.

In the nineteenth century the number of people, especially from the middle class, who had had some scientific education grew rapidly. Even the working class was hearing the distant rumble of scientific drums. The great Michael Faraday was the scientific working-class hero of the nineteenth century. The so-called lower class was breaking out of the cast of ignorance and subservience. It is significant that of the approximately 4000 members of the National Secular Society in the 1880s, nearly all were working men.

The academic world of England failed to realize what was happening. Until the second half of the century, neither Oxford nor Cambridge offered degrees in science.[6] The attitude of the ancient universities changed as the nineteenth century unfolded and as science became a socially acceptable profession. Respectability was bestowed on science by such events as the 1854 series of scientific lectures at the Royal Institution of Great Britain, given in front of the prince consort. Among the lecturers was Michael Faraday, who was a brilliant popularizer, and William Whewell, master of Trinity College Cambridge. All strongly advocated that science

---

[6]Neither did they admit women, Dissenters, Jews, or Roman Catholics. In England the first undergraduate laboratories in physics and chemistry were opened at University College London, a university founded in 1828 that was also the first to admit both men and women, regardless of creed, race, or class. As regards women, they were preceded by 2500 years by Pythagoras, who though he barred women from membership of the cult, allowed them to attend lectures.

be given a central place in the educational system. Support came from unexpected circles. A certain Reverend F. W. Farrar wrote an essay demanding the "immediate and total abandonment of Greek and Latin verse-writing as a necessary or general element in liberal education." Science was to be the main replacement.

In the 1860s it became apparent that Oxbridge was out of step with the times. Oxford took the lead in accepting physics as a subject for research at the university, and Manchester built university laboratories. In Cambridge the Cavendish physics laboratories were dedicated in 1874, and Clerk Maxwell became the first Cavendish Professor of Experimental Physics.

Across the English Channel, France had already introduced widespread scientific education in the eighteenth century. As early as 1828, the École Centrale des Arts et Manufactures had been established. Technical institutions, teaching both pure and applied science, were well established in eighteenth-century Germany, and after the turn of the century several of them attained university status. In the older German universities, in the first part of the nineteenth century, there were usually four faculties: philosophy, theology, medicine, and law. In the 1860s, science faculties began to appear, usually as an spin-off from the faculty of philosophy. Academic science would never look back, and in universities, technical colleges, schools, amateur scientific clubs, and popular lectures, science was propagated through the middle, and eventually the working, class.

America remained a scientific backwater during most of the nineteenth century. In the colonial period most American men of science saw England as the country with which they had contact, in the form of letters, visits, and articles sent for publication. Predictably, it was France that superseded England after the American Revolution. Benjamin Franklin, who had close connections with the French scientific world, was a familiar figure on the streets of Paris from 1776 to 1785. France maintained its attraction during the first half of the nineteenth century, but the influence of Germany grew rapidly in the latter half of the century. As in Europe, the change from amateur dabblers to professional scientists was primarily a nineteenth-century phenomenon, but America initially lagged behind Europe in the area of research facilities. A major milestone was the creation of the Smithsonian Institution, founded in 1846. By the end of the century, American science was taking off. There were a number of industrial research laboratories staffed by university graduates, and several scientists had attained world stature. J. W. Gibbs at Yale revolutionized chemical thermodynamics, and the experiments of Albert Michelson at the University of Chicago, on the speed of light, perturbed the world of classical physics.

In 1832 there was not a single observatory in the United States; fifty years later there were nearly 150, and by 1900 American astronomy had attained world-class stature. National prestige should not be underestimated as a motivation here. It was certainly a major factor in the construction of the, for those days, huge Pulkovo observatory near St. Petersburg in 1839, by order of the "Iron Czar," Nicholas I, the creator of the Russian secret police.

All over the Western world, science blossomed. But not everyone was filled with joy.

The Backlash

... the hideous new religion of Science ...

—Edmund Gosse, *Father and Son*

The predictable backlash against the scientific bulldozer came most markedly in the 1870s and 1880s, which saw a revival of fundamentalist Christianity. There was a revolt among many educated laymen against the, to some, menacingly spectacular growth of science and its apparent encroachment on all aspects of life. Queen Victoria led from the top: "Science is greatly to be admired and encouraged, but if it is to take the place of our Creator, and if philosophers and students try to explain everything and to disbelieve whatever they cannot prove, I call it a great evil instead of a great blessing." The formidable William Gladstone (he was reputed to chew each mouthful of food eighty times), then prime minister of England, also spoke for many when, in 1881, he said: "Let the scientific men stick to their science and leave philosophy and religion to poets, philosophers and theologians."

It was not only Darwinism that was perceived to threaten the mildly religious Victorian world. The upsurge of geology, especially the work of Charles Lyell, fostered antibiblical notions, such as the immense age and gradual development of the Earth. Investigations of life processes, inexorably cutting away the shroud of mystery that had singled out Life as something outside Newton's mechanistic universe, were deeply troubling—and still are, to some people.

Not only the religious establishment, but also many laymen were worried by the rise in atheism and the prospect of humanitarianism as a substitute for religious faith. It was common to blame the decline of religion on science, although among the clergy themselves there was just as much, if not more, concern about the so-called new criticism of the Bible, which tended to seek for rational explanations of the more unlikely episodes in the Scriptures. However, although there was a decrease in churchgoing during the century, it was not precipitous, and atheism was not a widespread social phenomenon, its image being well out of proportion to the number of its adherents.[7]

Notable voices were raised against what was seen as the arrogance of scientists. John Ruskin, the great (and neurotic) art critic, spoke out against the construction of the new physics laboratories, the Clarendon Laboratories, at Oxford. Specifically attacking Darwin, he declared that science should not be pursued for its own sake, and not even for its practical benefits, but as an activity subordinate to ethics. What this would mean in practice to men like Maxwell was not clear, but Ruskin's words were repeated frequently by antivivisectionists and those opposing vaccination. Ruskin's concern with ethical problems remains valid to this day, especially in the life sciences and medicine.

The distaste for science and the accompanying objection to vivisection reached greater heights on the Continent than in England, feelings typified by books such as Ernst von Weber's *Die Folterkammern der Wissenschaft* (The Torture Chamber of

---

[7]A high-profile atheist was Charles Bradlaugh, propagandist for scientific materialism, editor of the radical *Reformer*, and president of the National Secular Society. Bradlaugh gave popular antireligious lectures under the name "Iconoclast." For five years he unsuccessfully attempted to take his place in the House of Commons. He refused to take the religious oath. In the end he sold out, taking the oath and his seat.

Science) (1879) and the following passage from the novel *Gemma oder Tugend und Laster* (Gemma or Virtue and Vice), by Elpis Melna (1887): "What can you expect of a man who tramples on the most sacred laws of Nature, a man whose every last grand and noble feeling has died in him, who cannot even appreciate the wonderful spiritual qualities of a daughter worthy of worship, a man who seeks his pleasures in the lowest debaucheries and gory orgies, whose conscience and hands are forever smeared with blood. . . . In short what can you expect of a vivisector?" Not much, apparently.

Darwinism was more readily accepted by the Church in England than it was by the rigid Protestant sects of the New World, a pattern that is maintained by the almost complete absence of present-day controversy over evolution in Great Britain, in contrast to the continuing clashes between fundamentalists and secular authority in the United States.[8] Nevertheless, social Darwinism was more warmly embraced in the United States than in Europe. Herbert Spencer's banner "the survival of the fittest" proved a rallying point for moneyed laymen, tenured academics, and worldly clerics. Said John D. Rockefeller, "The growth of large business is merely the survival of the fittest," and his conviction was confirmed by Graham Sumner, a professor of political science at Yale, who diplomatically saw millionaires as "the naturally selected agents of society" (you never know who the university's next benefactor will be). The blessing of the Church was bestowed on mammon by Bishop William Lawrence of Massachusetts, who incredibly expressed the opinion that "godliness is in league with riches. . . . The race is to the strong." We could be generous and interpret that as a cry of woe, but it was not. The distinctly un-Christian Bishop might have applauded Spencer's opinion, borrowed from Malthus and expressed in Spencer's *Man versus the State* (1884), that it was not the business of the state to help the poor, whom he implicitly categorized as "unfit." Many concurred with Spencer's pseudo-Darwinian conclusion that the traditional role of women in society was the result of a process of natural selection: selection for reproduction. This bit of wishful thinking on the part of male society was rarely challenged by men, an honorable exception being the prominent music critic Ernest Newman, who, in an article on "Women and Music" (1895) wrote that men

> have started out with the theory of the natural inferiority of women, have assumed—like Rousseau, because it suited them to do so—that her "true" sphere was the home and her "true" function maternity, and then persuaded themselves that they had biological reasons for keeping her out of the universities and for denying her a vote.

Which did not prevent the sexologist Havelock Ellis, a supporter of women's rights, from declaring that "Nature has made women more like children in order that they may better understand and care for children." Social Darwinism was profoundly reactionary.

Darwin is, in the twentieth century, still a *causus belli* in the United States. In 1965 a courageous teacher in Little Rock filed a suit against the State of Arkansas

---

[8]In 1994(!), the broadcasting of a program in the United States on the significance of "Lucy," the 3.5-million-year-old "missing link" found in Africa by Donald Johanson, was, due to pressure, followed by a two-hour slot given to creationists to "balance" accounts. This is unthinkable in England.

on the grounds that her right to free speech had been violated by the local antievolution law. She won her case, but the verdict was reversed by the Arkansas Supreme Court in 1967. The case finally came before the United States Supreme Court, which in turn decided for the teacher. Before it came to its decision, there was an unprecedented approach to the Court by a group that included seventy-two Nobel laureates, who prepared a document defining the scope and nature of science. The kind of opposition they were up against can be judged from the following quotation from a letter to one of them, the physicist Murray Gell-Mann: "Ask the Lord Jesus to save you now! The second law of thermodynamics proves evolution is impossible." It doesn't.

Among American intellectuals, responses to evolution were perhaps conditioned by a strong streak of pragmatism. Unlike their European counterparts, they in general avoided appealing to vital forces or cosmic will. Rather, attempts were made to apply evolutionary ideas to down-to-earth problems in law, education, and social ethics. Everything was simply explained by the prosaic theme of man's continual adaptation to his environment, both biological and cultural—which was a healthy approach after the tiresome, and sometimes batty, European invocation of will.

The Achievements

All in all, there were four major *practical* results of nineteenth-century basic science:

1. The advent of electricity as man's most versatile source of energy and the main basis of the coming twentieth-century technological explosion.
2. The practical consequences of Maxwell's discovery of electromagnetic waves.
3. The advances in medicine consequent on the germ theory.
4. The development of the chemical industry.

The major breakthrough, in terms of its *intellectual* influence, was undoubtedly Darwinian evolution, which left its mark on social and political thought and infiltrated philosophy and literature. The achievements of Maxwell and Faraday were immense but cannot be said to have had a widespread *direct* effect on the way that people thought because, in contrast to evolution, most people saw only the *practical* implications of electromagnetism.[9] The theoretical aspects were not felt to be relevant to man's place on Earth, although the concept of fields and Maxwell's equations were to have a deep effect on twentieth-century physics. There is a similarity with the great twentieth-century breakthroughs in physics—quantum mechanics, particle physics, and relativity—which were to have little effect on the intellectual life of HMS. There is of course a difference in the perceived degree of incomprehensibility between nineteenth- and twentieth-century physics that puts this century's physics well out of the range of interests of most people. This is not to say that the advances of Einstein, Planck, and company had no practical or political consequences; the atomic bomb is not a trivial matter. But how many people's way of thinking has been affected by modern theoretical physics?

---

[9]The pointillist artist Georges Seurat was intrigued by Maxwell's electromagnetic waves, but I doubt if this affected his art.

There is one major achievement of nineteenth-century science—namely, thermodynamics—that neither had significant practical applications of which the public were aware nor penetrated the imagination of HMS. To HMS the subject seems abstruse, and irrelevant to daily life. Moreover, it lacks visual appeal. One can imagine genes and galaxies, and one can handle electronic equipment. The second law of thermodynamics has as much visual appeal as a hymn, and the only mark it has made outside science is the usually totally inaccurate use of the word *entropy* whenever the speaker wishes to lend universality and depth to an otherwise prosaic statement. The famous prediction of the eventual thermodynamic death of the universe raised some interest at the time but is not now a topic of conversation in pubs. Even in the Deep South, I doubt whether the second law of thermodynamics will raise enough ire to be repealed by a state senate.

Humanitarianism

It is natural in a book of this kind to present the nineteenth century as an era of scientific and technological advance. The wide sweep of history has not been given a hearing, but an unprejudiced eye must surely record that the other great triumph of the century, at least in the West, was the flowering of humanitarianism, typified by Wilberforce's fight against slavery and Salisbury's crusade against child labor. In 1844, Engels published *The Condition of the Working Class,* and whatever one can say about its eventual political fruits in eastern Europe, it was a plea for the downtrodden of the earth, those who were described in words by Dickens and Zola, and in images by Gustave Doré's drawings of the London poor. It would be gratifying to a scientist to link the progress of science with the progress of humanitarianism, and a look at the nineteenth-century globe reveals that very often the map of social injustice superimposes well on the map of scientific backwardness. But the correlation is a little too easy, and I make no attempt to encroach on the territory of the historian or the sociologist.

The nineteenth century saw the rise of Great Britain as a world power, and the British were well aware of their industrial and military strength and of their massive contribution to the advance of science. To give a taste of this national self-confidence, I quote part of a chauvinistic article that appeared in 1897, in the *Illustrated London News.* The underlying tone of the article reflects the materialistic and realistic frame of mind that typified much of European society, and that owed so much to the technological fruits of science. The quotation sums up, at first hand, the prestige of science, the associated belief in progress, and the British sense of world leadership at the end of the nineteenth century. Notice that only one non-British scientist—Siemens—is mentioned by name. Pasteur is apparently relegated to the status of a "Continental descendant of the great Jenner."

> The sum and substance of all the efforts of Victorian science has been to put the body and mind of man more at ease. Thanks to the Stevensons, the Brunels, and their Disciples, to Faraday, Siemens, Kelvin, Rowland Hill, and their followers, space has been almost annihilated, and the ends of the earth brought together. Simpson, Lister, the Continental descendants of the great Jenner, Parks and the pioneer band of English sanitarians have alleviated the sting of disease, and put our bodies more at ease. More than any nationality,

Englishmen have been busy to put our minds at rest by teaching us the nature of the world we live in, and making us less the creatures of chance. The seas have been charted, the depths of space fathomed, the heavens mapped, the earth and its plants and animals examined. Darwin, Lyell, Spencer and Huxley have replaced empirical theories by a history written in rocks, bones and living tissue.

It must surely have been that God spoke English. What the writer of the article did not know was that, as he wrote, yet another Englishman, working in the Cavendish Laboratories in Cambridge, was breaking into the unbreakable atom.

12

The Material Trinity:
The Atom

Colors, sweetness, bitterness, these exist by convention; in truth there are
atoms and the void.

—Democritus

When I was a young man, I also gave in to the notion of a vacuum and atoms;
but reason brought me into the right way.

—Leibniz, letter to Caroline, princess of Wales, 12 May 1716

In 1863, Ernest Solvay, a Belgian industrial chemist, announced a new process for
making washing soda. It made him a fortune, which allowed him to engage in phil-
anthropy and to express offbeat opinions, such as his belief in the evil of inherited
wealth and the undesirability of money in general. Among the projects to which
his own undesirable money went was a series of international meetings, the Solvay
Conferences. Einstein attended the first Solvay Conference on Physics, held in
Brussels in 1911, and was given the honor of being the final speaker. Madame Curie
was also there, but the most significant talks were given by two other participants:
Ernest Rutherford, who announced the discovery of the atomic nucleus, and Jean
Perrin, who gave the most convincing evidence until that time that atoms actually
exist.

For twenty-five centuries, the experimental evidence for the existence of atoms
was not strong enough to convince everyone of their reality. It was only in the early
years of the twenieth century that the distinguished French mathematician Poincaré, after hearing Perrin, could say, "Atoms are no longer a useful fiction. . . . The
atom of the chemist is now a reality."

In recounting the story of the particles that make up the atom, we will gloss over
two and a half millennia of the idea of atomism. All the meaningful advances were
made in the nineteenth and twentieth centuries. Nevertheless, we will go back to
the beginning and give the Greeks their due.

The Greek Guess

It is said that when, around about 1000 B.C., the Ionian Greeks came to Miletus, on
the coast of present-day Turkey, they came without women, ensuring their future
by slaughtering the local men and marrying the widows. The innovative recipe
worked; the community flourished. Up to about 500 B.C., Miletus was the wealthi-
est Greek settlement in the East, and home to an impressive array of philosophers,
including Anaximander, Anaximenes, and Thales, which says much for mixed

marriages. Leucippus, the first person that we know of to speak of atoms, was a citizen of Miletus but later settled in the Thracian town of Abdera, renowned for the stupidity of its citizens. There he took as a student Democritus, of whom Aristotle said, "He thought about everything."

One of the things that Democritus thought about was the possibility that the world was composed of tiny, discrete, indivisible particles—*atomos.* My guess is that he had been talking to Leucippus, but Democritus was a fragmenter by nature, believing that not only matter but both time and space were divided into discontinuous portions. He also, correctly, guessed that the seemingly continuous Milky Way was in fact a discontinuous collection of thousands of stars.

Democritus is usually given credit for being the father of the atomic theory of matter, but he did no experiments and had only the flimsiest evidence for postulating the existence of atoms. Democritus's theory was kept alive by the Roman poet Lucretius (see Chapter 14) in his masterpiece, *De rerum natura,* only one copy of which survived the Dark Ages, to be discovered in 1417.

The idea of atoms was inevitable because it is natural to ask how finely one can go on cutting up matter. Speculation along these lines led to some strange conclusions. Buridan, the fourteenth-century French philosopher, thought that: "One could have such a small quantity of a substance that it does not continue to exist for a noticeable time. This small quantity will continually tend towards its own destruction." This is taking an inferiority complex to absurd limits.

Averroës, the twelfth-century Muslim philosopher and physician, was much nearer the mark: "A line as a line can be divided infinitely. But such a division is impossible if the line is taken as made of earth." He was expressing the feeling that, although endless subdivision is an acceptable concept in mathematics, we cannot go on subdividing matter forever: sooner or later, surely there are irreducible particles. But since no one had seen them, the best that could be done was to venture, with Francis Bacon, that the atom "is a true being, having matter, form, dimension, place, resistance, appetite, motion and emanations: which likewise, amid the destruction of all natural bodies, remains unshaken and eternal."

The Smallness of Atoms

A woman is about 2 million times "taller" than the average *Escherichia coli* bacterium in her hamburger. By way of comparison, Everest is only about 5000 times as tall as the average man. A man is about 10 *billion* times "taller" than an oxygen atom. If the atom were scaled up to the size of a golf ball, then on the same scale a man would stretch from here to the Moon.

Even with the best optical microscope available, no one has the slightest chance of seeing atoms, and a conventional electron microscope would not help us. Even in this century there were scientists who, despite a great deal of circumstantial evidence, questioned the existence of distinct microscopic particles in matter. For these skeptics, atoms were just a convenient hypothesis; matter was a kind of infinitely divisible jelly.

The Atom Revealed

Why do we believe in atoms? Look at Figure 12.1. It is a schematic representation of an image produced by an instrument known as a scanning tunneling microscope, STM for short. It is produced by the surface of a piece of common salt,

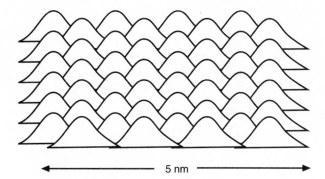

Figure 12.1. A very schematic representation of the surface of a sodium chloride (common salt) crystal as revealed by a scanning tunneling microscope.

5 nm

sodium chloride, magnified about 33 million times. It has a definite structure, appearing to be a very regularly stacked collection of objects. These objects are atoms. Believe me. The first photograph of a single atom was made by A. V. Crewe in 1970, using the STM technique.

Orderly stacking of atoms (or ions) is typical of metals and of all crystals. Who does the stacking? You will get rather different answers from a physicist and a Buddhist priest.

Since microscopes inform us what magnification they are working at, we can use images like that in Figure 12.1 to estimate the size of atoms. They turn out to have diameters ranging between about a tenth to a half of a nanometer.

What more can we say about atoms? Don't expect to turn the page and find an image of the atom's interior, taken through a super-STM or some other expensive toy. If current theory is to be believed, there is no possibility of our *directly* observing the interior of an atom, even with an as-yet-to-be-built supermicroscope. Nature doesn't let us get too close. Nevertheless, we know a very great deal about the structure of atoms.

The tale of the unraveling of the structure of atoms should be awarded the Detective Book of the Century award. The real breakthroughs all came in the late nineteenth and early twentieth centuries. We will go straight to that exciting period, one of the golden ages of science.

The Electron Revealed

When you switch on a light, an electric current flows through the filament or neon tube. This current is a flow of *electrons,* tiny negatively charged particles, and about 3×10^{19} (30 million million million) electrons flow through the fixture every second. What are electrons? The smallest stable components of ordinary matter. They are essential components of all atoms and were discovered in 1897 by J. J. Thomson, in the Cavendish Laboratories of Cambridge University. Thomson was, by all accounts, not a particularly skilled experimentalist, but he was good enough to get a Nobel Prize.

It had been known from the beginning of the eighteenth century that if electrodes were attached to a glass vessel containing air at low pressure, the air glowed. An interesting turn occurred in the investigations of this glow when technical advances made it possible to go to much higher vacuums. A key experiment was done in 1858 by Julius Plücker at the University of Bonn. Figure 12.2 shows the kind of

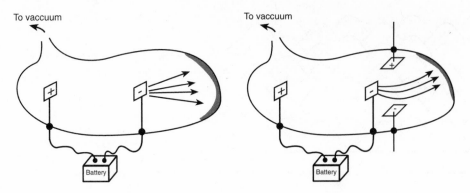

Figure 12.2. Cathode rays (electrons) leave the negative plate and strike the wall of the glass vessel. The position of the resulting glow can be altered by applying an external electric field.

apparatus he used. At moderate vacuums the gas in the bulb glowed, but as pumping continued and the pressure in the vessel dropped, the glow disappeared. Instead, Plücker observed that the glass next to the cathode gave off a green light. There were some puzzling features of this eerie glow. The position of the site on the glass from which it came could be altered by putting a magnet near the vessel. It looked as though something was coming out of the cathode and making its way to the glass. What were these "cathode rays"? Some said they were tiny bits of metal from the cathode. Heinrich Hertz (of kilohertz fame) thought they were radiation of some sort. Thomson solved the puzzle, his crucial experiments being those in which he examined the way in which electric and magnetic fields disturbed the path of the rays.

The path of the cathode rays was deflected by an electric field, which immediately suggested that the rays carried an electric charge. Furthermore, the deflection was away from the negatively charged plate inside the cathode-ray tube (Figure 12.2). This implied that the rays themselves were negatively charged. Thomson assumed that anything carrying a charge must be material, and he set out to find the mass and charge of the particles which he believed to constitute the rays. He did this by measuring the deflection of the rays in electric and magnetic fields of known strength. He realized that all he could find from his particular measurements was the *ratio*, e/m, between the charge and mass of the particles in the cathode rays. This he did, and although he hadn't determined separately either the mass or the charge of his particles, he ventured to propose that they were far smaller than atoms and were a universal component of all atoms. He was right. He had discovered the electron.

"The Saddest Words Are: 'It Might Have Been'"

Hertz thought that cathode rays were radiation and not particulate because when he, like Thomson, had tried to deflect them with an electric field, he had failed. He used too weak a field, and in consequence may have missed out on a Nobel Prize. A sadder case is that of Walter Kaufmann in Berlin, who, working at the same time as Thomson but using a different method, also measured the charge-to-mass ratio of cathode rays. His results were actually nearer the presently accepted value than Thomson's. Kaufmann reported his experiments but made absolutely no speculation about the existence of a subatomic particle. Francis Bacon remarked that people often reject new ideas just because they are new, but in Kaufmann's case there may be another reason: the influence exerted by the physicist and philosopher Ernst Mach. In the year that Thomson discovered the electron, Mach stated that atoms were a "mathematical model for facilitating the mental reproduction of facts"—in other words, a convenient fiction. Mach was a *logical positivist* (see Chapter 37), although this school of philosophy only flourished some twenty years after his death in 1916. The logical positivists were extremely suspicious of hypotheses and of concepts, such as atoms and electrons, that could not be experienced directly through the senses. Kaufmann's respect for Mach's philosophy cost him dearly.

Absolute Values

Thomson attempted to measure the charge of the electron but not very successfully. The man who did succeed, and also earned a Nobel Prize, was Robert A. Millikan, at the time at the University of Chicago. Taking his value for e and Thomson's value for e/m, the mass of the electron turned out to be about 1836 times less than that of the smallest known atom, that of hydrogen. Millikan published his work in 1911, the same year as Rutherford found the nucleus.

The Nucleus Revealed

Ernest Rutherford, First Baron Rutherford of Nelson and Cambridge, was a bluff, rugby-playing New Zealander, the kind of man one would expect to be suspicious of complex theories and who indeed expressed his wariness of theory publicly and bluntly. He was an experimental genius. The boy from South Island was awarded the Nobel Prize in 1908, knighted in 1914, elected president of the Royal Society in 1925, and made a baron in 1931. At the height of his success he declared his work to be of no practical value. In this, at least, he proved to be wrong. Our understanding of the nucleus and the atom are the foundation on which stand modern chemistry and much of physics.

Rutherford discovered the nucleus, and the emptiness of the atom. His experiment was not only one of the crucial experiments in our understanding of matter but also a classic example of simple, logical scientific thinking.

Rutherford's Experiment

Suppose you fired a revolver at a wooden fence, one of those continuous fences made of a layer of adjacent slats of wood. You would expect the bullets to pass straight through the fence, a few perhaps deviating very slightly from a straight path because of knots in the wood. Now imagine a set of iron railings, like those around Buckingham Palace, and suppose that a thin sheet of plastic has been fixed to them, to keep the wind off the queen's guards. Again fire your pistol. The great majority of your bullets will pass through the plastic, traveling in a straight line. However, a small proportion of bullets will hit one of the railings and will be deflected. Depending on exactly where the bullet strikes a railing, it will be deflected by a small or large angle. If you are really unlucky, a bullet might hit a railing straight on and bounce back to you. The difference between the two fences is that in the wooden fence, the material of the fence is more or less evenly spread out, while in the iron railings-plus-plastic-sheet the mass is very unevenly distributed.

A large proportion of the mass is in the narrow confines of the iron railings. A bullet hitting the plastic is hardly perturbed by the small amount of mass that it meets. A bullet hitting a railing encounters a massive object that can easily deflect it. This was the conceptual basis of Rutherford's experiments, carried out at the University of Manchester from 1909 to 1911.

In 1896 the Frenchman Becquerel discovered that uranium salts gave off invisible rays. Marie and Pierre Curie called this phenomenon *radioactivity*. The three French scientists shared the Nobel Prize in 1903. During a stay at the Cavendish Laboratory from 1895 to 1898, Ernest Rutherford discovered that certain radioactive substances violently ejected positively charged, relatively heavy particles, which he named *alpha particles*. It occurred to him that he could learn something about these particles, and about matter in general, if he used them as projectiles, examining what happened to them and their targets. In 1898 he moved to McGill University in Canada, and among the research projects that he started was a study of the path of α-particles (bullets) after they struck a piece of gold foil (a fence) about 0.00004 cm thick, so thin as to be semitransparent. When he came to Manchester in 1907, Rutherford asked one of his students, Hans Geiger (of Geiger counter fame), to continue this work. Geiger found that most of the α-particles that originated in a sample of radium went straight through the foil, but that a few were slightly deflected. Rutherford asked another student, Ernest Marsden, to find out if any particles were scattered through large angles. Marsden succeeded: about 1 in every 20,000 particles was deflected by 90° or more from the foil. This was totally unexpected because the gold foil was assumed to be a wooden fence, not a set of iron railings. The model of the atom currently in favor, which had been proposed by J. J. Thomson in 1908, was what could be called the Christmas pudding model. If, reasoned Thomson, electrons were components of all atoms, then, since matter is electrically neutral, there must be positively charged matter in all atoms. He supposed this positively charged matter to be *evenly* distributed in a ball, and the negatively charged particles, the electrons, were presumed to be scattered throughout this ball like raisins (Figure 12.3a).

Rutherford chewed over Marsden's results and concluded that, contrary to Thomson's model, the atoms in the gold foil had an extremely *uneven* distribution of mass (Figure 12.3b). The foil resembled a set of iron railings, not a fence. Nearly all the mass of each atom must be concentrated in a very small volume while the rest was spread out in a diffuse cloud. That is why the average α-particle "bullet" met little resistance on its way through the atom. The puny electrons were brushed aside by the massive α-particles, which weighed well over 7000 times as much as they did. Only those α-particles whose paths took them close to, or directly toward,

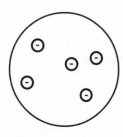

Figure 12.3a. J. J. Thomson's raisins-in-a-pudding model of the atom. The raisins are negatively, and the pudding positively charged.

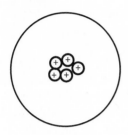

Figure 12.3b. Ernest Rutherford's model of the nuclear atom. The positively charged nucleus sits in a negatively charged swarm of electrons. In the figure the size of the nucleus is grossly exaggerated.

the tiny but massive central particles—the nuclei—were strongly deflected. The cause of the deflection was the enormous electrostatic repulsion between the positively charged α-particle and the positively charged nucleus when they were very close to each other.

In a historic paper entitled "The Scattering of Alpha and Beta Particles by Matter and the Structure of the Atom," Rutherford briefly described his experiments at a meeting of the Manchester Literary and Philosophical Society in 1911, and a few months later he addressed the Solvay Conference. He had discovered the nucleus. There have been few more fateful findings in science.

The discovery of the nucleus of the atom was announced at the same conference at which the French physicist Jean Perrin presented the strongest circumstantial evidence, until then, for the reality of atoms. You can find what Perrin did in most freshman physics textbooks.

Atoms Are Overwhelmingly Composed of Empty Space

Rutherford allows us to make this statement. Consider an atom of tungsten. Almost all of the mass of the atom is concentrated in a tiny central piece of matter, the nucleus. Around the nucleus buzz seventy-four bits of matter, all identical. These are electrons. You can think of the electrons as effectively forming a spherical cloud around the nucleus. The diameter of the atomic sphere is about 100,000 times the diameter of the central nucleus. Imagine a sphere with a diameter of 5 miles, representing an atom. The nucleus on the same scale would be the size of a tennis ball. The electrons don't do much to fill up the lonely spaces; they are individually smaller than the nucleus, and they are also very much lighter. You can see why Rutherford's α-particles so rarely hit a nucleus and generally went straight through the gold foil. The nucleus of a tungsten atom weighs over a third of a million times as much as a single electron. Thus nearly 99.8% of the mass of a tungsten atom is located in the tiny nucleus. This figure is typical of atoms in general.

The sheer concentration of mass in atomic nuclei is awesome. A teaspoon of *closely packed* nuclei would weigh about 500 million tons, which is more than the weight of the whole population of the world.

It is extremely hard to compress a lump of iron. Before Rutherford, we would have said that in such a lump the atoms are touching each other, like a pile of *solid* cannon balls, and that's why it is extremely difficult to squeeze a piece of any solid into a smaller space. And yet Rutherford showed that atoms are almost ghostlike.

The conclusion is that the pitifully thin cloud of electrons that surrounds an atom effectively "fills" the seemingly empty space around the nucleus, *as far as the intrusion of other atoms is concerned,* a little like the air forces of neighboring countries *effectively* filling their national air spaces. Recall Hamlet: "O! that this too too solid flesh would melt." It isn't all that solid; your flesh is well over 99% empty space. And so is every other object you look at.

What Are Nuclei Made Of?

Matter is almost always electrically neutral, but if electrons are negatively charged, then, as Thomson stated, atoms must have some other components carrying a positive charge equal in magnitude to that carried by the electrons. Rutherford had shown that these components are in, or constitute, the nucleus. The smallest atom known in the early 1900s, and today, is the hydrogen atom. A century of chemical research indicated that almost all other atoms had weights that were very nearly whole-number multiples of the weight of the hydrogen atom. In 1815 the English chemist Prout suggested that all atoms were constructed from hydrogen atoms. Thus the sodium atom, which weighs about twenty-three times as much as the hydrogen atom, was supposed to be built from twenty-three hydrogen atoms. In those days there was no way to test this hypothesis. The next step in the story belongs to two young men, one of whom only had time to make a single critical contribution to science.

In 1913 the Danish theorist Niels Bohr, aged twenty-eight, published a theory that, among other things, dealt with the frequencies of the light that could be produced by atoms. He predicted that these frequencies would depend on the magnitude of the positive charge on the nucleus of Rutherford's atom, so that experimental measurements of these frequencies could lead to a determination of the positive charge on any nucleus. The young English physicist Henry Moseley chose to examine the frequencies of the X-rays that are given off by any element when it is bombarded by a suitable beam of electrons. Bohr's theoretical relationship exactly fitted Moseley's results, and Moseley was thus able, in 1913, to give a list of the electric charge on the nucleus of the ten metals that he had examined. He noted two striking facts:

1. All the nuclear charges were very close to being whole number multiples of the charge on the electron. Thus, in units of the charge on the electron, the nuclear charge on the calcium nucleus was 20.00; on vanadium, 22.96, and so on. This is explicable if we accept that atoms are electrically neutral. If an atom has a whole number of electrons, the total charge on the nucleus must be a whole number of electron charges.
2. The order of increasing nuclear charge was, with one exception, identical with the order of increasing atomic weight.

Start with the lightest atom, hydrogen, and give its nucleus a charge of one, in units of the electronic charge. The next heaviest element is helium. Its nucleus has a charge of two units. After helium comes lithium; it has a nuclear charge of three units. And so on. Moseley had found a way to answer the question typified by: "How do you know there isn't an as yet undiscovered element with an atomic weight between those of hydrogen and helium?" There can't be because its nucleus would have to have a charge between one and two. We can count the elements.

Today there is a known chemical element for all the nuclear charges going from 1 to 110 (see Figure 13.1). There can be no other elements in this range, because an atom must have a whole number of nuclear charges. Any newly discovered, or produced, elements must have a nuclear charge of 111 upward. As Frederic Soddy put it, Moseley has enabled us to call "the roll of the elements." The numerical value of the nuclear charge is called the *atomic number*, since it is the ordinal number that we give to an element when all the elements are lined up on parade, arranged according to increasing nuclear charge.

In 1915 a British, Australian, and New Zealand expeditionary force, with the strong approval of Winston Churchill, landed on the Turkish coast, not far from Constantinople, in an attempt to unbalance the German high command. Gallipoli became a synonym for bloody military failure. Among the more than 8 million who fell in World War I, well over 200,000 fell at Gallipoli, among them signals officer Henry Moseley, aged twenty-seven.

Prout had proposed building up all atoms from amalgamations of different numbers of hydrogen atoms. Since the hydrogen atom had a single electron, it was natural to suppose that its nucleus, which had one positive charge, was also a single particle. This particle later became known as the *proton*. Now, following Prout, we might construct the helium nucleus out of four hydrogen atoms. Why four? Because the helium atom weighs close to four times what the hydrogen atom weighs. But this presents a nasty problem. Four hydrogen atoms would have the right weight, but they would have a total of four protons, which means *four* positive charges in the nucleus. This doesn't agree with experiment, which gives the helium nucleus *two* positive charges, that is, an atomic number of two. To get around this difficulty, you could suppose that there were also two electrons in the helium nucleus. This would hardly affect the weight, and the two negative charges would reduce the total positive charge to two, the right number (Figure 12.4).

That is how the atom was seen for about twenty years: as a tiny, massive nucleus containing protons and electrons, surrounded by other electrons. As often happens in science, the theory had a flaw.

The Material Trinity Is Completed

The dazzling complexity of the material world can, for almost all purposes, be reduced to a simple trinity: the proton, the electron, and the *neutron*.[1] The neutron, a component of the nucleus of every atom except that of hydrogen, was the last of the trinity to be discovered, in 1932. Had they all been a little younger, the scientists who uncovered the neutron might have met on the battlefields of World War II.

In 1930 the German scientist Walter Bothe bombarded the metal beryllium with α-particles. He detected what he thought to be an as yet unidentified form of radiation emerging from the metal, which sums up his part in the overall story. During

Figure 12.4. Two different models of the nucleus of the helium atom. Historically the first (and wrong) model was that on the right.

---

[1]Descartes, entirely without experimental or theoretical justification suggested that there were three "elements," basic forms of matter that differed in their size and motion.

World War II, Bothe worked on the Nazi's nuclear energy project. At the same time, in Paris, another participant in our story was doing work for the French Resistance.

In 1925, at the age of twenty-five, Frédéric Joliot began work as an assistant to Marie Curie at the Radium Institute in Paris. He was initially instructed by Madame Curie's daughter, Irène Curie. Frédéric detected in her an "extraordinarily sensitive and poetic creature," and they were married in 1926. From 1928 they signed themselves Irène and Frédéric Joliot-Curie. They were both eventually to be Nobel laureates, like Irene's parents, Pierre and Marie Curie. The Joliot-Curies read of Bothe's experiments and were curious about the nature of the novel radiation that he reported. They repeated his experiments and allowed the "radiation" from the bombarded beryllium to fall on materials having a high proportion of hydrogen atoms, for example, paraffin wax, in which over 50% of the atoms are hydrogen. They found that protons (hydrogen nuclei) were ejected from the target, but at such high speeds it was difficult to understand how they could have been given such velocities by radiation falling on them. They published their results on 18 January 1932.

Back in England, James Chadwick, a student of Rutherford's, read the Joliot-Curies' paper and tried out the effect of the mysterious radiation on a variety of targets. In contrast to those who came before him, he assumed that the so-called radiation was in fact a stream of *particles* and that the velocities of the atoms ejected from the targets could be explained by using Newton's laws of motion for two colliding bodies. Using the known masses and the measured velocities of the ejected nuclei, he deduced the mass of the enigmatic bombarding particles.

He concluded that the new particle had about the same mass as the proton. Furthermore, it was known that the "rays" were far more penetrating than beams of either electrons or protons, and it was therefore almost certain that the new particle was electrically neutral. When either protons or electrons are fired at matter, their paths can be affected by the electric charges on the nuclei and electrons of the atoms through which they attempt to pass. The uncharged neutron is blind to these charges and therefore penetrates matter relatively easily, even passing very close to nuclei without being disturbed.

Chadwick had identified the neutron. He published his results on 27 February 1932, a little more than a month after the Joliot-Curies' paper had come out. Again the discovery was made in the Cavendish Laboratories, and Rutherford and Thomson were still on the staff. Chadwick held that the neutron was nothing more than an electron and a proton going around together, but the neutron was slowly accepted as an independent particle in its own right during the 1930s. The basic components of the atom, and indeed its very existence, were thus established, after twenty-five centuries, in the thirty-five years between 1897 and 1932.

The neutron allowed a simple picture of the nucleus. There was no longer any need for electrons in the nucleus. The helium nucleus was not composed of four protons and two electrons but of two protons and two neutrons (see Figure 12.4). The mass comes out right, as does the charge. The atom had been completed. A decade later, James Chadwick headed the British group participating in the Manhattan Project at Los Olomos.

In 1942, Irène Joliot-Curie fled to Switzerland with her children. Frédéric Joliot-Curie joined the French Communist Party and organized the production of explosives for the Resistance. After the war he was awarded the croix de guerre. In 1939 the Joliot-Curies, worried by the rise of Nazism, had deposited a sealed letter with the Académie de Sciences. It contained the principle of the nuclear reactor and was not opened until 1949.

We can now summarize what could be called the British picture of the atom:

> Atoms are all composed of three kinds of particles. Except for the hydrogen atom. The nucleus contains two kinds of particles: protons and neutrons, except for the nucleus of hydrogen, which consists of a single proton. The proton and the neutron have nearly the same mass, but the proton is positively charged, while the neutron is electrically neutral. The mass of the proton is 1.67265×10^{-27} kg; of the neutron, 1.67495×10^{-27} kg.[2] Around the nucleus buzz the third kind of particle, the electrons. The electron has a mass of 8.5473×10^{-31} kg and carries a negative charge of 1.60219×10^{-19} coulombs, exactly balancing the positive charge of the proton. The number of electrons in an atom is equal to the number of protons, so that the atom has no net electrical charge. The diameter of an atom is about 100,000 times the diameter of its nucleus. The nucleus accounts for over 99% of the total mass of the atom.

Figure 12.5 depicts the nuclei of a few elements.

The nucleus and the electron are held together by electrostatic forces. To get some idea of their enormous power, imagine that you have taken a tablespoon of sugar and separated out the electrons from the nuclei. Now keep the nuclei in a box on Earth and take the electrons to the Sun, 93 million miles away. If the electrons are held in the Sun by Superman, then about 1 million people would have to stand on the box of nuclei on Earth to prevent it from moving toward the Sun. You can now understand why most macroscopic lumps of matter in the universe are effectively electrically neutral. Any attempt to remove a significant proportion of the charge from a neutral piece of matter is met by a stupendous force of attraction pulling the charge back.

The power of electrostatic forces poses a problem concerning the stability of the nucleus. We have said that any nucleus is made of protons and neutrons. Now the neutrons have no charge, but the protons are positively charged. How does a collection of protons keep together? Surely they would repel each other with such force that the only stable nucleus would be that of hydrogen, which consists of a single proton. There is a hint here that there are forces other than those that we have dealt with. There are, and we will meet them later on.

Hydrogen

Deuterium

Tritium

Lithium

12.5. The composition of the nuclei of some elements. Protons and neutrons are indicated by the letters p and n.

---

[2]If you want to get some idea of the smallness of these numbers, multiply them by 10^{27} (a billion billion billion) and you see that a billion billion billion protons, for example, have a mass of only 1.67265 kilograms.

Unstable Nuclei

A large number of nuclei are unstable. This is true of all the nuclei above atomic number 83 (bismuth), but it is also true of some lighter nuclei. Of the approximately 350 naturally occurring isotopes, about 50 are radioactive. Instability is manifested by the breakup of the nucleus and the violent ejection of particles or radiation. The particles are either the α-particle, which is simply a helium nucleus, consisting of two protons and two neutrons, or the β-ray, which is an electron. The radiation is γ-rays. The type of particles or rays ejected depends on the particular nucleus. Radioactive breakup gives a new nucleus, which in some cases is stable but in others may itself be radioactive. The same applies when this daughter nucleus breaks up, so that there are well-known cases of chains of radioactive breakup, until finally a stable nucleus is formed. The moment of breakup of a specific atom cannot be predicted, but for a very large collection of atoms we can say that half of them will disintegrate in a certain time. This *half-life* can be anything from a tiny fraction of a second to thousands of years, depending on the nucleus. We can do nothing to alter half-lives, which are immune to changes in temperature or pressure or to chemical reactions.

α-Particles, as befits such massive, charged particles, cause considerable damage when they strike matter, but they are also stopped after a short distance because they interact so violently with their surroundings. β-particles (flying electrons) are far more penetrating, being, for example, able to penetrate deep into the human body. γ-rays can easily pass through your body, and they can knock electrons out of atoms that stand in their way. This is the origin of their harmful medical effects, and the rationale behind their use in destroying tumors.

The classic destructive use of radioactivity is the atomic bomb, which is based on the finding, by Otto Hahn (1879–1968) and Fritz Strassmann (1902–1980) that when neutrons penetrate a heavy nucleus, the resulting nucleus can disintegrate to give more than one neutron, a process that Frisch termed *fission* (Figure 12.6). If this happens in a sample of suitable material, say uranium 235, each nucleus that disintegrates can be responsible for the disintegration of two (or more) other nuclei, which can result in the disintegration of four nuclei, which can end in a mushroom cloud. This is because the process of fission releases very large amounts of energy. It is a curiosity that the most destructive reaction in history might not have been discovered when it was, but for two women. Ida Noddack heard about some experiments in which Enrico Fermi had bombarded uranium with neutrons and had concluded that he had produced heavier nuclei. Noddack thought otherwise, correctly hypothesizing that the uranium nucleus had split in two. No one took any notice. When Hahn and Strassmann performed their experiments, it was the

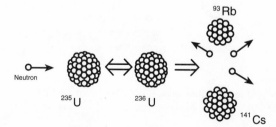

Figure 12.6. The neutron is absorbed by the uranium nucleus, and fission follows. Sometimes U-236 returns to U-235 and a neutron.

physicist Lise Meitner (1878–1968), a refugee from Hitler, and her nephew Otto Frisch, who realized exactly what the implications were.

Notice that, without being explicit about it we have been talking about the realization of the centuries-long dream of the alchemists—to transmute one element into another. Each chemical element is defined by the number of protons in the nucleus of its atoms, the atomic number. This number can change as a result of radioactivite decay, as Rutherford and Soddy were the first to state. Later on we will see that man can synthesize nuclei with higher atomic numbers than any found in nature, thus building previously unknown elements.

A Question of Priorities

I guess I'm just an old mad scientist at bottom. Give me an underground laboratory, half a dozen atom smashers, and a beautiful girl in a diaphanous veil waiting to be changed into a chimpanzee, and I care not who writes the nation's laws.

—S. J. Perelman

There will be readers who will protest that there are more than three elementary particles. What about the science section of the *New York Times*, which almost monthly reports successful or unsuccessful experiments costing millions of dollars, and revealing or attempting to reveal a new particle to be added to a long list of exotic elementary particles—quarks, leptons, muons, baryons, omega particles, neutrinos, positrons, gluons, and so forth? Those of these particles that can be made to appear in the laboratory are usually induced to do so by bombarding either nuclei or simple particles with a variety of projectiles, including protons. They generally "live" for less than a millionth of a second before being annihilated by collision with matter or by disintegration. A physicist will tell you that the motley crew of subnuclear particles partially listed here hold the deepest secrets of the universe. This may be true, but to understand the workings of the everyday world these particles are irrelevant.[3] To build a house, to understand the working of the liver, to design a new pharmaceutical on a computer, to launch a spacecraft, to design a fiber-optic information network or an artificial heart valve, to investigate the mode of action of a virus, you need know nothing about the nuclear physicist's particles. There is hardly any human activity, except writing articles-on-particles, for which one needs to know about the existence of any other particles but protons, electrons, and neutrons.

If we had never discovered the nuclear physicist's exotic particles, life as we know it on this planet would be essentially the same as it is today. Most of the particles resemble ecstatic happiness: they are very short-lived and have nothing to do with everyday life. Somewhat unfairly, it could be suggested that the motto of those who work on these particles be the smug statement of the English mathematician G. H. Hardy: "I have never done anything 'useful.' No discovery of mine has made, or is likely to make, directly or indirectly, for good or ill, the least difference to the amenity of the world."

---

[3]Except for positron emission tomography (PET), a method of scanning the cells of the body using positrons.

Research in particle physics is expensive. It costs hundreds of millions of dollars or more to construct the installations necessary to produce a new particle having a lifetime measured in millionths of a second or less. I quote the *New York Times*: "Big Science Squeezes Small-Scale Researchers" (29 Dec. 1992). The article below the headline discusses the plight of small-scale research projects that have to battle for government funds against the demands of particle-seekers. In the *New York Times* of 5 January 1993, a headline referred to a report published by a group of 315(!) physicists working with equipment costing about a quarter of a billion dollars: "315 Physicists Report Failure in Search for Supersymmetry." This kind of thing is red meat for anti-intellectuals, but it is no wonder that in some informed circles, both scientific and nonscientific, there is questioning of the justification for the high cost of searching for new particles in a world in which the budgets for research into ecological, medical, and other "relevant" problems are often insufficient. In order to accelerate protons to huge velocities, so that they can smash into each other and (hopefully!) produce what some believe is the final undiscovered particle, a vast installation, the Superconducting Supercollider (SCSC) was planned to be built in Texas. Cost? About $11 billion. The particle, known as the *Higgs boson*, will almost certainly have no practical applications. The U.S. Congress decided to stop funding the SCSC project. One congressman said, "It's good science, it's simply not affordable science."

HMS may suspect that among the motives for these gigantic ventures are national pride and scientific empire-building. He should also realize that there is a real fascination about particle physics, a fascination that I hope I can convey in Chapter 31.

"You've Come a Long Way, Baby"

Both Marie Curie and her daughter Irène were Nobel Prize winners. Marie's work on radioactivity was a central part of the story of the atom. Irène and her husband, Frédéric, discovered artificial radioactivity, the production of radioactive elements that do not appear in nature. Neither Marie nor Irène was a member of the French Académie de Science, despite their Nobel Prizes. They were women. Irène appeared twice before the Académie in an effort to convince it that women should be permitted as members. She failed. The British Royal Society has an equally proud record; it only legally allowed women members in 1923.

13

The Stuff of Existence

Human senses cannot penetrate to the insensible particles of bodies, and by
observing their Figure, Size, Connexion, and Motion, tell why rhubarb
purged, hemlock killed, and opium made man sleep.

—John Locke

We may not know exactly why rhubarb purges, but during the twentieth century
we have opened up the internal world of atoms and molecules. We can build molecules that nature never dreamed of, molecules tailor-made to meet our needs. The
evolution of alchemy into science accelerated dramatically following the discovery
of the basic structure of the atom and the incursion of quantum mechanics into
chemistry, a process initiated primarily in the 1920s and 1930s by the American
scientists Gilbert Newton Lewis and Linus Pauling. In this chapter we will briefly
outline the nature of the everyday material universe: the chemical elements and
the molecules that can be built from them.

Man's prehistoric fascination with naturally occurring materials, and their unexpected transformations, gave birth to *alchemy*, a 2000-year search for artificial
gold and for the panacea that would cure all ills, the philosopher's stone. We can
understand why the alchemists spoke of materials in the language of the occult. It
must have seemed magical to early man that one could take a heap of rocks and
conjure a shining, tough material from them, a material from which one could
make sharp, durable tools and weapons. Completely inexplicable changes of this
sort gave credence to the belief that substances could be transmuted; if rocks gave
iron and ice gave water, why (and this was the great dream), why couldn't lead give
gold? Alchemy, from the beginning, had both a practical and a mystical side, and
the straightforward descriptive entries in the notebooks of the alchemists were
mingled with a running commentary devoted to the signs of the zodiac and other
astrological or magical matters. The psychologist Jung saw alchemy as a sublimat-
ed search for spiritual wholeness, but though some of the alchemists were mystics,
many were involved as much in a material, as a spiritual, gold rush.

To the alchemist, nature seemed to exist in endless variety, but, as we have seen,
the instinctive search for simplicity led to early attempts to construct theories that
would account for the behavior of matter in terms of a small number of so-called *elements*. The best known of these efforts is the ancient Greek belief that earth, air,
fire, and water were the basic building blocks of the universe, but the Chinese had a
very similar tradition of five elements or "powers": earth, wood, metal, fire, and
water. Even while the sterile doctrine of the four Elements was being taught in European universities, the alchemists were building up a body of experimental
knowledge concerning the way that different substances reacted to each other, to
heat, and to corrosive solutions. Out of this cumulative and disjointed body of

practical experience grew a trail that led through Boyle and Lavoisier to Dalton and Rutherford and the modern concept of elements. It was this concept that allowed a rational explanation of the material world.

In the nineteenth century chemists finally acquired sufficient know-how to begin to confidently analyze material into its elementary components, and to start building molecules unknown in nature. This expertise was the basis for one of the first realizations of Bacon's dream of the application of science to the benefit of man. It was the nineteenth century chemical industry that gave the poor man colored fabrics.

During the nineteenth and twentieth centuries, it became clear that all matter, in its splendid variety, is constructed from eighty-three different chemical elements—nature's building blocks. To these naturally occurring elements, man has, up to the time of writing, added another twenty-seven new elements created in the laboratory.[1]

The Chemical Elements: Roll Call

All normal matter is made of protons, neutrons, and electrons, but the common currency of research, the basic units of matter consists of the elements, at least as far as the chemist, biochemist, geologist, metallurgist, physiologist, molecular biologist, medical man, and so on are concerned.

The first elements with which man played were carbon, from burned material, and a handful of metals, such as gold, which are loathe to combine with oxygen of the air and produce powdery oxides—which is exactly what iron does when it rusts. The economy of the Athenian polis, the birthplace of democracy, depended on slaves and on a shiny metallic element extracted from the mines near the city; silver was one of the earliest elements isolated by man. A more useful element, known to prehistoric man, was copper, which was employed in the manufacture of weapons and tools, especially when alloyed with tin—hence the Bronze Age. The Iron Age seems to have been largely based on iron that came from meteorites, since analysis of objects from that era shows the presence of relatively high percentages of nickel, typical of meteoric iron. Many iron ores are just dirty rust, which is basically iron in combination with oxygen. At some time in prehistory someone must have heated an iron ore in a wood fire. Luckily, at high temperatures, carbon has a greater affinity for oxygen than does iron, and in the heat of a fire it takes oxygen to its bosom, turns into carbon monoxide and dioxide, and leaves metallic iron behind. This simple chemical reaction was to turn England into the leading world power in the nineteenth century.

A dozen elements were known long before Christ was born (although the modern concept of an element was unknown): carbon, sulfur, copper, silver, zinc, tin, lead, gold, mercury, arsenic, antimony, and iron. All these species are elements because they are single substances, each composed of one single type of atom. For nearly 3000 years, from about 1000 B.C. to A.D. 1735, only three more elements were isolated: phosphorus, platinum, and bismuth. In the eighteenth century another sixteen were discovered, in the nineteenth century another fifty, of which

---

[1]The new elements are invariably radioactive and very short-lived. Their nuclei break up to give those of atoms of already known.

Humphry Davy found six important elements: boron, sodium, potassium, calcium, barium, and cadmium. Thus at the dawn of the twentieth century, all but two of the naturally occurring elements had been discovered. All the man-made elements were made in the twentieth century, most of them at the University of California, at Berkeley, by teams of scientists led by Segrè and Ghiorso. There are no elements that have been detected in the stars that do not exist on Earth.

Order Emerges

The 110 elements known at the time of writing are shown in Figure 13.1, in the periodic table of the elements. The table contains the name of each element, the conventional symbols used to denote the element, and the atomic number (the number of protons in the nucleus), conventionally symbolized by the letter Z. At the time of writing, some of the man-made elements have not yet been named.

The number of protons in the nucleus, Z, determines the chemical identity of the atom. Z gives the number of positive charges that there are on the nucleus. For an electrically neutral atom this must equal the number of negative charges, which can only be supplied by the electrons. In neutral atoms the number of protons in the nucleus thus determines the number of electrons.

When atoms meet each other there is a collision of electronic clouds. Sometimes the collision results in no change in the atoms. In such cases, if the collision is in a gas, the atoms just go their separate ways. This is what happens in a cylinder of helium gas, for example. In some cases, however, atomic or molecular encounters result in a drastic rearrangement of the electrons of the colliding species, and then we speak of chemical change. Whether a reaction occurs or not is almost entirely in the hands of the electrons; the natures of the nuclei, those tiny specks that are so far

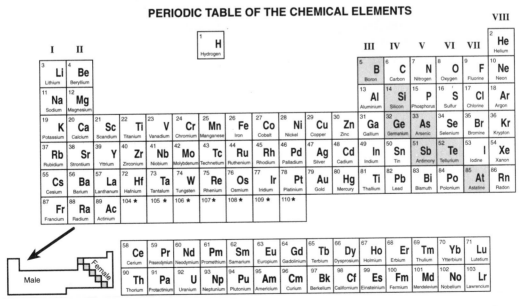

Figure 13.1. The periodic table of the chemical elements. The number in each square gives the atomic number of the element—the number of protons in the nucleus. The name and conventional symbol of each element is given. The elements marked with an asterisk have been discovered but not officially named. For the significance of the sexual classification see the text.

away from the borders of the atom, are not too relevant to the rearrangement of electronic clouds. *The chemical behavior of atoms is completely dominated by their electrons.* So true is this that the nucleus can be altered and the atom retains its chemical characteristics, provided the number of electrons remains the same. An example will illustrate the principle.

The hydrogen atom's nucleus consists of a single particle, one proton; the hydrogen atom has one electron. Some hydrogen atoms found in nature have nuclei consisting of two particles: one proton and one neutron. These atoms are given a special name, *deuterium*, popularly known as heavy hydrogen. Why "hydrogen"? Because the deuterium atom still needs only one electron to electrically balance the single proton, and therefore both "normal" hydrogen atoms and deuterium atoms both have one electron. It is this electron that determines almost completely the chemical behavior of the two types of atom. So similar is this behavior that we are entitled to call them both hydrogen.[2]

All known elements exist in two or more nuclear variations, all having the same number of protons and therefore the same number of electrons and the same chemical properties. These variations on a basic atomic theme are termed *isotopes*. They differ only in the number of neutrons in the nucleus. Thus hydrogen has three isotopes (Figure 12.5), namely, hydrogen (nucleus: one proton), deuterium (nucleus: one proton plus one neutron), and tritium (nucleus: one proton plus two neutrons). About 0.015% of the atoms in a sample of hydrogen obtained from natural sources are deuterium atoms. Tritium, which was first detected in 1950, also illustrates the fact that while the *chemical* properties of different isotopes are (almost) identical,[3] their nuclear properties can differ entirely: tritium is weakly radioactive, giving off β-rays (electrons). About one in 10^{18} hydrogen atoms is a tritium atom; at least that was the number before thermonuclear weapon testing began in March 1954. It then went up about a hundredfold but is now dropping back to its previous value. A historically significant isotope is uranium 235, which accounts for about 0.72% of the naturally occurring uranium atoms (uranium 238 accounts for most of the rest) and has been the basis of nuclear reactions producing huge amounts of energy.

The fact that the chemistry of an element depends on the number of protons in its nucleus suggests the possibility of transforming one element to another by pushing more protons into its nucleus, or knocking some out. This alchemy was first observed by chance by Rutherford in 1919 when he found that he could knock a proton out of the nitrogen nucleus by bombarding nitrogen with the nucleus of the helium atom (the α-particle). The first deliberate transformation was performed by Cockroft and Walton in 1932, at the Cavendish Laboratory. A beam of speeding protons smashed into a lithium (three protons in the nucleus) target and knocked out protons, producing helium atoms, which have two protons in the nucleus. Reactions resulting in nuclear change occur in nuclear weapons and in the stars.

---

[2]The behavior of hydrogen and deuterium can be very different, if their *nuclei* are involved. An important practical example of this is the use of heavy water in some nuclear reactors, based on the fact that deuterium nuclei absorb neutrons far better than the protons in normal water.

[3]The chemical properties do in fact vary slightly between different isotopes of the same element, a difference mainly manifested by tiny differences in the strength of the bonds to other atoms. Thus deuterium is a little more strongly bonded to other atoms than is normal hydrogen.

The Periodic Table

The periodic table of the elements (Figure 13.1) is constructed by arranging the elements in order of increasing atomic number, which means increasing number of protons in the nucleus, and therefore an increasing number of electrons in the electrically neutral atom. It is clear that the atoms of different elements have different weights.

Similarity in the properties of certain elements makes a periodic arrangement convenient, and was the spur for the first attempts to arrange the elements in some kind of order. Thus the elements helium, neon, argon, and so on, in the far right hand column, labeled VIII, are all gases occurring in nature as single atoms, which hardly react with other substances. This is why they are sometimes called the noble gases. All the elements in the column labeled I are soft, silvery metals that react vigorously with water and combine rapidly with the oxygen in the air.

A number of chemists, the first apparently being de Morveau in 1772, played around with the arrangement of the elements, but the present form of the periodic table is primarily the result of the work of the Russian Dimitri Mendeleev. Mendeleev felt himself to be a disciple of Newton. He wrote, in his *Principles of Chemistry* (1869) that it was necessary to harmonize chemical theories "with the immortal principles of Newtonian natural philosophy, and so hasten the advent of true chemical mechanics." In his 1889 lecture at the Royal Institution in London he tried, unconvincingly, to apply Newton's third law of motion to the behavior of atoms in molecules. Mendeleev arranged the elements in the order of their relative atomic weights, but this is pretty much the same order as that based on atomic numbers, the number of protons in the nucleus.

Mendeleev saw that if the elements that were known at his time were arranged in order of increasing atomic weight, then certain properties recurred at regular intervals: "Elements placed according to the value of their atomic weights present a clear periodicity of properties." Thus, in terms of atomic numbers, atoms 9, 17, 35, and 53 were all corrosive gases that interacted violently with metals and most other substances. He therefore arranged them beneath each other—see the column labeled VII. These elements are known as the halogens. (Element 85, astatine, is a man-made element, first produced in 1940.) He did likewise for all the other known elements. At the end he found that there were spaces in his table, which is not surprising since some of the elements had not yet been discovered. It was rather like laying out an incomplete pack of cards into four suits; you would know exactly where to leave holes. Most chemists were very skeptical about his table, which is difficult for us to understand because it looks so convincing. Once again, a new idea met sticky brains.[4] What eventually won everybody over was that when three of the missing elements (gallium, germanium, and scandium) were found, they had properties corresponding almost exactly to those predicted by Mendeleev, including their atomic weights. What Mendeleev had done was to look at the properties of the known elements above and below, and to left and to right of the "holes," and predict the unknown properties by taking a rough average of the properties of the hole's neighbors.

---

[4]Mendeleev himself was wary of new discoveries; he refused to believe in J. J. Thomson's discovery of the electron.

When Mendeleev was buried in St. Petersburg in 1907, his students carried a chart showing the periodic table in the funeral procession. Two decades after his death, quantum mechanics provided a very satisfying theory of atomic structure, in which the way the electrons are arranged about the nucleus inevitably leads to a periodicity in chemical properties.

Atoms must have a definite number of protons, and we know all the atoms from atomic number 1 to atomic number 110. The heaviest are all extremely unstable, and it is becoming harder and harder to create new nuclei. The largest nuclei are so unstable that they fall to pieces almost before we can look at them. The genesis of the elements will be discussed later, but there have been some funny theories in this field.[5]

The heaviest elements that we know, those above uranium (atomic number 93), have all been made by man. The method is basically to bombard known elements with a variety of nuclear projectiles and hope that all or part of the projectile will stick to the target nuclei long enough for us to separate, identify, and characterize a new element. A typically brutal synthesis was the bombardment of curium (Z = 96) with a massive projectile, the carbon nucleus (Z = 6), to give the new element nobelium (Z = 102).

Of the known chemical elements, about 80% are metals, but if you take a spoonful of earth you'll find that, on the average, it contains about 46% by weight of oxygen and about 28% of silicon. The dominance of silicon and oxygen is accounted for by the fact that earth, clay, and rocks contain very high proportions of these two elements. Aluminum, a very common constituent of clays, is the third most common element, at about 8%. Carbon (0.02%) and hydrogen (0.14%) are comparatively rare.

By comparison with the dead world of rocks, a woman contains about 62% by weight of oxygen atoms, 23% carbon, 10% hydrogen, 2.5% nitrogen, 1.4% calcium, 1.1% phosphorus, and smaller percentages of potassium, sulfur, sodium, iron, and cobalt. Very small amounts of other elements, such as copper, zinc, manganese, magnesium, and selenium, are necessary for the healthy functioning of the body. You can buy all these elements fairly cheaply from a chemical company. A human being costs about a couple of hundred dollars, retail.

By far the most abundant element in the universe is hydrogen, the main component of the stars.

Molecules

The elements are characterized by an extraordinary range of qualities, and that range becomes seemingly infinite once atoms combine with each other. Atoms combine, either naturally or at man's inducement, to form literally millions of combinations. When these combinations are reasonably stable, we call them molecules. Thus the nitrogen molecule consists of two nitrogen atoms that go around with each other in a stable pair, if no one interferes with them. Likewise, the sugar molecule is a combination of twelve carbon atoms, eleven oxygen atoms, and twenty-

---

[5]Sir William Crookes, whose belief in the spirit world we have encountered previously, believed that the elements all evolved by a process of Darwinian evolution from a primary material, which he called "protyle."

one hydrogen atoms and, as you know, if it is left in a dry container it appears to last forever. Molecules vary greatly in their stability under normal conditions, but the vast majority of them, say benzene, water, or the plastic polyvinyl chloride (PVC), can be kept in bottles for years without any, or almost any, detectable sign of change. Molecules have properties that, unless you have some chemical knowledge, are usually completely unpredictable on the basis of the properties of their constituent atoms. With a modicum of chemical knowledge you can make a reasonable guess at the chemical and physical properties of most molecules, although their biological effects may be totally unexpected. Well over 90% of all known molecules have been made by man.

The potential of atoms to produce new and varied properties in combination is illustrated by the fact that the following commonly occurring molecules, some of which are shown in Figure 13.2, are made only of combinations of one or more of the atoms hydrogen, carbon, and oxygen: polyethylene, water, cellulose, ozone, butane, glucose, benzene, carbon dioxide, acetone, alcohol (the kind in drinks: ethyl alcohol or ethanol), ether, cholesterol, formaldehyde, testosterone, progesterone, cortisol, phenol (carbolic acid), and antifreeze (ethylene glycol). Thousands of other molecules are constructed only from combinations of varying numbers of these three atoms. Add nitrogen atoms to the building blocks and you can construct most amino acids, TNT, urea, nitroglycerine, the cyanide ion, lysergic acid diethylamine (LSD), epinephrine (adrenaline), nicotine, morphine, and thousands more. Now throw in sulfur and phosphorus and you can build most proteins, DNA, penicillin, and so on. If you still doubt the extraordinary flexibility of atoms, you should reflect on the fact that about 99% of the living parts of any member of the animal or plant world are made solely from four kinds of atom: carbon, hydrogen, nitrogen, and oxygen.

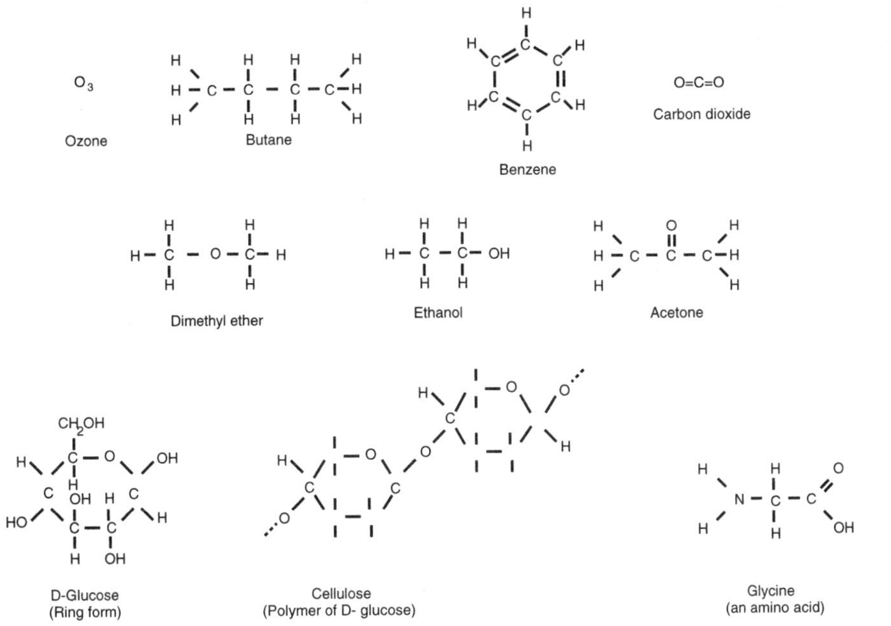

Figure 13.2. A tiny selection of important molecules that can be built from carbon, hydrogen, oxygen, and nitrogen atoms.

The preceding selection of molecules should convince you that atoms in combination usually have properties entirely unrelated to those of the isolated elements. A heap of carbon, when shaken with a mixture of hydrogen and oxygen gas, gives nothing but a heap of carbon in a mixture of gases; it doesn't behave like cholesterol or cellulose. The atoms have to interact much more seriously than forming a simple mechanical mixture. Chemical combination involves a rearrangement of the electrons of the species combining.

A further point that comes out of an examination of Figure 13.2 is that it is not only the types and numbers of atoms in a molecule that determine its properties. Look at dimethyl ether and ethanol, both of which have two carbon atoms, six hydrogen atoms, and one oxygen atom—C_2H_6O. They certainly have different chemical and physical properties. To start with, ethanol is a liquid and the component that makes whisky drinking worthwhile, but dimethyl ether is a gas. It is precisely this sensitivity of molecular properties to molecular structure that accounts for the incredible variety of matter. Perhaps the best-known example of properties depending on structure is the carbon atom, which occurs in nature in two forms: graphite and diamond. In graphite, which is a soft material, the atoms are arranged in *planar* sheets that can slide over each other; each carbon atom is joined to *three* others. In diamond each carbon atom is joined to *four* others, arranged at the corners of a regular tetrahedron, to produce a single immensely rigid, *three-dimensional* skeleton, similar to a system of girders and accounting for the diamond's hardness. Recently it has been found possible to synthesize a new form of carbon, a network of sixty atoms forming a ball-like structure (a "bucky ball") that has been christened *buckminsterfullerene*, after the architect Buckminster Fuller, the designer of the geodesic dome.

The Behavior of Atoms

Galvani and Volta found that they could obtain electric currents by a suitable arrangement of two metals and a damp solid. Unknowingly, they had provided an explanation for many of the chemical properties of matter. The title of the next section is a tribute to the connoisseur of the "electricity of women."

Atomic Sexuality

Divide the periodic table into three classes: male, female, and bisexual (Figure 13.1) The male elements are the *metals*; they are hard and flashy, and their atoms tend to give up electrons to the soft, subtler, female elements, the *nonmetals*. The bisexual atoms are the *semimetals*, which are prepared to accept or donate electrons. In the table they are located between the male and female elements. This classification is an oversimplification, but almost all classifications are.

Now consider potassium iodide. Potassium is a group I metal which is so reactive that, if held in the hand, it would soon eat through the skin. Iodine is a violet, iridescent, crystalline solid that, at room temperature, gives off a corrosive, choking violet vapor. When potassium and iodine react, they do so violently, giving a harmless white crystalline solid, potassium iodide, which, far from being dangerous, is a component of the "sea salt" sold in your supermarket. What happens is that potassium atoms (male) in contact with iodine (female) give up an electron to their partner. The initially neutral atoms are now both electrically charged. The potassium atom, having lost a single negative charge, now carries one net positive charge, which represents the fact that there are nineteen protons in the nucleus but only eighteen electrons in the atom. Conversely, the iodine atom now has one more electron than it has protons in the nucleus, so it carries one net negative charge. Charged atoms or molecules are called *ions*. Thousands of materials consist of assemblies of oppositely charged ions.

Potassium iodide is a white solid which, when examined by X-ray crystallography, is revealed as a very orderly array of ions, each ion of one kind having six close neighbors of the other kind (Figure 13.3a). It is the attraction between opposite charges that gives the crystal its stability, and we say that the crystal is held together by *ionic bonds*. Most substances that are of combinations of metals and nonmetals are called *salts* and consist of ions, held together by ionic bonds. The word *salt* the common name for one specific salt—sodium chloride—comes from the Latin *sal*.[6]

Notice that the addition of an electron to the iodine atom and the loss of one from the potassium atom change both of them from vicious beasts into sleepy pussycats. This is an illustration of the fact that although it is the nucleus that determines an element's *name*, it is the electronic cloud that determines how the atom *behaves*. Most metals have a strong tendency to give up electrons to nonmetals. When they have done so, they calm down and are much more stable. The nonmetals have a strong tendency to accept electrons, and they too usually change from highly active, corrosive substances into harmless, negatively charged ions.

Many salts break up (dissolve) in water, a process that liberates the ions from their neighbors. The free-floating ions can move in an electric field so that solutions of salts conduct electricity, just like electrons in a wire. That's why James Bond threw an electrical appliance into the bath in which the villain had fallen. He knew that the bathwater, which contained salts, would conduct electricity very efficiently. Pure water is a very poor conductor.

Back to Volta. The tendency for a metal atom to give up an electron varies from metal to metal. Put a copper atom in contact with a zinc atom. The zinc atom "pushes harder," and the result is that electrons try to pass from zinc to copper. It is this inequality in the "pushing power" of different metals that is the basis of the voltaic pile and of similar chemical batteries.

When identical metal atoms get together in a piece of metal, for a stag party, they have no one to whom to give their electrons to, but they release one or more electrons anyway. Thus a piece of metal is a collection of positive ions floating in a sea of electrons (Figure 13.3b). Sodium metal is an ordered array of positively charged sodium ions, each of which has contributed one electron to the negative sea.

○ Iodine ion

● Potassium ion

Figure 13.3a. The arrangement of ions in a crystal of potassium iodide.

---

[6]The Roman army was partially paid in salt, each soldier receiving his *salarium*, or salary, presumably in accord with Pliny's sentiments: "Heaven knows a civilized life is impossible without salt."

13.3b. A thin slice through a metal. The regularly arranged positive ions are held together by an electron sea.

Copper atoms give up two electrons each. Without these electrons, the positively charged metal ions would repel each other; it is the negatively charged electrons that are the mobile glue, sometimes called the *electron gas*, that holds them together by electrostatic attraction. We say that the atoms in a metal are held together by a *metallic bond.*

In 1900, only three years after J. J. Thomson's discovery of the electron, the German physicist Paul Drude suggested that some of the electrons in metals were free to migrate. The mobility of the electrons explains electrical conduction. If you put a metal wire between the terminals of a battery, electrons flow into one end of the wire and out of the other; the electron gas is moving.[7]

If we take a walk inside a metal, we find that the ions are arranged in *regular* arrays (Figure 13.3b). The arrangement of atoms in metals and crystals is in fact invariably orderly. Robert Hooke, in 1665, suggested, with sharp physical intuition, that the beautiful macroscopic shapes of crystals were a result of the orderly arrangement of their atoms. He stacked musket balls in different ways to help his speculations, but since nothing of importance was known about atoms, he could not progress.

Strong Bonds

Most substances are neither metallic nor saltlike; they are molecules of the kind shown in Figure 13.2. Take, for example, that media personality, the ozone molecule. It is built solely of three oxygen atoms. No ions here. The atoms in ozone are held together by *sharing electrons*. Most of the electrons in the ozone molecule spend their time wandering about in the space between the three nuclei, forming a cloud of negative electricity, which pulls the nuclei toward it (Figure 13.4a). Most of the molecules in your body are held together by such *covalent bonds.* You are most likely to find covalent bonds when you have atoms that have pretty much the same tendency to push (or pull) electrons. Thus, for example, the "pulling power" of the nitrogen and oxygen atoms is quite similar, and when they are brought together neither is the outright winner, so they share electrons, create a covalent bond, and form the nitric oxide molecule, NO.

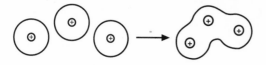

13.4a. The formation of a covalent molecule from three atoms. The electron clouds of the atoms merge to give one molecular cloud.

---

[7]When an electric current passes through a piece of wire, the electrical balance between positive ions and electrons is left unchanged. All that happens is that electrons leave one end of the wire but are replaced by electrons coming in at the other end. A wire carrying a current does not carry a *net* charge, which is why, as stated in Chapter 8, you can't feel an electric field outside such a wire. If you touch it, electrons will flow into you, but they are still being replaced at the end of the wire. A body with static electricity on it *has* got a net inbalance of charge, and the excess will flow into you if you touch the body.

Carbon atoms, when found in the natural world, are almost always either in the form of the covalently bound molecule carbon dioxide or in molecules in which two or more carbons form covalent bonds between themselves, giving chains or rings of carbon (Figure 13.2).

Gentle Bonds: Molecular Velcro

Ionic, metallic, and covalent bonds vary in their strength according to the particular atoms involved. A different and much weaker bond found in nature is named after a Dutch chemist. Van der Waals force exists between all molecules. It is a very weak attractive force that has its origin in the attraction of the positively charged nuclei of each molecule for the electrons of the other molecule (Figure 13.4b). The van der Waals' force is generally larger between larger molecules, which explains why it is important in the interaction between biological molecules such as proteins.

The last type of chemical bond that occurs in nature is also weak, but it controls all our destinies—the *hydrogen bond* (not bomb). Recall that a hydrogen atom is a proton with a single electron running around it. If a hydrogen atom sits on an oxygen atom, say, one which is part of a molecule, the powerful pull of the oxygen nucleus (it has eight protons) will partially shift the hydrogen's lone electron in the direction of the oxygen's electron cloud (Figure 13.4c). This partly bares the hydrogen's nucleus, the proton, which is of course positively charged. The net result is the creation of a spot of positive charge at the edge of the oxygen atom. This charge can attract the negative electron cloud of another molecule, thus gluing two molecules together. Usually hydrogen bonds form when the hydrogen atom is sitting between two oxygen atoms or two nitrogen atoms, or one of each. The glue is not very strong, but it is ideal if you need to have two molecules joined to each other in such a way that the connection is normally stable but can be broken nonviolently if need

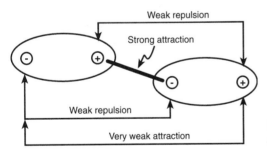

Figure 13.4b. Van der Waals force between two hydrogen atoms. The attractive forces outweigh the repulsive.

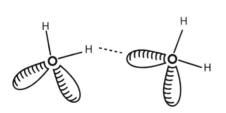
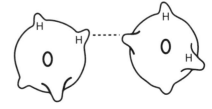

13.4c. The hydrogen bond between two molecules of water. Each oxygen atom carries two "balloons" of negative charge and two hydrogen atoms.

be. We will see that this is exactly the requirement of the DNA molecule if it is to fulfil its role.

Both the van der Waals force and the hydrogen bond are easily formed and easily broken. You can classify them as molecular Velcro.

The Semimetals

A major characteristic of metals is their ability to conduct electricity, some much better than others. The nonmetals do not have this ability; they have no "electron sea." Their tendency is to accept electrons, not give them up. The semimetals are neither here nor there. Take silicon, the symbol of the electronic age. Silicon crystals have a structure like diamond; each silicon atom forms four strong (covalent) bonds to neighboring atoms. This structure, like diamond, has no loose electrons left over, as there are in a metal. But silicon is a bigger atom than carbon and it has more electrons. It doesn't take too much to get a few electrons to break away and wander through the crystal. These electrons allow silicon to conduct electricity—not very well, but better by far than nonmetals. Now when an electron leaves an atom of silicon, it leaves a "hole" behind where it has been. This hole can also move. Well, it isn't really the hole that moves, it's an electron from a neighboring atom falling into the hole and leaving a new hole on the neighboring atom (Figure 13.5a). Since a silicon atom with an electron missing is positively charged, when a hole moves, so does a positive charge. But moving charge is what we call an electric current. Thus silicon has two ways to conduct electricity: the few free electrons can move, and the positive charge can move. Now we can start to play clever tricks.

If you look at the periodic table, you can see that the next element after silicon is phosphorus. Phosphorus has an atomic number of 15 compared with 14 for silicon, which means that it has one more electron. Suppose we put a phosphorus atom into a silicon crystal. It uses some of its electrons to join on to four neighboring silicon atoms, just as silicon does, but it has one electron extra, with nothing to do. This electron can wander around, and it increases the electrical conductivity of the crystal. In fact, if the crystal is "doped" with very small amounts of phosphorus, its conductivity depends almost entirely on the phosphorus atom's contribution to the "free" electron population. Such a crystal is called an *n*-type semiconductor, "n" for negative charge carrier, "semi" because it doesn't conduct anywhere as near as well as a metal.

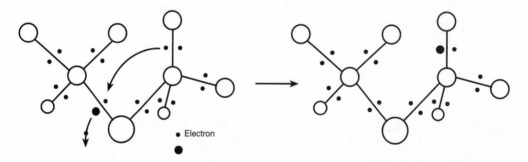

Figure 13.5a. Inside a silicon crystal. The two electrons involved in a bond between neighboring atoms are symbolized by two dots. An electron deserting a bond between two silicon atoms leaves behind a positively charged hole. Another electron can fall into this hole, with the result that the hole moves.

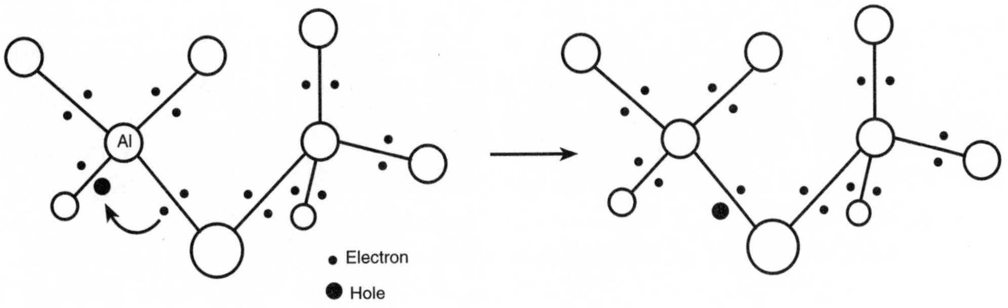

Figure 13.5b. An aluminum atom has invaded a silicon crystal but does not have enough electrons to form four bonds to neighboring silicon atoms. There is thus a built-in positive hole in the crystal into which an electron can fall.

Now look at silicon's other neighbor on the same row of the periodic table. Aluminum has an atomic number of 13 and therefore has one fewer electron than silicon. Put an aluminum atom in a silicon crystal, and it does its best to form four bonds to its neighbors, but it has one electron too few (Figure 13.5b). You can think of one of the bonds as being incomplete, lacking an electron. This missing (negative) electron creates a positive hole, and the number of these holes, and the electrical conductivity of the doped silicon, depends on the amount of aluminum in the crystal. This is a way of producing a *p*-type semiconductor, where "p" stands for positive.

Semiconductors formed the basis of the electronics revolution. The transistor is based on successive layers of p- and n-type silicon-based semiconductors, for example, *npn* and *pnp*. Subsequent technology has been marked by dramatic miniaturization of such devices and also the development of new semiconductors based on other semimetals, in particular gallium and arsenic. The whole field depends on the chemist's ability to produce very pure substances, because even minor amounts of impurities can result in different batches of semiconductors having widely varying and unpredictable conductivities.

In recent years, semiconductor devices have been built that, in response to electrical impulses, produce very pure light, effectively of one frequency. These "semiconductor lasers" are revolutionizing communications technology. There are now telephones that convert your voice to electrical impulses (that's not new), which are then converted to light by a minute semiconductor laser. The light, traveling along glass fibers, carries the message that, at the other telephone receiver, is converted back to electrical impulses and then to sound. We are in the age of optoelectronics, but only because we know how to handle the necessary materials.

The Creation of Materials

Man is now routinely making new molecules, unknown in nature, a capability that has altered the visual, tactile, and medical environment of much of mankind. There is no possibility, within the confines of this book, to survey in depth the nature of materials and our ability to conjure with them, but we will very sketchily summarize the main types of synthetic materials to be found in our world.

Ceramics

A rough definition of a ceramic is a tough, durable substance, resistant to environmental attack and composed of a blend of metallic or semimetallic elements and a nonmetallic element, which is usually oxygen. A well-known ceramic is Carborundum, the common name for silicon carbide. Here the semimetal is silicon and the nonmetal is carbon. Another example without oxygen is silicon nitride, an extremely hard-wearing material that is used for turbine blades.

Almost all ceramics up to modern times were based on silicon and oxygen, because these two elements form the basis for clay and sand. There is an atomic pattern that recurs in many different types of ceramics (Figure 13.6). The arrangement of the atoms is based on tetrahedra consisting of a silicon atom surrounded by four oxygen atoms. The radius of the silicon atom is about three and a half times as large as that of the oxygen. We thus have a collection of footballs and tennis balls, and what happens in large three-dimensional structures is that the footballs form tetrahedra and the tennis balls creep into the central cavities. If you look at Figure 13.6, you will be able to appreciate that in such large structures every oxygen belongs to two tetrahedra, so that there is a continuous chain of links running through the material.

In general the strength of ceramics comes from their interlinked array of atoms, rather like a pylon or a bridge, in which one can continuously walk along strong interconnections. Compare this with, say, oxygen gas, where the only strong bonds are between discrete pairs of oxygen atoms, and the force holding individual oxygen molecules to each other is very weak. Similarly, a crystal of sugar is a collection of discrete molecules that falls to pieces in water, each molecule floating around on its own.

Disordered Solids

If you heat a crystal, the atoms, or ions, vibrate more and more energetically until, if the temperature is high enough, they shake the ordered structure to pieces and

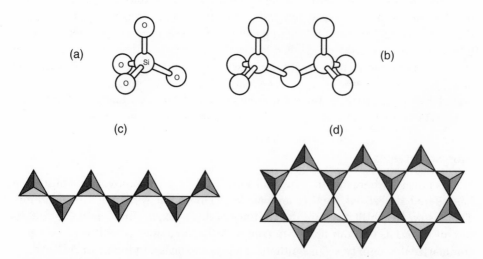

Figure 13.6. (a) The building block of many important silicates. (b) Two linked blocks. (c) A linear chain of blocks. (d) Another common pattern.

the individual atoms start wandering where they will. You have made a liquid; the orderly arrangement of the solid has disappeared. Now, suddenly lower the temperature. The melt solidifies and, before the atoms can find their way to the positions that they had in the ordered structure, they are trapped, wherever they are. The resulting solid has not got a regular structure. The solid is *amorphous*, from the Greek for "without form." Almost all types of glass are amorphous.

As we saw earlier, glass has found an important new use with the coming of fiber-optic communication. These fibers obviously must have a high degree of transparency, otherwise the light signals would die out too quickly. All glass absorbs light to a certain extent, but it is a measure of the state of glass technology that a pane of standard window glass absorbs as much visible light as a 10-mile-thick block of the glass used in optical fibers.

The Infinite Variety of Carbon

In 1846 the German chemist Johann Gmelin first used the term *organic chemistry* for the study of carbon compounds. Most university chemistry departments have professorships in organic chemistry. Why not a professor of sodium chemistry? The reason is twofold: the enormous importance of organic compounds and the tremendous number of carbon compounds. The carbon atom has the ability to easily form a strong covalent bond to other carbon atoms so that molecules with chains or rings of carbon atoms are found everywhere. Some simple examples were shown in Figure 13.2, in which all the larger molecules shown have a number of carbon-hydrogen bonds. The vast majority of carbon compounds contain such bonds. Of the 15 million or so known molecules, some 12 million are organic. And the number is growing. Many chemists spend the whole of their professional lives making new organic molecules.

Molecular Monsters

One particular class of carbon compounds plays a starring role in the living world and in the material environment of much of mankind, namely, *polymers*. A polymer is a chain, produced by linking together small molecules, known in this context as *monomers*. The monomers in a chain may be identical, as in a simple necklace, or similar. Monomers are usually small carbon-containing (i.e., organic) molecules. Polymer chains can reach enormous lengths, containing thousands of atoms. The synthesis and industrial production of man-made polymers are of major economic significance, but by far the most important polymers occur naturally.

The most abundant organic polymer on earth is almost certainly cellulose (Figure 13.2), a polymer of the sugar β-glucose, which is the tough substance that forms the outer cover of the plant cell. Cotton is almost pure cellulose; wood is about 50% cellulose. We make about 200 million tons of paper a year from cellulose. Glucose is also the basis of starch. In animals, glucose is stored in another polymeric form, as glycogen, which is a heavily branched molecule containing anywhere from about 2000 to 600,000 glucose units. A macroscopic sample of any polymer is invariably a mixture of chains of different lengths.

The normal functions and structure of the cell are almost entirely in the hands of a group of polymers: the proteins. Proteins consist of chains of amino acids. Twenty amino acids are used by the cells of the human body, and in any given protein the same amino acid may appear many times. Different proteins contain different numbers and differently ordered combinations of amino acids. The number of possible proteins is vast. The number of different proteins in the human cell is well over a thousand.

The proteins have an astonishingly varied range of tasks. Some of them act as *structural* components, holding the cell in shape. Other proteins are *antibodies* that protect, or should protect, the body. Fibrinogen is a protein that is essential in blood clotting. Muscles depend for their contractile action on two proteins, myosin and actin. There are proteins whose job is to carry materials from one place to another, like the hemoglobin molecule that ferries oxygen from the lungs to the rest of the body and depends for its action on the presence of four iron atoms. All *enzymes* are proteins. Enzymes catalyze (speed up), the reactions which, at the most basic level, are life.

From man's point of view, the most important polymer on earth is the DNA molecule, the molecule that carries genetic information from one generation to the next.

The Importance of Shape

When a small molecule, such as a drug or a hormone, acts on a living cell, it initially binds to one or more very specific sites on the cell surface or within the cell. This binding is usually by van der Waals (Velcro) forces, which, because they are weak, allow the small molecule to break away relatively easily at a later time. A crucial factor in successful binding is a good match between the shape of the absorbing site and the shape of the absorbed molecule. For example, the action of all enzymes in the living cell depends on their coming into contact with other molecules and interacting with them. This interaction does not occur anywhere on the enzyme molecule but at specific sites, usually termed *active sites*. Like a lock on a door, active sites are shaped so as to accept only one specific molecule (key) or only molecules containing a small group of atoms arranged in the right shape. This ensures that a given enzyme speeds up only a certain type of reaction and does not act as a general accelerator of cellular reactions. Complementarity of shapes is a device extensively used by nature to ensure the smooth functioning of living matter.

Shape has life-or-death implications when it comes to the design of new pharmaceuticals. The nineteenth-century German scientist Paul Ehrlich described the action of drugs as being a "lock and key" phenomenon, in which only the correctly shaped key would fit the lock. His idea has been extremely fruitful in studying the reaction between large molecules such as proteins and smaller, biologically active molecules such as drugs and hormones.

In studying the shapes and interactions of proteins and other large molecules, increasing use is being made of visual representations of the molecules on a computer screen. The evolution of very fast computers with large memories has revolutionized this field. It is possible to "put a protein on the screen," rotate it, focus in on certain parts of it, and thus examine the detailed shape of the protein and of its active site, and use the information as a guide to how the protein functions. This kind of approach is being used more and more to examine the interaction of drugs and small, biologically significant molecules with proteins. The drug need not be synthesized in the initial stages of the research—it can be built on the computer screen and then manipulated on the screen to see how well it fits into the active site of the protein.

Man-made Polymers

The first man-made polymer was produced by Leo Baekland in 1905. He took formaldehyde—the material used for preserving medical specimens—and phenol, commonly called carbolic acid, and heated them together. He got a hard black material which he called Bakelite. He filed a patent application in 1907 a day before a certain James Swinburne arrived at his local patent office to apply for the same patent. Bakelite was a common plastic in the 1930s and 1940s, but the limelight was taken by nylon in 1934. Nylon made its popular debut in 1938 in Dr. West's Miracle Tuft Tooth Brush. Our daily surroundings, if we are city dwellers, are now replete with polymers, from polyethylene to polyvinyl chloride.

In the days of Bakelite the creation of polymers was not far removed from cookery—a bit of this, a bit of that, heat, and stir. Polymer chemistry and physics are very sophisticated disciplines today, and we have reached the stage at which new polymers are being tailor-made for specific purposes.

Liquid Crystals

If you have a digital watch, the display is almost certainly made of liquid crystals, a class of compound that was discovered in 1888. Liquid crystals can flow like liquids, but their molecules are normally arranged with a high degree of order, like solid crystals. The simplest possible kind of liquid crystal consists of long, thin molecules, that above a certain temperature behave like a normal liquid, the molecules being completely free to move as they wish. When cooled, the liquid suddenly changes into a form that, although still fluid, has its molecules all roughly lined up in the same direction, like part of a logjam floating downriver. It's as though a plate of cooked spaghetti changed from the usual chaotic mess to a plate of approximately aligned strands of spaghetti—aligned but still sloppy enough to pour off a slanted plate, thus behaving like a liquid. When all the molecules in a sample of material are roughly parallel, the sample will exhibit different physical properties from the same sample in the liquid state, in which the molecules are tumbling around freely. One of these properties is the way that the sample reflects and scatters light, and it is this property that has been used in liquid crystal displays. In your watch the numbers change as a result of the action of tiny electric fields that are capable of changing the orientation of liquid crystal molecules so that they line up with the direction of the field. Such crystals are man-made, but there are natural substances that are liquid crystals. A very important type are the *phospholipids.*

Phospholipids are the main constituents of the membranes of living cells. They are long floppy, molecules that naturally tend to align themselves with each other. In fact, many phospholipids, when shaken with water, have the tendency to spontaneously form closed, hollow spherical aggregates, termed *vesicles* or *liposomes.* Such spheres have fascinating physical properties, but you might care to reflect that mere shaking can induce the formation of a structure that resembles large portions of the membranes of living cells. No "vital force" was needed to construct this highly organized entity. Real cell membranes consist of double layers of phospholipids of this kind, with a variety of other molecules embedded in them.

Theory Pays Off

Chemistry handles the visible cloth from which the universe is made. Only in the investigation of the fundamental particles of matter has chemistry nothing to say. But it is chemistry that has shown us what the Earth is made from, what living material is, and (in the guise of biochemistry and molecular biology) how the cell works. Its methods have enabled us to analyze and purify the materials that have made possible the communications revolution: microchips, semiconductor lasers, fluorescent screens for monitors, and so on. Modern medicine depends on a multitude of substances, from anesthetics to antibiotics and hormones, that are either synthesized or purified by the skills of the chemist. Chemistry enables us to build materials that nature could not, on her own, produce; materials that have colored our buildings and our clothes, have alleviated our ills, and eased the daily grind.

Chemistry has been called the central science, and no other discipline has contributed so much to the understanding and betterment of our condition, even though the products of man's ingenuity have sometimes turned around and bitten him.

In the past two chapters we have seen how man's curiosity about the nature of the material universe, and his power of reasoning, have led from alchemy to the nuclear atom and onward to science-based technology. Before we get carried away by modern man's ingenuity, it is fitting that we now pause, turn off the TV, and look back over the centuries at two of our scientific ancestors.

interlude
Honor Thy Father and Thy Mother

Science and the search for order in the universe have their roots in the ancient world. In the following pages we look back at two of our intellectual ancestors, one a mystic, the other a materialist; one a charismatic leader and the other a tough-minded poet. They represent two essential ingredients of the human quest for knowledge.

14

Scipio's Dream

He sat on a rubbish heap for faulty geometrical figures,
Nearly straight lines, kinky circles, bent rectangles, 3.7-sided triangles
Their skinny, useless shapes clinked and sagged under his weight
Like a heap of ten thousand metal coat-hangers.

—Adrian Mitchell, *Pythagoras on Elba*

Those who are not permanently hunting for food or money have time to hunt for meaning. Philosophers, introspective adolescents, French film directors, poets, and HMS all seek an underlying or overbearing order. The search for natural order in the physical world, the drive to detect pattern, may in the end be a consequence of the computer architecture (the way our brains are wired) or of the computer software (the way our brains are programmed). Undeniably, our attempt to understand the cosmos has been driven by far more than the wish to improve our standard of living.

The search for order in the wider sense, the "What-does-it-all-mean?" syndrome, is at least partially motivated by the awareness of suffering and the smell of mortality. The two quests, for natural order as exemplified by science, and for metaphysical order exemplified by religion, have often overlapped. Order in the natural world hints at an overall design and by inference an overall Designer, so that belief in a deity is bolstered by the symmetry of the Tyger. The intertwining, or even identity, of the two quests was rarely questioned before the seventeenth century. Newton took the ordered motion of the planets as evidence of the existence of the Creator. He agreed with the psalmist: "The heavens declare the glory of God, and the firmament sheweth his handiwork."

The yearning for the scientist's Grand Design and the mystic's Ultimate Reality is unlikely to fade. Scientists, often driven by irrational rather than rational belief in the unity of nature, have striven to unite and simplify. This is not a recent phenomenon, associated only with Oersted and Faraday, Maxwell and Einstein. The simplifications, and unifications, of myth and religion have been associated with every stage of human history, from the caveman's wall paintings onward. Simplification was raised to the status of a principle by the fourteenth-century Franciscan philosopher William of Ockham: "What can be explained by the assumption of fewer things is vainly explained by the assumption of more things." Ockham's razor still guides those seeking to account for natural phenomena.

The traceable roots of both Western science and philosophy lie in Greece and in the Greek settlements of Asia Minor. The French philosopher Renan called the marvelous flowering of Greek thought "the only miracle in history." There, in the Greek world, the first known Western philosophers appear: Thales, Anaximander, and Anaximenes. There, reason separated itself from myth and mysticism, but that separation was rarely complete, and the mingling of the mystic and the mathematical is nowhere more intimate than in the shadowy, enigmatic figure of Pythagoras of Samos. He has a right to be seen as the father of the Western search for order. His spirit is still with us.

The Godfather

Pythagoras was born about 580 B.C., about the time that Nebuchadnezzar II sacked Jerusalem. The first definite sighting of him that we have is in Samos, a lush island just off the coast of present-day Turkey, with a tradition of wine making initiated by the god Bacchus in person. Pythagoras's whereabouts before Samos are unknown; the grapevine has it that he was born in either Tyre or Sidon and spent many years in Egypt and Babylon delving into the occult and mathematics. He seems to have made a reputation as a thinker in Samos, but he next appears in Croton (now Crotone) on the sole of the Italian peninsula, where he joined an existing Greek colony and quickly set up a tightly knit, secretive cult characterized by asceticism, vegetarianism, a belief in the transmigration of souls, and a ban on bean eating. "To eat beans," said the Pythagoreans, "is to eat one's parents' heads." The rationale was that life was breath and since beans induce flatulence, to break wind after eating was proof of having eaten a living soul. Shelley wrote a piece in praise of the "Pythagorean diet."

The similarities between the Pythagoreans' lifestyle and that of the Buddhist priesthood of India is striking. There is an ancient rumor that Pythagoras traveled as far as India. He was fifty when Buddha was about thirty.

There were Mafia-like aspects to the Pythagoreans. Their leader was a charismatic, well-traveled, snappy dresser: white robe and trousers and a gold coronet. He is said to have sacrificed an ox (not a horse) every time he proved a new mathematical theorem. One Hippasus of Tarentum was reputedly drowned at sea for blabbing out the secret rules by which the cult was governed. After Pythagoras's death, his followers infiltrated local (southern Italian!) politics and apparently controlled dozens of city-states for about half a century.

Popular history has not been kind to Pythagoras. Today almost the only response that his name invokes is "the square on the hypotenuse." He would have been gratified to find that there is a pleasant, white-walled red-roofed port in Samos named after him: Pithagorio. But his legacy is far larger than that. His statue, alongside that of Aristotle, Euclid, Ptolemy, Christ, and an assortment of saints, stands above the massive west door of Chartres cathedral. His spirit pervades science and philosophy. Newton revered Pythagoras. Let us begin to see why.

The Legacy of Pythagoras: The Numbers Game

Pythagoras said "All is number" and "Number is the ruler of forms and ideas, and the cause of gods and demons."

Divide the infinitely long number 11111111 . . . by 9. The answer is 12345678901234567890123. . . . Or divide 1010101010 . . . by 9 and you get 112233445566778900112233. . . . So what? Trivially amusing? It would not have been to Pythagoras—nothing to do with numbers was trivial; they were a bridge to the deepest, hidden secrets of the universe, a path to what Plato was later to see as the reality behind the shadow.

The Pythagoreans were obsessed with numbers. With your sixth-century B.C. eyes, look at what the Pythagoreans saw when they looked at the first four integers, 1, 2, 3, and 4:

| 1 | 2 | 3 | 4 |
| Point | Line | Plane | Solid |

A straight line can be drawn between any *two* points, a plane through any *three*. The simplest solid with plane sides must have a minimum of *four* corners. One can build a temple to the Gods with lines, planes, and solids. Is it so strange that the Pythagoreans regarded the numbers 1 through 4 as being special?

Take their sum, 1 + 2 + 3 + 4 = 10. Arrange ten points in a triangle:

```
            *
          *   *
        *   *   *
      *   *   *   *
```

The central dot together with four dots along any side makes 5, which, if you are a Pythagorean, represents earth, water, fire, air, and the soul. The three points at the corners of the triangle enclose a hexagon, and 6, the sum of 1 + 2 + 3, represents life, although I don't know why. Together with the central dot the hexagon gives seven, a number of mystic significance to many civilizations and representing intelligence and light, the prerogatives of Athene. (The Japanese had seven gods responsible for happiness.)

The infinite series of numbers 1, 3 (=1 + 2), 6 (=1 + 2 + 3), 10 (=1 + 2 + 3 + 4), 15 (=1 + 2 + 3 + 4 + 5), and so on, were dubbed *perfect numbers,* of which 10 was regarded as by far and away the most significant. It was given a special name, the *tetraktis,* τετρακτυζ, and was sacred. Pythagoras swore by it.

The supposed occult significance of numbers has occupied man from prehistory. The Chaldeans held 4 and 7 in special regard. The Chinese constructed magic squares, a game overlooked by the Greeks. Relax in your favorite chair and improvise: One is one, the only, the Godhead, the cosmos . . . me, numero uno. Two is ying and yang, heaven and hell, man and woman. Three is the Trinity, the triangle, the male genitals . . . and so forth. All of which is harmless rubbish and which, for a Pythagorean, misses the point. Take the Trinity. If there was for Christianity a purely religious significance attached to 3, it has long been lost. It is the personae of the Trinity—Father, Son, and Holy Ghost—that are the objects of veneration, not the number 3 in itself. With the Pythagoreans it was the *numbers* that were in danger of canonization.

The Game Turns Serious

Inevitably the religiously inspired manipulation of numbers led some of the Pythagorean cult in the direction of mathematics. As an example, go back to the perfect numbers 1, 3, 6, 10. . . . The next three in the series are 15, 21, and 28. Now note that the sum of any two successive perfect numbers is a perfect square. For example:

$$1 + 3 = 4 = 2^2$$
$$3 + 6 = 9 = 3^2$$
$$6 + 10 = 16 = 4^2$$
$$10 + 15 = 25 = 5^2 \text{ etc.}$$

Those with high school algebra might try proving this result, which is for us a mild curiosity but for the Pythagoreans was a hint of cosmic order. *And it was really cosmic order that obsessed them, and that in different ways has obsessed man ever since.*

Take the perfect number 28. Its *factors* (the numbers by which it can be exactly divided) are 1, 2, 4, 7, and 14. Add them together. You get back 28! By the tetraktis! This must be another type of perfect number, one that is the sum of its factors. If you get your kicks from solving puzzles, you could try looking for other such perfect numbers.[1] An easy one is 6, which, as we have seen, is a perfect number in the sense that it is the sum of 1, 2, and 3, but is also perfect in the new sense since 1, 2, and 3 are also its factors. Another two are 496 (the sum of all integers from 1 to 31) and 8128 (the sum of all integers from 1 to 127). Thus 8128 is the sum of its factors: 1, 2, 4, 8, 16, 32, 64, 127, 254, 508, 1016, 2032, and 4064. I warn you that up to 19 February 1992, only thirty-two perfect numbers of the "factor" kind had been found. The largest has 455,663 digits, and you will not therefore be surprised to learn that five of the known numbers were unearthed by the use of Cray supercomputers.

The Pythagorean concept of *means,* which remains a part of elementary mathematics to this day, will have special significance for us shortly. Among other types of mean the Pythagoreans defined the *arithmetic mean* (or average) of two numbers, which is given by $(x + y)/2$; and the *harmonic mean*, which is given by $2xy/(x + y)$. The arithmetic mean of 6 and 12 is $(6 + 12)/2 = 9$; the harmonic mean is $(2 \times 6 \times 12)/(6 + 12) = 8$.

The later Pythagoreans were interested in mathematics, as distinct from mystic numerology, but they remained essentially a religious cult with beliefs that often echo older religions. We remember their math because we use it; we tend to forget the, for us, wasted time they spent trying to find a number or group of numbers to associate with material objects and abstract concepts. To us these associations are meaningless. The number 1 represents intelligence; 2 is opinion, because you can jump from each of its component "1s" to the other $(1 + 1)$. Even numbers were declared to be feminine, and odd were masculine. In a male-dominated society this inevitably led to the doctrine that even numbers were evil and odd were good. Certain numbers were considered to have aphrodisiac properties and were written down and swallowed—with unrecorded effects. Multiplication?

The Pythagoreans appear to be the first to have stated that mathematics was an activity that could be divorced from the practical world. Mathematics undoubtedly arose from practical needs such as commerce and the construction of buildings. It was the Pythagoreans who saw that there was an abstract entity behind the socially

---

[1]For the foolhardy, here is a helpful hint: if 2^P-1 is a prime number, then $(2^P-1) \times (2^P-1)$ is a perfect number. Thus for P = 5, $2^P-1 = 31$, which is a prime, leading to 496, a perfect number. Prime numbers of the form 2^P-1 are known as *Mersenne primes* after the remarkable Marin Mersenne, mathematician, theologian, estimator of the speed of sound, and friend of René Descartes.

useful instrument. This was one of the critical turns, if not *the* critical turn, in the history of mathematics, and it had a deep influence on later Greek thinkers, Plato and Euclid in particular. It is here that we see the beginning of what is today called basic science, the study of nature, and knowledge, for its own sake. Plato, a firm Pythagorean, specifically stated that arithmetic should be studied for its intrinsic value, not for the purposes of commerce.

A major discovery of the Pythagoreans was that √2, the square root of 2, could not be written down as *a/b*, where *a* and *b* are whole numbers. √2 is the number which when multiplied by itself gives 2. It can be proved that there is no way of writing this as the ratio of two whole numbers, in the way that we can write 1.5 as 3/2. The Pythagoreans called √2 and similar numbers *irrational,* as we still do, and the existence of such numbers disturbed them since they saw the cosmos as based only on *rational* numbers, which could be written as *a/b*.

Despite the mathematics, the real influence of Pythagoras over the centuries is rooted in the mystic, some would say metaphysical, content of his thought: his search for cosmic harmony.

The Music of the Spheres

Take a guitar—there must be an acoustic guitar left over from the sixties, or a high-tech electronic brain blaster belonging to you or your son. Pluck the thickest string. If your house is in order, you should hear the note E. Measure the length of the string, which should be about 25.8 inches. Now place an index finger between the fourth and fifth frets, the strips of metal crossing the fingerboard (Figure14.1a). Press down firmly on the string and with a finger of the other hand pluck it again. You should be listening to the note A. Measure the length of string that vibrated— the length between the fifth fret and the bridge of the guitar. It's about 19.3 inches.

The ratio of the two lengths that you have measured is 4 to 3. The two notes that you have played, E and A, define a musical interval of a *fourth*, because you have played the first and fourth notes of a major scale.

You are repeating an experiment performed by Pythagoras. You may not be amazed by the fact that a pleasant musical interval is related directly to a simple numerical ratio. He was apparently thunderstruck. Follow him further: play the "open" string again, the E. Now put your finger between the sixth and seventh frets. Play and measure. The second note that you have heard is a B, and you have played an interval of a *fifth*. The ratio of string lengths is 3 to 2.

Transport yourself back again to the sixth century B.C., to a time long before the

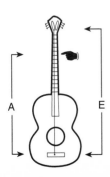

Figure 14.1a. Pythagoras discovers cosmic harmony.

frequencies of notes could be measured, before psychological or physiological explanations could be generated to account for the scales that please our ears. For the
Greeks, the striking link between music and numbers was astounding, beautiful in
its simplicity and a reflection at the deepest level of the harmony of nature. It was
without doubt a major source of Pythagoras's belief in cosmic order and its numerical basis. Let us go a little deeper.

Earlier in this chapter we talked about *means*. Pick up the guitar again. Let the
length of the open E string be 12 units (Figure 14.1b). If you play a length of 6 units,
you get an interval of an octave—you are playing the next highest E. Now notice
that, as we found earlier, the *arithmetic mean* of 6 and 12 is 9, and the *harmonic
mean* of 6 and 12 is 8. Play lengths corresponding to 9 units and 8 units. You are
again playing the A and the B. The arithmetic and harmonic means evoke harmonious sounds. To Pythagoras this was an example of a far deeper order, an order
that encompassed Heaven and Earth.

It is to Pythagoras that we trace the first use of the word *cosmos* both to label the
universe and to imply beautiful and harmonious order. It was he who proposed
that every planet emitted its own musical tone, the combined sound being the
"Music of the Spheres," which, according to his disciples, only Pythagoras among
mortals could hear. There is an echo here of the ancient Hindu belief that, apart
from audible sound, there was an "unstruck sound," inaudible to man.

Pythagoras, more than anyone else, has the right to be considered the father of
the Western way of seeking order in the universe. His influence stretches from ancient Greece to the modern world.[2]

Pythagoras and the Ancient World

Pythagoras's fingerprints can be detected all over the Greek and Roman world.
Plato was influenced by the Pythagorean doctrine of the transmigration of souls. In

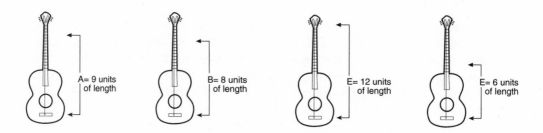

Figure 14.1b. The mathematics of music.

---

[2]Pythagoras said, "A friend is another self." Like most denizens of Soho coffeehouses in the
1950s, I played the guitar. On 1 June 1957, The Vipers, the group of which I was a founder-
member, reached number eight on the British Hit Parade. The Vipers were later helped by George
Martin, who went on to greater things as the Beatles' arranger. Late one night in the Two Is, in Old
Compton Street, the Royal Shakespeare Company actor Mike Pratt pointed out to me that the
fourth and fifth chords that were part of my limited musical vocabulary had been the origin of
Pythagoras's dream of cosmic harmony. Mike died, cruelly early, long ago. I record my gratitude
to him for initiating my interest in Pythagoras, thus allowing me to sense, if not hear, the music of
the spheres.

his *Phaedo*, where he tells the unforgettable story of Socrates' last day, he invents (for he was not present) a discussion on immortality in which Socrates' companions are nearly all Pythagoreans. At the end of the *Republic* Plato introduces the music of the spheres. Later in his life Plato fell for numerology and the nonmathematical significance of numbers. Perhaps Pythagoras was indirectly responsible for the motto reputed to have been chiseled above the door of Plato's Academy: "Let no one ignorant of mathematics enter my door." Almost certainly it was the Pythagorean doctrine that the sphere was the perfect solid body that induced Plato to declare the Earth to be spherical.

Euclid was a mathematical fan of Pythagoras in the third century B.C. Ovid, in his *Metamorphoses,* portrays Pythagoras expounding transmigration to the legendary king of Rome, Numa Pompilius, who predated Pythagoras by a century! Cicero's *Republic* contains a section entitled *Somnium Scipionis* (Scipio's Dream), in which the elder Scipio Africanus appears in a dream to his grandson and presents a panorama of the universe, heavily dependent on the philosophy of the Pythagoreans. Transmigration, immortality, and the music of the spheres all make their appearances.

And then came the Barbarians.

Pythagoras Survives the Dark Ages

Thrown out of southern Russia by Attila's Huns, the Ostrogoths fled westward and in A.D. 452 invaded the Italian peninsula. The Roman Empire in the West had run its course. From A.D. 493 Theodoric the Goth ruled the kingdom of Italy. The Dark Ages were to last for several hundred years.

These particular Barbarians were not all that barbaric. Like the Romans that they had conquered, they were Christian, albeit adhering, like Isaac Newton, to the Arian heresy. They sensed that the Romans had something to teach them and, in view of the fact that Theodoric had been educated in Constantinople, it is understandable that he employed Roman advisers. One such was Anicius Manlius Torquatus Severimus Boethius, who was dubbed "the last of the Ancients." It was he who would carry the Pythagorean torch, almost alone, through the millennium that led to the Renaissance.

Boethius's (unrealized) ambition was to translate all the works of Plato and Aristotle into Latin. Pythagoras left no writings; much of what we know of him is contained in Plato's writings, particularly *Timaeus*. Through Plato, Boethius met Pythagoras.

The translations of Boethius had an enormous influence on medieval learning. With the birth of the universities in twelfth-century Europe, it was his texts that were used, almost exclusively, to teach arithmetic and music. He himself wrote a five-book treatise, *De institutione musica,* which was the standard European text on music for a thousand years, from the fifth century onward. Boethius's treatise was firmly based on the scale that Pythagoras had constructed, built on simple ratios of string lengths.

The curricula of the medieval universities typically included seven subjects classified into the *quadrivium* (arithmetic, geometry, astronomy, and music) and the *trivium* (rhetoric, grammar, and dialectic). Note that music was classified with the sciences, another legacy of Pythagoras. We see the Pythagorean connection very clearly when the music syllabus is examined in detail. Instrumental and vocal music were regarded as the least prestigious branch of music. Of higher importance were the study of the "harmony" of the body and spirit of man, and here we detect the ancient belief in four *humors*, which only when they were perfectly (harmoniously) balanced allowed man to be perfect. Music reached its highest peak in the music of the world, *musica mundi*, which subject included the study of the four elements (earth, water, fire, and air), the movement of the heavenly bodies, and the dance of the four seasons—subjects not on the curriculum at Juilliard. Music, math, and nature mingled—this is pure Pythagoras. It was largely through Boethius's text that the message of Pythagoras was preserved for some 500 years until, from the eleventh century onward, original Greek texts reached Europe via the translations of Arab and Jewish scholars.

Boethius was unlucky in his choice of patron. Theodoric was an unpredictable, violent man. For obscure reasons, he had Boethius imprisoned in Pavia, and after nine months of uncertainty had him horrifyingly tortured and then clubbed to death. Yet the spirit triumphed. During his incarceration, Boethius penned *De consolatione philosophiae,* in which the personified figure of Philosophy solaces the author in his misery. This monument to man's dignity in the face of the slings and arrows was probably number two on the best-seller list in Europe for a thousand years, and was translated by both Alfred the Great and Queen Elizabeth I.

Pythagoras Corrected

To be proved wrong 2000 years after your death is a fate that has befallen some of the best people. Pythagoras did not escape. He, or at least his school, was responsible for generating a series of relationships between the musical note emitted by a plucked string, the length of the string, and the *weight* hanging on it. These results were certainly not the outcome of experiments, but in the great medieval Scholastic tradition they went unquestioned and untested until, after two millennia, a musician actually hung weights on strings, plucked the strings, and listened. He didn't hear the notes that he was supposed to, and his own experiments sparked off a reinvestigation (or investigation) of the physics of vibrating strings. The musician was Vincenzo Galilei, a virtuoso lutenist, and it was Boethius's book that had aroused his curiosity. I would like to think that it was his experiments that fostered an interest in science in his son, Galileo Galilei. Thus would the link be made, through Boethius, between the Pythagorean origins of order and the seventeenth-century scientific revolution.

Newton and the Ancients

Newton firmly believed that his work was merely a resurrection of knowledge already bestowed on the sages of old by the Almighty. While Sir Isaac was loathe to give credit to his contemporaries when they propounded views similar to his own, he was wildly generous to the ancient Greeks, the ancient Hebrews, the ancient anybody who by some stretch of the imagination had made a pronouncement resembling something that he had said.

By a deviously twisting argument, worthy of a Jesuit, Newton attributed to Pythagoras a statement of the law of universal gravitation. In bestowing on the Sun the power to hold the planets on their courses, Newton was doubtless aware of the second-century B.C. Pythagorean belief that identified the Sun with Apollo Musagetes, the "conductor" of the Muses who sang the music of the spheres. And when he famously split the light of the Sun into a spectrum, Newton counted seven colors, because there were seven notes in the Pythagorean scale.

It is ironic that the mystic side of Newton, which was probably of more importance to him than anything else, has been widely ignored, and yet if one looks underneath his scientific writings one finds that they are permeated by the belief in a harmonious, ordered cosmos that was the visible representation of an underlying profound mystic order. The belief in an all-embracing scheme, of which modern unified field theories are typical, has been the common ground of scientific thought for centuries. The spirit of Pythagoras infused the lives of Galileo, Newton, Kepler, and a host of their successors. As the late Paul Feyerabend remarked, when Copernicus set out to order the heavens he did not consult his scientific predecessors but went back to a "crazy Pythagorean"—Philolaus. The search for a mathematical framework on which to hang the visible world is in no small measure the legacy of Pythagoras, the man whom the philosopher Whitehead regarded as the founder of European philosophy and European mathematics.

The Poets Abandon Pythagoras

Pythagoras spoke to scientists and poets alike. Aeschylus, at least according to Cicero, was a follower of Pythagoras. Chaucer knew of *Scipio's Dream,* as did the author of the medieval *Roman de la Rose.* The music of the spheres was a common literary device up to the Enlightenment. Lorenzo, in *The Merchant of Venice,* declares: "There's not the smallest orb which thou behold'st, / But in his motion like an angel sings." And the Elizabethan Sir John Davies, in *Orchestra,* puts into Antinous's mouth the sentiment that: "all the world's great fortunes and affairs / Forward and backward rapt and whirled are, / According to the music of the spheres."

It is difficult to know when ideas such as the music of the spheres changed from being objects of belief to being literary conceits. When Wordsworth speaks of "harmony from Heaven's remotest spheres," we know that he does not believe in literal harmony, as did the Elizabethans.

The Romantics' vision of the universe was anti-Newtonian. The Pythagorean vision of mathematically based cosmic order was abandoned by the Romantic poets. Wordsworth retreated into subjectivity—a retreat that still shapes Western poetry. The poets have staged an almost wholesale withdrawal to the inner world, despairing by default of a universal framework and leaving the search for cosmic order in the outside world to the scientists. No contemporary poet would undertake to emulate Lucretius's delineation of the universe, and few poets today would wish or dare to write of the harmonious cosmos. But here is one of the few exceptions:

(*The next total eclipse of the Sun, visible from London, will be in the year 2142.*)

> I shall not see it,
> nor the last child born,
> nor even yet his son.
> Still it will happen.
> This is a calculation
> you can count on; not

like the railway timetable
that can be disrupted,
cancelled, or run late.

It does not need us,
and we cannot interfere.
If, by that time,
there should be no one
left alive down here
to witness, even so
it will take place.

And knowing this we are
at once diminished, yet
at the same time strangely
reassured—stirred by
the synchromesh of cosmic
gears, as if believing
what has order must have
purpose,—as if we sensed
the music of the spheres.

Tony Lucas, *Total Eclipse*

The Atomist's Reply

Thou great Lucretius! Thou profound Oracle of Wit and Sence!

—Thomas Shadwell, *The Virtuoso*

Since none of Pythagoras's writings have survived, we only know of him at second-hand. But we do have the text of a poem about the universe, penned by a Roman 600 years after Pythagoras lived. It was written by a very different man from Pythagoras, a man for whom the mysticism of the Pythagoreans was symbolic of all the superstition that, usually in the guise of religion, was deluding man.

Lucretius was an atomist and a materialist, no less awed by nature than Pythagoras but seeing it with cooler, dare one say braver, eyes. Because his poem was written in ancient Rome, it is not read much these days, but it deserves to survive. It represents a tradition in science that is no less central than the Pythagorean belief in order. Lucretius is the voice of all those who are prepared to challenge authority and face the cosmos as it is, even if that means a cosmos without faith.

Lend Me Your Ears

The reality of previous generations is problematic for many of us, rather like the difficulty that teenagers have in believing that their parents have a sex life. The average schoolboy sees the ancients as a bunch of remote bores, dressed in bedsheets and talking Elizabethan English—but they often speak to us as directly as our contemporaries.

Here a short, bitter poem from Italy:

I hate and I love,
And if you ask why,
I have no answers, but I discern,
Can feel, my senses rooted in eternal torture.

Translator: Horace Gregory

The pain transcends the centuries. The poem behind the translation was written about 2000 years ago in Latin, by Catullus, a poet's poet, passionate, jealous, sarcastic, and hopelessly in love and hate with Clodia, the flighty wife of the stodgy governor of Cisalpine Gaul. Like the majority of present-day poets, he had difficulty making a living from poetry, and at one stage found work on the staff of Memmius, a dry, egotistical civil servant. To Memmius was dedicated one of the greatest texts to survive from the classical world: *De rerum natura* by Lucretius. Lucretius speaks our language.

Titus Lucretius Carus was an out-and-out materialist. For him, not only objects but also ideas and sense impressions were material. Thus I can see my dog because he sheds an excruciatingly thin layer of atoms (note the word) that penetrate my eyes and end up somewhere in my brain. The explanation has distant echoes of the corpuscular theory of light, the "atoms" of light being reflected by an object into the observer's eyes. With suitable modification it is also not a bad description of how I *smell* my dog. Because man had constructed the idea of the gods, they existed somewhere—because ideas are material. However, in complete variance with contemporary dogma, Lucretius avowed that the Gods had no interest in us whatsoever.

Lucretius was one of the earliest publicists of the idea of atoms. He believed, as few did in his day, in the experimental method of determining facts; he was endlessly curious about nature and courageously antiestablishment. And he knew how to write.

In 60 B.C. Lucretius published his masterpiece, "On the Nature of Things," in the six books of which he explained the universe. He leaves almost nothing untouched: how the universe was formed, how the stars and the planets move, the origins of man and speech, the nature of politics and religion, the discovery of fire, the beginnings of music, and so on, and so on. He frequently goes into great detail. There are seventeen listed reasons why the soul must die when the body dies.

Almost everything that he wrote was wrong. There just weren't enough facts available to him. He was a propagandist for atoms, but he had absolutely no evidence for their nature or even their existence.

Why concern ourselves with a would-be scientist? Partly because we owe him a debt. He was a poet, a thinker standing out against the irrational, but not the mysterious; against superstition, but not against awe. He is the counterpart of Pythagoras.

Lucretius's writings, like those of Catullus, are powerful and direct. They speak to the feeling, rational man, to HMS. Here is a loose modern-language version of the famous lines in which he extols his spiritual father, the Alexandrian Epicurus, who died 200 years before Lucretius wrote[3]:

When human life was still oppressed, suffocated with the dead weight of Re-

---

[3]Those with a morbid turn of mind might be interested to know that Lucretius's Elizabethan translator, Thomas Creech, hung himself. He was given to toying with a rope while reading *Biathanatos,* John Donne's sympathetic study of suicide.

ligion, which had come down from the Heavens, terrifying and hovering over mortal men, it was then that a Greek had the courage to raise his eyes, mortal eyes, and look Religion boldly in the face. He was cowed neither by the reputations of the Gods, nor the thunderbolts of Heaven; their threats only sharpened his resolve to smash down the gates that guard the secrets of the Natural World. And his driving energy carried him through to victory. He passed beyond the fiery regions surrounding the Earth and roamed in thought through the immeasurable universe. And he returned in triumph, bringing back the spoils of victory, explaining what is possible and what is not, what governs the potentialities of matter, and what are the deep and inherent limits that dwell in all things. And Religion was trampled underfoot. One man's victory put us on a level with the Gods.

Lucretius made little impression on the Romans, probably *because* he admired Epicurus. In later centuries, Epicurus's philosophy was often wrongly taken to be an enticement to la dolce vita, a hedonistic response to the gods' indifference to man. The Romans saw it correctly, as a search for peace of mind by living the quiet life, devoid of excessive desires. Horace called it the happiness of a pig in a sty. This philosophy, with which Lucretius sympathized, was averse to excessive, or even moderate, wine, women, and gladiators, and didn't at all suit the Roman mentality, which was basically action-oriented. In contrast, Lucretius's writings reveal him as a man to whom melancholy was familiar. Like Camus, he seemed to see man standing alone, but proud, in a hostile universe.

Democritus, and his conviction that there was "nothing but atoms and the void," was the other great influence in Lucretius's thought. Reading Lucretius, one senses that he would have made a name as a physicist had he been born a few centuries later. He was an observer, and he had the curiosity, intuition, and skepticism of a good scientist.

Lucretius was effectively forgotten for 1500 years, until the same spirit with which he swept aside obscurity and superstition, mysticism and rites, tore through Renaissance Europe. Lucretius was one of the sources of the atheism that, not always overtly, underlay the thinking of many of the new men who overthrew the moldy dogma of the Middle Ages. He spoke to a generation only too willing to discard the inheritance of Aristotle, Christianized by Albertus Magnus and Thomas Aquinas, and fossilized in a way that Aristotle himself, the probing, questioning inventor of logic, would probably have hated.

Lucretius was widely read in the Enlightenment. The philosophes sympathized immediately with his rejection of mysticism and his vigorous attack on religion. They saw in him an echo of their own Newtonian belief that observation and reason would provide an understanding of the workings of the universe.

Exactly 1600 years after Lucretius published the never-finished *De rerum natura*, Giambattista della Porta founded the first recorded scientific society, in Naples in 1560. And in 1960, men sent up the first weather satellite and Lt. Don Walsh together with Jacques Piccard descended 35,800 feet into the Pacific near Guam.

If we have journeyed outward to the planets and downward to the floor of the ocean, if we think the way we do, a way so different from that of pre-Renaissance man, it is partly because of men like the pre-Renaissance poet Titus Lucretius Carus.

V

Cracks in Newton's Pedestal

A nineteenth-century scientist would find this part deeply subversive. For two centuries, Newton's mechanical universe dominated the way that both scientists and laymen thought about the workings of nature. Newtonian-Galilean mechanics hinted strongly at a deterministic framework for the cosmos, a conclusion that had deep psychological, moral, and religious implications. But the seeds of scientific heresy were germinating in the quiet fields. The nature of light and of heat— problems that occupied Newton, Hooke, Huygens, and their contemporaries—were to eventually split asunder the analogy between the universe and a well-oiled clock.

It is a mistake to think that the downfall of the Newtonian universe started in the twentieth century, with the development of relativity and the quantum theory. The rot started long ago, and in fact there is almost nothing in this destructive part that falls outside the realm of classical physics. Newton would not be amazed if he read this part, but he might be very perturbed by some of its conclusions. He would find a fundamental law of nature that should follow from his laws but doesn't, and he would find that in practice there are many situations where his laws should give an answer but are completely impotent.

15

Making Waves

My mother taught me to enjoy light waves, my father to enjoy sound waves. My mother was the eldest of four sisters born at the turn of the century in Gomel, in what is now called Byelarus. Her earliest memory was of being hidden under an inverted tin bath in the house of an old colonel, during an attack by cossacks engaged in their national sport of slaughtering Jews. Years later, in London, I found out that when my father went to work in the East End, and I had gone to school, my mother would often take the Underground to central London and wander around the pricey, fancy art galleries. Once, as a university student, I cut through Dover Street, near Piccadilly, and unexpectedly caught sight of her through the window of a dealer. She looked small, shy, and shabby amongst the real and would-be connoisseurs. She could never afford originals, but she brought home catalogues and purchased postcards in the big public galleries like the Tate. She had almost no formal education, but her tastes were extraordinarily sophisticated: Kandinsky, Klee, Miro, Chagall, Marino Marini. And she loved Vermeer and Turner. Color fascinated me as a child.

My father, on the other hand, had very conventional visual tastes. He loved music, whereas my mother was tone-deaf. He felt uneasy with Chagall's blue cows floating above the roofs of Vitebsk. I remember as a boy arguing with him when he told me that colors vanished in the dark and would not exist at all if there were no animals to see them. He quoted Bacon: "All colors will agree in the dark." He listened to Bach in the dark, and I discovered the Chaconne in D Minor. My mother picked up reddish golden autumn leaves and put them by his bedside.

Light and sound, color and music, Chagall and Monteverdi, were once all considered to be manifestations of *waves*. That sound is a wave phenomenon is indisputable, but light is a far more complex creature, as it showed at the beginning of the twentieth century when it came out of the closet and revealed a particulate nature. Newton had always suspected that light was corpuscular; Huygens said it was a wave phenomenon. Neither offered convincing experimental evidence for their guesses, although experiments in the nineteenth century strongly supported the wavelike nature of light. However, as a result of certain observations in the twentieth century, Einstein went against convention (as usual) and opted for particles. The position at the moment is that it is very convenient for purposes of calculation to regard light as wavelike in some cases, but the particle-picture seems to be closer to reality and can also account for phenomena that are inexplicable in terms of waves. The reasons for all this will become apparent later. In this chapter we will speak the language of waves.

The Messenger of the Gods

Luz vertical,
luz tu
alta luz tu
luz oro;
luz vibrante,
luz tu.
Y yo la negra, ciega, sorda, muda sombra horizontal.

—Juan Ramon Jimenez, "Luz Tu"[1]

There are few words more evocative than *light*. The word itself has become a metaphor for happiness and understanding. In Francis Bacon's *New Atlantis* (1627), the remote island kingdom of Bensalem engaged in external trade in only one commodity: light. The export-import agencies dealt in the light of understanding—the "knowledge of causes and the secret motion of things."[2]

Light is a basic component in the functioning of this planet. The light of the Sun maintains life on Earth. The mental world of most of us is based heavily on the content of light signals, and light acts as a time cue (zeitgeber) for the biological clock in mammals.

Man manipulated light long before the nature of light or vision was understood. The art of constructing optical instruments was highly developed by the seventeenth century. As for vision, it was believed for centuries that light issued from the eye of the observer, a Pythagorean idea that was taken up by Plato.

Newton was deeply interested in the nature of light and particularly in color. He made notes on nearly forty recipes for mixing paints, mainly reds. His best-known experiment was the splitting of a ray of sunlight by allowing it to fall on a prism. "Light," he wrote, "consists of Rays differently refrangible," meaning that they were bent by different angles when they passed through his prism.[3] To this statement we can trace the beginning of the modern technique of *spectroscopy*, the analysis of light signals into their components—and by "light" I mean any kind of radiation. It is to this ability to *separate the different frequencies in electromagnetic radiation* that we owe much of modern physics and chemistry. It is this that allows us to detect the chemical elements in the stars, to analyze complex mixtures of materials, to measure the vibrations and rotations of molecules, to probe the electronic cloud of atoms and molecules and to detect tumors by magnetic resonance.

Light, in fact any electromagnetic radiation, has many *wavelike characteristics*; sound *is* a wave phenomenon. All waves have certain common properties, and it is to these that we now turn.

---

[1]Vertical light / you, light / tall light / golden light / vibrating light / you light. / And I, the black, the blind the deaf, the dumb horizontal shadow. (Translation: Eleanor Turnbull)

[2]It was Bacon's fictional Salomon's House on Bensalem, the original model for the modern research establishment, that was almost certainly in the minds of the founders of the Royal Society.

[3]When Goethe read of Newton's experiment, he interpreted it as showing that light was in fact homogeneous and it was only the interaction of light with matter that produced colors—they were not there to start with. This was a perfectly acceptable interpretation, but it runs into difficulties with Newton's demonstration that a second prism, placed in the path of the light emerging from the first, can recombine the colored rays to give white light.

Waves: Periodicity

The beating of our hearts, the circling of the Sun, the dance of the seasons, menstruation, metronomes, the tides, marching feet, Gene Krupa, and Ravel's *Boléro*—all these are reminders of natural or man-made *temporal* periodicity. Wallpaper patterns, tiled roofs, Greek decorative patterns, Arabic tiles at the Alhambra or Isfahan, the scales of a fish, the arrangement of atoms in crystals, are all examples of *spatial* periodicity. Our through-space interaction with the universe is primarily mediated by periodic movement—sound waves and light waves. Both sound waves and light waves are scientific constructions. If our sense organs could directly detect light and sound *as waves*, it would not have taken man so long to show their wavelike nature.

If we define a wave as a *moving periodicity* (Figure 15.1a), then there are three simple characteristics that are possessed by all waves: the *velocity*, the *wavelength* (or its frequency), and the *amplitude.* Two of these properties can be seen very clearly for ripples spreading across a pond. If we fix our eyes on one crest, we can define the velocity of the wave as the velocity of the crest. The wavelength is simply the distance between two neighboring crests. A quantity that is directly related to the wavelength is the number of waves that pass a given point in one second. The number of ripples that pass a reed in a pond in one second is the frequency of the wave. For waves traveling at the same speed, the frequency increases as the distance between crests, the wavelength, decreases. If one crest passes by every second, we say that the wave has a frequency of 1 hertz, written 1 Hz, which some people would write as 1 cycle/second.

The frequency, v, wavelength, λ, and velocity, v, of a wave are related by: $v = v\lambda$. Velocity = frequency x wavelength, which is the same as saying that the speed of a train is equal to the number of carriages which pass you in one second, times the length between the front ends of two adjacent carriages (or any two similar points on adjacent carriages). In a vacuum all electromagnetic waves (visible light, UV, X-rays, etc.) travel at the "speed of light," c, which is about 300 million meters per second, or about 186,000 miles per second.

To and Fro versus Up and Down

Put a cork in a pond, on a windless day, and then throw a stone into the pond and watch the cork as the waves pass it. The cork moves up and down. All the movement is perpendicular to the direction in which the waves appear to move. For this reason, waves in water are termed *transverse waves* (Figure 15.1b). If you rhythmically wave one end of a rope while someone holds the other end, you are creating transverse waves. In this case it is very clear that the rope is neither moving toward nor away from you. Its only movement is *perpendicular* to the motion of the wave, as you can check by tying a red ribbon on the rope. In transverse waves in matter, the movement of the wave in a given direction is not a movement of matter but a

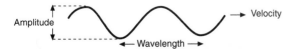

Figure 15.1a. A simple wave. This could be a rope or a line drawn on the surface of the sea.

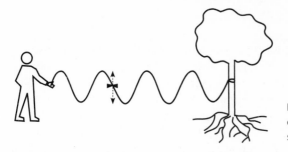

Figure 15.1b. A transverse wave. The ribbon oscillates vertically. A cork far out to sea behaves similarly provided the waves are gentle.

movement of *form;* the movement of matter is only transverse to that direction. Watch a "Mexican wave" in a football stadium. Light waves are also transverse waves, but they are harder to visualize than waves in a rope or in a pond.

In contrast, sound waves in air are *longitudinal.* If you could observe the air molecules in the path of a sound wave, you would find that they oscillate back and forth, but this time *in the direction* that the sound moves (Figure 15.1c). Nevertheless, *on the average* they stay where they are because, although longitudinal waves carry molecules in the air forward, they carry them back again. You can see what a longitudinal wave looks like by holding one end of a "Slinky" (those tightly coiled metal springs sold in toy shops) and letting the spring hang down vertically. Now move your hand up and down, and you will see vertical longitudinal waves. Obviously, since the spring remains in your hand, there is no *net* movement of metal in the direction of the wave. Simple waves do not permanently displace matter.

Velocity

A baritone and a soprano produce notes with different frequencies. Can we choose the velocity of a wave in a given medium, as we can choose the frequency? Not really. The velocity is overwhelmingly determined for us by the medium, although it is mildly sensitive to changes in conditions. The speed of sound waves in dry air at 20°C is about 1125 feet per second. Changing the air pressure has no significant effect on this value. Temperature has a small effect. At 0°C the speed is 1086 feet per second and at 100°C it is 1266 feet per second, whether you are singing bass or soprano.

The speed of sound depends on the medium through which the waves are traveling. The "stiffer" the medium, the faster the wave. The speed in water is about

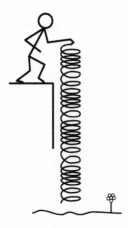

Figure 15.1c. The wave travels along the axis of the Slinky. At any given point above the ground, the closeness of the coils varies periodically.

1500 feet per second; in iron it is about 5160 feet per second. Similarly, the speed of sound through the ground is much faster than it is in air. That's why the Lone Ranger's faithful Tonto put his ear to the ground to get early warning of the galloping hooves of the bad guy's horses.

The velocity of electromagnetic waves *in vacuum* has been measured for wavelengths ranging from about 6 meters to less than 100 million millionth of a meter (10^{-14} m), and found to be effectively the same, as Maxwell predicted. This independence of velocity on wavelength is not true for light traveling through matter.

The velocity of electromagnetic waves in a given material medium is slower than in vacuum and depends slightly on the frequency of the wave, which is the cause of the splitting of "white" light into different colors (frequencies) by prisms, lenses, and raindrops.

The velocity of a wave is not dependent on the velocity of its source. For example, if I wave my hand in water, I create waves that move forward at a velocity of about 1500 feet per second. If, while moving my hand back and forth I also walk forward, the wave front will still travel at the same speed. The disturbance that I create, once it has established itself in the water, is no longer affected by my hand. However, although there is no affect on the velocity of the waves, my walking forward means that I will partially catch up with the first wave by the time that I make the second one. I am compressing the waves, so that their peaks are closer together. I have shortened the wavelength of the wave. If I continually walk forward *at the same speed as the waves*, then I will be moving forward on the crest of the first wave, and every new wave will be produced at the wave front, which never escapes from me. This extreme bunching, when it occurs in air, is termed a *sonic boom*, which is heard when a plane travels at the speed of sound.

Frequency

Light waves and radio waves exhibit a range of frequencies; otherwise we would see only one color and hear only one radio station. The light that comes from the stars, the Moon, and the Sun is a mixture of different frequencies. It is rare that we see light or hear sound that is of one frequency. Try joyously singing two different notes. If you sing middle C you are producing waves having a frequency of 256 Hz. Now listen to this note when it is made by a tuning fork. It sounds different. The note that you are singing is not pure. All musical instruments, including the human voice, emit notes contaminated by other frequencies. It is this contamination that allows our ears to differentiate between instruments and gives them their character.

Light from almost any natural source is a mixture of very many frequencies. The invention of the laser allows us to produce light that is almost pure (monochromatic) in that it contains effectively only one frequency, which also means one wavelength. Yet even a laser doesn't produce absolutely monochromatic light. Thus a krypton laser produces light with an *average* wavelength of 605.78021 nanometers and a spread of about plus or minus 2 nanometers. By ingenious instrumental tricks, this 4-nanometer spread in wavelength can be reduced by a factor of 100 million to give almost pure light.

The Electromagnetic Palette

Until the nineteenth century, what was called "light" was radiation with a very limited range of frequencies, the range detectable by the human eye. Maxwell predicted that there was radiation that the human eye failed to see. His equations predict that it is possible to generate electromagnetic waves of almost any frequency that, when in a vacuum, all travel at the speed of light. As it should do, theory resulted in previously disconnected and puzzling phenomena dropping into place. Thus Maxwell's theory made sense of some observations made by the astronomer William Herschel in 1800. He attempted to measure the heating effect in the different colors of the visible spectrum by placing the bulb of a thermometer in the patches of different color obtained when sunlight passed through a prism. He noticed that when he placed the bulb of the thermometer in the dark area just outside the red end of the spectrum, the mercury also rose. He concluded that the cause was invisible radiation. Since the frequency of this radiation was apparently lower than that of red light, he called it *infrared* (IR). In 1801, Johann Ritter had also detected "something" just outside the other, violet, end of the visible spectrum. He found that photographic salts were blackened by these invisible *ultraviolet* (UV) *rays.* Maxwell's theory shepherded IR and UV into the electromagnetic fold. In 1886, Hertz created radio waves. Marconi was waiting in the wings.[4]

The year after Hertz died, Roentgen discovered electromagnetic radiation at much higher frequencies than either radio or light waves. This radiation is still called after him in some countries but is more commonly known by the name that he gave it: X-rays. The English satirical weekly *Punch* was highly amused by the penetrative properties of X-rays. The following uninspired verses were part of a poem it published on 25 January 1896:

> O, Röntgen, then the news is true,
> And not a trick of idle rumour,
> That bids us each beware of you,
> And of your grim and graveyard humour.

> We do not want like Dr. Swift,
> To take our flesh off and to pose in
> Our bones, or show each little rift,
> And joint for you to poke your **** in. . . .

(*Nose* is a four letter word.) In 1901 Roentgen was the recipient of the first Nobel Prize awarded in physics.

In 1900 the French scientist Paul Villard detected γ-rays from radioactive material. They were shown to be electromagnetic radiation, with frequencies even higher than those of X-rays. A whole universe of waves was opening up.

The properties of natural and man-made electromagnetic waves have revolu-

---

[4]Even long after "the radio" became a standard household item, it aroused amazed hosannas, being described by one author, in 1938, as, "the miracle of the ages. . . . Every vision that mankind has ever entertained, since the world began, of laying hold upon the attributes of the Almighty, pale into insignificance besides the accomplished fact of radio."

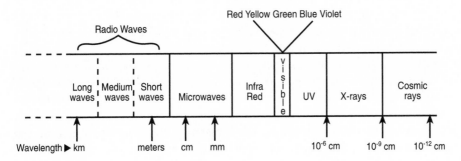

Figure 15.2. The known range of electromagnetic waves.

tionized science and technology. The known spread of frequencies is shown in Figure 15.2.

Walk out on a dark night, far from the city lights. Darkness is the name we give to the inability of our eyes to detect any other than visible light. But the whole universe is teeming with electromagnetic radiation. We see only wavelengths between about one six millionths of a centimeter and one 12 millionths of a centimeter, which, as you can see from Figure 15.2 is but a tiny fraction of the known range of radiation. Radiation over the complete range of frequencies shown occurs in nature; yes X-rays, microwaves, and radio waves are detected coming from space. They are in no way evidence of intelligent life. Nature has her ways, and in fact our Sun sends out IR, visible, UV, X-rays, and γ-rays, although the last two fail to penetrate the atmosphere.

Man creates a wide variety of radiation.[5] Some of it is involuntary, as in the case of the IR radiation arising from the jiggling of the molecules in our body, and which can be detected and photographed with suitable equipment. Deliberately created radiation such as that from lightbulbs, neon lighting, and lasers all result in the end from electrons moving, either between or within atoms, and thereby creating electromagnetic waves.

Cat Whistles?

As for sound, human ears can respond to frequencies in the range of 16 to 20,000 Hz. Dogs hear up to about 50,000 Hz, cats to 65,000 Hz, and moths have a hearing range from 3000 to 150,000 Hz. By way of comparison, the range of the piano is 30 to 4100 Hz.

The human ear is remarkably sensitive. Not only can it distinguish something like 400,000 different sounds, but it can detect sounds so quiet that the vibratory movement induced in the eardrum is not much more than the width of a calcium atom.

Intensity

Velocity and frequency (or wavelength) are not enough to completely characterize a wave; we also need to specify its intensity. When waves rock a boat, we feel intuitively that the force they exert is related to their height, the distance between the

---

[5]He is not unique in this respect. There are several organisms that produce light, such as the glowworm, bioluminescent bacteria, and certain deep-sea creatures.

peaks and valleys of the wave, *as measured perpendicular to the wave.* We could measure this distance, the *amplitude,* by watching a cork bobbing up and down in the water and noting its maximum and minimum height.

If I walk in the sun and feel its heat on my shoulders, I know that the heat originated 93 million miles away and reached me across an effective vacuum. Waves undoubtedly transfer something from place to place: *waves transfer energy,* as Joshua demonstrated at Jericho. We live in a huge energy bath.

In the case of water waves, the energy transferred to a floating cork from the point at which a stone strikes the water is connected to the *amplitude* of the vertical motion of the cork (Figure 15.1b). It can be shown that the amount of energy transferred by a water wave, or any wave, is proportional to the *square* of its amplitude. If the amplitude is increased, say, threefold, then the energy transferred increases by a factor of nine.

Waves have curious characteristics that arise from their periodic nature. Two of these have played an outstanding part in the general history of science: interference and the Doppler effect.

Interference

The great Dutch scientist Christiaan Huygens (1629–1695) realized that two waves can mutually interfere in such a way as to annihilate each other. Suppose you simultaneously, and with the same force, throw two identical stones into a pond, at separate, closely spaced points (Figure 15.3). Waves spread out from the two points. Both waves will have identical speeds, frequencies, and amplitudes. Take a point P, equidistant from those at which the stones entered the water. The two waves pass through P like a pair of synchronized swimmers. What one does, the other does. If a crest of one wave passes through P, then a crest of the other wave will pass through at exactly the same time. If a valley of one wave passes through, a valley of the other wave accompanies it. In short, the two waves reinforce each other; they exhibit what the aficionados call *constructive interference.* To rub in the superiority of the expert to the layman, they will also tell you that the two waves are *in phase.*

Now consider the point Q. We choose it so that *exactly* as the crest of one wave passes, the valley of the other arrives. Put a cork at that point. It goes neither up nor

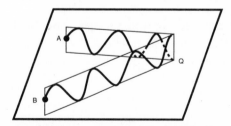
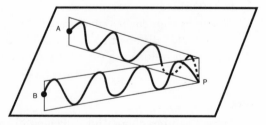

Figure 15.3. Two identical stones land in the water simultaneously at points A and B, which are equidistant from point P. At P the waves arrive in phase: when one wave is at its peak (or trough) so is the other. This is known as *constructive interference.* At point Q, the distances from A and B are such that when the peak of one wave passes through it coincides with the trough of the other. The waves cancel each other out. This is *destructive interference.* We can find a whole range of points similar to P or to Q. This is the basis of the next figure.

down as the waves pass it, because there are no waves at this point. They have canceled each other out—they are exactly *out of phase.* The name of this phenomenon is *destructive interference.*

Interference can be demonstrated easily for water waves because they are obvious to the eye. But what about electromagnetic radiation? Maxwell had claimed that light consists of waves. If light could be shown to exhibit interference, that would be convincing evidence that Maxwell was right. Which brings us to Thomas Young, who preceded Maxwell.

A Famous English Experiment

I learned history when the map of the world was heavily splashed with red—representing the British Empire. History, as taught at my first school, was a succession of dates on which "great events" occurred: 55 B.C., Caesar lands; 1066, William of Normandy lands; 1215, Magna Carta; followed by a string of dates on which English armies won battles abroad. Science was taught on the same principle: large numbers of clever Englishmen and important (mainly English) experiments. The great event method is a terrible way of learning history but a fairly good way of learning science. One of the great events in science was undoubtedly an experiment performed by Thomas Young.

Young (1773–1829) was a man of diverse talents who read at two years old. His hobby was Egyptology, and he was one of the team that deciphered the Rosetta stone. He was trained as a physician, and his medical interest in the senses resulted in his discovery of the cause of astigmatism in 1801, and the correct suggestion that we have three types of color detectors in our eyes. His particular interest in light led him to perform one of the more famous experiments in scientific history.[6]

Young guessed that light was wavelike and that the waves were not longitudinal but transverse; the vibrations, whatever they were, were perpendicular to the direction of the light ray (Figure 15.1b). You can replicate his experiment by shining a strong, reasonably monochromatic light at a screen pierced by two slits, each not much more than about a thousandth of an inch across, and spaced about one-hundredth of an inch from each other (Figure 15.4a). Observe the image produced on a second screen, placed about 1 yard behind the first. You should see not two lines of light but a roughly elliptical patch crossed by parallel dark lines. This is what Young saw, and he realized that he had demonstrated the wavelike behavior of light.

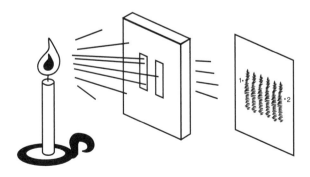

Figure 15.4a. Young's experiment. The light waves from the two slits give alternate bands of dark and light on the screen.

---

[6]If there are any engineers reading this and you have ever used Young's modulus in the study of the elasticity of a material—it's the same man.

The explanation of the dark lines exactly parallels that for the interference of water waves (Figure 15.3). At points like P on the second screen, the two light waves coming from the slits will be in phase. The waves reinforce each other and give a bright area; this is constructive interference. If we now move to a point like Q, we see that the two waves travel different distances, and we can account for the dark line by supposing that the point Q is so situated that the two waves are exactly out of phase at that point, annihilating each other; this is a case of destructive interference. If we move to other bright areas, we can explain their brightness by supposing that, although the two waves travel different distances, they arrive in phase again. If you reduce yourself to sub-Lilliputian dimensions and go for a walk along the second screen—from point 1 to point 2 in Figure 15.4b—you will find that the light intensity on your path falls and rises periodically. On the basis of Young's experiment, it is difficult to quarrel with his conclusion that light is wavelike.

Newton believed that light consisted of corpuscles, from which he deduced that light could be attracted by gravity: "Do not bodies act upon light at a distance, and by their action bend its Rays, and is not this action (*caeteris paribus*) strongest at the least distance?" But no one had seen light bend in response to gravitational pull. And now Young's classic experiment appeared to utterly destroy the corpuscular theory of light. How could one possibly look at the striped pattern on the second screen and explain two corpuscles of light producing darkness? Two waves might be able to annihilate each other, but how can two bullets? (The final hope of the revolutionary physics student facing the firing squad?)

A dramatic example of constructive interference is the laser. Traditionally, light is produced by heating or burning something, say, a lamp filament or oil. The light originates in the movements of the electrons in the atoms of the hot body, and a simple picture sees the electrons oscillating and thereby producing electromagnetic waves. Now it is highly improbable that all the electrons responsible for radiation will be oscillating in phase. In general each electron will be "doing its own thing" quite independently of the others. Thus the waves emitted by the body are not in phase; they are a disorganized jumble of waves. Furthermore, the light is a mixture of frequencies, so that the waves don't "fit together." If you stood in the path of such light and could see the individual waves, their peaks would not pass you together. If, however, they all did, if all the waves had the same frequency and were in phase, you would expect to feel a very strong effect. It is as though many people were simultaneously shaking you such that they all pushed and pulled in the same direction together. That's how laser light acts; all the waves in the light are effectively of the same frequency, and they all travel in phase. The effect of this army of simultaneous shakers can be dramatic. Lasers can burn through steel.

As obvious as the implications of Young's experiments appear, his conclusions

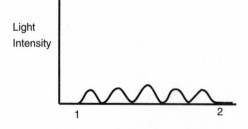

Light
Intensity

1   2

Figure 15.4b. The variation of light intensity in walking along the line from point 1 to point 2.

about the wave nature of light were greeted with ridicule by his compatriots. Who was an upstart physician to argue with Newton? The strength of the paradigm prevailed. In France, primarily because of the confirmatory experimental work of Dominique Arago and Augustin Fresnel, Young's ideas were quickly accepted, and the doubting English scientific establishment eventually had no choice but to follow suit.

Notice that in the explanation of Young's experiment it was assumed that light emerging from a narrow slit spreads out slightly. If this were not so, then the image of the two slits would simply be two narrow lines of light. This spreading out of light when it passes through a narrow orifice is termed *diffraction*. In fact it occurs for orifices of any size but is not usually noticeable, being manifested by a slight fuzziness at the edge of shadows. Diffraction is a property of all waves, and you can observe it for water waves.

In Young's experiment the spacing of the pattern of lines that he saw on the screen depended on the wavelength of the light and also on the spacing of the slits. In fact the best, most readily detected diffraction patterns are produced when the spacing between the slits is roughly the same as the wavelength of the light. This fact is the basis of one of the most powerful methods we have to determine the shapes of molecules and the internal arrangements of atoms in crystals. We jump forward a century.

The Microscopic Young Experiment

Without *X-ray crystallography* and related diffraction techniques, we would never have known the structure of the DNA molecule, and modern molecular biology would have been a stunted flower. We owe the idea to Max von Laue (1879–1960). Aware of Young's experiment, von Laue realized that the wavelength of X-rays was of the same order of magnitude as the spacing between the nuclei of atoms in crystals or molecules. Since the atoms in all crystals are arranged in very regular arrays, von Laue guessed that they might diffract X-rays to give simple patterns that would depend on the spacing of the atoms. He and his collaborators passed X-rays through a crystal of copper sulfate and obtained complex diffraction patterns. The "slits" in crystals are usually complicated, and so is the connection between them and the form of the patterns. Only in the case of certain simple types of crystal, such as sodium chloride, where atoms are arranged in very simple ways, is the analysis of the patterns in terms of the structure of the crystal relatively easy. The early analysis of the patterns given by many crystals was usually long and tedious, but today they can be analyzed rapidly by computer techniques so that even for crystals of large complex molecules such as proteins, we can work back from the diffracted pattern to find the arrangement of atoms in the crystal.

The method should not be confused with the medical use of X-rays to examine the interior of the body. This use is based on the fact that X-rays usually penetrate noncrystalline matter in *straight* lines just as light penetrate glass, so that we are essentially looking at X-ray *shadows*. Diffraction would be a nuisance for this purpose, since we rely on the X-rays traveling in straight lines to give sharp shadows.

If, as Young seemed to have proved, light was indeed wavelike, then certain interesting phenomena could be explained or predicted. One of these was to be the tool by which, in the twentieth century, it became clear that the universe is expanding at a frightening rate.

The Redshift

When, in 1924, it was proposed that our universe is expanding, the conclusion relied on work published in 1842 by the Austrian physicist Christian Doppler (1803–1853). The well-known *Doppler effect* comes into play whenever a source of waves and an observer are moving relative to each other.

Let's throw another stone into the pond. On the surface of the pond there is a stationary pond skater. As the wave passes, he takes out his stopwatch and counts the number of times per second that a crest pushes him upward. He is measuring the frequency of the wave, *as he sees it.* Now suppose that we repeat the experiment but that this time we are dealing with a cowardly pond skater who, on seeing the stone strike the water, skates rapidly away from the stone. Cowardly, but still cooperative enough to take out his stopwatch and measure the number of crests that lift him upward each second. Since he is running away from the source of the wave, fewer waves will pass him in a second than will pass his static friend. You can see this very easily if you think of a special case: a surfer riding on a wave. He is traveling at exactly the speed of the wave, and therefore no waves will pass him; and *for him*, according to the definition of frequency, the frequency of the wave is zero. In general, if you move away from a static source of waves, you observe a lower frequency than if both you and the source are static. If you think for a moment, you will see that if *you* are static and the source is moving away from you, you will also observe a lowered frequency. All these closely related effects are included under the heading of the Doppler effect. The example of the Doppler effect usually given to undergraduates is that of an ambulance or anything with a horn or siren on it. As such a vehicle passes you, the pitch of the note that is heard suddenly drops as the source of the sound moves away from the listener.

If a source of light of known frequency is moving away from you, the frequency, as you observe it, will be lower than when the source is static. Since red light is at the low-frequency end of the spectrum of visible light, we can say that the light we are observing displays a *redshift.* Such shifts have been observed in the light emanating from certain heavenly bodies. As early as 1814, Fraunhofer observed the light of the Sun and detected a large number of different frequencies. When he did the same thing with light from the stars, he found very many frequencies that were almost identical to those from the Sun, but slightly red-shifted. He did not realize the huge significance of this discrepancy, which in the twentieth century was a key factor in the development of our theories about the birth of the universe.

Polarization

More "proof" that light is wavelike and, more specifically, that it is a transverse wave, comes from experiments which showed that light exhibited a property that was only known to be associated with transverse waves: the property of *polarization.*

Take hold of one end of a rope and ask a patient friend to take the other end. Move your arm up and down. You will produce a transverse *vertical* wave along the rope. Now change to a side-to-side motion of your hand. You will produce a transverse *horizontal* wave. In fact, you can produce a wave in any direction that you like, which is what we mean when we say that transverse waves can be *polar-*

ized. This is something you can't do with a longitudinal wave, where the motion is always along one line, the direction of the wave. Recall the Slinky.

Normally the light we experience, from either natural or artificial sources, is unpolarized. Perhaps a better word would be *multipolarized*, since such light consists of a mixture of light in very many planes. Each radiating atom gives radiation in an arbitrarily oriented plane. However, there are natural and artificial substances that can polarize such light. What they do is to cut out all wave motion that is not in a certain plane. Some crystals have this property. They have a plane, which we will refer to as the *polarization plane*, such that unpolarized light shone onto the crystal emerges polarized in that plane. You can do much the same by waggling your rope through a slot (Figure 15.5). Only a waggle in the plane of the slot will pass through.[7]

If you vertically waggle a rope through a vertical slit, the wave will pass through, but if you now put a second slit, oriented horizontally, after the first slit, you will kill the wave. Similarly, you can extinguish a beam of light by passing it through two polarizing crystals, with their polarization axes arranged perpendicularly. Later on these simple conclusions will lead us to doubt whether we understand reality.

Movers and Shakers

Electromagnetic waves interact with matter, which accounts for photography, the mechanisms of sight, Sun-induced skin cancer, bleaching by the Sun, the mutation of genes, the greenhouse effect, the destruction of the ozone layer, and photosynthesis. All these phenomena have something in common.

Water waves move things like ships and pond skaters. Sound waves move eardrums and shatter crystal decanters. What do electromagnetic waves move? The answer follows directly from their nature: the varying electric field acts on electric charges; the varying magnetic field acts on magnets. The electric interaction is far stronger and is the only one that we will consider, although magnetic interactions cannot be ignored in certain important areas of medical instrumentation and scientific research. Since all atoms contain charged particles, the door is open for electromagnetic waves to act on matter. The net result of absorbing light is always seen in a rearrangement, however small, of the molecule's electron cloud. If this rearrangement is sufficiently drastic, the molecule may change shape or even fall to pieces. UV light can do this to a large number of molecules, which brings us to the ozone layer.

Much attention has been given recently to the depletion of the ozone layer and the consequent increase in the quantity of UV light reaching the Earth's surface. The frequency of UV light is higher than that of visible light. Very crudely put, it

Figure 15.5. The polarized light passes through the first slit but is "killed" by the second.

---

[7]Strictly speaking, if the waggle is at any angle to the slot except perpendicular to it, part of your waggle will pass through, but the rope on the far side of the slot will only waggle in the plane of the slot.

can shake the electrons even more violently than visible light, so violently, in fact, that the electronic "glue" holding the molecule together can be disrupted and the molecule falls apart. The resulting fragments are usually very chemically reactive. In the living cells of humans, the best scenario that can be hoped for, following this molecular disruption, is that the two fragments should join together to reform the shattered molecule. The worst scenario is that the active fragments attack other cell components. This can happen when sunlight falls on the skin and can be the first stage in the development of malignant melanoma.

The molecules used in pressurized spray cans include a group called CFCs for short. CFCs contain chlorine atoms bonded to carbon atoms, and the UV light in the Sun can break this bond and liberate free chlorine atoms. These attack and disrupt the ozone molecule. Now atmospheric ozone absorbs much of the UV light of the Sun, and the depletion of the ozone layer in the Southern Hemisphere, caused by CFCs and other gases, has given rise to what has become known as the "ozone hole" through which UV light streams almost freely. The number of cases of skin cancer has risen significantly in Australia. It should be pointed out that nature itself is not always on our side in regard to the atmosphere. Volcanic eruptions can result in the emission of harmful chemicals that circle in the stratosphere for three years or more.

From what was said earlier about laser light, it is understandable that when laser light falls on matter it zonks the electrons around so violently that there is a strong rise in temperature, which explains why lasers can burn their way through metals.

For well over a century, the light generated by heating different materials has been used as a means of identifying them. In 1860 a German chemist noticed that when he placed a small quantity of common salt on a platinum wire and put the wire into the flame of a Bunsen burner, the flame burned yellow. Why a Bunsen burner? Because he was the man who invented it: Robert Bunsen (1811–1899). Piped gas, generated from coal, had become newly available as a consequence of the development of the German chemical industry. Bunsen put other salts in the flame. Copper salts burned green, potassium salts gave lilac, lithium gave a red flame. His colleague Gustav Kirchhoff (1824–1887) suggested that they let the light from the incandescent salts fall on a prism. The emerging light was thus split into a spectrum of frequencies, each spectrum being characteristic of the substance they were heating (Figure 15.6). These spectra were one of the first pieces of evidence for the internal structure of atoms, the once supposedly indivisible units of matter. Bunsen and Kirchhoff had no means of answering the obvious questions: Why these frequencies? Where does this light really originate? But their work provided a powerful empirical method of identifying the components of a sample of matter, although it was primarily limited to substances containing metal ions.

In 1860 Bunsen and Kirchhoff discovered a new element, cesium, which they

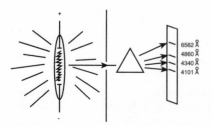

6562 Å
4860 Å
4340 Å
4101 Å

Figure 15.6. The wavelengths of the visible light emitted by hydrogen atoms that have been heated by an electric discharge. Many more wavelengths are emitted, which are invisible to the human eye.

first detected by its unfamiliar spectral frequencies and subsequently isolated from their sample by chemical methods. In 1861 they found rubidium, another metal.

Analysis of the light from the stars provides a means of detecting chemical elements in the heavens. It is true that we can land on some cool heavenly objects, such as the Moon, and actually take samples to analyze, but only spectroscopy reaches to the farthest galaxies.

More about Waves: Wave Packets

Unlike a particle, a wave has no position. You cannot point at a specific spot on a wave and say, "That's where it is." However, interesting things happen when you start to mix waves of different frequencies.

Figure 15.7a shows what happens when two waves of slightly different frequencies are mixed. As you can see, the intensity of the joint wave varies in a regular way. Anyone who has played the same note on two different strings of a slightly out-of-tune guitar will have heard "beats," a kind of throbbing the frequency of which increases as the difference between the two frequencies increases. Figures 15.7b and 15.7c show what happens when several waves of different frequency are mixed. The result is known as a *wave packet*. It is a traveling disturbance that, compared with a pure wave, is comparatively localized; at any given time it has a "position," which can be defined by the point of its maximum intensity.

The spatial spread of a wave packet depends on the range of frequencies used to make it. If all the waves that contribute have pretty much the same frequencies, then the wave packet will be fairly long, as in Figure 15.7b. If a wider range of frequencies is used, the packet is narrower, as in Figure 15.7c. If an extremely wide range of frequencies is mixed, the packet narrows so much that we can attribute to it a very well defined position, as though it were a particle. These are not idle games. They touch on the very nature of reality and will be of deep significance when we come to talk of quantum mechanics.

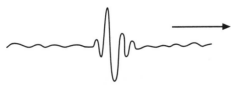

Figure 15.7a. The superposition of two waves of different frequencies produces "beats."

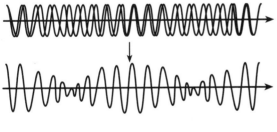

Figure 15.7b. The combination of many waves of slightly different frequencies produces a broad traveling wave packet.

Figure 15.7c. The combination of many waves of widely varying frequency produces a narrow wave packet.

The Problems Start

So far, so good. Huygens's waves have triumphed over Newton's corpuscles. But in the hour of Maxwell's tour de force, doubt was seeping through the Victorian floorboards. Maxwell had shown that light was a wave, but a wave in what?

Water waves travel in water; sound waves travel in gases, liquids, or solids. In what do electromagnetic waves travel? Unlike any other kind of waves that we know of, they can travel in a vacuum. What is waggling in a vacuum? It would in a way be easier if light consisted of corpuscles. Bullets have no trouble going through a vacuum. But Young seemed to have killed that possibility. Hertz said of Maxwell's equations that "Maxwell's theory consists of Maxwell's equations." In other words, use the equations, don't ask too many questions. Maxwell's laws allow us to calculate the strength of the electric and magnetic fields for any system, including light waves. They give the right answers, and it is of no practical importance whether we have a reassuring physical picture of electromagnetic waves. Logical positivism ignores awkward questions.

The idea of waves-in-nothing was deeply disturbing to the scientific establishment. We noted previously that the "stiffer" the medium through which sound waves travel, the faster waves travel through it. It is possible to quantitatively relate the velocity of a wave to the stiffness of the medium; if this is done for light waves it turns out that they must be traveling through a medium that has the stiffness of steel. Where is this medium? Why aren't we and the planets continually hacking our way through it? In order to salve their consciences, the physicists invented a universally present medium, dubbed it the *ether,* and hemmed and hawed when they were asked to explain its properties. HMS might well regard this as a step of dubious morality. No one had the slightest evidence for the ether.[8] Moreover, the required properties of the ether were a bit too much to swallow. Apart from having the stiffness of steel, it was also supposed to penetrate transparent substances, so as to provide a medium in which light could waggle. Not too credible.

Scientists realized that *if* the ether existed it might provide an answer to an old problem: Is there an absolute space, a framework fixed in the universe to which all motion should be referred? If the ether was motionless, then perhaps the Earth's motion through the ether could be measured.

The attempt was made in 1887 in Cleveland, by two Americans, Albert Michelson (1852–1931), born in Prussia, and Edward Morley (1838–1923). They argued that if we were drifting through a sea of ether, like a fish in water, then, if light was a wave in the ether, a light beam shining in the direction of the Earth's motion should behave slightly differently from a beam traveling at right angles to the motion. Using extremely sensitive methods, they found no difference whatsoever. This was one of the great null experiments of all time. If the ether was there, it gave no sign of it.

The result induced instant migraine in the physics community. Michelson regarded the experiment as a failure, and Lord Kelvin (1824–1907) pronounced it a real disappointment. It is an interesting sidelight on the dynamics of scientific respectability that when the experiment was repeated thirty-five years later by D. C.

---

[8]Descartes had postulated an omnipresent ether, not to carry waves but to satisfy his belief that there could not be a vacuum in nature. The idea was rejected by Robert Boyle, since he could find no experimental evidence for it.

Miller, his positive result was rejected. By that time the special theory of relativity was so firmly established that the existence of an ether was no longer compatible with accepted theory. Anyway, his results later turned out to be experimentally suspect.

Despite the fact that Maxwell's equations worked for most purposes, the physicists knew that all was not well with the world of physics. But the ether was not the only problem with light.

The Problems Continue

Why is it pleasant to kiss a lipsticked mouth but not a red-hot poker? Both are red. The answer becomes evident in a darkened room. All visible light *originates* in very hot bodies, but what we normally see is *reflected* light. Lipstick is not a source of visible light; it's not hot enough.

The light given off by a body depends on its temperature. Heated bodies give off a range of electromagnetic waves, due to the motion of the charged particles from which they are made. At normal temperatures this radiation is rather low-frequency; if you like, you can imagine that the movements of the particles are leisurely. Almost everything in your surroundings is at about room temperature. These objects are giving off a range of frequencies, mostly in the IR, which is invisible to the naked eye. If your temperature goes up, the distribution of frequencies will change slightly. The light at the lowest frequencies will decrease in intensity and the higher frequencies will increase in intensity. In general the distribution of frequencies looks like a hill (Figure 15.8a), and as you warm up, the peak of the hill moves toward higher frequencies. You are unlikely to get so hot that the hill starts to encompass visible frequencies. The Sun, of course, is hot enough, and gives off not only visible light but also radiation of much higher frequencies, in the UV. The Earth, on the other hand, has an average surface temperature of about 8°C and the top of the frequency hill is in the IR range of radiation.

In the ninteenth century a peculiar fact emerged about the radiation produced inside a closed heated container. The way that this radiation is analyzed is to construct a hollow cavity, which can be maintained at various temperatures, and drill a tiny hole in it to allow a beam of radiation to escape. This radiation is termed *blackbody radiation.*

We can construct a graph showing the distribution of the intensity of the different frequencies that are detected in blackbody radiation (Figure 15.8b). This graph,

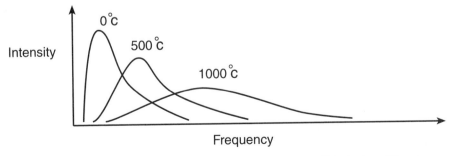

Figure 15.8a. The distribution of the frequencies of the radiation emitted by a body at different temperatures. As the temperature increases, the radiation moves to higher frequencies.

of course, depends on the temperature at which the cavity is maintained, but the intriguing finding is that, for a given temperature, the curve is quite independent of the material from which the container is made; it depends only on the temperature of the container. This rather out-of-the-way observation eventually produced the greatest revolution in science since Newton. The problem was that no one could explain the shape of the curve that showed the distribution of the intensities of the different frequencies. It was not that there was a poor fit between observation and Maxwell's theory—there was no fit.

Here we have an example of the periodical impotence of established science, the failure that smashes the paradigm and prepares the ground for the next revolution. It was evident, as the nineteenth century drew to a close, that, as far as blackbody radiation and the problem of the ether were concerned, Newton and Maxwell were of no help; there was something very wrong with the basic physics. There was far more wrong than the scientists of the time could have known, but before a new paradigm was constructed, Newton ran into more deep trouble. To appreciate his posthumous tribulations, we must first say a word about molecular motion.

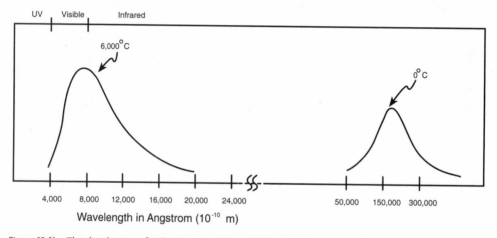

Figure 15.8b. The distribution of radiation emitted by a blackbody at two temperatures.

16

The Ubiquity of Motion

Shake it, baby, shake it!

—Marvin Gaye

The Ubiquity of Motion

In 1827, as Beethoven lay dying, the Scottish botanist Robert Brown observed an aqueous suspension of pollen grains under the microscope and found that they were in continual motion, not in smooth waltzlike trajectories but in sudden, break dance-like jumps in all directions. He concluded that this motion was displayed only by the male sexual cells of plants. (In those days women moved more decorously than men.) Later he found that pollen from long-dead plants exhibited the same phenomenon, as did unambiguously nonliving systems, such as tiny specks of dust derived from crushed stones. Brown was, of course, not observing molecules, but unknowingly he had stumbled across the visible consequences of molecular motion.

Brown was not the first to observe what has since become known as *Brownian motion*, but he was the first to study it seriously. He volunteered no explanation. It was another fifty years before the correct explanation was put forward—that the pollen grains were being battered on all sides by the endlessly scurrying, but invisibly small, molecules of water. Pollen grains are far smaller than the average dust particle; they are small enough to be detectably pushed around by molecular bombardment. In a dense crowd you would not expect to be simultaneously and symmetrically bumped into by all your neighbors. Similarly, the chances are that a pollen grain in water will be hit unevenly, in the sense that at any given moment its surface will not be subjected to equally strong simultaneous molecular blows from all sides. Any imbalance in the blows will push the grain into movement. That is why it jumps around.

Brownian motion provided the first direct evidence of the immortality of molecular motion: *Brownian motion does not die out.* If you observe a suspension of fine particles under the microscope and preserve the sample, your descendants to the nth generation will see exactly what you see. The inference is that the water molecules are exhibiting perpetual motion. Furthermore, even the average length of the jumps exhibited by the particles will remain the same—provided the temperature at which the observations are made is always the same. This strongly suggests that the average strength of the blow delivered by a water molecule is unaffected by the passage of time, which in turn implies that the average speed of the water molecules, at a given temperature, is constant. A similar statement was made about the average speed of our molecules-in-a-box, and you can look upon Brownian motion as very strong circumstantial evidence for the validity of that statement, which I previously asked you to take on trust.

The central idea of this chapter is that *all* the microscopic components of matter are jumping around; that gases, liquids, and solids are characterized by ceaseless microscopic motion.

In a gas, molecules move in straight lines until they bump into each other. All molecules except those few that have only one atom, like helium, also exhibit *internal* motion; they can rotate and vibrate, twist and shake. This may seem to be an out-of-the-way professorial bit of information, but we will see that the greenhouse effect is largely due to the existence of internal motions in the carbon dioxide and methane molecules.

Gas molecules are free to move where they will, which is consistent with the fact that a sample of gas released into a container will immediately spread out and fill the entire space. In liquids the molecules attract each other enough to keep very close together. However, they still move about, but in a manner resembling people in a dense crowd. Their motion is akin to the mixing of the tightly packed participants in a crowded cocktail party.

If molecules stopped moving in liquids we would all drop dead. Within the living cell, materials have to be moved from place to place; otherwise, substances coming into the cell would not reach the molecules that they have to meet for the life process to continue. The movement of molecules *in* the cell is in general simply a result of their inherent inability to remain stationary. The movement of molecules *into* the cell is not always so simple, and may involve their being carried by other molecules.

In solids, atoms or molecules generally occupy permanent sites from which they rarely stray, but they can be crudely envisaged as a gospel choir, each singer swaying, wobbling, and bobbing up and down, but remaining tethered to their places as a consequence of the close proximity of their neighbors. Motion in solids is thus almost entirely *vibrational*. You can't see or feel these vibrations; they are too small.

Movement of molecules from place to place, so-called *translational* motion, occurs continually in gases and liquids but can occasionally occur in solids, although it is generally very slow. One manifestation occurs in old color films. The differently colored dyes in a cinema film were originally confined to certain areas. However, in old films the borders between these areas are sometimes blurred because, over the years, dye molecules slowly migrate across the film, smearing Jeanette MacDonald's lipstick and Gable's razor-sharp moustache. This is an example of *diffusion*, the slow spreading of one substance through another by molecular motion.

Molecular motion is universal. It underlies almost every kind of change found in the inanimate and animate world. The explanation of molecular motion, how molecules move and what the consequences of that motion are, has grown into a sophisticated and highly successful theory. Theory grows from observation, and although it is not easy to see molecules, there is a wealth of readily observable macroscopic phenomena that depend on microscopic molecular motion. It was from the study of these phenomena that the unifying theme of molecular motion emerged. The first such study was conducted Robert Boyle. In Chapter 1 we saw that if a sample of gas is kept at constant temperature, then the product of its volume and the pressure exerted on it remains constant, no matter what the pressure. The correct model for the explanation of Boyle's Law was given by Daniel Bernouilli, who guessed that the molecules are in motion. The discovery of Brownian motion subsequently supported his guess.

What Molecular Motion Can Do

Mark off, in your imagination, a square inch of skin. Every second this area will be subjected to 2×10^{24} blows from the molecules in the air. This incredible and ceaseless bombardment affects every surface exposed to the air. The human body is not sensitive enough to detect the tiny individual blows delivered by gas molecules, but nevertheless a very significant observable effect arises from this bombardment.

Imagine a vertical metal plate welded to a wheeled platform standing on a railway track. The plate is perpendicular to the direction of the tracks. Fire a few shots at the plate from a revolver. The effect of each impact is a ping and a very slight jerk forward of the wheeled platform, which then comes to rest after each shot. (In a frictionless world, and in a vacuum, the plate would of course roll on forever, condemned to do so by Newton's first law of motion.) Now take a submachine gun and fire a prolonged, continuous, very closely spaced stream of bullets at the plate. *The platform accelerates smoothly.* According to Newton's second law, acceleration implies that there is a net force acting on the body. What is happening is that the storm of very closely spaced impacts appears to create an effectively *constant* force on the plate. If the points at which the bullets strike are fairly equally distributed over the whole surface of the plate, then the force is evenly spread out over the whole area. In this case we can work out the force due to the bullets striking a unit area (say 1 square inch) of the plate. The definition of pressure is force per unit area. The ballistic bombardment results in pressure.

Now replace the bullets with molecules and it is believable that what we call air pressure is merely the macroscopic effect of a microscopic molecular bombardment. The atmosphere presses on everything, not because of its "weight" but because of the ceaseless movement of its component molecules. The first person to differentiate between the weight of a sample of air and the pressure that it exerts was the Frenchman Pierre Gassendi (1592–1655), the atomist and mechanist who is attacked, along with Descartes, in *Gulliver's Travels*. Gassendi realized that the weight of the air in a container had nothing to do with the pressure that it could exert. A 50-liter compressed air cylinder usually contains about 200 gram of gas, but the internal force acting on 1 square centimeter of the container is about 5 kilogram weight. If the cylinder is heated, the weight of the gas remains unchanged but the pressure rises. The explanation of gas pressure in terms of molecular motion was first clearly stated by the Swiss Daniel Bernoulli, one of seven distinguished Bernoullis.

The Inquisition Pursues the Inquisitive

The Bernoullis were originally based in Antwerp, but when that town was taken by the Catholic Spanish forces of the duke of Alva, the Protestant family prudently packed their bags and fled to Switzerland, to avoid the Inquisition. They settled in Basel in 1583 and produced three generations of mathematicians, physicists, and astronomers holding academic positions in Basel, Berlin, and St. Petersburg. Daniel (1700–1782), perhaps the most distinguished member of the clan, taught mathematics, physics, botany, anatomy, and, in the end, philosophy at the University of Basel. In Basel, the Bernouillis were free to worship as they wished and to publish their scientific thoughts.

In 1738, Daniel Bernoulli published his most important and influential book, *Hydrodynamica.* In it he suggested that invisible microscopic bombardment ex-

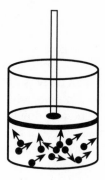

16.1. Why Bernoulli's piston did not fall.

plained some commonly observed macroscopic phenomena. In particular, he asked why it was that, even in the absence of significant friction between cylinder and piston, the piston in Figure 16.1 didn't immediately fall to the floor. His explanation assumed that the molecules (he called them particles) were in continual motion. If the piston fell, the molecules would be crowded closer together and the number of hits per second on each square inch of the piston would increase. The pressure of the enclosed air on the walls of the container, and on the piston, would increase and resist the fall of the piston. Think of 1000 dim-witted houseflies in Carnegie Hall, bumping into the walls. Now reduce the hall to the size of a cigar box. There would be a huge increase in the number of collisions per second on each square inch of wall. In other words the fly-pressure on the walls would increase tremendously. Bernoulli showed that Boyle's law followed from his model of a gas. His explanation is better than Newton's because Bernoulli's model of moving as opposed to static molecules can be used as the basis of explanation for many other properties of gases.

Establishment Inertia, Again

> New Opinions are always suspected and usually opposed, without any other reason but because they are not already common.
>
> —John Locke

Bernoulli's correct hypothesis about the origin of gaseous pressure, which was the first example of modern theoretical atomic physics, was published in 1738 and ignored until 1821. Only then did one Herepath dig it up, in the year that Napoleon was buried. The eighty-year neglect of Bernoulli's hypothesis is yet another reminder of the all-too-frequent conservatism of the scientific establishment. Thus it was that, when an important equation in the theory of gases was formulated by the Englishman John Waterston, and submitted to the Royal Society, it declared the manuscript to be "nothing but nonsense." Poor Waterston decided to publish his paper in an obscure railway magazine of which he was the editor. Not surprisingly, the paper did not attract a huge amount of attention. Eventually its merit was recognized and, in a pitiful attempt to make amends, the Royal Society published the paper in 1892. This was an example of scientific pie in the sky—Waterston had died nine years earlier.

The Jimmy Connors Effect

If the concept of perpetual molecular motion in gases is accepted, a whole range of phenomena can be rationalized. We concentrate on one, avoiding math, but showing that extremely simple ideas can explain a phenomenon that puzzled generations of wise men.

A tennis ball hits a vertical wall. It leaves the wall with (roughly) *the same speed* it had when it struck. Now replace the wall with Jimmy Connors, making his famous two-handed forehand drive. As the ball hits the racket, it meets the strings coming the other way. The ball leaves the racket at a *greater speed* than that which it had when it hit the strings (Figure 16.2). Now change the tennis court into a closed container, convert the tennis ball into a molecule, and let a piston play the part of the tennis racket.

Molecules in a container into which a piston is moving forward will rebound off the piston at higher speeds than that at which they hit. This is the Jimmy Connors effect, which results in a rise in the average speed of molecules in the container. But we know that increased average molecular speed corresponds to a rise in temperature, and this is exactly what is observed. If you compress a gas, its temperature rises. Those of you who grew up with a bicycle will remember that when you vigorously pumped up a tire the pump heated up. That was the Jimmy Connors effect. Molecular motion is not a fiction.

The Newtonian Picture Becomes Blurred

It should be noted that we have not the slightest possibility of describing the motion of a single molecule in a gas, in the way that we can describe the path of the Moon. We can't see molecules, and they make too many collisions per second. What predictive limitations arise from our ignorance of the movement of individual molecules? How can we quantitatively predict anything about a collection of trillions of wildly careering particles?

Our present, and probably eternal, inability to predict the motion of single gas molecules can be (very imperfectly) compared to our inability to predict the life span of a given individual. This latter limitation doesn't worry life insurance companies. They will give you a good estimate of the *probability* of a certain client dying reaching, say, age forty-seven. They can't point to a specific Joe Bloggs and tell you that he *will* be battered to death at the age of forty-seven, but, after examining his medical and other records, they can say that he has a 22% *chance* of dying

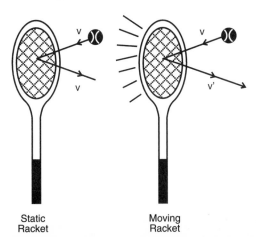

Static
Racket

Moving
Racket

16.2. The Jimmy Connors effect.

before he is forty-seven; that is the best that they can do, but it is good enough—life insurance companies don't often go broke.

For many systems, probabilities can be predicted with great accuracy. No one, except Fat Louey, can tell you that the first card that you are going to select from an unmarked pack of cards will be the king of clubs, but the *probability* that you will do so is exactly 1 in 52. We often use the language of probability when we speak of the properties of large collections of molecules. No one can predict the velocity of a given molecule in a gas at a given time, but you can get an excellent estimate of the average velocity and of the *probability* that a molecule has, at any time, a given velocity, say 2.65 or 4.76 times the average velocity. These estimates come from the kinetic theory of gases, a theory to which many made contributions, none more than the great James Clerk Maxwell. This theory deals almost entirely in terms of probabilities. The reason that it works is the same reason that insurance companies remain solvent. They have enough customers. When you deal in large numbers, probabilities are almost the same as certainties. I wouldn't bet my life on the toss of a single coin, but I would, with great confidence, bet on heads appearing between 49% and 51% of the throws of a coin if the number of throws was 1 billion.

There is a subtle and fateful change in going from the language of Newtonian laws to the language of probability, and it is a change that will hound us on and off until the end of this book. In the meantime, we ask another question about molecular motion and get an awkward answer.

Perpetual Motion

Most moving objects in the familiar, commonsense world eventually come to rest, and if they don't do so of their own accord we can, in principle, stop them dead in their tracks. Thus water roaring over Niagara Falls can be collected in a bucket and will in a few seconds come to complete rest—except for the microscopic motion of its molecules. We all eventually come to rest—except for the microscopic motion of the molecules of our skeletons. Under what circumstances does this molecular motion cease completely? Under no circumstances!

It is easy to stop the *translational* movement of molecules. Freeze a bottle of olive oil, and the molecules stop moving from place to place. Any *rotational* movements they may have had in the liquid state are also frozen out. But they don't stop moving. They *vibrate* in complicated patterns. The amplitude of the vibrations can be reduced by lowering the temperature. Can the amplitude be reduced to zero, so that the vibrations are killed? No! At the lowest temperatures attainable in the laboratory, *or theoretically attainable,* all molecules are still vibrating. We say that they have "zero point energy." For reasons that will appear later, *there is no way in which molecules, or atoms, can be stopped from vibrating.*

Molecular Motion, Violent Encounters, and Chemical Change

Molecular motion is a prerequisite for chemical interaction between molecules. The piece of magnesium metal in a flashbulb cannot burn if oxygen molecules do not somehow arrive at its surface. The fact that molecules usually have to meet to react is trivial, but molecular motion has another indispensable role in most chemical reactions.

Chemical reactions almost invariably involve the rearrangement of molecules.

For example, if you put a match to a mixture of hydrogen and oxygen, you will get a jolting explosion, and the net result of the reaction will be the production of water. In the process, the atoms of the oxygen molecules and the hydrogen molecules become rearranged to give molecules of water, which, as we know, consist of two atoms of hydrogen linked to one atom of oxygen. Now you can mix hydrogen and oxygen gas and nothing at all will happen. You can warm up the mixture by placing the container in boiling water. Nothing happens. We know, or can guess, that molecules of hydrogen and oxygen are bumping into each other at a staggering rate, and yet water is not formed. They have to meet to react, but meeting is not enough. That is because they are not bumping into each other sufficiently violently. It is rather like two Rolls-Royce Silver Clouds hitting each other head-on when both are traveling at 2 mph. At worst the chauffeurs will vibrate slightly. Nothing gets broken, even if the champagne slops around a little. This is what usually happens when two molecules in the air meet at room temperature. The way they vibrate may be slightly altered, but nothing much else occurs. If you want a visual model you can think of a "molecule" made of a few metal balls held together by short springs, bumping into another such "molecule" at moderate speed. Plenty of wobbling, but no real damage.

When two cars, each traveling at 100 mph, collide frontally, there is major structural damage. When two molecules smack into each other at high enough relative velocity, there is a breaking of existing bonds and the formation of new bonds. The molecules that come out of the encounter are not those that came in. If there is no reaction, we can guarantee to produce one by raising the temperature. For a pair of any given makes of car, there is a minimum relative velocity below which no damage occurs on contact. Something similar is true for reactions between simple molecules. There is a certain critical relative velocity, specific for each reaction, below which no reaction occurs.

In general, raising the temperature encourages chemical reactions to go faster. That's why you put food in the refrigerator, to slow down the chemical reactions—usually induced by bacteria—that we call decay.

Motion is the key to reactions of another kind—reactions between *nuclei.* In the Sun, radiation is generated by the collision of protons. Protons, being positively charged, will repel each other, and at close quarters this repulsion is so strong that the chance of two nuclei getting close enough to "touch" is very small. This disinclination to get to grips can be overcome by giving them high velocities. In the Sun the temperature varies from about 15 million degrees Celsius at the center to about 6000°C at the surface. At the higher temperature the average speed of the protons is over half a million kph, about 1000 times the average velocity of an oxygen molecule at room temperature. Two protons approaching one another at these speeds have sufficient momentum to overcome their electrostatic repulsion and smash into each other, forming a deuterium nucleus and two unstable particles—a neutrino and a positron. This is the first step in the production of the solar energy which I hope is warming you now. An average proton in the Sun can rush around colliding for a billion years before it reacts with another one. That is why the Sun is burning slowly,[1] which suits us fine.

---

[1]Per kilo, a human being produces 10,000 times the energy produced by the Sun in the same time. But (big but) the mass of the sun is 10^{19} times that of the total mass of the population of the Earth, so the Sun produces 10^{15} times the heat energy produced by the whole population of the Earth in the same time.

For the kind of temperatures prevailing on Earth there is a huge range of rates at which chemical reactions occur. Some reactions appear not to go at all. The nitrogen in the air doesn't burn in the oxygen of the air, even when I light a match. You can imagine the consequences if it did; the first man to have made fire would also have been the last. On the other hand, put a small piece of phosphorus in a jar of chlorine gas and a violent reaction occurs, lasting no more than a few seconds. Life, at least in the form that we know it, would be impossible if all chemical reactions were either very slow or very fast. In fact, many of the reactions that take place in the living cell take place extremely slowly when we try to reproduce them in the test tube, even though we know that the rate of collisions between molecules is colossal. What is it about the internal environment of the cell that makes the difference? There was a time when the answer to that question would have included the "Life Force," the magic ingredient that was supposed to distinguish living matter from dead. The real answer is that all living cells contain *catalysts*.

A catalyst is a substance that speeds up the interactions between molecules or atoms. Often it does this by holding down one of the reacting molecules while the other one attacks, rather like a "heavy" holding someone by the arms while his boss delivers blows to the victim's belly. The interaction with the catalyst alters the fine details of the molecule's structure enough to either break it on the spot or soften it up to attack by molecules that would not otherwise be able to react with it at that temperature. The essential feature of most catalysts is that they are built in such a manner that their shape, or part of their structure, is adapted to the shape of one or both of the molecules between which they are attempting to induce a reaction. There are something like 1000 catalysts in each living cell. We call them *enzymes*, and they speed up the reactions that keep life going.

Another example is the catalysis of the reaction between hydrogen and nitrogen, which is carried out industrially to produce ammonia. This reaction can be speeded up enormously by the presence of a piece of platinum metal. The reaction takes place on the surface of the metal. The German chemist Fritz Haber developed this process, which saved the German cordite (explosive) industry in World War I.[2] The British munitions industry at the time was saved by Chaim Weizmann, later to be the first president of the State of Israel. He invented a novel process for manufacturing acetone. In 1934 Haber died on his way to Israel to take up a position as head of the Sieff Institute, later to be called the Weizmann Institute.

It has been estimated that about a quarter of the GNP of the United States is based on the sale of products the manufacture of which involves catalysts.

Chemical change is impossible without molecular motion, but molecular motion has another string to its bow: it lies behind the idea of heat.

---

[2] Haber strongly objected to the Nazis' requests that he dismiss Jewish employees in his research institute. In a letter of protest he wrote, "For more than 40 years I have selected my collaborators on the basis of their intelligence and their character, not their grandmothers."

17 | Energy

A man meets a woman on a bitterly cold, windy Chicago street corner. Her hands are freezing, his are warm. He presents her with a bunch of faded flowers and then, humming, "Your tiny hands are frozen," warms her hands with his. There have been three very different kinds of transfer between them. One was material and tangible: the flowers. The other two were intangible and involved no material transfer: the heat that went from his hands to hers, and the information that went from his mouth and eyes to her brain. There are very many intangible transfers between animate and inanimate objects, transfers that appear at first to have nothing in common but in fact are closely linked.

Throw a ball. Come to rest after the ball leaves your hand. The ball is moving; you are not moving, but you were. An observer could summarize what he saw by saying that *motion* had been transferred from you to the ball. A falling column of water impinges on the blades of a turbine in a hydroelectric station, and the wheel turns. Again, motion appears to have been transferred from one body to another. When the attentive wimp warms his companion's cold hands, nothing material is transferred, but something has obviously moved from one body to another. It was only in the middle of the nineteenth century that the transfer of heat and the transfer of molecular (or atomic) motion would have been jointly recognized as involving the transfer of *energy*.

Energy, like force, is a concept that unifies a host of phenomenan. In discussing energy it is convenient to use the concept of work, so let's immediately define work as force times distance. Suppose you take an apple that has a weight of 2 newtons. This is the force that you have to exert on the apple to hold it up. Now lift it to your mouth, a distance of about half a meter. The work that you have done in lifting the apple is 2 newtons x 0.5 meter = 1.0 *joules.* The joule is a unit of work, named after the nineteenth-century Lancashire scientist James Prescott Joule (1818–1889). We all have a feeling for the meaning of work, but what about energy?

Energy, like work, is an abstract, man-made concept. You can't see energy or work. Beware of common usage here. You can see someone work, and you may be able to see the results of his work, but you cannot see the work itself. Energy can be thought of as something that enables work to be done. Thus a waterfall has the ability to do work by virtue of the fact that the water is moving. This is true of all moving bodies, and we call that ability *kinetic energy*. A body moving with a velocity v has kinetic energy equal to one-half of the mass of the body times the square of its velocity: $KE = (1/2)\ mv^2$. In the case of the waterfall, the kinetic energy of the water can be converted into the kinetic energy of a spinning turbine, which in turn is converted into the electric current produced by the dynamo, which can be used to run an electric motor, doing work.

On the other hand, an immobile boulder perched on the edge of a cliff has the *stored* ability to do work. It too could fall on a turbine blade, not a particularly elegant arrangement but sufficient to prove that its stored energy can be turned into work. We call stored energy of this kind *potential energy*. In countless cases a body is not moving but has the potential to do work, provided that it can move in response to the forces acting on it. In the case of the boulder the force that in the end produces the work is gravity, as it is in the case of the waterfall. The potential energy of the boulder, its ability to do work, clearly increases with its initial height. As the boulder falls, it is continually losing potential energy and speeding up, thus gaining kinetic energy.

There are many kinds of potential energy. A twisted rubber band in a toy airplane can potentially do work if it is released. The stored chemical energy in your muscle cells can do work if released. So can the stored chemical energy in petrol. There is no single simple expression for potential energy except in the case of a body that can do work by virtue of the fact that it can fall. In this case, the work, in joules, that it can do by falling through a height difference of h meters is given by mgh (m x g x h), where m is its mass in kilograms and g is the acceleration due to gravity, which appeared in Chapter 4. On the Moon g would have to be replaced by $g/6$.

The language of daily life assumes that we need energy to perform work and that, as we do work, energy gets used up. We cannot use the same coal twice. It looks as though energy—kinetic or potential—and work are different faces of the same thing. This statement includes any kind of potential energy, including chemical potential energy.

Suppose we lift a boulder onto a ledge. In doing so we do work and also lose part of the store of chemical potential energy contained in certain molecules in our muscles. So kinetic energy, all forms of potential energy, and work are all interconvertible. Daily life tells us that we have to extend this statement.

Consider a coal-fired railway engine. When the engine is moving, work is being done, since the engine is certainly exerting a force and thereby moving something. This time the immediate source of the work seems to be the heat of the burning coal. It appears as if *heat* can also be changed into work. Are heat and work also the same kind of thing? Are they two faces of energy?

The Return of the Fat Sumo

Two hundred years ago, heat was believed to be an invisible indestructable fluid called *caloric*, a word coined by Lavoisier. Caloric is another example of a theory that fits *some* of the facts. The particles of caloric were supposed to repel each other. This explained why heat had a tendency to spread from one body to another. It also explained why a body expanded upon being heated; the more particles of caloric in the body, the greater the internal repulsive force pressing the body outward, and so on. The fat sumo was alive and well.

The first real advance in burying Lavoisier's caloric was made by the man who married Lavoisier's widow. The highly colorful, and moderately obnoxious American, Benjamin Thompson, Count Rumford, appears to have lived about ten lives (some simultaneously), as scientist, diplomat, spy, philanderer, philanthropist, military adviser, creator of the Royal Institution, and others. Thompson observed that heat was produced during the boring of cannon barrels—for the Bavarian armed forces. It was clear to Thompson that as long as the drill was turning in the block of metal, heat was being produced. Caloric was supposed to be conserved, and yet here was an endless stream of caloric coming out and none going in. In the paper that he presented to the Royal Society, Thompson suggested that heat was a form of motion. Humphry Davy added to the evidence by showing that if two pieces of ice were rubbed together, they melted. No one took much notice.

Around about 1840, James Joule set out to prove quantitatively that mechanical work could be converted into heat and that work and heat were therefore equivalent, two aspects of the same thing. His apparatus is shown in Figure 17.1. The falling mass is subject to a force—its weight, which is mg (see Chapter 4). As the mass falls, the paddles stir the water. The mass falls through a distance, h, say. From the definition of work as "force times distance," this means that the work done by the attraction of the Earth is mg x $h = mgh$ joules. You may be unused to the idea of the Earth doing work, but if you stick to the scientific definition, you will see that a mass is being moved through a distance by a force. That's work!

The mass comes to rest on the floor. When it was suspended above the floor it had the *potential* to do work; on the floor that potential has gone. We can say that it has lost its potential energy. But if you measure the temperature of the water at the end of the fall, as Joule did, you will find that it has risen. Joule concluded that mechanical work, done by the falling weight in turning the paddles, could be converted into heat. The farther the mass falls, the greater the rise in water temperature: a certain amount of work was equivalent to a certain amount of heat. Another way of looking at it is that we started with a certain amount of (potential) energy and ended up with the same energy in another form: heat. Joule had shown quantitatively that work and heat were two sides of the same coin. In modern nomenclature, 4.184 joules of work is equivalent to 1 calorie of heat.

Once again the concept of moving molecules simplifies everything: the moving paddles hit the water molecules and, like a tennis player returning a ball, they increase the speed of the molecules that they hit. The Jimmy Connors effect is working. (The blows will also increase the violence of the internal vibrations of the water molecules and the vibrations of the atoms in the paddles.) We know that increased motion is associated with a rise in temperature, which is exactly what is observed. All that Joule had done was find a complicated way of transferring the motion of the falling mass to the water molecules. Nothing material has gone from the mass to the water; only energy has been transferred.

In fact any collision of macroscopic moving objects results in the production of heat, because part of the translatory motion of the moving bodies is converted into the random motion of their atoms. Joule spent his honeymoon in Switzerland,

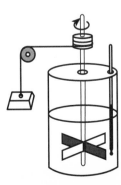

Figure 17.1. Joule's experiment.

where it occurred to him that water falling over a waterfall should warm both the Earth and the water slightly when it hit the Earth. He spent part of his honeymoon confirming this prediction. Mrs. Joule's comments have not been recorded.

Joule presented the results of his paddle experiments to the British Association in 1843. The reception veered from skeptical to frigid. He tried again two years later, this time adding his idea that water at the bottom of Niagara should be warmer than at the top. No one cared. In 1847 he returned, but this time there was a genius sitting in the audience. William Thomson, twenty-six years old, realized what his elders failed to see: that Joule had made one of the great advances in the understanding of nature. Thomson, later to become Lord Kelvin, sparked the interest of the scientific establishment.

The stubbornness of the scientific world in the face of Benjamin Thompson's observations and Joule's experiments were not the only examples of mental inertia in the embarrassing history of heat. The equivalence of heat and mechanical work was also stated by the German physician Julius Mayer (1814–1878) in a paper that was rejected as worthless by the German scientific journal *Annalen der Physik*. A modified version of the manuscript was published by another prestigious journal, but it left the readers cold. The acceptance of Mayer's ideas was probably hindered by the obscuring influence of *Naturphilosophie* in his writings and the fact that he did not have Joule's clear conception of the meaning of energy.

In 1847 the distinguished physician and scientist Hermann von Helmholtz sent a manuscript to *Annalen der Physik* presenting his conclusion that heat and work were equivalent and that energy was conserved in any process. You guessed it: rejected. The author subsequently presented his ideas at a scientific meeting in Berlin and paid for the article to be published privately. It was this article that was the major influence in persuading the rest of the scientific world that heat and work were interconvertible and quantitatively related. The idea of the conservation of energy was to become a cornerstone of science.

To get a feel for the units of work and energy, lift Newton's apple to your mouth. You are doing very roughly 1 joule of work, which is equivalent to a little less than quarter of a calorie of heat. To warm an average cup of water to boiling, you need about 16,000 calories. To get this amount of heat by doing mechanical work, you would have to do work equivalent to that needed to raise 64,000 apples to your mouth, or a 13-ton barbell from the floor to hip height.

Heat and Motion

The perceptive Robert Hooke guessed the truth: "Heat is a property of a body arising from the motion or agitation of its parts, and therefore whatever body is thereby touched must necessarily receive some part of that motion, whereby its parts will be shaken." Although the exact connection between molecular motion, heat, and temperature took a long time to gel, when it was finally established it provided simple, transparent explanations of countless phenomena—such as making a cup of tea.

Fill the container with water, plug in, and switch on. Electrons begin to flow through the heating coil. They bump into the vibrating atoms from which the wire is made. The blows that the atoms get from the rushing electrons induce them to vibrate more violently. This enhanced movement conveys itself to the atoms in the ceramic coating of the wire. The atoms of the ceramic beat upon the metal base of the water container, thereby increasing the vibrational motion of the metal atoms. You can guess the next step: the enhanced vibrational activity of the metal of the container is passed on again, this time to the molecules of water. The blows that they receive affect them mainly by increasing the average velocity with which they move and by increasing their internal vibrations.

There are two effects that can be easily measured. The water expands. This is because the increased motion of the molecules tends to push them farther away from each other. They need more lebensraum, more elbowroom. The other effect is that the temperature rises. How do I know? I can put my finger in. The enhanced molecular motion of the water molecules is now transferred to the molecules of my skin and progressively works inward until it reaches nerve endings, the molecules of which also begin to vibrate more vigorously. I hope that I'm not yet at a temperature where I'm doing physical or chemical harm to my cells. Somehow the increased vibration is converted to an electrical signal to my brain, and I "feel that the temperature of the water has risen." The phrase does not indicate what actually happens in my brain; I am merely using conventional language to describe the sensation. Of course I could put a mercury thermometer into the water. Again the heightened movement of the water molecules is conveyed by collision to the glass tube and via the glass to the mercury of the thermometer. The mercury atoms consequently also need more lebensraum, and the mercury expands, rising up the stem of the thermometer.

Notice that we have described the heating of water without mentioning the word *heat*. What has happened to the water is that it has "augmented its *molecular motion* by being at the receiving end of a succession of collisions, starting with the galloping electrons in the heating element." In everyday, macroscopic language: "The water absorbed *heat* from the heating element."

To heat an object is simply to make its atoms or molecules move more violently, whether by placing the object in contact with a hotter body or by exposing the object to radiation. No substance is transferred in either case.

"Heat" is a macroscopic concept. Almost every time we use the word *heat* we can replace it with the idea of molecular motion. However, normally it is far more convenient to use the macroscopic concept of heat rather than the microscopic concept of molecular motion.

The realization that heat and work are aspects of energy, and that work can be converted into heat, is one of the unifying principles that underlie the complexity of nature.

The Conservation of Energy

In Joule's experiment, when the mass can fall no farther it can do no more work. It had the ability to do work before it fell, and it has now lost that ability; we say that the potential energy of the mass is lost. On the other hand, the water has gained energy, the molecules are moving faster; they have increased their kinetic energy. Joule believed that the loss of potential energy of the mass was equal to the gain in the kinetic energy of the water. He was very nearly right. In fact the string connecting the mass to the paddle passes over a wheel, and a small amount of heat is produced by the friction between the axle of the turning wheel and its suspension. The loss of energy of the mass was in fact equal to the total amount of heat produced by friction and by the movement of the paddles. *No energy was lost in the process.* If we want to be pedantic, we have to admit that the water that heated up in Joule's experiment also heated the air surrounding the apparatus. A nit-picking summing up of the experiment is that the loss of potential energy of the mass is exactly equal to the *total* heat released into the environment. Nothing has been lost; energy is conserved.

The idea of energy as being something that was conserved was formally stated in the 1860s by the German physicist Rudolf Clausius (1822–1888). Today we speak of the first law of thermodynamics. In its most general form, the law says that there is a certain amount of energy in the universe and that we can never increase or diminish it. We can transform it from one form to another—work to heat, or potential energy to kinetic energy, or chemical energy to heat, and so on, but we never gain by these processes; we are merely changing currency (with no commission costs). The law is based on experience, and on the exact definition of what we mean by energy. It was not derived from a simpler law. All that we can say about it is "That's the way the universe is," as far as we know.

This is our third conservation law, after those for charge and mass. In each case the laws put restraints on the way the universe behaves. In each case the law is deduced from the behavior of nature. We noted that because of Einstein we had to modify the law of conservation of mass to take into account the interconvertibility of mass and energy. Obviously the same applies to the conservation of energy. If we want to be correct, we should speak of the conservation of mass plus energy, but since the conversion of one to the other is not something that happens in everyday life, it is far more convenient to use the two laws separately. We will run across situations where that cannot be done.

If the first law says that the amount of energy in the universe is constant, then why should we worry about how much energy we use? When a car engine works, part of the energy of the fuel is turned into heat, which warms the air. Why not take all that heat and reuse it? Better, consider all that kinetic energy in the molecules of the water of the ocean. Since Joule took the potential energy of a body and used it to raise the kinetic energy of water molecules, why can't we reverse the process? Let's turn the kinetic energy of the molecules in the ocean into work.

This brings us to the next chapter, the contents of which are really bad news for Newton.

18

Entropy: Intimations of Mortality

So far as scientific evidence goes, the universe has crawled by slow stages to a somewhat pitiful result on this earth, and is going to crawl by still more pitiful stages to the condition of universal death.

—Bertrand Russell

Open a can of petrol in a closed room. The room is also thermally isolated; heat cannot enter or leave. The smell of petrol will slowly spread through the room. The mechanism is simple: molecules leave the can and meander across the room. Why do they leave the can? Because, in the can, in the space above the liquid petrol, they are moving. Some of them will, *purely by chance,* find their way out of the neck of the bottle. It is absolutely essential to realize that they do not *purposefully* make their way to the opening; remember the blind, deaf, radarless, unsmelling flies of Chapter 1? In fact, some of the molecules that have left the can may find their way back, not because of a homing instinct but purely by chance. Molecules have no will. It would be a scary world if they did.

In time the petrol will distribute itself evenly over the whole room. This process is spontaneous—nothing pushes the molecules around. They are condemned to eternal motion. Because this motion *always* seems to result in their spreading out, never in their all going back to the can, the spread of petrol through the room is termed an *irreversible process.*

A large number of randomly moving molecules would be expected to end up *uniformly* distributed inside a container.[1] At any given time, we do not expect *exactly* the same numbers to be in each half of the room; there will be small fluctuations in the numbers as the molecules rush around. All the same, we do expect the numbers to be very nearly equal, and *on the average*, taken over a period of time, we can confidently predict them to be equal. This prediction is based on experience. You wouldn't expect all the molecules in the air of the room in which you are sitting now to accumulate on the other side of the room, thus leaving you in a vacuum and wantonly snuffing out your life. And yet *there is nothing in Newton's laws that says that this cannot happen.* In principle it could. Why doesn't it?

Why do we never find that all the petrol molecules have returned to the can? Most people would instinctively say, "Because it's very unlikely, isn't it?" Others might reply, "Yes, but why?"

---

[1] We can neglect the fact that, because of gravity, there would be slightly more molecules in the bottom half of the room than the top. This increase in density nearer the Earth is displayed by the atmosphere. However, the density of molecules would be equal in the left and right halves of the container. We also discount cases where the molecules condense to form a liquid, for example, the perfume in a closed bottle or the petrol in a closed can.

Where do we proceed from here? We could go along the path labeled "Newton" and try to *prove* that the molecules never return to the bottle by using the laws of motion. Instead, we'll try something more intriguing; we will follow the path labeled "probability."

We have taken a fatal step here. We are going to use probability to explain the workings of nature. This is not the way that Newton spoke. The implications for our understanding of reality are not trivial, and it is advisable for us to stop for a moment and define what we mean by probability. We need to do this because probability is another of those words that is shared by science and everyday language.

If I state that the probability of throwing a "four" with a fair die is one in six, what do I mean? What is a "fair" die? I could define it as one in which the chances of throwing a "four" are one in six! Another way to say what I mean is to claim that for a physically symmetrical die there is no reason to suppose that the die will fall on one side more or less than any other. We believe this because we don't, or shouldn't, believe in occult influences. I do believe that a perfectly symmetrical die should fall on any of its faces with the same probability.

In a simple system we may know enough to state probabilities exactly. Thus, with a pack of unmarked cards the chance of picking out a given card is *exactly* 1 in 52, provided the man holding the cards is not a cardsharp. It turns out that the probabilities associated with the behavior of molecules can often be stated exactly. We are going to use probability to see why the petrol vapor will eventually spread evenly throughout a room. We go back to our molecules-in-a-box.

Suppose that there is *one* molecule of oxygen in an otherwise empty box. What are the chances, at any given moment, of finding it in the left-hand side of the box? A gambler would say "fifty-fifty," by which he means that there is an equal chance of finding it in the left-hand side or the right-hand side. We can say that the probability of finding it in the left-hand side is one-half.

Now suppose that there are two molecules in the box, and assume that neither molecule affects the behavior of the other. This is a crucial assumption. Thus if you put a man into an empty room, the chances are "fifty-fifty" that he would be found in the left-hand side. If you put a heterosexual man and a heterosexual woman in the room, the chances are very high that they would be found in the same side of the room. This is not true of, say, nitrogen molecules; they behave effectively independently of each other as far as their location is concerned. What is the probability that both molecules will be found in the left-hand side of the box? We can assume that each molecule individually has a probability of $1/2$ of being in the left-hand side. The rules of probability are quite clear on how to continue: if two independent events each have probabilities of one-half, then the chance of the events occurring together is $1/2 \times 1/2 = 1/4$. This is exactly the same reasoning that makes the chance of throwing two "heads," in two throws of a coin, equal to $1/4$. The chance of throwing three heads, one after the other, is equal to $1/2 \times 1/2 \times 1/2 = 1/8$. We can write $1/2 \times 1/2 \times 1/2$ as $(1/2)^3$. In a box containing three molecules, the chance of finding all three on the left-hand side (or the right-hand side) is also $(1/2)^3$, that is, one in eight. The generalization is obvious. If there are n molecules in a box, the chance of finding them all in the same half of the box is $(1/2)^n$. This becomes a very small number for even moderate values of n. Thus suppose that there were only ten molecules in the box; the chance that they would all be on one side is $(1/2)^{10}$, which is equal to a little less than $1/1000$. The chance of finding all twenty molecules in a box on one side is a little less than one in a million.

Now let's consider the chances that you are going to be suffocated because all the gas molecules in the room in which you are sitting spontaneously collect in the other half of the room. (Remember that Newton's laws of motion do not forbid this happening.) The number of molecules in a modestly sized room is about 10^{28}. It follows that the chances of all the molecules being on one side of your room is about $(1/2)$ multiplied by itself 10^{28} (10,000 trillion trillion times) times. You would appear to be safe: you can trust the calculations. They also indicate why it is that, once the petrol vapor is spread evenly throughout the room, unlike the genie, it doesn't go back in the bottle; it's just too improbable.

What is the probability of finding some other arrangement of air molecules besides the extreme case of all being in one half of the room? To answer this we consider a much simpler case: the box with ten molecules in it. We work out the probability that there will be nine molecules in one half and one in the other; eight on one side and two on the other, and so on.

The probabilities, as gamblers should know, are exactly the same as throwing a given number of heads in ten throws. They are plotted in Figure 18.1, where it can be seen that the maximum probability is associated with 5/5, that is, equal numbers of molecules in each half. This is intuitively expected. If you had a box with many gas molecules in it, you would *expect* to find the gas equally distributed throughout the box. The least probable distribution is that in which all the molecules are on one or the other of the two sides. Again, this is a reasonable result. To bolster our belief in our new method of approaching the world, let's consider a cocktail party problem: two sets of guests socializing.

Mingle!

We go back to our box and place twenty molecules in it, ten oxygen on the left-hand side, ten nitrogen on the right. Again, for later purposes, I want the box not only to be closed but also to be thermally isolated. Obviously, the separation of the gases will not last, because the molecules are in motion. We can guess that the most probable distribution is that in which there are equal numbers of molecules on each side of the box, and we will suppose that this is the case. But within the limitation that there are ten molecules on each side, there are several possible distributions of oxygen and nitrogen molecules. In the initial state the separation is complete: ten oxygen on the left and ten nitrogen on the right. Without going into the math, I ask you to believe that the most probable distribution of molecules is that in which each half of the box contains five nitrogen and five oxygen molecules. Again, with-

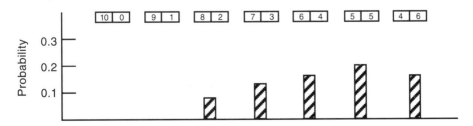

Figure 18.1. The probabilities of finding different distributions of ten molecules in a box. Two of the probabilities are too low to appear on the scale of the figure.

in the restriction that there are equal numbers of molecules on either side of the box, the least probable distribution is when all the molecules of one type are on one side and all those of the other on the opposite side. This confirms us in our belief that we will not be sitting in a recital in a concert hall and suddenly find that all the oxygen molecules have left the singer and concentrated around the audience. It is possible, just not very probable.[2] The probability of finding all ten oxygen molecules together on one side, right or left, is about one in a hundred thousand.

When mixing is complete, experience shows that the system has no tendency to stray away significantly from that condition. We say that the system is at *equilibrium*, a term we apply to any system that shows no spontaneous urge to change. We can say of our system of two gases that it started off from a state that had less than maximum probability and arrived at a state with maximum probability—the *equilibrium state*.

Time Makes Its Appearance

> There is a road that turning always
> Cuts off the country of Again.
> Archers stand there on every side
> And as it runs time's deer is slain
> And lies where it has lain.
>
> —Edwin Muir, *The Road*

We expect the spreading of perfume through a room and the spontaneous mixing of two gases to be *irreversible*; we never observe the evenly spread or evenly mixed molecules reverting to their original arrangements. Likewise, we never observe the Worcestershire sauce in the Bloody Mary unstirring itself and separating. We believe that such processes go only one way, and our belief is based on 10,000 years of experience. *The concept of irreversibility drags time into our picture of the world.* When we say that the spread of perfume is irreversible, we are picturing in our minds a process that develops with time. Suppose we take a video of the process of two gases mixing, which could be done for chlorine and oxygen since chlorine is colored green. If we run the video in reverse, we would immediately realize that time was going backward. Time could not, in our experience, possibly be running forward if molecules spontaneously unmixed, or if a perfume returned to its bottle.

Why Is Time a One-Way Road?

Now consider a box in which all the molecules are in one half of the box. They will, of course, spontaneously spread out. If we could observe a film of this process, there is an excellent chance that the molecules will move *progressively* through more and more probable states—moving from left to right in Figure 18.1—until it arrives at the most probable state. You will never observe the opposite behavior, a set of evenly distributed molecules spontaneously working their way through a series of less and less probable states until they arrive at the least likely state, that in which all the molecules are on one side of the box. In making this statement I am

---

[2]It would be a God-sent response to the wish of Coleridge: "Swans sing before they die—t'were no bad thing, / Should certain persons die before they sing."

assuming that you have a frame of reference, an everyday world in which the concepts of "before" and "after" are agreed by everyone.

Unfortunately, *there is no law based on the behavior of individual molecules that indicates that mixed gases cannot spontaneously unmix.* And yet there seems to be a natural direction for spontaneous processes. No one in the history of the Earth has observed spontaneous unmixing of the molecules in the air. Mixing processes of this kind *always* go in one direction only. Any sane scientist would tell you that there *must* be a physical law behind this universal directionality. And there is another point: *the direction of spontaneous, irreversible processes is always the same as that in which we think that "time" develops.*

Are we justified in jumping to the thought-provoking statement that *the direction of time is determined by probability*?

Let us pause to see where we are. What we have suggested is that, in describing a basic property of nature, we can dispense with physical laws of cause and effect, such as the laws of motion. We are also hinting that the direction of spontaneous processes is determined by probability, and that the direction of time is tied to the direction of increasing probability. Is it? Are the laws of motion impotent when it comes to describing the natural direction of microscopic physical processes?

We have betrayed mechanics and taken up a passionate liaison with probability. Yet maybe we have given up those laws too quickly; maybe there is something missing in the probabilistic explanation of the way in which large groups of molecules behave. There certainly is. Our assumption that a system moves from less probable to more probable states is all very well; it fits in with the facts, but it doesn't say how, or why, the system does this. A childless couple living in a country where the average number of children per couple is 2.2 is in a less probable state than a couple with 2 children. The probabilistic approach is to say that they will *probably* proceed to have one child and then a second one. But it doesn't tell us how they will make these steps. The frustrating fact is that there is no way of using our probabilities for molecules-in-a-box, to prove that the system must go "forward" rather than "backward." *Probability in itself gives no preferred direction for a system to evolve.* This may be hard to accept, since we might expect that a system *would* move from a less probable to a more probable state. In general we are right; it does, but in the case of molecular motion we don't know *why*. And for small numbers of molecules it is easy to see that the system can obviously move from more probable to less probable. An extreme example: If you have two molecules in a box and they are in different halves of the box (the most probable arrangement), it would be no great surprise to find both of them in the same half (less probable) a second later.

Newton Disappoints

Let's give mechanics another chance. Let's make a videotape of the spreading of the petrol vapor and run it backward. In our case it will have to be an imaginary film, since we can't see the molecules. If the reversed video shows a process that contradicts Newton's laws, then we have to assume that only the forward process is possible. The opening scene in our box office sellout is a box containing an evenly distributed collection of molecules. As the reversed film proceeds, the molecules drift toward one half of the box, finally leaving a vacuum at the other half. Is this impossible; does it defy the laws of motion? The disturbing answer is that, at every stage of this process, *each individual molecule behaves according to Newton's laws.* There is nothing happening to *any given pair of colliding molecules* that would induce a scientific observer to say "This is unnatural."

According to a therom of Henri Poincaré, a system obeying Newton's equations of motion will, if left to its own devices, inevitably return again and again to any state that we care to specify. I want to specify the unmixed state of two gases. Poincaré assures me that this state will return. Why doesn't experience confirm that prediction?

Neither Newton nor Poincaré helps us. The reverse video leaves their laws inviolate, in other words the laws are obeyed by the system even if we know that it is going backwards in time. Nevertheless there is something terribly wrong about the reverse film. The *total process* which it depicts *is* unnatural—gases do not unmix—or collect in one half of a container.

Our conclusion at this stage is that *there is something about the behavior of the mechanical universe that is not contained in Newton's laws.* The laws do not contain within them anything that predicts the irreversibility of processes such as those that typify the natural world.

We have looked at the behavior of a gas from two completely independent points of view: mechanical laws and considerations of probability. We believe that the mechanical laws must be obeyed, but they are obeyed whether the processes through which the gas goes run backward or forward. It is only probability that appears to control reality. For the scientist this is a frustrating situation—the laws that control the movement of individual bodies should, hopefully, be sufficient to explain completely the behavior of collections of such bodies. And yet they seem to be insufficient. The direction of mechanical processes involving molecules is, in the real world, governed by a different principle: they generally move from the least probable state, through increasingly probable states, to the most probable state.

Let us consider another example of irreversibility, one that threatens us with extinction. Place some cold water in a perfect thermos flask, one that neither lets in nor lets out any heat whatsoever. Next throw in a heated lump of iron. Close the flask. Open it after a half an hour. The contents, iron and water alike, will be at the same temperature. The iron has cooled, and the water has warmed up. We can say that heat has passed from the iron to the water. Or, in the microscopic version, we can say that the average molecular motion of the iron atoms has decreased and that of the water molecules has increased. Why? Why shouldn't the water have cooled down and frozen, and given part of its molecular motion to the atoms of the iron, warming the lump of iron to an even higher temperature than it had before? The first reason is that all our experience since childhood proclaims this process to be idiotic; but that is not a demonstration that a scientific law is being broken. Maybe the first law of thermodynamics has been violated? No! The first law merely says that energy is conserved. It is conserved in the thermos, even if the water got colder and the iron hotter. All that would have happened is that energy would have gone from the colder body to the hotter body. The total amount of energy is neither decreased nor increased. Admittedly such a process contradicts our common sense, but that does not explain why it doesn't happen. If we take a microscopic viewpoint, the passage of heat from a hot body to a cold body just involves the loss of kinetic energy (motional energy) by the atoms or molecules of the hot body and the increase in kinetic energy of the constituents of the cold body. This comes about by collisions. If a fast-moving body hits a slow-moving body, then the faster body usually slows down and the slower body speeds up. One might think there was no alternative but for heat to pass from hot to cold. The problem is that if you took a (hypothetical) video of the process and ran it backwards, each individual microscopic collision would satisfy Newton's laws and you could not tell that anything was wrong until you suddenly realized, looking at the *whole* system, that in the cooler part of it the molecules were actually slowing down and in the hotter part they were getting faster. Of course it's wildly improbable, but it contradicts no law of nature that we have spoken about up to now. A slowly moving body hitting a faster-moving body *can* give up part of its kinetic energy to the faster body and slow down, if they hit at the right angle. So what's going on?

Equilibrium

The common factor that connects all our examples is that we started with systems that were, in our experience, inherently unstable and they all moved spontaneously to states in which there were no further macroscopic changes: the petrol vapor remains evenly spread, the two gases remain perfectly mixed, the iron sits silently in the water, both of them at the same unchanging temperature. The systems evolve spontaneously but eventually show no tendency to change: we say that they have reached *equilibrium*. Even if they fluctuate slightly about that state, they show no macroscopic changes whatsoever.

There is something else that the systems have in common. They are all what are technically known as *isolated* systems, where the word has this very specific meaning: they are closed to the transfer of matter or heat, in or out. They were deliberately chosen this way because there is one mightily important isolated system: the universe. It cannot exchange heat or matter across its borders because by definition it includes everything that exists. What we have learned about little boxes applies to the cosmos:

Isolated systems move toward equilibrium and stay there.

So the entertaining thing that we have learned is that the *universe* is spontaneously moving toward equilibrium. In simple words, we are going to end up in a universe where there will be no further macroscopic changes. This condition, of course, precludes life of any kind. This is too dramatic a conclusion to be left like that.

The Problem of Time

We see that the innocuous molecules-in-a-box, with which we started this book, have led us to a dismal prediction and to this nasty problem: what physical law determines the spontaneous direction of natural processes? Where is the direction of time built into the laws of science?

If science is supposed to give a rational explanation of the physical universe, then it must surely be able to account for the passage of time. Put another way, if our description of the universe is to be crystallized in a number of basic physical laws, such as the laws of motion and Maxwell's laws, then one would hope that the use of these laws would allow us to predict unequivocally the direction in which *all* natural processes proceed with time.

Our hopes are not fulfilled. The laws of motion and Maxwell's laws are unchanged by reversing time. Any process described by these laws can be run backwards without violating the laws. In summary, we can say of nature that if you run the film backward it may look ridiculous, but everything will proceed strictly according to the laws of nature. Except one.

There is one law in nature that ensures that processes like the mixing of gases, the spreading of Chanel No. 5, the passage of heat from the warm hands of the flower-bearing man to the cold hands of his beloved are all irreversible. They only go one way: they only go forward with time, and in doing so they obey the second law of thermodynamics.

The Second Law

There are a number of equivalent ways of phrasing the law, the simplest of which is the statement that:

Heat does not flow spontaneously from a colder to a hotter body.

The more combative among you will declare that I have been cheating because what I have done is to observe what happens in nature and then declare that it is a *law* of nature. From which it follows that what happens in nature can be explained by science, by the second law of thermodynamics in fact. You are right. It *is a* law based purely on observation—as all fundamental laws are.

The second law of thermodynamics is on a standing with Newton's and Maxwell's laws. It is one of the cornerstones of the scientific attempt to explain the universe. It is the only law that gives a direction to time and thus "explains" why spontaneous natural processes are irreversible. Incidentally, neither the laws of quantum mechanics nor relativity can do this because they, too, are unaffected by reversing time.[3]

The law was not derived from fancy theory; it did not come out of a professor's study. The opposite is true. It was put on the road by a French army engineer who had made a study of steam engines, particularly those being built in early nineteenth-century England. A logical analysis of the way such engines operate led Sadi Carnot (1796–1832) to formulate a law that looks very different from the way we have stated the second law but can be shown to be completely equivalent:

It is not possible to construct an engine that operates in cycles and whose only effect is to convert heat into an equivalent quantity of work.

I am going to leave this definition, since it is not as obvious as the previous one. You might see other definitions of the second law in your wanderings; they can all be shown to be equivalent. For now, we have everything we need in the previous definition and our ideas on probability, but I stress that both forms are based on observation of the way the world behaves. Carnot's law is not derived from other laws or from theoretical axioms. In the end, all our physical laws have come from some kind of observation of the "real" world, including the laws of motion and Maxwell's laws. We do not have to be ashamed of this.

The second law does not allow natural spontaneous processes to reverse, even

---

[3]One of the forces within the nucleus, the *weak* force (see Chapter 31), is not time-reversible, but it is not considered to interfere with the workings of the atomic world.

though they would not be breaking any other law. Because the second law has time built into it, any derivation of the law from the other laws, which are time-reversible or time-independent, is extremely suspicious.[4]

Entropy

Here is another statement of the second law:

The entropy of the universe is continually increasing.

Entropy is a convenient concept, allowing us to use short sentences instead of paragraphs. Changes in entropy can often be measured, which allows us to use it to make *quantitative* predictions about the behavior of systems. We can readily make the transition from probability to entropy because entropy is probability in disguise.

Which is more probable, to find a room with normal air in it or a room with all the oxygen on one side and the nitrogen on the other? We already have two answers, one based on our experience and the other on considerations of probability. Both lead us to expect the mixed state. We can say of the unmixed state that it is more *ordered* than the mixed state. Thus if your girlfriend's T-shirts are mixed up with her sweaters and you unmix them, you are bringing more order to her cupboard. We can extend this idea to the case of the molecules clustered in one corner of the room. They are more ordered than if they were all over the room, just as picking up the scattered trash in a public park and bringing it to one corner is an act that increases the order in the park. Recall the water and the iron. At the beginning, one part of the system, the iron, contained all the violently moving particles. This is an ordered state; it is as though someone has taken kinetic energy from one part of the system (the water) and pushed it into the other part (the iron). The kinetic energy, like the paper in one corner of the park, is ordered. When the temperature is the same everywhere and the average kinetic energy of all the molecules throughout the system is equal, the system is less ordered than it was; the paper is all over the park.

Notice that in all the preceding examples the ordered state is less likely (unless you have an orderly girlfriend). In general, *order is less probable than disorder*. But you knew that.

The terms *order* and *disorder* are qualitative, but it is possible to give a number to the disorder in a system. We can say not only that the disorder in a system has increased; we can say by how much. The clue lies in the probabilities that we attached to different states of a spreading gas and to a system of mixing gases. Both systems went from order to disorder, and on the way we were able to give clear estimates of the probabilities of the intermediate states through which they passed. This can be done for any physical system in theory, and for very many in practice. In general we do not use probabilities but a very closely related quantity: entropy. In thinking of entropy, think of order and disorder:

The statement that the entropy of the universe is continually increasing is equivalent to the statement that the disorder in the universe is increasing.

---

[4]That profound Austrian thinker Ludwig Boltzmann (1844–1906) thought that he had derived the second law from the kind of probabilistic arguments we have been using, but there is a flaw in his reasoning.

The T-shirts and the sweaters are mixing more thoroughly; the paper is being blown all over the park.

Why use entropy when we can use probabilities? Because for all but simple systems it is impossible in practice to measure the probabilities that we need, but it has been shown that we can measure a quantity that is closely related to these probabilities. That quantity is entropy, and to measure *changes* in entropy it is quite often sufficient to measure the amount of heat entering or leaving a system, and the temperature at which this process takes place. We will not go along this path. The central point to be grasped at this stage is that in *naturally occurring spontaneous processes* (for example, the spontaneous mixing of two gases) we can make three equivalent statements:

- The universe has changed to a state that is more *probable* than that which preceded it.
- The *disorder* in the universe increases.
- The *entropy* of the universe increases.

In all our examples we can say that the entropy of an isolated system increased. It stopped increasing only when the systems reached equilibrium. There were no states more probable than the equilibrium state, and so that is the state of maximum entropy, for isolated systems. If you prefer, you can say that in an isolated system the equilibrium state is the state of maximum probability.

Clausius Coins a Bon Mot

To summarize some simple facts that Newton never knew:

1. Isolated systems spontaneously move toward states of higher entropy (probability) and stop changing when they reach an equilibrium state, which is characterized by maximum entropy (maximum probability).
2. This spontaneous process, for isolated states, is irreversible. The system never stops in any state that is not the most probable state, that is, the equilibrium state.
3. Because the universe is an isolated system, the preceding conclusions apply to it. The total entropy of the universe is increasing continuously, thus hastening not the day of judgment but the night of silence.

Clausius, who had invented the term entropy in 1865, summarized the heart of thermodynamics in the famous sentence:

The energy of the universe is a constant: the entropy of the universe always tends to a maximum.

The Autonomy of the Second Law

No one has yet succeeded in deriving the second law from any other law of nature. It stands on its own feet. *It is the only law in our everyday world that gives a direction to time,* which tells us that the universe is moving toward equilibrium and which gives us a criteria for that state, namely, the point of maximum entropy, of maximum probability. The second law involves no new forces. On the contrary, it says nothing specific about forces whatsoever.

It might seem possible to explain the unidirectional progress of spontaneous processes by supposing that there is a *mechanism* that drives systems from states of low probability (say, all ten molecules on one side of the box) to states of higher probability (nine on one side, one on the other). No one has found a logical basis for such a supposition.

Our molecules-in-a-box model suggests that the second law is a law of large numbers. Take a box with only four molecules in it. "Equilibrium," the most probable state, occurs when there are two molecules on each side. The probability of finding this state is 3 in 8. This is not very much greater than the chance of finding three on the left-hand side and one on the right—namely, two in eight. If a given observation showed the box to be "at equilibrium," then I would be prepared to bet on finding the system well away from equilibrium (three times as many molecules on one side as the other) at the next observation. This would contradict the second law, which says that when you get to equilibrium in an isolated system, you stay there. In fact, if someone gave me good odds, say 200 to 1, I wouldn't mind putting money on finding all four molecules on the left-hand side; the probability of finding that state is 1 in 16.[5] Thus for small numbers of molecules, we can say that the second law holds very weakly. Systems don't move inexorably to maximum entropy (probability) and stay there. In fact, even in systems containing enormous numbers of molecules we don't expect the system to remain *exactly* at equilibrium—because of fluctuations. For huge numbers of molecules these fluctuations are, percentage-wise, very small compared with those in systems with very few molecules.

The second law predicts that the available energy of the universe, that which is available for work, is being continually *degraded* into heat, becoming unavailable for work. This can also be seen as a move from order to disorder. Thus the atoms of the falling boulder are all moving downward together in the same direction, despite their random vibrations. Like the dancers in the Rio carnival, they wobble and twirl within the boulder, but their collective movement is forward and can easily be converted into mechanical work. When the boulder hits the ground, the column of dancers disperses. Each atom and dancer goes her own way, the King of Disorder rules. The boulder and the ground are static, the atoms in the ground vibrate more vigorously, the heat oozes away, the atomic and molecular motions are completely random, not directed along one line. And even if the boulder hits a turbine blade and thereby does some useful work, the blade and the boulder still steal some of the kinetic energy of the boulder and warm up. Useless heat. The second law is a law of irreversibly lost potency. Ashes to ashes, available energy to heat. As the blues man says, "There are thirteen goin' to the fun'ral, 'n only twelve men comin' back."

The reason we can't use the kinetic (heat) energy of the water molecules of the ocean to get work (the opposite of Joule's experiment) is that to run a machine (a car or a steam engine) heat has to flow from a higher to a lower temperature. If you have a bathful of water at the temperature of the room, you can't get the heat to flow out of the bath. There have been plans to use the temperature difference between equatorial and polar seas, but not practical plans.

What does all this mean for the universe?

---

[5]If you want a really good bet, try putting a few dollars on the chance of finding three molecules out of four on *either* the left-hand or the right-hand side. The combined probability is one half.

The Big Bang

The second law implies that the universe, throughout its history, has been continually drifting toward equilibrium, as does any isolated system in which changes occur. *This drift does not preclude local islands of decreasing entropy* (increasing probability), but the overall entropy increases inexorably. It follows that the universe started in a state of comparatively low entropy—it was highly improbable. We will return to this topic, but one point is worth making here, especially since the molecules-in-a-box model has been so prominent.

Don't be tempted to picture the universe as a huge box, full of molecules, for which the state of highest entropy, the equilibrium state to which we are journeying, is that in which all the molecules are spread out over the box, like perfume in a room. The molecules in our box were presumed to have no forces between them, and in fact these forces are so weak for most gases that we can ignore them for many purposes. How do we know, qualitatively, how to take the presence of forces into account when asking what equilibrium state we expect? It is not a bad principle to trust the world when it comes to asking in which direction entropy leads us. To appreciate that forces can make a difference to the equilibrium state of an isolated collection of molecules, you just have to consider a thermos flask partly filled with water. The molecules are certainly not spread out evenly throughout the bottle; most of them are in the liquid water, and some are in the space above the liquid. (This would be true even if there were no air in the bottle.) It is the comparatively strong forces between water molecules, as compared to the molecules in air, that pull them together. The system is certainly at equilibrium—it will not change if left alone. Its state of equilibrium is very different from that of air in a box, although if you cool the air enough you will find that the equilibrium state at low temperatures will also consist of a mixture of liquid and gaseous air, the mixture you get in a commercial container of liquid air. What happens at low temperature is that the molecules are moving so slowly that the weak forces of attraction between them succeed in holding most of them together as a liquid, against their tendency to rush off into the space above the liquid.

All this is relevant to the fate of the universe. The force that we forgot is gravity, which must be taken into account when considering a system containing as much mass as the universe. The upshot is that the equilibrium state of the universe will not be a state in which matter is evenly distributed. Rather, it is more likely to be typified by the gravitationally induced clumping together of all the matter in the universe into one horrifying black hole. So the fate of our universe may be to end up as a clump of matter at equilibrium, in which no further change takes place. Depressing, but notice that we slipped in the phrase "local islands of decreasing entropy" a couple of paragraphs back. Where and what are they?

Life versus the Second Law

The most spectacular exception (but only apparent!) to the second law is a living organism. There are two puzzles: the origin of life and the maintenance of life. Both appear to be wildly improbable. Is this a sign that the supernatural is at work? I am not being flippant; there are those who would answer yes.

When a living organism dies it decomposes, which just means that the extraordinarily complex molecules of which it is made break down into far simpler components. There are few systems more ordered than the *living* cell and its components. Just ask yourself what the probability is of finding all those atoms linked up to give the highly complex molecules of the cell, and the probability of finding those molecules arranged in the way that they are. The process of decomposition goes the way that the second law predicts: the complex, highly ordered system breaks up into a highly disordered mess. The structure of the cell collapses (becomes disordered), and the molecules that it contains disintegrate, destroying the miraculous order and leaving the fragments to wander about in a disordered manner. Some of the small molecules are incorporated into low-entropy large molecules in the bacteria that cause most decomposition, but the overall result is still that the entropy (alias disorder, alias probability) of the universe increases. That's what the second law predicts for any real process. And if they are left without food, the bacteria will also die, and their molecules will be degraded by the atmosphere. Living matter would much rather be dead. Living matter is a curiosity.

So, the (entropic) question that has to be answered with respect to the origin of life, is:

Is it possible to defy the second law and *spontaneously* create an *ordered,* low-entropy, low-probability system out of a disordered, high-entropy, high-probability system, that is, the primeval oceans, which contained no large molecules or highly ordered structures?

The second question arises from the stability of most organisms. They show no tendency to change, at least over time periods short compared with their lifetime. We could say that they show all the signs of equilibrium. And yet, as we just remarked, they are not really at equilibrium; equilibrium for an organism is attained when it dies and spontaneously disintegrates into a final, genuinely unchanging, high-entropy, unordered state. A living organism is apparently not in its state of highest entropy and yet it manages *not* to move in that direction, at least in the short run. So we have a second question:

How do living organisms avoid spontaneously increasing in entropy until they reach their state of maximum entropy? In other words, how can life maintain itself in a state of low entropy? How does an improbable state avoid becoming more probable?

First of all, let it be clear that, contrary to the claims of the creationists, there is nothing in the laws of thermodynamics that prevents the spontaneous creation of organized (entropy-poor, low-probability) systems on the primeval Earth. The statement that entropy has to increase in a spontaneous process applies only to an *isolated* system. The Earth is an *open* system, since both matter and energy can arrive and leave the Earth. Huge amounts of energy stream toward us from the Sun, and we radiate comparable amounts of IR light back into space. The arrival of matter is in amounts too small to be of any great significance, but the exchange of energy with the rest of the universe means that the Earth, and its flora and fauna, is at the very least what is called a *closed* system, one that can exchange energy but not matter with its surroundings.

There is no theoretical reason why entropy-poor, ordered, systems cannot be formed in an *open* or *closed* system, and we will subsequently illustrate this claim. Unfortunately, pseudoscientific babble from a variety of sources has suggested that science is on the side of the creationists, since the second law supposedly precludes the spontaneous formation of organized systems. Ergo, life required a super-natural hand.

The creationists have also carelessly dragged the second law into the question of evolution. Evolution has, on the average, led to increasingly complex living forms. The complexity of the higher apes, let alone man, far outdoes that of the unicellular organisms that were presumably the first, or near-first, life-forms. This can be interpreted in terms of a spontaneous evolutionary decrease in probability. (Remember that complex, highly organized systems are *less* probable than simple, comparatively less organized systems.) The creationists, those who deny evolution, have jumped on this as proof that evolution could not have happened since it would have required a spontaneous decrease in entropy (probability). The flaw in the creationist argument is again simple: the Earth is not an *isolated* system; a *closed* system can spontaneously produce islands of order, that is, low entropy. Science does not disprove the creationist thesis, but it does not support it either. It just makes it look infantile. Regarding the Creation, I feel that it is necessary to give at least one example of the spontaneous decrease in entropy (that is, the creation of order) in a comparatively disordered system.

The Spontaneous Creation of Order

We have seen that natural processes spontaneously move from *more* ordered to *less* ordered, which appears to rule out the spontaneous creation of ordered systems and force us back to magic if we want to create life. There is, however, a way out, and it is suggested by simple experiments that can be performed with simple materials by a layman. What is common to these experiments and makes them so relevant to the nature of living matter, is that (1) they start with a disordered mixture of molecules and *self-organize*, (2) the ordered state survives only as long as either energy or material is continuously fed into them, and (3) the systems are not isolated.

Dynamic Equilibrium

Mechanical equilibrium is often observed and readily comprehended. A child's swing at rest and a motionless table are at mechanical equilibrium, by which we mean that if we do not exert forces on these bodies they will maintain their position. But the equilibrium of these bodies is of a special kind. It is a *stable* equilibrium; a small disturbance will cause a small change in the state of each system, which moves to a new equilibrium state. Thus if I push gently on the swing, it will move to a new position, out of the vertical, where it will remain as long as I maintain my pressure. Furthermore, if the small disturbing forces are removed, the systems return to their original positions. This is obviously not true of all systems. Thus a stationary billiard ball resting on a table will not return to its original position if it is temporarily subjected to a force—it is in a *metastable* state.

Now consider an adult human being. In the healthy state, the organs of the body function from day to day, from year to year, without any significant change in shape, size, or function. If the body begins to overeat, its size and the size of its organs can be affected, but the system is in stable equilibrium in the sense that small disturbances, which could mean one too many hamburgers, bring it to a new equilibrium position, specifically, a state in which the total mass of the body is greater.

Now there is something different about the stable equilibrium of living organisms and the stable equilibrium of chairs. The equilibrium of the chair is a state it can maintain without any help from its environment. To maintain their state of equilibrium, living organisms need an input of material and energy, without which they drift rapidly away from their equilibrium toward another, highly disordered, equilibrium state—a state of decomposition preceded by death.

The maintenance of life is an example of a *dynamic equilibrium*, the system being maintained in an equilibrium state removed from its "real" equilibrium, by the input of energy. If you are not convinced that this can happen, I ask you to consider a very simple physical system in a similar state.

Picture a horizontal tube containing helium gas. The tube is divided into two equal halves by a partition that contains holes a little larger than the atoms of helium. The partition is there to stop *bulk* movement of gas from one side of the tube to the other; it's like having two rooms connected by a very narrow door. At equilibrium, as we well know, there will be the same number of molecules on either side of the tube, a state we can refer to as "normal." Now, wrap one side of the tube in a heating coil through which you pass a steady current so that the temperature of that side is higher than that in the other. Maintain this temperature difference at a constant value. You will find that a new equilibrium will be established in which there are more molecules on the cold side of the tube than on the hot side. Provided you maintain the temperature difference, the number difference will remain unchanged. The point is that you have managed to keep a system in an "unnatural" state, and the way you are doing it is by putting energy through the system: heat energy flows into the hot end of the tube, from a source of heat, and flows out at the other end. The moment you stop this flow, the system starts to revert to its original equilibrium state—it "dies." In terms of the probabilistic models that we have been playing around with, the state of *dynamic equilibrium*, with its unsymmetrical distribution of molecules, is more ordered and less probable than the normal state—it has a lower entropy. We agreed that in nature systems tend to move in the direction that increases the entropy of the universe, and here we have a system that has moved from a more probable state (before heating) to a less probable state (during heating), but there is no paradox. The second law says that the entropy of the *universe* increases in all natural processes, and although the entropy in the tube may have decreased, it can be shown that the entropy of the tube plus its surroundings, increases because heat is spontaneously flowing from the hot heating coil to the cooler air. The dynamic equilibrium is maintained by a flow of energy through the system.

Such a system is too simple to provide a convincing model for life, but it shows that molecular ordering, and a lowering of the entropy of the system, can be created from disorder by nonsupernatural means. There is no need for a life force.

Dissipation

Life is a curiosity. It built itself out of a high-entropy (disordered) world and it maintains its low entropy (order), against the overall tendency of the universe to go to ever higher total entropy. Life's motto is "Ordnung must sein!" Except for the living world and some of the artifices that it constructs, the universe appears to be universally disordering itself. Living organisms are islands that swallow matter and energy from their surroundings and use them to keep themselves in a state of ordered dynamic equilibrium, a state that collapses and becomes disordered when they die. Structures that use the resources of their environment in this way have been called *dissipative structures* by the Belgian Nobel Prize winner Ilya Prigogine (1917–), a theoretician who has been largely responsible for initiating the study of such structures. So even the holiest of us, the nondrinking, nonwenching saint, is a dissipative structure. And so is Earth, which receives a continuous input of energy from the Sun. This energy comes from one direction only (at any one time) but radiates in all directions. We are in approximate dynamic equilibrium, the temperature remaining stable within fairly narrow limits. If the energy source were cut off, it would result in a drop in temperature of at least a couple of hundred degrees as we rushed toward a new equilibrium.

There is, as your priest will tell you, a price to pay for dissipation. The heated tube of gas described earlier illustrates a basic fact about dissipative structures. They may hold their own entropy constant, but they *increase* the entropy of the surroundings. In the end we can't win; the total entropy of the universe increases.

By the generality of its implications, and their doomsday aura, the second law of thermodynamics, or at least the vague statement that entropy condemns the universe to death, has attracted the attention of nonscientists, many of whom were, or are, otherwise completely uninterested in science. Oswald Spengler (1880–1936), another of the prolific Continental school of obscurantists, wrote in his best-selling *Decline of the West* (1918), apropos the second law, "Something Goetheian has entered into physics—and if we are to understand the deeper significance of Goethe's passionate polemic against Newton in the *Farbenlehre* we shall realize the full weight of what this means. For therein intuitive vision was arguing against reason, life against death, creative image against normative law." Which is about the level of stuff that the second law has inspired among those who don't understand it. Henry Adams expressed a more human and understandable sentiment when he wrote to William James: "The universe has been terribly narrowed by thermodynamics."

The prediction that the universe will edge inexorably closer to dead equilibrium speaks of a time possibly billions of years in the future, yet we sometimes feel as if we have to cancel next month's tickets to the Met. There is something repellent about that final silent sea, a sea alive only in the sense that the kinetic energy of its component particles will never cease. There are a variety of scenarios that offer a way out.

The thought of the Supreme Architect galloping over the hill to rewind the spring of the great clock is comforting, but you wouldn't get very good odds on it at the local betting shop. So perhaps we should try those cosmological theories that see the universe eventually collapsing to the size of a golf ball, only to reappear like a phoenix, born again to repeat its history in a series of endless cycles. This is the Big Crunch, followed by another Big Bang—and who says that our Big Bang was the first? Unfortunately, cosmology offers no clear way out of the grip of the second law. The problem with thermodynamics is its generality—the first and second laws don't ask anything whatsoever about the detailed nature of a system.[6] If the universe is an isolated system, which it clearly is, then the second law leaves us with

---

[6]There is a third law. The second law allows us to calculate only *differences* in the entropy of a system when something happens to it. The third law allows an absolute estimation of the entropy of a system.

no alternative: we are heading for equilibrium however many cycles of bang and crunch we go through.

Some scientists have suggested another loophole. What if the laws of thermodynamics are valid only in certain parts of the universe? There is as yet no evidence for this hypothesis, which echoes one explanation of suffering—that God is in control of only part of the universe—and it isn't our part.

Another lifeline is the fluctuation theory. As we saw, even at equilibrium there are fluctuations that take the system away from equilibrium, which is where we want to be taken, O Lord. Even in a universe that *on the average* was at equilibrium, there could conceivably be one or more places where there were significant deviations from equilibrium, in other words, where things can happen! Or is that just wishful thinking?

Entropy for Beginners

The reader should be on guard against books in which entropy appears without a definition, or with a faulty definition, and in which it is used in relation to undefined systems to give heartwarming results. I won't take up space unpicking these balls of wool, of which perhaps the outstanding example is Jeremy Rifkin's *Entropy: A New World View* (1980), in which can be found the fourth law(!) of thermodynamics: "In a closed system, the material entropy must ultimately reach a maximum." The term *material entropy* is meaningless; it has not the slightest connection with entropy, or indeed science. The "law" is the brainchild of an economist, Nicholas Georgescu-Roegen. It is a child that should have been left on a bleak Thracian mountainside, a fate that should also have befallen the essays written in the late nineteenth century by the American historian Henry Adams in which he maintained that history should be "scientifically" based on the laws that governed energy and entropy. The objection to such texts is not that they challenge established science. We are all for that, I hope. The problem is that they use very well-defined concepts in situations to which they do not apply, and in a way that reveals that the authors have misunderstood the meaning of both entropy and the second law. Entropy is one of the buzzwords in the realm of pseudoscience, being blamed for crimes that it never committed and praised for deeds it never did. One of the problems with people who write books like Rifkin's is that if you question their concepts, they turn their back on you, face the audience, and proclaim that once again the blindness of conventional science is revealed. They are sometimes right, but never when their "innovations" are based on the gross distortion of well-defined concepts.

What Use Is Entropy?

Because we are concerned more with ideas than with their application, we will not discuss the practical use of the second law. A law that deals in the entropy of the universe might seem somewhat removed from the marketplace. It isn't. Chemists, chemical engineers, mechanical engineers, and biologists are among the people who use the law at their places of work. It can help us to decide, for example, whether or not a certain potential manufacturing process for a chemical will be economically feasible.

We seem to be making excellent progress. We have Newton and Maxwell to cover the mechanical and electromagnetic aspects of matter, and the second law to help us deal with huge assemblies of molecules. Are we at last in a position, except for some niggling problems with light, to explain *and predict*, the course of nature? No, often we are not. In the following chapter we are going to be humiliated.

19

Chaos

I tell you: one must have chaos in one, to give birth to a dancing star.

—Nietzsche, *Thus Spake Zarathustra*

Is prediction possible? When the astrologer confidently tells you that you will meet a tall dark man this year, she is merely exploiting the fact that about 5% of the population fall into that category, and if you leave the house there is almost no chance that you will *not* meet a tall dark man within a year. But what about prediction in the inanimate world, or the world of nonhuman life-forms where cold physical law operates?

Newton and Maxwell gave an unambiguous answer: given the initial conditions of an inanimate system, and assuming that the laws apply to them, the future of the system can be mapped out exactly. And, if need be, as in the examples of the preceding chapter, we can call on the second law to help us out. *In classical physics the present determines the future.* It is said of classical physics that it is *deterministic*.

So that there is agreement about what we mean when we say that a theory is deterministic, we will define our terms. A theory may be said to be deterministic if, using only the theory and a complete description of the state of a system, every subsequent state of the system is logically inevitable. Newton's equations are certainly deterministic in this sense. A body acted on by forces has no choice at all about its behavior; if Newton's laws hold, its future is absolutely certain. Another way of expressing the creed of extreme determinism is to say: For every event there must be a cause, and so, given the conditions preceding the event, and the laws of nature, every event must in principle be predictable and in practice be *inevitable*. To be a little more pedantic, which never hurts in this kind of argument, I should replace "cause" by a "set of causes," the complete collection of circumstances which guarantee that an event will occur. Thus the predicted path of an artillery shell depends, among other things, on atmospheric conditions, the positions of all the heavenly bodies, and so on. And on the fact that there is not a faulty detonator in the shell.

We can see planets and missiles, but are we so sure that the microscopic world is a deterministic system? It certainly is if we agree that the same laws of motion that Newton declared to hold for the solar system also hold for molecules. When we wrote $\mathbf{F} = m\mathbf{a}$, we didn't put upper or lower limits on the size of m, the mass of the body. Newton's laws, *if* they hold for molecular motion, tell us that the movement of a molecule after a collision, either with another gas molecule or with the wall of the container, is *completely* determined by the laws of motion.

In Newton's world the future state of *all* particles in the cosmos, great or small, is determined *only* by their present conditions and those laws. We should be able to predict exactly the future of any physical system—given complete knowledge of its present.

The Inexorable Gears

The great eighteenth-century mathematician Laplace dreamed up an ambitious project. Since, said he, Newton's laws completely determined the motion of all bodies, if we could at any given moment measure the position and velocity of every particle in the universe, we could use the laws of motion to determine their future movement completely. Thus the future of the universe is in principle calculable from its present state, just as the future position of the planet Mars is completely predictable given its position and velocity at any given moment. More than this, we can use the laws, together with the present state of the universe, to calculate where everything *was*, and how fast it was moving, at any time in the past. And, says Laplace, there is no room for uncertainty, either in the past or in the future; the birth, life, and death of every physical object, including humans, can be unambiguously plotted. The future is completely predetermined.[1]

Laplace had taken Newton's science and turned it into philosophy. The universe was a piece of machinery, its history was predetermined, there was no room for chance or for will. The cosmos was indeed an ice-cold clock. Maxwell leaves this conclusion unchanged. He merely added additional deterministic laws to the game.

All we have to do is include the human brain in this strait-jacket and we have a depressing bottom line: the whole history of the cosmos, *including human history,* was fixed at the moment of Creation. There could have been only one possible history for the cosmos—the history that happened—and there is only one future history, which, if I had enough information and a big enough computer, I would be able to calculate. *Everything is predictable.* To which Newton would cry: "Wait, I didn't mean that at all!"

What we intend to show here is that *even in the deterministic Newtonian world, prediction can be impossible.*

Determinism or Free Will?

The determinism that the *Principia* engendered was at the core of the thread of ironclad certainty that ran through science until the nineteenth century. It supported the Deism of the Enlightenment and created a breed of materialistic determinists such as d'Holbach and La Mettrie. Although determinism was not invented in the seventeenth century, from Newton's time until the end of the nineteenth century few scientists doubted that the *inanimate* world was deterministic. But what about man?

In its extreme form, the doctrine of determinism includes man. If man is a part of nature, is he not also bound by its unbreakable laws? If so, this has the profoundest implications for man's self-image and his attitude toward religion. Are we destined

---

[1]If we could predict the future, would it be avoidable? This is the oft-considered paradox of the time traveler who jumps into the future and sees his own life story. If I perform a Laplacian calculation that shows that I am going to drink a Heineken tonight, can I perversely choose to drink a Budweiser instead?

to do what we must do? Is my every thought fixed by the movement of molecules and charges in my brain, movements and charges that, in turn, are completely predestined by the previous physical history of my body? *Is there no free will?* Are we just immensely sophisticated automata?

So thought the English philosopher Thomas Hobbes (1588–1679) who published his treatises *De corpore* and *De homine* (On Matter and On Men), in 1655 and 1658. Newton was still at school, but Galileo's mechanics had reached Hobbes, and he fell for the same deductive reasoning that he had found in Euclid. Deeply influenced by mathematics and mechanics, Hobbes stated that philosophy should only be concerned with the action of contiguous moving bodies on each other and that the results of such action were to be obtained from the laws of dynamics. Hobbes realized that the mentally directed actions of humans were different in nature from the movement of inanimate bodies, but he deemed them not different enough to take them outside the realm of mechanics. Hobbes provoked fierce criticism, and twice in his life he chose discretion rather than valor and slipped across the Channel to live in France. It is good to know that he was clear-minded and active to the end, dying at the age of ninety-one. Paradoxically, his agnostic negation of free will was backed by religion.

Religion and Science Agree

Any system of belief, like Christianity, that includes an all-knowing deity can run up against the following dilemma: if God knows and can foresee *everything*, then he knows exactly what is to come and he cannot be mistaken. "Que serà serà." Boethius, the sixth-century thinker, explains: "If God beholdeth all things and cannot be deceived, then what He forseeth must inevitably happen. Wherefore if from Eternity He doth foreknow not only the deeds of men but also their counsels and their will, there can be no free will." As Karl Popper comments, an omniscient God would be impotent, since his actions would be completely determined by his prior knowledge of them.

The religious problem of free will was intensified by Galileo's and Newton's deterministic physics, although Newton himself makes it clear that he leaves room for God to modify the operation of his cosmos, thus breaking the iron chain. But did he consider the paradox of the all-knowing God?

Man Is Not an Exception

Unlike Hobbes, many who were prepared to accept determinism in the nonliving world made an exception of man, but since it has now been demonstrated beyond any reasonable doubt that living material, including the human brain, is constructed from molecules that obey the same laws as any other molecules, there seems reason to suppose that the behavior of human beings can be explained on the basis of *deterministic*, chemical and physical processes. Max Macho might think that he *chose* to chat with that attractive barmaid, but he had no choice; it was predetermined. It was in their stars. At least that's what he's trying to tell her.

It looks as if our personal lives have already been punched out on the pianola scroll. Like the intricate clockwork dolls that amused the baroque courts of Europe, the only tune we can play is the tune that we will play; there is no free will.

This chilling conclusion in fact preceded modern biochemistry and biophysics; as we saw, several Enlightenment thinkers viewed man as a machine. Diderot expressed the consequence in general terms: "Look closely and you will see that the word liberty is a word devoid of meaning; that we are no more than fits in with the general order of things, the product of our organization, our education and the chain of events." Voltaire, as usual, speaks with more bite than most:

> It would be very singular that all nature and all the stars should obey eternal laws, and that there should be one animal five feet tall which, despite these laws, could always act as suited his caprice. It would act by chance, and we know that chance is nothing. We have invented this word to express the known effect of any unknown cause.

Note the Newtonian assumptions (and the average height for a man). The scientific backing for determinism sharpened the philosophical debate on ethics and particularly on punishment, a debate that continues to this day. Diderot, like Hume, concluded that the concepts of vice and virtue were meaningless; there was no point in apportioning praise or blame for any action, since all actions were inevitable. Criminals should not be punished but should be eliminated, preferably on the public square. In this century Bertrand Russell expressed related sentiments: "If, when a man writes a poem or commits a murder, the bodily movements involved in his act result solely from physical causes, it would seem absurd to put up a statue to him in one case and to hang him in the other." A deterministic universe in which all effects come inexorably from previous physical causes has no room for morality; it is a universe with facts but not values.

Is there an escape from this claustrophobic, flypaper world, or are we really trapped by unbreakable chains of cause and (completely predictable) effect? I don't want to believe that my actions are controlled by a rigid chain of cause and effect going back to the Big Bang; but do I have a choice? And if I don't believe, is my disbelief inevitable? This conclusion has been strongly rejected by most thinkers. A common attitude is to accept the determinism of the natural world but to exclude man from the dictates of physics. This liberates history, which becomes unpredictable. There could have been a multitude of histories. This conclusion may be an illusory victory of wishful thinking.

The Impossibility of Prediction

The idea that, in principle, the future of the universe is completely determined by its present state is usually associated with Laplace, but in fact it had been previously put forward by the Croatian scientist Rudjer Boskovic in 1758 and had also been hinted at by Newton in the *Principia*. Boskovic, however, was the only one to qualify his statement, admitting that "we do not know the number, or the position and motion of each of these points."

Boskovic's comment kills the project. It is clear that no one will ever be able to do Laplace's calculations. Go back to Chapter 1 and you will see that if we want to "do a Laplace" on a spoonful of air we will need to know, at a given moment, the position and velocity of something like 10^{20} molecules. One minor difficulty arises here: How can anyone possibly obtain this information? Furthermore, we saw that, on the average, molecules in the air bump into each other and change velocity about 6000 million times per second. This means that we would have to complete our 10^{20} measurements in one 6000 millionth of a second to have any hope of measuring the position and velocity of most of the molecules before they were changed by collision. To be honest, the problem is worse than I have presented it. Even for a simple gas like helium, in which each molecule is a single atom, we need, for each atom, three numbers to specify its position and another three to specify its velocity, because velocity includes a specification of direction. For a calculation on a sample of helium containing 1000 atoms we need 6000 pieces of data. At the present time, the task of measurement alone is well outside the capabilities of mortal man.

But supposing it were not. How are we going to do a calculation that involves registering and then manipulating data on 10^{20} molecules? At present, no way. To complete our calculations before the future caught up with us, we would need to repeat the calculation about 6000 million times per second to keep up with the changes induced by collisions. Not even Aristo the Human Calculator, whom as a child I saw in a fair, could cope with that.[2]

But the problem of prediction is in fact far deeper than Laplace or Boskovic knew. To begin to see why, let us lower our sights and tackle what appears to be a far simpler problem. The success of the Newtonian-Galilean laws of motion lay in their ability to explain and predict with great accuracy the movement of some mechanical systems. If we take the classic Newtonian system, the Earth and the Moon, we can solve the equations of motion *exactly*, using the law of universal gravitation to tell us what force to take. Any two bodies can be treated in the same way. Thus one body could be the Earth and the other a cannonball. The expressions for the paths and velocities of the bodies involved can be expressed in analytical form, which means as simple mathematical expressions. Thus if I throw a body vertically into the air with a velocity of v_0 meters per second, then its velocity, v_t, after t seconds is given by the simple expression: $v_t = v_0 - gt$ meters per second. And if necessary I can take air resistance into account.

Now let us do one experiment too many. Take three bodies, say Mars and its two moons, Phobos and Deimos. Can we work out the paths of these bodies? Yes and no. We can predict the paths of the moons quite well, but not exactly. The strange truth is that we cannot solve the equations. We can solve them *approximately*, by first assuming that the planet is a stationary object around which the moons revolve *independently*, not interacting through their mutual gravitational attraction. Any attempt to treat the problem in its entirety has in the end to resort to a computer to produce approximate paths. No analytical expression can be found. This is not a weakness in our mathematical techniques. Now that we have started sowing doubt, let's keep up the good work.

Let's take the most famous example of Newton's success and niggle away at it. The path of the Moon is described by a simple (analytical) mathematical expression, which appears to work beautifully and has been working for more than 300 years. Or has it? The model used to calculate the Moon's motion, involves two bodies: the Earth and the Moon. The laws are the laws of motion and the law of universal gravity. But we have neglected the fact that the Moon is attracted by the Sun. We feel justified in this because the calculated and observed paths of the Moon are ef-

---

[2]I have simplified the problem by assuming that all molecules collide simultaneously at intervals given by the average time between collisions. By definition of an average, a substantial number of molecules will collide within a shorter interval.

fectively identical, but in principle we should take into account the influence of the Sun and work out a new more correct equation for the Moon's path. We should be treating the problem as that of the motion of (at least) three interacting bodies.

In 1889, Henri Poincaré startled the world of the mathematical physicists by showing that *in principle* there is no analytical solution to the three-body problem.[3] The best we can do is to obtain approximate answers. For a problem involving three, or more, interacting moving bodies, there is no closed solution, no simple mathematical expression. We are lucky that in many cases a problem that really involves the interaction of three or more bodies can be simplified by taking physical facts into consideration. Thus the path of a baseball is influenced by the Earth, the Moon, the Sun, the planets, and so on. But the influence of the Earth so outweighs that of the other bodies that we can ignore them for a problem that involves a path of a couple of dozen yards. We assume that the Earth is stationary and that the only moving body is the baseball. Likewise the path of the Earth around the Sun is given to an excellent approximation by ignoring the effect of other planets and the Moon. But we have no exact solution.

Is Determinism Dead?

Not yet. At this point we can say that although nature may obey deterministic equations in principle, *in practice* it is impossible to find an exact solution to the general problem of the movement of three or more bodies.

We have run into two types of difficulties. One is a matter of *principle*: Poincaré's proof that the three-body problem has no analytical solution, which can sometimes be surmounted by simplifying the physical problem and getting an analytical, and approximate, solution as in the cases of the baseball and the Earth. The other problem, which is *practical*, is that typified by Laplace's problem: an enormous and inaccessible quantity of data.[4] Despite our inability to find exact solutions to these two types of problem, *we still believe that the systems are governed by deterministic laws*. We believe that Newton's laws hold. We can say that although determinism holds in principle, practical determinism is somewhat shaky. Nevertheless, since we have better computers every month, there seems to be no significant limitation on our ability to obtain extremely good approximations to any problem involving the classical laws of motion.

At this point we can say that not only is determinism unchallenged but, for any systems in which the quantity of data is manageable, determinism (possibly helped by a computer) allows us to make very good predictions. This is an illusion.

The Uncertainty of Measurement

Suppose that you are godlike and could know the position and velocity of every molecule in the box at a certain time. Concentrate on one molecule. In principle,

---

[3]The three-body problem is not a problem if all the bodies are nailed to the wall; it is a simple matter to calculate the forces acting on each body, from the law of universal gravitation. It is only when at least two of them can move that we get into trouble in trying to predict their paths.
[4]The three-body, or many-body, problem is not applicable to our simple model of a gas since, between collisions the molecules are supposed to be traveling in the absence of mutual forces.

we can work out its future path because we have the deterministic laws of motion to help us. The three-body trap doesn't worry us because in our model there are no forces between the molecules between collisions, and it is the forces that usually create the difficulty. Ignore for the moment that you couldn't possibly have all the knowledge that you need, and suppose that you have actually performed the calculation of the molecule's path for the next second. You can tell *exactly* where it will be in the box at the end of the second. Now someone comes up to you and tells you that the initial direction that you were given for the molecule's motion was very slightly inaccurate. What can you say about the final position of the molecule after one second? Obviously, it will not be where you calculated it to be. Will it be *near* where you calculated it to be? Very unlikely, even if the initial data that you used for its direction were in fact very close to the truth. In one second a molecule in the air makes several billion collisions; a tiny change in the direction of our molecule at the beginning of its journey may only slightly alter the way in which it makes its first collision, but that will slightly alter the direction and speed with which it carries on after the collision, *and* slightly alter the subsequent movement of the molecule that it hits. After very few collisions our molecule, and the molecules in its vicinity, will have entirely different positions and velocities from those that we predicted. After 4 billion collisions the molecule is likely to be in an entirely different place from that originally calculated. It is easy to believe that the effect of an extremely small change in initial conditions will have an effect which is out of all proportion. What we are seeing is an extremely simple example of the fact that:

There are systems in which the outcome of a series of events is very sensitive to the initial conditions. So sensitive that the behavior of the system may be unpredictable in practice, even if it is predictable in theory.

It was Poincaré who first pointed out this sensitivity in gaseous systems. He was also the first to realize that the difficulty of making meteorological forecasts was a consequence of the same sensitivity. An extraordinary illustration of this sensitivity is provided by the calculations of the physicist Michael Berry, who considered a collection of oxygen molecules at atmospheric pressure and room temperature. He placed an electron at the edge of the known universe (about 10^{10} light-years away) and asked: After how many collisions would a given molecule miss a collision that it would have had if the electron were not there? Now the electron is supposed to act only via its *gravitational* field, which as you know must be so incredibly small that any right-thinking scientist would completely ignore it. Bad mistake! After fifty-six collisions the molecule misses a collision. I find this result to be almost incredible, but you can see why any attempt to predict the microscopic future of molecular systems requires a macroscopic amount of optimism.

The next step is to realize that I may never be able to specify the initial position of the molecule, not because of instrumental limitations but because of the way the world is. Suppose I use a ruler in order to help me specify where the molecule starts. I choose a tiny point at the center of the molecule and line it up with a point on the ruler. Is this possible in principle? Yes, but only if I'm prepared to approximate. Let us see why.

The ruler is marked with numbers, the integers, 1, 2, 3, . . . If I want to specify a point, I can estimate where it is with respect to the two nearest integers. But how many points are there between zero and one, or between one and two? An *infinite* number. Some are rational; they can be expressed as the ratio of two integers, for example, 1/2, 9/74, 260/513. But there are, nestling between the rational numbers, an infinite number of *irrational* numbers: √(7/9), √(9/11), and so on. These numbers cannot be expressed as the ratio of integers, and they present an insurmountable difficulty. When an irrational number is expressed in decimal form (in order to allow me to say exactly where I want to put a molecule, and give the information to a computer), the row of digits, which is randomly ordered, stretches to infinity. Thus √(7/9) is 0.293972367 . . . and so on, *forever.* The number of such irrational numbers being infinite, I am almost certain of putting the molecule on one of them. If I want to put the number into a calculation, I have to cut it off somewhere. If I cut off the number, say √(7/9), after 9 digits, as I have in the example, I will get a different answer to my calculation of the molecule's path than if I had taken 12 digits, or 27 digits, or . . . I can never get an exact answer for the path. Computers have a limited number of places to which they can work. A long multistage calculation on two different computers can predict entirely different behavior for a physical system. Now, suppose that the number that I start with is an irrational number, like the square root of two, which cannot be expressed as a decimal with a finite number of digits. There is no finite computer that can accommodate such a number, and anything less than that could lead to trouble. If I cut off the infinite row of integers in the decimal after 100 places, and my calculation involves many steps, I could get a very different answer from that which I would get if I took 101 places. Problems that on the face of it should be soluble turn out to be insoluble.

The problem of predicting the molecule's flight is an example of two principles that are gathering in importance in modern science: the impossibility of completely specifying the initial state of a system (in our case the position of a molecule) and the extreme sensitivity of the future development of a system (the path of the molecule), to its initial state (the position of the molecule).

It is important to realize that our inability to predict the behavior of such systems is *not* a refutation of determinism. The path of a molecule, in our Newtonian model, is completely determined by the laws of physics. *Predictability* is the problem, not determinism. We just don't have an accurate enough description of its initial state, and if we were to improve our measurement devices to excruciatingly high standards, we would still run up against the problem of irrational numbers.

An important objection can be made to our pessimistic analysis of the motion of a molecule in a gas. It is true that the behavior of a single molecule is unpredictable, requiring an impossible accuracy in the initial conditions of all the molecules in the box; but the behavior of the whole sample of gas is very predictable, and this is of far greater practical import. Thus there are laws that tell us how the pressure of the gas changes if we heat it or compress it. The results of such calculations are quite insensitive to small errors in our initial conditions, for example, in the assumed pressure and volume in the box. An error of one millionth of a percent in the initial position of one molecule in the gas will give a completely different final position after one second. A 10% error in the initial volume will cause only a 10% error in the calculated value of the final pressure of the gas after it has been heated. Thus, all is not lost—there are predictable quantities in nature.

Chaos: The Ubiquity of Unpredictability

> Chaos umpire sits
> And by decision more embroils the fray
> By which he reigns.
>
> —John Milton, *Paradise Lost*

We have looked at a system that is deterministic and whose development is extremely sensitive to its initial state. In addition, *in practice* it is impossible to specify the initial conditions with sufficient accuracy to avoid considerable unpredictability in the development of the system. These are the characteristics of what are called *chaotic systems*. Chaotic systems occur all over the place.

Ecological Chaos

We will take two basic systems that have had an important role in the history of the development of chaos.[5] I am going to slip in a simple equation here. The equation is based on a naive model for the number of animals in a given locale, in our case the number of rabbits in Sherwood Forest. Suppose that we start off at a certain time with N_0 rabbits and that on the average each rabbit gives birth to λ offspring a year, so that the value of λ is a measure of the breeding rate. For convenience in our calculations, we scale N_0 so that it has a maximum value of 1. We can always multiply our results at the end to get the real numbers.

All births occur at the same time of the year, and the parent rabbits die soon afterwards. N_1, the number of rabbits in the *first* generation is given by λN_0, which is just the average number of offspring per rabbit times the number of parents. All the first generation grow up fit and well, because this is a rabbit's paradise; there is an ample supply of food for everyone, and there are no predators. The number of rabbits in the *second* generation will be λN_1, which is the number of rabbits in the first generation times the average number of offspring that they have. Notice that this is a *linear* relationship; the number of rabbits in any generation is directly proportional to the number in the previous generation. However, life is not like this. In the real world there is a limited amount of food in the forest, and, because there is competition for food, the weakest rabbits perish and we have to reduce the estimated number of rabbits surviving in each generation. We do this by multiplying the ideal number of rabbits in any given generation by a factor of less than one. How do we choose this factor? Since it is less than one we can write it as (1 - something). One simple way to decide on the value of "something" is to realize that the more rabbits there are in the forest, the tougher the food situation becomes, and the smaller proportion of those born will survive. So we want a "something" that gets larger as the number of rabbits increases. Taking all this into account, for the first generation, instead of λN_0 rabbits, we write λN_0 multiplied by (1 - N_0), which is written $\lambda N_0(1 - N_0)$. Notice that the more rabbits there are at the beginning, the bigger N_0 is and the smaller (1 - N_0) is. This makes sense since the more rabbits there are initially, the more the pressure there is on food, and the smaller the fraction that will survive in the first generation.

---

[5] I was first made aware of the fascination of chaotic systems by an article on the pendulum written by David Tritton, which was published in the *European Journal of Physics* in 1986, and which I recommend.

So, according to our admittedly simple model, the number of rabbits surviving in the second generation will be $N_2 = \lambda N_1 (1 - N_1)$, and so on. This is *not* a linear relationship; the number that survive in a given generation is not directly proportional to the number in the previous generation. It is this nonlinearity that will produce the unexpected effects that await us.[6] (By the way, the equation is known to professionals in the field as the *logistic map*.)

It is extremely informative, from the point of view of learning about the nature of chaos, to see what happens to this population of rabbits under a variety of circumstances. We will start all our ecological experiments with the same number of rabbits, 0.2. As we noted earlier, this doesn't mean one-fifth of a rabbit. For convenience we have just scaled the number of rabbits so that the maximum number is one. Thus we start with $N_0 = 0.2$ and vary λ, the average number of offspring per rabbit. We plot the number of rabbits that survive in each generation.

1. $\underline{\lambda = 2.}$ Each rabbit on the average produces 2 offspring. The results are shown in Figure 19.1a. As you can see, the population in successive generations grows but levels out at a constant value of 0.5. This leveling out could have been expected on two counts. First, that's how the mathematical equation behaves, independently of whether it refers to rabbits or company shares. Second, at the beginning of the experiment there is more than enough food in the forest and the population grows, but it cannot increase indefinitely because the food is limited. Where is the chaos in Figure 19.1a? There isn't any, yet.

2. $\underline{\lambda = 2.9.}$ Figure 19.1b shows that the population again settles down to one value, a higher one than previously. Notice, however, that the population oscillates for many generations before it settles down. The larger number of offspring than in the previous case gives a baby boom that depletes the food supply faster and results in a lower number of survivors in the following generation. This is still not chaos; there is nothing unexpected here. Forward.

3. $\underline{\lambda = 3.}$ Figure 19.1c shows that the population oscillates for many generations but appears to very slowly approach a constant value. The baby booms are bigger and the following shortage of food more severe than in the previous case.

4. $\underline{\lambda = 3.2.}$ Figure 19.1d reveals something entirely new, although one might have guessed what was going to happen after seeing the last two graphs. The population oscillates forever. It is as though the time during which there are oscillations has been getting longer as λ increases and has finally become infinite. The population settles down to *two* fixed values. We say that there has been a *bifurcation,* a splitting into two. After a generation with a high population a small generation follows, forever. We say that the population has a *period of two.* For values of λ up to about 3.4, this behavior continues, alternate generations of rabbits have only one of two possible sizes, but the difference between the values grows if λ is increased.

5. $\underline{\lambda = 3.5.}$ Again something new happens. The population again oscillates forever but settles down to four fixed values: the population has a *period of four.* Every fourth year the value for the population returns. Each of the two previous values

---

[6]If you multiply out the expression, to give $N_2 = \lambda N_1 - \lambda N_1{}^2$, you can see that there is a term in the square of the previous generation. That is why this is not a linear expression. A linear expression would include only N_1.

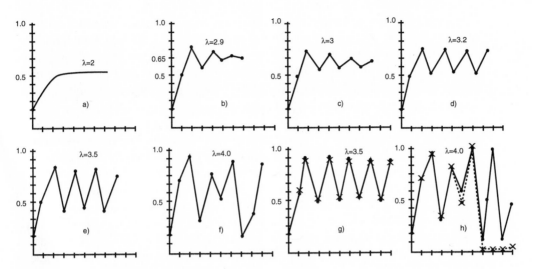

Figure 19.1. The population of rabbits in a forest. The initial population is 0.2, and λ is the average number of offspring (see text for explanation). In graphs (g) and (h) the crosses and/or dotted lines indicate the population changes for an initial population of 0.202.

(for λ = 3.2) splits into two: there are two bifurcations. It begins to look as though we may be heading for a doubling process as λ increases. There is, but we will not work our way along a series of values of λ. If we did, we would find out that as λ increased, for certain ranges of values of λ the value of the population would settle down to oscillations between 8 values, then 16 values, then 32. . . . . The results are plotted in Figure 19.1e, and you should read the caption at this point.

6. **λ = 4.** There is absolutely no pattern discernible in Figure 19.1f. The size of the population of one generation seems to have no connection with that of the previous generations. We have reached a state of permanent instability. Is this chaos? Let's see if it obeys one of the essential requirements.

We have said that one of the basic characteristics of a chaotic system is that its development with time is extremely sensitive to initial conditions. We can test this by doing all our experiments again but slightly changing the initial rabbit population. Let's make a 1% increase, from 0.200 to 0.202.

From Figure 19.1g, we see that this change makes almost no difference to the behavior of the populations for λ up to 3.5, a range in which the behavior was nonchaotic. We have plotted the results up to the tenth generation, but after a few generations there ceases to be any significant difference between the populations based on the two different initial values. In fact, the results in the nonchaotic range, far from displaying oversensitivity, show almost complete insensitivity to the initial population. No matter what initial population is chosen, after a few generations the numbers settle down to values determined only by λ.

The moment we go to λ = 4, chaos takes over. After the first few generations, in which it appears that nothing untoward has happened, the population numbers for N_0 = 0.200 or 0.202 diverge. There is no connection between them. The small change in initial population has a very strong effect on the number of rabbits observed in following generations (Figure 19.1h). This is a sign of chaos. How sensitive is the pattern of population? Instead of making a 1% change in the initial number of rabbits, we make a change of 1 in 1000 and put $N_0 = 0.2002$. Predictably, the population remains close to that for $N_0 = 0.200$ for more generations than in the case of $N_0 = 0.202$. But at the tenth generation the pattern also breaks away completely, again confirming our statement that in a chaotic system the evolution of the system is very sensitive to the initial conditions.

In practice these results simply mean that if you are asked to predict the way in which the rabbit population changes in a forest, then for the first few generations it doesn't matter much if you make a slight error in estimating the initial population, and in the long run it doesn't matter at all—*provided you are in the nonchaotic range.* If you aren't in that range, then there is little hope for you, unless you have a cast-iron method of counting rabbits exactly. You can solve the equations using the observed value of the breeding rate, λ, but that doesn't help, because the values that you get for the successive populations are so sensitive to the initial population.

We see that a nonlinear system, as shown by our example, can be perfectly well behaved for some conditions but switch over to chaos at others. And, by the way, the scheme given in Figure 19.1 contains undisclosed riches, but you will have to go elsewhere to enjoy them.

I return to a vital point: the system is deterministic, it is not outside the laws of nature. In this case, the initial information involves only integers (the number of rabbits) or rational numbers like λ. If you have the *exact* initial information, you can always calculate the results exactly, even in the chaotic range. If you haven't, it doesn't matter much provided the system is in its nonchaotic mood. Otherwise you would be well advised not to start the calculation.

Again, distinguish between chaos and unpredictability: chaotic processes are by definition unpredictable *in practice* because of their oversensitivity to initial conditions, but unpredictable processes need not be chaotic. Consider a series of four flips of a coin. The final result cannot be predicted. We can only state probabilities, for example, the chance of throwing three heads and one tail, is one in four. But the process of flipping a coin is not chaotic, because the final result of flipping four coins has nothing to do with the initial conditions. Each flip is absolutely independent of the previous flip. This is not true of the flight of our molecule. The path of the molecule after each collision depends on the positions and velocities determined by the previous collision, and this chain of cause and effect goes back to the initial condition of the molecule as it set out on its odyssey. Similarly, in the region in which the rabbit population behaves chaotically, the successive populations do depend on the initial population—very sensitively.

Attractors

We are going to put some foxes into Sherwood Forest. For simplicity we will suppose that the only factor that controls the number of foxes is the number of rabbits, because the rabbits are their only food. It is also presumed that in the absence of the foxes, the breeding rate of the rabbits is such that they would settle down to a single value for their population; λ is 2, say. We will not write down the equations this time; they are known as *Volterra's equations.* Qualitatively, it is clear what will happen.

Suppose that we have introduced a small number of foxes. They find plenty of rabbits to eat, they reproduce, and their cubs likewise enjoy easy hunting. The fox

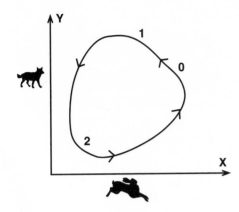

Figure 19.2a. Rabbits and foxes. The populations at successive times are given by a succession of points that when connected make up the graph.

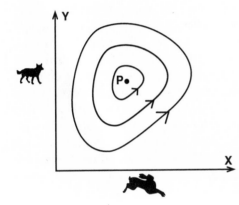

Figure 19.2b. The population graph for a number of different starting populations.

population grows at the expense of the rabbits, who are defenseless. In dealing with chaos, you will find that pictures are invaluable, so let us make a graph (Figure 19.2a). Along the horizontal axis (x) we plot the number of rabbits; along the vertical axis (y), the number of foxes. The zero marks the initial populations. Now since the number of foxes increases, we move upward on the y axis, and we also move left along the x axis, because the number of rabbits is decreasing. We move away from the initial population at point zero, along the arrow and toward the point labeled 1. The rest of the curve can also be understood qualitatively: the number of rabbits decreases and the number of foxes increases, but gradually the foxes start to find it hard to find rabbits—they have eaten most of them. The foxes start to get short of food and some die of starvation because, as Darwin says, only the strong survive. The number of foxes decreases, and because it does so the rabbits find it easier to live a full and rewarding life. In short, as the number of foxes falls, the number of rabbits grows. We are at point 2 on the curve. The continuation is inevitable: as the rabbit population grows, the foxes find hunting easier and, after reaching a minimum, their population grows again and they start decimating the rabbits. We are back to where we started. Both the populations oscillate when plotted against time.

In the model system the cycle goes on forever. Furthermore, different sets of initial populations can give different cycles (Figure 19.2b). A more realistic model of the predator-prey system reveals two important differences from the simple Volterra model (sometimes known as the Lotka-Volterra model). The plot does not return

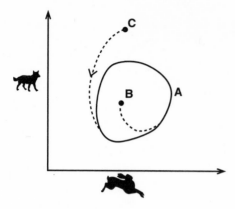

Figure 19.3. A slightly more realistic model of the animal's histories gives a graph that converges onto a single "attractor" no matter what the starting populations.

on itself except for certain initial conditions (Figure 19.3) Only plot A in the figure shows this behavior; any starting population numbers that correspond to a point on this plot will result in the populations following the plot and returning forever. Any other starting populations that lie off this plot will result in populations that follow a spiral path. Starting points like B will result in plots that spiral outward until they reach plot A, which they will then follow. Starting points like C, which lie outside plot A, will result in a plot that spirals inward until it reaches plot A. Plot A can be called a *limit cycle*, since in the limit of long times the other plots all end up there. It is also very natural to call plot A an *attractor*. The plots that we have drawn are deterministic, and there is nothing chaotic about them. Their evolution is almost the opposite of chaotic—they are very insensitive to initial conditions, since they all predictably arrive at plot A.

There is one further point before we carry on to chaos. Going back to the simple situation illustrated in Figure 19.2b, we realize that by suitably choosing the initial populations we can get smaller and smaller cyclic plots. There must be a point, P, in the interior, which is the limit where the plots collapse. If the initial populations correspond to this point, they will stay there. The foxes won't increase because there is no surplus of rabbits; the rabbits won't decrease because there are not enough foxes to reduce their numbers. Point P is also an attractor in its modest way. We could call it a *steady-state* attractor, because the state of the populations never changes.

The value of this example is that we have introduced the idea of an attractor, a cycle to which the system will gravitate whatever the initial conditions; or a steady-state point at which the system starts and where it remains. But where is chaos in all this?

First we must remind ourselves that all these plots are of mathematical expressions, like those for the rabbits alone. It has been shown that great numbers of very diverse types of mathematical equation, when plotted, demonstrate the presence of attractors. Wherever you start plotting the equation, it develops in such a way that it ends up being confined to a certain path, like path A in Figure 19.3. We now extend the idea of an attractor.

We add a third beast to the forest, a group of stoats, say. We set up a model for the changes with time in the population of all three beasts, but now we would need *three* axes on our graph and we might expect a series of plots in the three-dimensional space, each plot representing the changes in all three populations from generation to generation. We would hope that in one of these plots the behavior of all three populations would be cyclic, going through the same succession of numbers again and again. This plot would be a three-dimensional attractor, and in principle there could also be a point attractor in there somewhere, a set of starting values for fox, rabbit, and stoat that would stay unchanging.

In general, for a three-animal forest, we can picture the attractor as an imaginary three-dimensional surface. As the system develops in time, it traces a line on the surface of the attractor. Now such a line can wander in a closely wrapped coil of enormous length until it returns on itself. In our three-beast forest, this would mean that if the initial populations were represented by a point outside the surface they would eventually zoom into it and then move over the surface, running over a very large range of sets of values until they started repeating their tour. This is still not chaos. The path taken by the point is not very sensitive to the initial conditions. Visually what this means is that if you take two points, representing slightly differing sets of initial populations, they will follow each other as they move over the surface. Both paths are completely predictable, and they do not diverge. Next stop, chaos.

There are attractors that have an *infinite* surface, and yet, as we will see later, this surface can be packed into a finite space. Such surfaces can often be regarded as being infinitely folded upon themselves, *not* a series of separate surfaces. The difference can be appreciated by taking all the paper on a new toilet paper roll, stretching it out, gluing the two ends together and then crushing it up. The surface is not infinite, but you get the idea; it is continuous, and in principle a journey on the surface can cover every point in the surface.[7] It is very different from the Sunday *New York Times*, which is a stack of independent sheets.

There are sets of starting values for the three animals which generate such infinite attractors and for which two closely spaced points that start moving on the surface can easily get separated in a very short distance. The two points represent slightly different initial populations of the three animals. The two paths, showing the development of the three populations with time, are extremely sensitive to their initial conditions; the slightest alteration in this position can result in the system following an entirely different path. This is the hallmark of chaos—we cannot know the initial conditions accurately enough to predict the system's future. The points may occasionally come close together after being separated. This means that starting with slightly different populations, they diverge greatly but can come close to each other after a number of generations and then diverge again. What is the nature of these new, infinitely surfaced, attractors? They are *fractals*.

Fractals

One of the most delightful meetings in science has been between chaos and fractals. An important idea that links the two concepts is that of an infinitely long line contained within a finite space. The classic example is the Koch snowflake, the construction of which is described in the caption to Figure 19.4. Two points are worth noting: the shape of the line looks the same at all magnifications, and the length of the line can be increased indefinitely while still remaining within the circle. After 100 steps performed on a triangle with 1-centimeter sides, the length of

---

[7]If you twist the strip once to make a Möbius strip, you can travel over both sides.

Figure 19.4. A Koch snowflake. Each successive shape is obtained by mentally dividing each straight segment into three and then building an equal-sided triangle on the center segment.

the line is about 93 million kilometers. After 240 steps, the line has a length of over 10^{11} light-years, about the distance of the farthest object visible in the universe—with the corrected Hubble telescope. And the line never crosses itself. The concept of an infinitely long line within a finite area may be qualitatively justified if you accept what we were taught at school: a line has no width.

Now imagine a *surface* that never crosses itself, never ends, and can be extended indefinitely but never leaves a finite portion of space, just as the Koch snowflake sits within a finite area. Since a surface has no depth, presumably this is also allowed. If you progressively shrink yourself, examining the surface in finer and finer detail, what you see is basically the same at all magnifications, like a wanderer inside the snowflake. Such a surface is a *fractal*, a term invented by Benoit Mandelbrot, who is the best-known name in this field.

A line is one-dimensional, an area is two-dimensional. If you allow the Koch snowflake to go on forever, although you will still have a line, there is a feeling that this infinitely closely packed object is an honorary area. This statement can be quantified: one can define a "fractal dimension" for fractals, and the Koch line has a dimension of 1.2618, more than a line but less than an area. For most of us the concept can only be appreciated intuitively via the imagination, but fractal dimensions have their importance in solving certain problems. As you might anticipate, fractals constructed in three dimensions have fractal dimensions that vary between 2 and 3; they are, in a way, more than surfaces but less than solids.

Many physical processes, when modeled in mathematical terms, give chaotic behavior, which is demonstrated by the fact that they have attractors that are fractals. In other words, if you plot the history of the system, it wanders over an *infinite* surface, and the history of two systems starting off close to each other can diverge to an extent out of all proportion to their initial differences. That is chaos—an extreme sensitivity to initial conditions. These fractal attractors are the so-called *strange attractors.* Two points placed in close proximity on a fractal can easily get separated very quickly in the infinite maze through which they travel. We have a situation in which a chaotic system moves through a series of states that appear completely random, never repeating in our lifetime, always finding a new place on the infinite surface of the strange attractor, and yet limited to a certain range of values for whatever parameters we are measuring. If you traveled along the edge of the infinite Koch snowflake, you would go through an infinity of positions, but you would never leave the very limited area enclosing the snowflake.

Fractals are shapes that can be generated by mathematical expressions or by geometrical operations—as for the snowflake. It is obvious why the Koch snowflake looks the same at all magnifications—it was built that way. However, not all fractals have this property of self-similarity on all scales, and in fact many of the fractals associated with chaotic natural phenomena have a different geometry at whatever magnification you look at them. The study of fractals has attracted popular attention because of the amazing forms that can be generated by the mathematical expressions that give fractals. I well recall the strange, entirely mathematically generated landscapes that Benoit Mandelbrot brought to my office, landscapes that were a cross between Swiss calendar photographs and the dream landscapes of the surrealists. And if you ask the computer to zoom in on the detailed structure of these scenes, it reveals more and more structure, whatever magnification you go to.

Practical Chaos

Of what practical significance is the study of chaos? The distinguished Belgian physicist David Ruelle, the "inventor" of the strange attractor, has said, "The physics of chaos, however, in spite of frequent triumphant announcements of 'novel' breakthroughs, has had a declining output of interesting discoveries." The initial excitement has cooled slightly. The realization that nature is replete with nonlinear systems and that some of them display chaotic behavior encouraged those in the field to think that chaos theory could be fruitfully applied to an enormous range of problems. It is possible that it can, but although it has had some fascinating successes in some fields, the complexities of many systems frustrate the easy application of theory. But the general feeling is that it will find more and more applications. The scientifically important fields in which attempts have already been made to use the theory include weather prediction, astronomy, certain chemical reactions, the nature of turbulent flow, medical statistics, electronics, mechanical engineering, and the behavior of biological systems. It is even intruding into economics, where I suspect that it has a natural home. The movement of some asteroids (see Chapter 33) has been shown to be chaotic—their orbits are unpredictable. The same is true of the planet Pluto, and in a very muffled way of the whole solar system. The physical reason is the fact that the paths of all the bodies in the solar system are affected by several other bodies. We have seen that an attractor can be localized in a region of space. Our luck is that the attractors for the chaotic motion of the planets are quite limited in extent. In the past, some asteroids, which circulate mainly in a dense belt between Jupiter and Mars, seem to have had large attractors; in other words, their orbits were "wild." The spatial spread of the attractor depends on the forces on the asteroids, and those whose orbits were in a certain relationship to that of Neptune displayed chaos dramatically enough to have left their companions, most of whom continue to circulate in reasonably settled paths. Their original orbits are empty, and the asteroid band has gaps in it that were previously a mystery but are no longer.

There is a modest number of convincing applications of chaos theory to real systems, including cardiac rhythms and certain chemical reactions. The theory is here to stay, and as computers become more powerful, it will be possible to tackle an increasing number of problems using models that are nearer to reality than our rabbits in Sherwood Forest.

The Significance of Chaos

The problems for those who believe in scientific predictability have mounted in this chapter. If we cannot solve the three-body question, if we cannot specify the state of some simple systems accurately enough to prevent chaotic solutions, what hope have we of predicting the behavior of complex systems?

Does this mean that determinism is dead? It may be, but if it is, chaos didn't kill it. The fact that we have no hope of predicting the behavior of chaotic systems does not mean that they are not deterministic. If I can put it this way: we may never be able to predict the behavior of a chaotic system, but the universe "knows" what its future is. The fact that I can't measure a length doesn't mean that that length has no objective existence (at least that's what most nonphilosophers believe). Nevertheless, *in practice*, for chaotic systems determinism is a fiction. The main consequence of the development of the theory of chaos is the collapse of *practical* determinism and the realization that there are insoluble problems within the framework of deterministic science. They are not theoretically insoluble, if we still believe in the laws of motion and in cause and effect. But for many real systems, we can never specify the initial conditions accurately enough to obtain a believable prediction. This finally puts an end to the seventeenth-century belief that all aspects of nature could be, in principle, the subject of quantitative analysis and reliable prediction. The scientist has been humbled, and not for the only time in this century.

Some have seen chaos as a road by which to escape the mechanistic universe of Hobbes and La Mettrie. This seems to me to be grabbing at straws. Unpredictability and determinism are, as we have seen, completely compatible bedfellows. It is also difficult to see how chaos provides support for the operation of free will. The argument is that the brain may include chaotic processes in its working, thus freeing us from determinism. It doesn't—chaos can be deterministic. At least that is what the present physicochemical state of my brain cells has forced me to write.

The unwillingness to include man in a deterministic universe, the hope that free will really operates, is symptomatic of our wish to be qualitatively different from the rest of the universe. This is one of the drives behind creationism. Unfortunately, the more we learn of biology, the more it seems that man is just a particularly complicated example of organized matter, firmly within the animal kingdom. Or maybe you believe that Bonzo has an eternal soul.

vi

Life

Life is anything that dies when you stomp on it.

—Dave Barry

The most complex systems of molecules in the universe are probably on this obscure planet. Living cells are by far the most organized form of matter that we know. Here is matter that can reproduce itself and, in the case of man, examine its own structure and workings, and ask why it exists. It is possible that you hold the vitalist belief that there is a *qualitative* difference between living and nonliving matter; that the laws of physics and chemistry, on their own, plus a little feedback theory, cannot explain the phenomenon of life. I am not on your side, but I am not going to unleash a stream of polemics. What I ask is that you, likewise, have patience and accept that science has found rational reasons for a host of once mysterious phenomena associated with life, and that these findings have proved themselves in combat.

All living forms are partially the products of their genes. We are going to examine the genetic basis of inheritance. We will be discussing great science, absorbing science, useful science—and disturbing science.

As well as delving into the nature of life, we will be reviewing the current state of research into the origin of life. Here we are on shakier ground than any we have trodden on to date.

20 | The Slow Birth of Biology

Sir Nicholas is discovered practicing swimming on a table:
Longvil: Have you ever tri'd in the Water, Sir?
Sir Nicholas: No, Sir; but I swim most exquisitely on Land.
Bruce: Do you intend to practise in the Water, Sir?
Sir Nicholas: Never, Sir; I hate the Water, I never come upon the Water, Sir.
Longvil: Then there will be no use of Swimming.
Sir Nicholas: I content myself with the Speculative Part of Swimming; I care not for the practick.

—Thomas Shadwell, *The Virtuoso*

For centuries no area of science flaunted such a small ratio of meaningful "practick" to empty theory, as biology. The early history of biology was a catastrophe because of man's inability to put "practick" before authority and "speculative" thought. Observation was downgraded to an irrelevancy. It is difficult to excuse the immense durability of Aristotle's descriptions of the form and functioning of animals and plants. There is very little in his writings on biology that has stood the test of time, and much of what he wrote could have been shown to be wrong by simple use of the naked eye. But no one dared to seriously challenge the master until the late Middle Ages. Unfortunately, he was not the only sacred bull.

On the kitchen wall of my home hangs a large, luxuriant drawing of a lavender plant. Beneath the proliferating roots are the words *Mattioli, Venezia 1565*. Pierandrea Mattioli (1501–1577) published a best-seller in 1544: the first translation into a modern European language of the classic *De materia medica* of Dioscorides.

Dioscorides, a first-century A.D. surgeon in the service of Nero's armies, used his spare moments to make drawings of plants and to list their real or supposed medicinal effects. The resulting herbal, written in Greek, was to influence European botany for the next 1500 years. It became a holy text, repeatedly copied. Like the story whispered from ear to ear at a party game, the illustrations drifted farther and farther away from the originals, and for fifteen centuries people were reading and teaching from books that contained drawings of plants removed from the reality revealed to them by their own eyes. Dioscorides was the standard text in many European universities until the middle 1500s. Only then did three Lutheran Germans produce a trio of herbals that were illustrated by accurate, and often very beautiful, drawings. The accompanying texts, however, still borrowed heavily from Dioscorides. The most successful of these herbals was Leonhard Fuchs's *Historia stirpium* (1542), which came out in a German translation the following year. The fuschia was named after Fuchs.

The fossilization of botany was paralleled by the stagnation of zoology. For centuries Aristotle was the accepted textual authority, but it was the bestiary of Physiologus, a second-century Greek, that was the prototype of the immensely popular medieval bestiaries. Physiologus's descriptions of animal behavior were rather low on fact. Everything was buried in a morass of mythology, theology, and popular beliefs. The animal kingdom was apparently created to provide Aesop with material for his tales. Every beast illustrated a moral lesson.

A totally unreliable bestiary does not threaten the health of the average armadillo, but a surgeon could be lethal if he worked from a Renaissance text on human anatomy.

Of Apes and Pigs and Amateur Surgery

If you are in Uppsala, visit the seventeenth-century anatomy lecture theater in the Gustavatium. The benches are arranged in elliptical tiers, to form a steep funnel at the bottom of which stood the table upon which the dissection was performed. Climb the narrow steps up to one of the upper benches and imagine yourself in the world of a seventeenth-century student: the crowded, exclusively male gathering looking down at the waxen, gashed corpse as its inner architecture was revealed by the dissector, to the accompaniment of a commentary read by the professor. Usually it was a male corpse, since most of the corpses were those of criminals. Often the dissection took longer than a week and so winter was the preferred season because it was colder and the nauseating stench that rose through the theater was reduced.

When I sat in that theater and looked down to the table, I knew that blindness, literal and metaphorical, had prevailed. First, it was impossible at that distance for the student to see anything but the coarsest detail. Second, in the seventeenth century, the commentary of the learned professor of anatomy was taken from an unreliable text written fifteen centuries before the dissection.

Galen (c. 130–200), a Greek born in Pergamos, learned much of his anatomy as a physician to gladiators. He settled in Rome and became the court physician to the philosophical emperor, Marcus Aurelius. He turned out torrents of treatises on everything from physiology to philosophy.

Galen was worthy of his reputation as the greatest physician after Hippocrates (c. 460–377 B.C.). In accord with twentieth-century trends, he favored prevention over cure. He was a brilliant experimenter, among other things demonstrating the pulse that spread along the arteries from the heart. Dissection of the human body was forbidden in Rome. Galen had to content himself with the dissection of pigs and Barbary apes, and the examination of a couple of human skeletons that he came upon by chance.[1] The extraordinary fact is that the anatomy of two animals was regarded as part of the definitive source of information on human anatomy for about 1500 years. In the dissecting theaters of sixteenth-century Padua and seventeenth-century Uppsala, the dissector did his work while the professor stood to one side and read from second-century Galen. Galen, like Dioscorides, had emphasized the importance of observation and the danger of being overly influenced by existing texts. To no avail. He survived through the Dark Ages, and, like many other

---

[1]The dissection of human bodies for teaching and anatomical research was practically unknown in Europe before 1300, when it was revived by the Italian Mondino de Luzzi. The ancient Greeks had performed occasional human dissections, but the Church effectively shut down such work for centuries. Medieval Islam also forbade dissection of the human body, as does orthodox Judaism today.

Greek authors, his reputation blossomed at the time of the Renaissance. He was first published in a printed Latin translation by the famous Aldine Press in Venice, in 1476. The text spread over Europe, and in most universities it was considered near heresy to question anything that Galen wrote. Any differences between the anatomical findings of a dissector and Galen's text (and there were many) were pushed under the carpet.

One of the very few to publicly revolt was the grandiloquently named Theophrastus Philippus Aureolus Bombastus von Hohenheim (1493–1541), known as Paracelsus. He was a wanderer for most of his life, a practicing physician without an official qualification, a professor at the University of Basel. His early success depended on the patronage of the humanist Desiderius Erasmus and the famous printer Johann Froben, both of whom were apparently grateful patients. His self-confidence flourished, and in 1527 he threw a copy of Galen into a student's bonfire. From now on, he announced, he would teach only what he had learned from his own patients. That year his main backer, Froben, died. The establishment took its revenge, and Paracelsus had to leave town. He was pushed out of practices in other towns and died in Salzburg in 1541. He believed that, at the Creation, God had supplied a remedy for every ill, and the only reason for a physician failing to find a cure was his own ignorance. He spoke in the language of astrology; he left no school.

Vive Leonardo!

One man might have saved anatomy at that time, the man who in his letter of self-recommendation to Lodovico Sforza wrote, "My work will stand comparison with that of anyone else." It could, but Leonardo da Vinci was not systematic, and in any case his anatomical drawings did not reach the printing press in his time. Dr. William Hunter, the distinguished eighteenth century anatomist who introduced the dissection of cadavers into British medical education, said that "Leonardo was the best anatomist at that time in the world. He certainly knew more than the doctors." His anatomical drawings are amazing. He invented the cross-sectional method of presentation. He made the first drawing to show the proper relationship between the small and large intestines, and he made wonderful drawings of the veins of the liver and of a dissection of the heart, "That marvelous instrument invented by the Supreme Master." Leonardo rarely mentioned God directly.

Da Vinci wrote, "The more thoroughly you describe the more thoroughly you confuse. It is necessary to draw." The seeing eye of the artist was often in those days more reliable than the tradition-bound scholasticism of the learned doctors. Da Vinci's drawings of flowers were miracles not only of art but of accuracy, as were Dürer's. But even da Vinci was under Galen's spell. His drawings of muscles and bones were the fruit only of observation, but the drawings of some of the deeper organs of the body were a cross between what the eye saw and what Galen's books described.

The Truth at Last

The man who finally saw through his own eyes, and not those of others, was Andreas Vesalius (1514–1564), who established a reputation when dissecting for medical students, while he was still a teenager. In 1537, at the age of twenty-three, he was already given the chair of surgery and anatomy at Padua. Early in his career,

Vesalius realized that there was a gap between what he saw and what Galen wrote. He entered a period of intensive activity, it being rumored that his enthusiasm for dissection reached such heights that he dissected bodies before they were dead, like those overzealous waiters who remove your plate before you've finished eating. He employed artists, some from the school of Titian, to prepare anatomical drawings and woodcuts. Before he was thirty, Vesalius had published one of the most handsome and influential texts ever printed: *De humani corporis fabrica* (On the Structure of the Human Body) (1543). By 1600 it had become the standard textbook throughout Europe, and thus it was that, in the sixteenth century, the life sciences finally made a serious break with the error-ridden past. The living eye had vanquished the dead word—although not everywhere. Avicenna's translations of Galen into Arabic were still the basis of medical practice in late-nineteenth-century Teheran.

Appropriately, the *Fabrica* came out in the same year as Copernicus's *De revolutionibus orbium cœlestium*. One author had described man better than he had ever been described; the other had moved man from the center of the solar system. Naturalists began to examine nature through their own eyes, unprejudiced by medieval herbals and bestiaries. One of the great questions that they asked was this: Is there any order, any sense, in the multitude of living forms?

The Spectrum of Life

A few years ago, ten trees in the Brunei rain forest were sprayed with insecticide and the dead insects collected. In round numbers, 20,000 insects were found belonging to 2500 different species. In the whole world, about 1 million species of insect have been described, but entomologists believe that there are at least another 5 million to be discovered, and it is estimated that in all there are between 10 and 100 million living species as yet unnamed. What is a species? Is a poodle a different species from a dachshund, a cat from a puma? Is every mongrel a new species?

Ideally, systematic classification should allow biologists to immediately place a species in the animal or plant hierarchy, simply from the name of the species. On what principles do we carry out this classification? How would you classify a collection of coins of different years? You could, for example, group all the coins of any given year together, or you could group all the nickels together, all the dimes together, and so on. The classification depends on what is most suitable to you. There is no best way. Is there a choice in classifying living forms, or is there one "true" way?

On my sixth birthday my father bought me a Bible and a pair of boxing gloves. The gloves were "because you're a Jew." He should have said that about the Bible, not the gloves, but my father was a realist. At the age of seven, if I had been asked to classify the human race I would have started off by asking if you were a heavyweight, a middleweight, a bantamweight, and so on. After that would have come color. Thus I had a binary classification system and, if pressed, would have classified Joe Louis as *Heavyweightus blackus*. If you think this system is idiotic, I suggest that you compare it with those used to classify animals up to the seventeenth century. Aristotle divided the animals into those that had red blood and those that were bloodless. In many bestiaries, the only attempt at classification was to list the animals alphabetically under the names they had been given for centuries. The order differed in different languages. So how do you classify animals and plants?

The man who first laid down rational general principles of classification was the Swedish naturalist Carolus Linnaeus (1707–1778). He used two concepts to guide him: the idea of distinct species and the concept of a hierarchy of living forms. The supposed hierarchical structure is a cornerstone of the theory of evolution.

The idea of species had previously occurred to the Englishman John Ray (1627?–1705), for whom a species was *a collection of individual animals or plants that reproduced among themselves to give individuals similar to themselves*. Reproduction had also been put forward by Aristotle as one possible criterion for grouping animals. Ray believed that species were fixed, or very nearly so, but not everybody did. Today the concept of species seems so obvious that it is difficult to understand why it took so long to be established, until we recall that, apart from the confusing results of crossbreeding, those reading medieval bestiaries and folktales found that there were beasts that gave birth to utterly different beasts and animals that were generated spontaneously. The mutability of species was taken for granted, but not in the evolutionary sense. Animals didn't change and adapt, they just changed. In such a world, Ray's definition of a species was controversial.

Linnaeus accepted Ray's concept of species, but he went further, envisaging an animal and plant world united by common characteristics. The one on which he laid special emphasis was the sexuality of plants; he was the Georgia O'Keeffe of his day, and the inventor of the commonly used signs ♂ and ♀ for male and female. He saw animal genitals everywhere in the plant world, and his descriptions were not inhibited by the prim conventions of the Dutch town of Leyden, where, in 1735, he published his seven-page summary of a classification scheme for plants: *Systema Naturae*. The organizational framework that Linnaeus introduced was to classify plants into *class, order, genus, species,* and *varieties.* The basis of the classification scheme, for flowering plants, was the form of the stamens, the *male* "genitals." He observed the number of the stamens and their length and grouped flowering plants into twenty-three classes. The remaining, nonflowering plants, such as mosses, were put into a single class, the Cryptogramia. Classes were divided into orders depending on the structure of the stigmas, the *female* "genitals," of the flowers. The calyx, the supporting structure for the flower, was referred to as a nuptial bed, so that the poppy, for example, was described as having "twenty or more males in the same bed with the female." This might have been considered as merely "naughty," but one wonders what his readers made of his comparisons of certain structures in the flower to *labia minora* and *majora.* To name a class of flowers Clitoria was asking for trouble and did not escape the censure of solid citizens, but Linnaeus could point to more blatant examples of the description of nature in suggestive terms. Consider the following verse from a poem on the lamprey, by William Diaper (1686?–1717):

> At length with equal Hast the Lovers meet,
> And strange Enjoyments slake their mutual Heat.
> She with wide-gaping Mouth the Spouse invites,
> Sucks in his head, and feels unknown Delights.[2]

---

[2] I would like to put in a plea for the little-known Diaper, a country parson who, if not a major poet, is often a lively and fresh alternative to the stereotyped nature poets of his time. Diaper wrote it as he saw it.

Having named the main *classes* of plants, Linnaeus, after some experiments, set-tled on a binary system of nomenclature for individual *species*. By 1758 he had named 4236 species. Then he turned to the animal kingdom. A century later, Agassiz and Brown had classified 129,370 animals and plants. We still use Linnaeus's binary system of Latin names.

The whole living world is classified into five *kingdoms*: the Monera, which are single cells without nuclei and include bacteria and some algae; the Protista, which are also single-celled but have nuclei; the Plantae, the higher plants; the Fungi, plants such as mushrooms that do not have the capacity to use sunlight as a direct source of energy; and Animalia, the animals. The kingdoms are split into *classes,* as we have already seen for plants. The mammals, mammalia, are one of the classes of the Animalia. Classes are split into *orders*. The rabbits and tailless hares belong to the order Lagomorpha. Orders are split into *families*. Thus the family Leporidae contains those members of the Lagomorpha that have elongated hind limbs, long ears, and short, curved tails. Families are divided into *genuses*. Rabbits belong to the genus *Oryctolagus*. Genuses are divided into *species*. The different species of rabbits are differentiated from each other on the basis of their average size, their color, length of hair, and so forth. Most common or garden rabbits are labeled *Oryctolagus cuniculus*. The official name consists of the genus, written with a capital letter, and the species, written with a small letter. There is no need to write more; the professional biologist will know from the genus where to find the organism in the tree of life. This is the great advantage of systematic classification.

There was no place for evolution, or transformism as it was then called, in Linnaeus's scheme. The hierarchy of living forms, so strongly suggested by his classification, did not in his eyes imply that lower forms had developed into higher. As he said, there are as many species as God created. But the doctrine of evolution was stirring in the brain of an aristocratic French contemporary of Linnaeus.

Buffon

As a youngster, Georges-Louis Leclerc, Comte de Buffon (1707–1788) was a dueler, and like Tycho Brahe he had to leave his university for this reason. He did not like Linnaeus and was not averse to saying so.

Unlike Linnaeus, Buffon saw the natural world, animal and plant, as an *everchanging* unified whole in which the difference between forms was often difficult to perceive. This was the basis of his objection to the rigid classification into unchanging classes, and the background to his belief in evolution. Of the forty-four volumes of his enormously popular *Histoire naturelle,* thirty-six appeared in his lifetime. This is a treatise written by a man in love with the natural world, but a man who had the clarity of thought of a mathematician. (I like his way with words: "The ass is not a marvelous production.") The drawings in his *Histoire naturelle* were to inspire a series of sketches of animals by Picasso.

Buffon objected violently to the concept of families of species, deriding "those naturalists who, on such slight foundations, have established families among animals and vegetables." He was referring obliquely to Linnaeus but was more specific and combative elsewhere: "This system, indeed, namely Linnaeus's, is not only far more vile and inferior to the already known systems, but is further exceedingly forced, slippery, and fallacious; indeed I would consider it childish."[3]

Buffon estimated the age of the Earth to be 74,832 years[4] and used this time scale to construct what amounts to an evolutionary history of the Earth. Instead of the seven days of Genesis, Buffon proposed seven lengthy *epoches*. Animals emerged in the fifth epoch, the continents separated in the sixth, and man appeared in the seventh. This elongated time scale is not part of the St. James version. When the first volume of the *Histoire naturelle* was published, it was predictably attacked by a committee set up by the theology faculty of the University of Paris.

Buffon issued a thoroughly devious retraction. But the seeds of evolution were stirring. The parts were slowly coming together: the hierarchical structure of the natural world as reflected in Linnaeus's classification; the work of the Danish Nicolaus Steno (1638–1686), who explained the stratification of rocks by a process of successive deposition (over a period of seven days!); and the realization, by Buffon himself, that the animal world was involved in a competition for the inadequate food resources of the earth. Classification was the necessary framework for the theory of evolution, although today we would say that evolution is the true basis for classification.

For many years classification was effectively a dead subject; the general outlines of a system had been constructed by Linnaeus, and there was not a lot more to be said. However, recently the subject has been given a shot in the arm by the advances in molecular biology. The comparison of the DNA, the collection of genes, of different species has allowed family connections to be revealed. Species that are closer in the evolutionary hierarchy will tend to have more closely related DNA maps. Similarly, analogous proteins in different species can be compared—they always differ somewhere in their amino acid sequences, and the differences tend to be smaller the closer the species. Taxonomy, the study of classification, has moved from the macroscopic to the molecular level.

But we have run ahead of our story: before molecular biology, it was the optical microscope that was at the leading edge of biological research.

Macroscopic to Microscopic

The early single-lens microscopes of the Dutchman Antony van Leeuwenhoek (1632–1723) were capable of revealing the tiny inhabitants of pond water.[5] His best lens gave a magnification of about 500. His assistant, Johann Ham, managed to see single cells—human spermatozoa. Seeing them swimming about vigorously, van

---

[3]Compare the gentlemanly Robert Boyle, in the preface to *The Sceptical Chymist*: "A man may be champion to truth without being an enemy to civility, and may confute an opinion without railing at them that hold it."

[4]There is evidence that Buffon believed the age of the Earth to run into several million years. He was wise enough not to make this guess public. His published estimate was off by a factor of roughly 50,000. It was also about thirteen times longer than the then widely accepted figure of Archbishop James Ussher who, in 1654, announced that, after years of research, he had fixed the date of the Creation as 26 October 4004 B.C., at the distinctly civilized hour of nine o'clock in the morning.

[5]Leeuwenhoek was a friend of Jan Vermeer, who used light in a somewhat different way.

Leeuwenhoek was so impressed that he declared them to be the real creators of new life, rather than the egg. Nevertheless, the concept of the cell did not occur to him. When he communicated his findings to the Royal Society, his description of "little animals" in water created a sensation. Leeuwenhoek and his compatriot Jan Swammerdam (1637–1680) were largely responsible for the now defunct *preformist* theory, which states that the sperm contained a minute replica of the grown organism. A variation of the theory places the replica in the ovum rather than the sperm. Swammerdam was the first man to see red blood corpuscles, but he also failed to be led to the conclusion that life came in small packets.

In England, Robert Hooke constructed a compound microscope, with two lenses, and produced exquisite drawings of parts of insects and other natural objects. It was he who discovered the multi-lens structure of the fly's eye. His work was frequently met with the same skepticism that met Galileo; it was imputed that he was studying optical illusions. He did not escape the barbed pen of Thomas Shadwell, who in the play *The Virtuoso* (1676) was referring to Hooke when he wrote: "A Sot, that has spent £2000 in Microscopes, to find out the Nature of Eels in Vinegar, Mites in Cheese, and the Blue of Plums, which he has subtilly found out to be living Creatures." Hooke was the first to employ the term *cell* in the context of living matter. He examined a cross section of cork and described the array of roughly circular holes as "cells," using the word in analogy to prison cells. He invented the nomenclature, but he too went no further with the idea. Marcello Malpighi (1628–1694) used both single-lens and two-lens microscopes. He was the founder of microscopic anatomy, and his work on the lungs, in which he discovered capillaries, was the real beginning of our understanding of respiration. He drove himself hard: "I have sacrificed almost the whole race of frogs, something that did not come to pass even in Homer's savage battle between frogs and mice." Like the work of Hooke and Galileo, his optical devices were held by his contemporaries to be falsifiers of reality. It is pleasant to think that in 1684, when he lost his microscopes in a fire at his house, the Royal Society had some lenses made for him and sent to Bologna. Again, Malpighi missed the significance of the cell.

François-Vincent Raspail (1794–1878), political activist, friend of Flaubert, and unqualified but successful medical practitioner, prepared the way for the modern concept of the cell theory, stating that "the plant cell, like the animal cell, is a type of laboratory of cellular tissues." He realized that the membrane around the cell acted as what he called a "sorter," selectively allowing only certain substances to pass. Even so, the origin of the cell theory is conventionally attributed to the German scientists Theodor Schwann (1810–1882) and Matthias Jakob Schleiden (1804–1881), both unusual characters.

Schwann was the founder of the mechanistic school of medicine. In contrast to the overwhelming feeling at the time, Schwann considered that living organisms were not produced by a directing will, aiming at a predetermined goal, but rather by the laws of chemistry and physics. He saw all organisms as being built from cells, a "cell" for him being "a layer around a nucleus," the whole structure being covered by a membrane. The cell nucleus had been discovered in 1830 by the Scots botanist Robert Brown, who we met in connection with Brownian motion (see Chapter 16). Schwann also made the pivotal suggestion that organisms develop by the differentiation of the original clump of undifferentiated cells. This is the basic axiom of embryology, which should be compared with Leeuwenhoek's preformist theory. Schwann's whole scientific career lasted only five years, from 1834 to 1839, after which he forsake materialistic rationalism and drifted into mysticism.[6]

About the same time that Schwann was singling out the cell as the central concept in biology, his countryman Jacob Schleiden was bringing out similar statements, particularly in an article written in 1838, and in a subsequent textbook in which he writes that the cell is "the foundation of the vegetable world." Schleiden, like Schwann in his early days, adopted a straightforward physical and chemical approach to living matter, rejecting the vague wafflings of *Naturphilosophie*.[7]

The cell theory needed the prestige of another German, Rudolf Virchow (1821–1902), to firmly establish it as official dogma. In 1860, Virchow coined the classic motto: *Omnis cellula e cellula*, All cells arise from other cells.

The cell is the universal unit of living matter. *The concept of the cell was the great turning point in the history of biology*. It was the microscope that was the indispensable instrument needed to detect objects that are generally no more than one five thousandth of a centimeter across.

From the middle of the nineteenth century, the cell theory established itself firmly at the center of the biological sciences. Different types of cells were characterized, their internal structure examined as best as could be, by the optical microscope, and their components separated out. This work was to lay the foundations for the astonishing advances in cell biochemistry and molecular biology that were, in the twentieth century, to sweep biology into the realm of the exact sciences. Biology, chemistry, and to a lesser extent physics have met within the living organism.

The metamorphosis of biology into a science came to full fruition with the elucidation of the nature of genes and their mode of action. The story began in a quiet Austrian garden.

---

[6]Schwann realized that alcoholic fermentation was caused by yeast cells, but he came under an extraordinary attack from Wöhler on this count. Wöhler wrote a satirical article suggesting that yeast cells in sugar solution change into small animals, which eat the sugar and expel carbon dioxide from their urinary organs and alcohol from their anus. Their bladders are shaped like champagne bottles.

[7]Schleiden was an individualist. At a time when the tide of anti-Semitism was rising in German universities, he wrote a complimentary article on the Jews, singling out their role in conveying knowledge to the West in the Middle Ages.

In a Monastery Garden

He's the spitting image of his dad.

—About one-quarter of the visitors to maternity hospitals.[1]

Gregor Johann Mendel (1822–1884) bred pea plants in a monastery in Brno, Austria, because he was interested in inheritance. Mendel concentrated on the inheritance of a very small number of characteristics, and those one at a time. He started off with plants that had been inbred so as to give a uniform population. Thus Mendel took plants with long stems, that had been inbred for at least two generations until all offspring had long stems. Similarly, he bred "pure" short-stem plants. Only then did he cross short-stemmed and long-stemmed plants, taking the pollen from one plant and transferring it to the other. In a large set of experiments he observed stem length in the progeny of his pure plants, and found that *all the stems were long*, even though one of the "parents" was short-stemmed. Then he took that first generation of long-stemmed plants and crossbred them among themselves, again documenting stem length among their progeny (Figure 21.1). He found that both long- and short-stemmed plants appeared in the second generation. He carried out experiments of this kind to observe other single properties: color of seeds, seed shape, pod color, pod shape, flower color, and flower position on the stem. He observed something like 28,000 results of crossbreeding.

What would *we* expect from these experiments? We know that inside almost every living cell there is a well-defined, roughly spherical body called the nucleus and within the nucleus there are elongated bodies called chromosomes. The chromosomes are basically strings of genes, the entities that control the different physical characteristics of our bodies. Genes are passed from generation to generation, and in the case of sexual reproduction the offspring receive half of their chromosomes from their mother and the other half from their father.

In peas chromosomes come in pairs, each member of the pair carrying the same kind of genes but not necessarily identical genes. For example, both chromosomes of a pair may carry a gene for color, but one may be the gene for yellow peas and the other the gene for green. In human beings each chromosome may carry about 2 to 3000 different genes.

In the normal human cell there are twenty-three pairs of chromosomes, forty-six chromosomes in all. In the female, each member of a pair, as seen under the microscope, is apparently of equal length. In the male, one of the pairs consists of a long chromosome (labeled the X chromosome) and a short one (the Y chromosome). In the pea plant there are seven pairs of chromosomes in each cell.

---

[1] Assuming that about half the babies are female, and about half the visitors to a male birth think that junior looks like his mother.

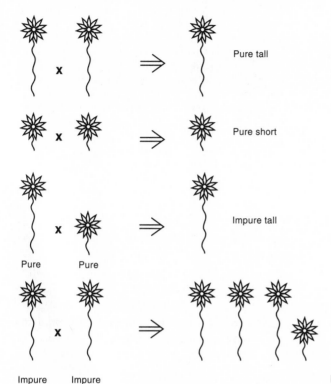

Pure tall

Pure short

Impure tall

Pure   Pure

Impure   Impure

Figure 21.1. Mendel breeds peas.

One of the genes in the pea plant controls the length of the stem; another controls the color of the seeds. When crossbreeding takes place, the genes of the parents are passed to the cells of the offspring. It is the way that they mix that is the clue to Mendel's findings.

To simplify our explanation we will look at a hypothetical plant that has only one pair of chromosomes in each cell (Figure 21.2). We will concentrate on the inheritance of color. We assume that the gene that controls color has two forms, two *alleles*, as they are called. One allele results in a yellow pea, the other in a green pea. In the case of "pure" yellow plants, the pair of chromosomes that carries the color genes each has an identical "yellow" gene. Likewise for the "pure" green variety (Figure 21.2).

The reproductive cells—the *gametes*—of the plant, like the sperm and the ovum of humans, contain only half the number of chromosomes of the normal cell, but they can contain either chromosome from each pair in the normal cell (Figure 21.2). In our hypothetical case, since we have only one pair of chromosomes, the gametes of a "pure" yellow plant all have a yellow gene. Now the single pair of chromosomes in the normal cell are not necessarily identical, even if both carry a yellow gene: for example, one chromosome may carry the short-stem gene and the other the long-stem gene. Thus in the present hypothetical case all gametes carry a yellow gene, but there are two types of gametes.

Next take the pollen from a pure yellow plant and pollinate a pure green plant. In principal the reproductive cells in the receiving plant will also be of two kinds, but in the case of a pure green plant every gamete will have a green gene. Since the meeting of gametes is random, any one of the two types of pollen has the same chance of meeting any one of the two "receiving" gametes. The consequences are

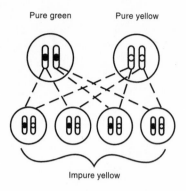

Pure green   Pure yellow

Impure yellow

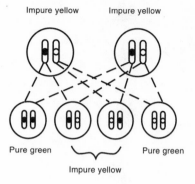

Impure yellow   Impure yellow

Pure green   Pure green

Impure yellow

Figure 21.2.
The inheritance of color.
For example, a pure green
plant crossed with a pure
yellow plant gives all
yellow offspring, but they
contain (recessive) green
genes.

shown in the figure, from which it is evident that there are four possible combinations—two things with two things. In other words, there are four types of fertilized egg, each containing a pair of chromosomes.

What will be the color of the seeds of each of these eggs when it develops into a plant? Mendel would have guessed that they would all be yellow. This is because all of them have a yellow gene in their cells, and if there is a yellow gene and a green gene, the yellow gene is *dominant* and forces the plant to be yellow. Thus, even though the cells of a yellow plant may contain green genes, they are not evident in its color. We say that they are not *expressed*—the green gene is *recessive*. All of us carry some recessive genes.

Now try crossing two of these first-generation yellow plants, two plants that both contain one green and one yellow gene (Figure 21.2). The gametes of both plants are now divided into those that contain a yellow gene and those that contain a green. We combine them as before, two objects with two objects at random. Mendel found that three-quarters of the progeny were yellow and one-quarter were green. We can get this result if we assume that in a second-generation plant carrying two yellow genes, the plant will be yellow; if there are two green genes, the plant will be green. If there is a yellow and a green, the yellow dominates and we get a yellow pea. As you can see, three out of the four fertilized combinations have at least one yellow gene, giving a ratio of 3:1, for yellow to green, as Mendel found. The fact that a real pea plant has another six pairs of chromosomes is irrelevant to the result; it is only the chromosomes carrying the color genes that control the color of the next generation. Thus the mating of two yellow plants can, if they are not "pure," give some green progeny—a scandal in the family.

Mendel knew nothing about genes, but by attributing the inheritance of single characteristics to a "factor" that could take on a dominant or a recessive character, he could explain the statistics that emerged from his painstaking work. The founder of genetics published his work in 1865, in a minor scientific journal, and also wrote to some eminent biologists. You've guessed it—his work was either dismissed or ignored. Mendel was crushed; he gave up trying to publish his results.

Mendel had read Darwin's *Origin of Species,* but, strangely, did not appear to see the relevance of his own work to evolution. Darwin is known to have seen a book published in 1881, citing Mendel's work, but he never referred to Mendel. Mendel's work appeared to have died. And then, in 1900, three scientists independently published results that completely confirmed Mendel's work. De Vries in Holland, von Tschermak in Austria, and Correns in Germany had carried out their work unaware of each other and of Mendel. All three went to the library to write their papers, and all three discovered that Mendel had scooped them. All three acknowledged Mendel's prior claim, but he had died sixteen years before. In 1904, Walter Sutton, in the United States, realized that genes (he didn't use that word) and chromosomes came in pairs, and he voiced the suspicion of several other investigators that genes, which were invisible under the microscope, were located on chromosomes, which were visible. Genetics was blossoming. Mendel had created one of the central fields of modern science. His sole honor in his life was to be made abbot of the monastery, not the last time that a scientist ended up as an administrator.

The eminent statistician and geneticist R. A. Fisher has suggested that Mendel thought up his theory before he did his experiments, and gently cooked his results so as to conform to his expectations. There are reasons for supposing that exact ratios of 3:1, for example, should not be observed. If Mendel did massage his results slightly, his confessions must have been rather unique for a monk.

Interesting Complications

We have considered an extremely simple case of heredity, the inheritance of one characteristic, determined by one gene (for color) with two alleles (green and yellow). Inheritance is often more complicated than this, and the study of inheritance at the genetic level is a major discipline. Here we will content ourselves by pointing out some of the simpler basic ideas that are part of the geneticist's armory.

We supposed color to be simply determined by two alleles, one of which, the dominant allele, overrides the action of the recessive allele. There are cases when this dominance is incomplete, and the various characteristics expressed in the organism are a mixture of those due to the two alleles. In the case of the pea, this would mean that we could breed yellowish green peas.

Another complication is that there are characteristics that depend on more than one gene. Whatever part of our intelligence is inherited, it is almost certain that it is not due to a single gene, and neither is skin color.

In the case of man, as compared with pea plants, there is an additional complicating factor that has to be taken into account. Humans exist in two sexes. As we mentioned earlier, in the cells of the male there is one pair of chromosomes that is not symmetrical. One chromosome, the Y chromosome, is considerably smaller than its partner, the X chromosome, and carries less genes. This asymmetry holds for most animals, although there are a few in which it is the male who has two complete, X, chromosomes, and the female who has one asymmetrical pair. A disagreeable consequence of the chromosomal difference between the sexes can be appreciated by considering the case of a gene that is defective.

Suppose the gene responsible for directing the synthesis of a certain protein is located on the X chromosome, which implies that for a woman the gene is on both members of the X pair (Figure 21.3). Now suppose that a certain woman has a defective gene on one of her X chromosomes. This gene is incapable of directing the synthesis of the protein in her cells. However, we have supposed that she has a corresponding normal gene on the other X chromosome, and therefore she will produce the protein. If a man has a defective gene for this enzyme on his X chromosome, there is a chance that he will *not* have a normal gene on his Y chromosome because the Y chromosome is much smaller and has far fewer genes; it may not even carry that gene. This man will suffer from whatever symptoms follow from a

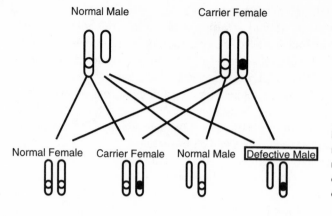

Normal Male   Carrier Female

Normal Female   Carrier Female   Normal Male   Defective Male

Figure 21.3. Sex-linked inheritance. The recessive (black) gene in question does not appear on the short male Y chromosome but is carried by the female on one of her X chromosomes.

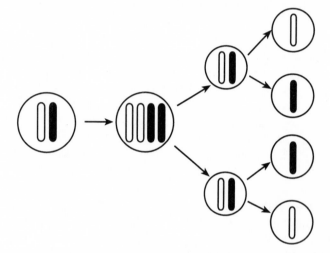

Figure 21.4. The production of male gametes, called *sperm* in the case of man. The organism in this illustrative case has only one pair of chromosomes.

lack of that protein. Notice that the woman is a *carrier*; although not suffering from the disease herself, she is carrying a defective gene that she can hand on to her children. If it goes to a male child, he may suffer from the deficiency disease. If it goes to a female child who is unlucky enough to inherit a defective gene from her father, then the child will have two defective genes and will certainly suffer from the disease. Hemophilia is a classic case of a "sex-linked" genetic disease, and the classic example is Queen Victoria's progeny. She was a carrier, not suffering from the disease herself but passing on the faulty gene to some of her offspring.

There are a variety of abnormalities associated with a small percentage of humans who carry more than one Y chromosome. The best known are XYY individuals, having forty-seven chromosomes and being popularly dubbed supermales. Such men appear to have an above-average chance of being violent.

Another complicating factor in heredity, one that occurs throughout the animal and plant kingdoms, is *recombination* or *crossing-over*. It plays a vital part in heredity.

The Gene Blender

It is obvious that sexual reproduction mixes the genes of a given population. This is one of the things that worry the minuscule brains of racists. However, genes get

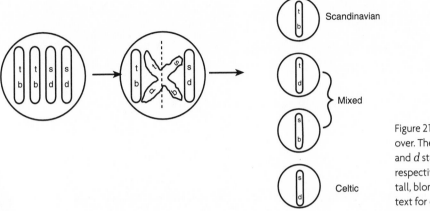

Figure 21.5. Crossing over. The letters *s, t, b,* and *d* stand, respectively, for short, tall, blond, and dark; see text for details.

mixed up in another way even before fertilization of the egg. They get mixed up in the cell of the parents, during *meiosis*, the process that results in the formation of the gametes from their precursor (germ) cells. Let us take a simple example and follow the process of sperm production in an animal with only one pair of chromosomes (Figure 21.4).

In the first stage of meiosis, each chromosome is replicated to give four chromosomes in the cell. These crowd together to form a tetrad, which subsequently, via two cell divisions, breaks up to give four sperm cells, each containing one chromosome. If the tetrad merely breaks up to give the original four chromosomes that made it, then there are no surprises. The chromosome in each sperm is exactly the same as one of the chromosomes in the original germ cell, which means that they are the same as one of those in every cell of the father. We have singled out two genes to illustrate what happens. One is a gene for hair color (dark or blond) and the other for height (long or short). In our example, then, father's sperm will contain chromosomes of two types: tall, blond-haired sperm (Scandinavian) and short, dark-haired sperm (Celtic), so to speak. If there is what is termed *crossing-over* the situation is more interesting.

In crossing-over, two chromosomes temporarily join together and when they separate, swap pieces of each other, as can be seen from Figure 21.5. This can result in new combinations of characteristics being exhibited in offspring. In our case, as you can see, two new kinds of sperm are produced: tall, dark-haired (Scandinavian Celts) and short, blond-haired (Celtic Scandinavians). It is because this kind of shuffling around that a couple can have two children who look very different from each other.

The last complication that needs mentioning is the most newsworthy: genes can undergo changes in their chemical structure. *Mutations* are, as we will see, one of the reasons life developed from a primitive, single-cell organism into a higher form of life: the drunken, beer-bellied soccer fan.

22 | Evolution

For when he takes his prey he plays with it to give it a chance.
For one mouse in seven escapes by his dalliance.

—Christopher Smart, "Jubilate Agno," (*My Cat Jeoffrey*)

One day in 1956, during the Suez crisis, I was summoned to the office of Professor J. B. S. Haldane, the famous geneticist, biometrist, and ex-Marxist. I was an undergraduate at University College London, and Haldane was for me an awesome personality. His room was littered with books, random objects, sheets of notepaper, and what seemed to be fossil bones. There was a box of germinating potatoes beneath his desk.

"Sit down, Silver."

"Thank you, sir."

"I'm told you're a leader among the students." He went on to outline a plan for bringing down the government of Anthony Eden, whose policy in regard to the Suez Canal he objected to. The plan involved thousands of students, armed with pennies, phoning government offices in Whitehall from public telephones. "When the phone rings, replace the receiver immediately. That way you can use the same coin repeatedly." The idea was that this continual tying up of the phones in Whitehall would induce the government to resign. He elaborated, and I sat there amazed at this zany, professorial plot coming from someone who was one of my intellectual heroes.

When the plan had been fully explained, Haldane chatted with me on science. It was, of course, more of a monologue than a chat. At one point I ventured a standard question: "I've always wondered if life has existed long enough for one-celled organisms to evolve into human beings." He smiled from under his bristly eyebrows, "It took *you* exactly nine months."

Was there time for an amoeba to be transformed into Frank Sinatra? Is Darwinian evolution credible?

There have been two major battles concerning living matter. One has been over the question of the uniqueness of living matter: are the living and the nonliving worlds separated by what in the nineteenth century was called "vital force," an undefined, occult "something"? The second, and longer-lasting, altercation, has been over the reality of evolution. The two controversies are connected. In both cases the argument is between those who believe that life and, more specifically, the emergence of man, can be explained by the laws that govern the behavior of the inanimate world, and those who don't.

The belief that living creatures contain molecules that cannot be made from nonliving precursors, or that living matter contains a "vital force," died down (but not out) in the nineteenth century. Evolution remains an area of controversy, although the arguments against it have the nature of a desperate, hopeless rearguard action. In this chapter I will attempt to demonstrate open-mindedness, but there will certainly be those who will condemn my scientifically conventional conclusions as wrong-minded at the best and satanic at the worst.

Those who see Old Nick's hand behind the theory of evolution still take or threaten court action over the place of evolution in the school syllabus. On the strictly scientific front there are sharp differences of opinion between professional research workers as to the exact *mechanisms* of the evolutionary process in specific cases. For example, there is no consensus on the cause of the so-called Cambrian explosion, the sudden emergence, about half a billion years ago, over the comparatively short period of about 20 million years, of a great variety of bigger animals. Nevertheless, the scientific world overwhelmingly believes in the principle of evolution, and it is the current, broadly accepted picture that will be presented here. The fact that all the details have not yet been filled in is a reflection of the infinite variety of nature, not a sign of a basic weakness in the theory.

The Evolution of Evolution

First, forms minute, unseen by spheric glass,
Move on the mud, or pierce the watery mass;
These, as successive generations bloom,
New powers acquire, and larger limbs assume.

—Erasmus Darwin, "The Temple of Nature"

"Evolution" invokes Darwin. And yet there are few ideas that can be ascribed in their entirety to one individual.[1] Darwin was preceded by Jean-Baptiste de Lamarck (1744–1829), who was preceded by Erasmus Darwin (1731–1802), who was preceded by many eighteenth-century scientists, amateur biologists, and writers, including Diderot. No one who observed the animate world could fail to notice the similarities between different species, and to realize that even in the absence of the fossil record there was a rough hierarchy of living forms, a steadily increasing complexity that was bound to suggest the idea that the simpler were the biological antecedents of the more complex. As far back as 1699, the English physician Edward Tyson (1650–1708) had published an account of the dissection of an orangutan and had pointed out that chimpanzees were anatomically closer to man than to the other primates. Linnaeus, in 1735, commented, "It is remarkable that the stupidest ape differs so little from the wisest man, that the surveyor of nature has yet to be found who can draw the line between them." It was he who gave us the name *Homo sapiens,* making us another species in the animal kingdom, but he also avoided suggesting, or did not dare see, the evolutionary link between man and ape.

Over a century before Darwin published his ideas, Benoit de Maillet had proposed that birds had evolved from flying fish and lions from sea lions. In 1751, Maupertuis suggested that new species could be formed since, "the elementary particles" which form the embryo "may not always retain the order which they

---

[1]As an ex-rugby player, it is clear to me that the invention of the game by the immortal William Webb Ellis catching (rather than kicking) the football, and then running with it, is perhaps the only example in human history of a stroke of genius unaided by any other mortal.

present in the parents." The endlessly curious Goethe suggested that all plants were descended from one primitive plant, to which he gave the rather inelegant name of *Urpflanze*. There were others with similar ideas, including Charles Darwin's own grandfather, Erasmus Darwin, a doctor and inventor, friend of Benjamin Franklin and of Josiah Wedgewood. However, Erasmus Darwin based his concept of evolution on a mistaken belief in the inheritance of acquired characteristics, a theory that implies that environmentally induced changes in the parent are passed on to its offspring. Incidentally, the term *Darwinism*, as applied to evolution, was first used in reference to Erasmus Darwin. It was Charles Darwin who summed up his own and others' ideas to propose a convincing theory of evolution.

A man who undoubtedly had a major effect on Darwin was the Reverend Thomas Malthus (1766–1834), who in his seminal *Essay on the Principle of Population* (1798) propounded the thesis that great masses of humanity perished because they failed to compete for limited available food. This was survival of the fittest. Listen to Darwin:

> In October 1838, that is fifteen months after I had begun my systematic enquiry, I happened to read for amusement *Malthus on Population,* and being well prepared to appreciate the struggle for existence which everywhere goes on from long-continued observation of the habits of animals and plants, it at once struck me that under these circumstances favorable variations would tend to be preserved and unfavorable ones destroyed. The result of this would be the formation of new species.

Malthus himself had already considered, in the spirit of natural selection, the lowered chances of survival of physically handicapped members of the American Indian community. Darwin took Malthus and applied his analysis of human history to all God's creatures: "It is the doctrine of Malthus applied with manifold force to the whole animal and vegetable kingdoms."

Darwin was a candidate for Holy Orders when, twenty-two years old, he set sail on the HMS *Beagle* in December 1831. We probably have to thank the great Swedish botanist Linnaeus for the fact that the commander of the *Beagle* was prepared to take a scientist aboard. Almost exactly a century before Darwin stepped on deck, the young Carolus Linnaeus, then a medical student, had joined an expedition to Lapland to examine the local way of life and to bring back biological specimens. By the time Darwin sailed, it was common practice for naturalists, as they were called, to travel the world on sailing vessels.

Darwin's five-year journey to South America changed his life, revolutionized biology, and united man with the natural world. When he set out, he believed in the immutability of species; when he returned, after observing and collecting a huge number of biological and geological specimens, he had become convinced that the only rational explanation of the facts was that species evolved into new species. He believed, as most open-minded scientists still do, that all life forms developed from one or a few very primitive organisms, although he did not speculate about their origin.

Darwin was convinced of the reality of evolution but had to find a mechanism for it, and this he did in the theory of natural selection, sometimes termed the survival of the fittest. Had he lived today, he would have published his ideas immediately, to establish priority—and get tenure. In fact, he worked for nearly twenty years collecting new facts, thinking, and analyzing, until one summer day in 1858 he received a letter from Alfred Russel Wallace, a naturalist working in the Far East. While going through an attack of malaria, Wallace lay down to think about "any subjects then particularly interesting me." It occurred to him that "the fittest would survive." Wallace, incidentally, had also read Malthus. His letter contained a short description of his ideas on evolution. They were practically identical to Darwin's. Darwin had to publish or perish, but his sense of fairness made the thought of publishing before Wallace repugnant: "I would far rather burn my whole book than that he or any other man should think that I had behaved in a paltry spirit." In fact, Darwin had a copy of a letter that he had written to Asa Gray at Harvard in 1857, and also possessed the 1844 version of a manuscript that he had originally prepared in 1839. These documents established his priority. His friends convinced him to present them, together with Wallace's letter, at a meeting of the Linnean Society in London on 1 July 1858. Neither author was present, and no discussion followed the reading of the communication. The annual report of the president of the society for 1858 included the following words: "This year has not been marked by any of those striking discoveries which at once revolutionize, so to speak, the department of science on which they bear." Oh well.

The date 24 November 1859 saw the publication of *On the Origin of Species by Means of Natural Selection, or the Preservation of Favoured Races in the Struggle for Life.* A few days before the book came out, Darwin wrote to Wallace: "God knows what the public will think."

Wallace must be given his due, but Darwin not only preceded him, he provided the bulk of the massive body of biological and geological evidence that gave the theory of evolution its credibility. For this reason, Darwin is considered to be the authentic father of the doctrine of evolution through natural selection.

Darwin had hoped to gloss over the inflammatory question of man's place in the evolutionary ladder, which he barely mentions in his book, but it was inevitable that most of the fuss around the book was centered on the question of the origin of man: Was he part of the evolutionary saga or wasn't he? One has the feeling that if Darwin had unequivocally excluded man from his theory, the book might have passed almost unnoticed by the general public. As it is, there are still those who would like to burn it.

The Darwinian Evolutionist's "I Believe"

Those who believe in Darwinian evolution generally believe that life on this planet originated by chance, although that is not an official part of the theory of evolution. They all believe that the first recognizably living forms were tiny unicellular organisms living in water, and that every single species, extant or extinct, is connected by an unbroken chain of intermediate species to these first life-forms. All species, here or long gone, belong to one biological family. In Darwin's own words, they can be arranged on "the great Tree of Life, which fills with its dead and broken branches the crust of the earth, and covers the surface with its ever-branching and beautiful ramifications." Historically, the reasons for believing in this scenario were first and foremost the possibility, for any given species, of finding close relatives. The relationship may express itself in a similarity of structure and function. The notorious example, of course, is *Homo sapiens* and his similarity to the higher apes, but a child at the zoo can sense the kinship of the tiger and the puma to her pet cat. That these similarities exist, that rough plausible chains of structural evolution can be constructed by examining living species, is undeniable. It is not true that all *living* species can be arranged in a branching tree. In the course of evolution some of the branches have ended up "dead and broken"; there are species that have developed from a common, long-extinct ancestor. This is where the fossil record often fills in the gaps. The remains of prehistoric species have been dug out of layers of rock that can be dated. In broad outline, the older the rocks in which a fossil is found, the simpler the species. Taken as a whole, the fossil record, which stretches over a geological time scale of billions of years, reveals an unmistakable, if incomplete, progression of living forms, a slow evolution from simple to complex. How is the evidence to be interpreted? Those who believe in the literal truth of the book of Genesis believe that every individual species, plant and animal, was created in one gigantic fait accompli, on Wednesday, Thursday, and Friday of Creation week. The Darwinian theory of evolution provides a rather different explanation.

The Bare Bones of Evolution

At the root of the modern theory of evolution is the genome, the collection of genes that determine the physical characteristics of every individual living organism. Differences in the genomes of parents and their offspring can result in minor or major physical differences between consecutive generations.

The mechanism of evolution is based on the appearance, in every new generation, of variants—individuals or groups of individuals of the same species that differ genetically from each other to a lesser or greater degree. A huge amount of experimental evidence shows that there is a wide range of variants in any given generation, although most of those variations are not particularly dramatic. Thus even in a fairly homogeneous isolated society like a remote tribe, there are variations in height, age of balding, facial structure, and so on. Some variations may be due to environmental causes such as better diet, but there are always variations that can more or less confidently be ascribed to genetic differences.

There are two methods by which species can produce variants: mutation and sexual reproduction. In the early history of life, mutation may have been the only method. Some basic points:

1. A "mutation" or "mutant" originally meant any variant, an organism or small group of organisms that differed in some observable way, in either form or function, from the majority of the members of the previous generation of a given species. With the birth and advance of molecular biology, it became clear that variations in the form, appearance, or functioning of the members of a single species are (except in the case of noninherited disease and obvious environmental factors) caused by variations in DNA composition. (An "obvious environmental factor" could be, for example, the company barber responsible for the crew-cut appearance of enlisted men.) It also became apparent that some mutations were created by chemically and physically caused damage to the chemical structure of DNA. We shall adopt the current usage, which is to confine the term *mutation* to a change in DNA caused by an external agent or an internal re-

arrangement of the DNA molecule. Variants caused by sexual reproduction are not termed mutants. Nonbiologists often overlook the role of sexual reproduction, unjustifiably assuming that only mutation can create the wide range of variants that is the basis of natural selection. The mixing of genes that occurs during fertilization is by far and away the main cause of the variants that occur in any one generation in species that reproduce sexually.

2. Over a geological time period, multitudes of variants have appeared.
3. In the natural environment of a colony of a given species, there will occasionally be cases in which one or more variants have characteristics that enable them to survive or to multiply more efficiently than all other variants present.
4. The advantage possessed by such a "superior" organism, and which ensures its surviving longer and/or breeding faster than other variants, normally results in that organism becoming the majority variant after one or more generations, if it is not threatened by subsequent, even more superior, variants. *This is natural selection, the survival of the fittest.*

This, in the barest possible outline, is the mechanism of evolution: natural or induced variations in every generation, plus natural selection of the fittest. Darwin did not have the benefit of our knowledge of the molecular mechanism of evolution; this is how he expressed his general conception of evolution in a critical passage from Chapter 4 of his book:

> Can it, then, be thought improbable, seeing that variations useful to man have undoubtedly occurred, that other variations useful in some way to each being in the great and complex battle of life, should sometimes occur in the course of thousands of generations? If such do occur, can we doubt (remembering that many more individuals are born than can possibly survive) that individuals having any advantage, however slight, over others, would have the best chance of surviving and of procreating their kind? On the other hand, we may feel sure that any variation in the least degree injurious would be rigidly destroyed. This preservation of favorable variations and the rejection of injurious variations, I call Natural Selection.

It is essential to realize that evolution is an automatic process; it involves no "will" on the part of the organism involved, and the occurrence and type of a given variation, unless human beings deliberately intervene, is governed by chance. There is no instinctive drive in a living organism to create "higher" forms of life. It is the combination of chance variation and the environment that gives evolution a direction.

A well-known example of the evolutionary process, which has taken place within your lifetime, is that of the adaptation of certain bacteria to antibiotics. In a colony of bacteria dosed with penicillin, it is quite possible that every single bacterium will perish. However, it is difficult to find a colony that is genetically homogeneous, because mutations result in bacteria whose genetic makeup differs from that of the majority. There could be many kinds of mutants, and it is possible that most of them will perish along with the majority variant, but it is evident from medical records that over the years certain strains (mutants) have flourished even in the presence of penicillin. These are the mutants whose genetic makeup confers on them the ability to resist the fatal action of penicillin that kills off their less talented brethren. We have selected a new "species" that, because it is fitted to its environment, will become the dominant mutant in the colony as long as the threat of penicillin is present.

The appearance of a completely penicillin-resistant type of bacterium need not be accomplished by a single mutation. It is quite possible that none of the mutants in a given generation is particularly resistant but that there are a few that, though all but wiped out, are sufficiently resistant to be the major species after the massacre. These mutants are now the origin of further mutants, some of which may well be *less* resistant and some more. And so on. Mutations are chance events, and there is frequent misunderstanding of this point. The bacterium does not know in which "direction" it is supposed to mutate; it has no will; it does not size up the situation, gird its genes, and head for the pass. If it is unlucky, none of the mutants will stand up to penicillin, and that colony will perish.

Was There Enough Time?

If *Homo sapiens* had developed from one-celled organisms over a period of a few thousand years, one could be forgiven for saying, "This couldn't have happened, that fast, by a chance series of variations." Sexual reproduction doesn't usually result in marked differences between generations, and as for mutations, the change per mutation would have had to have been too big to be believable to create a man from an amoeba in a million years, let alone a few thousand. Evolution is not a science fiction film; ants don't mutate into elephants in one step. But we are not talking about a million years. Evolution has been taking place over a time scale that is so huge as to be beyond our ability to grasp.

Was there enough time? If one looks at human history, the living species that are familiar to us appear to have changed negligibly over a period of several thousand years. Is such a slow rate of change sufficient to get from one-celled organisms to man over the span of time for which life has been estimated to exist?

Furthermore, both in nature and in the laboratory, when we do observe mutational changes in species, these changes are small, at least in comparison with the differences between a cockroach and a gorilla. Was there enough time for a succession of such *small changes* to result in a cockroach, let alone a man?

How much time was available? This was a problem that worried Darwin, because in his day the estimated age of the Earth was embarrassingly short. In the eighteenth century Buffon deduced a figure of 74,832 years, not exactly what Darwin wanted to hear. In 1854 the German scientist Helmholtz calculated that the Sun could not have been shining for more than 20 million years. This helped, but not enough; 20 million years was far too short for Darwin's purposes.

Darwin had allies in the geologists, in particular James Hutton (1726–1797) and Charles Lyell (1797–1875). Hutton, in his *Theory of the Earth* (1795), had concluded that the rock structure of the Earth had required an immense stretch of time to form, and that the Earth revealed "no vestige of a beginning—no prospect of an end." Lyell warily avoided discussing the origins of the Earth, but it was clear from the text that he believed the age of the Earth to be immense. Darwin should have listened to Kant, who in 1754 published an essay with a most un-Kant-like title: *Whether the Earth Has Undergone an Alteration of Its Axial Rotation*. His estimation of the Earth's age was in the range of hundreds of millions of years. In the end,

Darwin sidestepped the problem, not hazarding a guess at the date of the Creation.

We now believe that the Earth came into existence about 4600 million years ago. The earliest signs of life that we have date to about 3.6 billion years ago, with animals emerging perhaps over 1 billion years ago. For much of life's history, the organisms that flourished were unicellular or very simple, and such organisms usually have a very small "turnover" time. A colony of bacteria can easily produce a new generation within twenty-four hours, as against twenty-five years for humans. Thus the number of generations between the first recognizably living organism and *Homo sapiens* can be *very* conservatively estimated at 1,000 million. Now, even before sexual reproduction weighed in as a contribution to the production of variants, each generation almost certainly produced more than one new mutant. Simple organisms, and even such complex forms as insects, exist in huge numbers, and this means that very large numbers of different mutants could have occurred in a given species in quite a short time. Some of these must have been steps "up" the evolutionary ladder. And think of how many generations there have been!

We have not formally demonstrated that there was time for man to have evolved, but we can see that the number of possible steps between the earliest and most recent species was unimaginably large. This in itself does not prove that evolution took place, but it allows us to believe in its possibility.

We could shorten the time estimated for the evolution of man by supposing that among the mutations occurring in any one generation, there are always one or more that result in a *major* change in the form and function of the next generation. This is a highly unlikely scenario. First, we have almost no evidence for such mutations. Second, the very small amount of evidence that there is shows that an organism that is drastically different from the rest of its species is unlikely to survive as an individual or to breed. As we see with humans, very significant genetic changes are almost always classified as "defects," because they make life harder, not easier.

Man-made Evolution

The evolutionary process is not unique to nature. Man has carried out processes that mimic natural evolution. The best known is the breeding of animals for a specific purpose, the commonest examples being the breeding of racehorses and show dogs. All the elements of evolution are present. Horses that are fast runners are selected for mating, and the successful runners among their progeny can expect a similar fate. The selection process and the characteristic (speed) for which the selection is performed are, of course, entirely controlled by man, who in this case is part of the horses' "natural" environment. The difference between this environment and nature is that nature automatically selects for survival, while in the case of human selection the "losers" (the slower horses) are not doomed to become extinct. The same is true of dog breeding, in which it is the look of the dog that is the selection criterion. The dog world is in no danger of being overtaken by fancy-looking poodles; they do not compete for food with the average mongrel, who hopefully has a good home.[2]

In both these examples of animal breeding, man has dipped into a heteroge-

---

[2]Animal breeding was another of Darwin's triggers: "I was brought to my view by consideration of that which was practised in domestic animals, artificial selection." Letter to Sir Charles Lyall, 25 June 1858.

neous genetic pool and the characteristics sought (i.e., speed and good looks) are at least partially under genetic control. If they were not, it would be a fruitless exercise to attempt to encourage the desired characteristics by breeding.

An example of man-made evolution at the *molecular* level is provided by the work of Spiegelman in the late 1960s, a condensed account of which is appended to this chapter. In short, molecules of ribonucleic acid (RNA) were isolated from a virus. These molecules can act as a template, on which copies of the RNA can be assembled by the linking together of smaller molecules present in the same solution (Figure 22.1). Some of these copies are defective—a mutant has been created. These mutants will themselves form templates for a further generation of RNA, which will also include some mutants. In the course of the experiment, the RNA mutated to give a succession of different forms. Each form copied itself at a different rate. After seventy-four successive generations, the specific RNA that survived was that which copied itself fastest. The "fittest" RNA had survived. There will doubtless be useful applications of this type of procedure, and much work has followed Spiegelman's experiments, but the point for us is that the mutations were completely random; no one told the mixture to try to produce faster breeding mutants, no *will* was involved. In fact, at any particular stage there may be no new mutant formed that is copied faster than the fastest mutant at the previous stage. And yet the overall result was that the system produced "species" that bred faster than both their ancestors and their less well-adapted companions. That's how evolution works—by chance, not by will. I stress this because there are, among those who have accepted the phenomenon of evolution, many who have supposed that will, or something akin to it, is the driving force. It is difficult to see how a system of molecules in a test tube can have will.

The Superfluousness of Will

The will to survive is a universal human characteristic. It is also, apparently, a characteristic of nonhuman life-forms, "apparently" because there is very serious doubt that the patterns that such forms exhibit in life-threatening situations are the product of will. There is no evidence that anything except man demonstrates will.

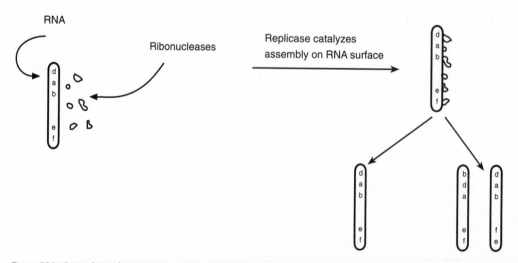

Figure 22.1. Spiegelman demonstrates molecular evolution; see text for details.

The fact that an animal fights for its life is not evidence of a will to survive. It is evidence of Darwinian evolution. Those foxes that live today are overwhelmingly the descendants of those whose lifesaving behavior favored their survival. Foxes whose behavior did not help them survive in life-threatening situations were mutants that were weeded out by natural selection. Their unsuccessful genes were not handed on.

The concept of will as a driving force inherent in human beings has appeared in many guises. George Bernard Shaw invoked the *life force,* and he was not the first or last writer to seek an undefined spiritual fuel for the living world. In the preface to *Back to Methuselah,* Shaw writes: "The will to do anything can and does, at a certain pitch of intensity, . . . create and organize new tissue to do it with." Elsewhere in the same preface he typifies Darwinism as a "rotten myth." One can sympathize: creation myths are usually replete with charismatic gods or goddesses, supernatural transformations, and all the brilliant scenery of an Italian opera. And here comes Darwin with his prosaic mutations.

Shaw echoes Nietzsche's concept of *Will* (with a capital *w*), a kind of animate force directing nature. Nietzsche was not a Darwinian: "The influence of the environment is nonsensically over-rated in Darwin . . . the essential factor in the process of life is precisely the tremendous inner power to shape and create new forms, which merely *uses, exploits* 'environment.'" Schopenhauer saw unconscious Will as the ultimate reality, a purposeful, irrational force that manifests itself in the phenomena of the natural world. He became intoxicated with his own ideas and ended up believing that it was will that drove the evolution of the cosmos, the geological history of the Earth and biological evolution.

The hidden assumption behind the myth of the will or the life force is that the living nonhuman world is somehow *consciously* pressing in the direction of perfection, or at least better adaptation to its environment. This concept is wrong. Evolution is based on variants, including mutants, and it stretches credibility to accept that the molecular processes responsible are under either conscious or unconscious control. The fact is that many mutations result in an organism becoming *less* well adapted to its environment. The death force? It is far simpler to assume that those who are better adapted, adapt better.

The fact that the formation of mutant genes in nature is a question of chance does not mean that all types of mutation are equally probable, like all possible sequences of three cards drawn from a full pack. Mutation involves a chemical change, and such changes do not all proceed at the same speed. Certain mutations will be favored because they occur through relatively easy chemical changes.

It is worth stressing yet again that within the range of variants found in a given population, one cannot automatically expect a preference for those that help the species to survive better. Thus if a certain mutant has properties that confer on it a greater chance of survival than its parents, that is pure chance—it has nothing to do with purpose.

How Is Evolution Studied?

Evolution can be studied in the natural world and in the laboratory. Under laboratory conditions it is advantageous to work with organisms that breed rapidly, thus allowing new variants to develop into significant populations by reproducing for several generations in a short time. Bacteria and certain insects are convenient from this point of view. Darwin himself took the natural world as his laboratory, observing animal form, function, coloration, and distribution. He began by observing living species rather than fossils, and as a result of his (and Wallace's) theories, the way in which anatomists looked at animals changed.

Of course, similarities between the forms of different species had been noted before Darwin. Lamarck had used such similarities to arrive at a primitive theory of evolution, and he must be given credit for emphasizing the vital significance of fossil remains, which are, after all, the only evidence we have for the temporal ordering of living forms over the past few billion years. Rocks originating nearly 4 billion years ago are the first to show any remains of living organisms, and these fossils are of microscopic forms of life. It was another 2 billion years before vertebrate fossils began to be laid down, and land animals only appear in the fossil record in rocks formed less than about 600 million years ago. The progression of forms is very clear.[3]

After *The Origin of Species,* no one could compare two similar living forms without looking for the origins of the more complex species in the anatomy of the simpler, or searching for a joint origin for the two species. Anatomy, first external and then internal, remained for many years the main lens through which evolution was scanned, and under anatomy must be included the structure of fossils.

The tools that have become available to study evolution have multiplied since Darwin's day. Under the microscope, anatomy becomes morphology, and from comparing thighbones, evolutionary biologists turned to comparing the structure of cells and their components. With the coming of molecular biology, the investigation went down to the most intimate level at which one can expect differences between species. Since genes control the nature of all the molecules produced by a given species, we can attempt to find evolutionary connections between species by examining variations in the structure of the molecules that they possess. It is known that the molecular makeup of a protein molecule in one species may differ in detail from that of a molecule performing the same function in another species. Thus the hemoglobin molecule, which latches on to oxygen, differs in its exact structure between different mammals. This variation does not apply to simple molecules; you wouldn't expect the water molecule in the cells of catfish to be different from that in the blood of zebras, and the structure of the small molecule glucose is the same in yeast and in man. However, for large molecules, particularly proteins, it has been found that the detailed composition of molecules with the same function can differ between species because there are parts of such molecules that are essential and parts that are not. Thus enzyme molecules, which are invariably very large, have regions known as "active sites" (see Chapter 13), which are responsible for the action of the enzyme. Changing the nature of the site destroys the enzyme's ability to perform its task, but there are other regions that are far less sensitive from this point of view and can be slightly changed without affecting the action of the enzyme, just as the performance of a concert pianist is, or should be, unaffected by his or her having a haircut. Enzymes that have exactly the same func-

---

[3]Possibly the first person to sense a correlation between geology and paleontology was Robert Hooke who, in the seventeenth century, suggested that fossils be used to establish the chronology of rock formations.

tion in different species almost always have similar active sites but differ in the amino acid composition of the rest of the molecule. *In general these differences are smaller the nearer we consider two species to be in terms of evolution.* This is exactly what we would expect. Thus the enzyme superoxide dismutase appears in the cells of higher animals but also in organisms down to the level of bacteria. The amino acid sequence in the enzyme differs from species to species, but the general similarity is unmistakable and the differences tend to be smaller, the smaller the evolutionary distance between any two species.

A powerful means of linking species is DNA analysis. Our DNA contains all the information responsible for our basic physical appearance. This being so, we might expect that, parallel to the visually obvious similarities between certain species, we will find similarities between their DNA, and that it might be possible to trace the evolution of DNA just as we trace the evolution of wings or eyes. Much work of this kind has been done. Much remains to be done, but the similarity of the DNA of species that have been related on other criteria is unmistakable. The general picture emerging from this research is entirely consistent with the premise that relatively small changes in DNA form the basis for the evolution of the marvelous variety of living forms.

We see that not only is there a more or less continuous evolution in the visible forms of animals and their organs, but there also is a trail left by the molecular variations that we find in chains of related species. The creationists will have to explain why Satan took a freshman course in molecular biology.

Evolution is studied in far subtler ways than the simple observation of macroscopic or microscopic variation. For example, animal behavior, a fascinating and complex subject, is revealing some extraordinarily interesting behavioral determinants of the selection process. An area that has attracted much attention in recent years is sexual preference. Why does an animal prefer to mate with one member of the opposite sex, and not another? Among humans the situation is complicated by considerations that are foreign to the rest of the living world. Mercenary female chimpanzees are not attracted to wealthy, cigar-smoking males. But animals do show preferences, and many of these can be shown to be genetically determined by the genetic constitution of *both* male and female. This is clearly a selection process in the Darwinian sense. It is too specialized a subject for us, but it does hint at the complexity of the evolutionary process, which tends to get simplified in popular science texts to taller giraffes winning out against shorter giraffes in the fight for leaves on trees.

Cherchez la Femme

Among the components of animal cells are small structures called *mitochondria*, whose functions are discussed later (see Chapter 26). Mitochondria are unusual in that they are one of the very few cell components not produced on orders from the DNA in the chromosomes within the nucleus; they contain their own DNA. Furthermore, when cells divide and proliferate, their mitochondria reproduce independently; their DNA replicates and is inherited entirely independently of the nuclear DNA in the chromosomes. This means that mitochondria have their own proliferation mechanism, a kind of segregated priesthood. Now mitochondria are too big to get inside sperm, which means that the origin of all the mitochondria in every cell of your body is the handful of mitochondria that were in your mother's egg when it was fertilized. What this means is that *your mitochondrial DNA is a gift from your mother;* your father had nothing to do with it. Furthermore, since she got her DNA from her mother, and so on, it is apparent that you can trace the history of your mitochondrial DNA through the female line of your family.

Analysis of the mitochondrial DNA of different human races not unnaturally shows close similarities within each race and greater diversity between different races. By comparing the nature of these similarities and differences, it has been inferred that all the mitochondrial DNA was derived ultimately from a common source—Africa. This research is part of the ongoing search for the origin, or origins, of modern man. Its conclusions have, however, been challenged, especially by those who rely more on the evidence of comparative studies on fossils. The skeptics present evidence supporting their claim that modern man appeared in several places on the globe, more or less independently evolving from more primitive forms. Both sides claim to be right, but neither side has the slightest doubt that *Homo sapiens* is the end product of evolution.[4]

Appendix: Molecular Evolution

The bacterium *Escherichia coli* is a common resident in commercial foods prepared under unhygienic conditions. These bacteria themselves can become infected by a virus, the Qß bacteriophage. This virus, like all living forms, contains the genes essential to its reproduction. However, viruses are extremely simple forms of life with a very limited, if efficient, range of structures and functions. Since the physical characteristics of an organism are genetically controlled, the rather short strip of RNA that acts as Qß's genetic material contains only four genes. The RNA strand is composed of a chain of small molecules called *ribonucleases*. In principle a solution of these could be assembled to give a strand of RNA by using an already existing strand as a template. This can be done using the good offices of an enzyme—"replicase" (Figure 22.1). One of the virus's genes provides the cell with the information needed to synthesize this enzyme.

It is possible to disintegrate the Qß virus and isolate its RNA and the enzyme replicase. Spiegelman found that if a sample of this viral RNA is placed in a solution containing a mixture of the right ribonucleases (the building blocks of the RNA) and a touch of the enzyme replicase, new molecules of RNA are manufactured (Figure 22.1). However, replicase has a bad work ethic; it frequently produces imperfect copies—mutants. These mutants differ from the original RNA in the exact sequence of the different nucleosides in the strand. It is as though the sentence written with nucleosides had had its words scrambled, so that, for example, a section of RNA might "read" . . . bda . . . instead of . . . dab. . . . Now this first generation of new RNA, mutants and all, can be replicated again, using replicase and more nucleosides. However, not surprisingly *each mutant is reproduced at a different rate,* depending on its constitution.

---

[4]Just to complicate matters, there has been a claim, by Sir Alister Hardy in 1960, that we are descended from aquatic apes.

Spiegelman made a mixture of the original RNA, nucleosides, and replicase and left it for twenty minutes. During that time a variety of mutants were produced. Suppose that one particular mutant, say mutant A, is copied faster than the others and than the original RNA. Its proportion will increase at the expense of the other mutants. After twenty minutes Spiegelman transferred a small portion of the mixture to another test tube containing fresh nucleosides and replicase. The replicase will continue to produce new copies, favoring mutant A, the proportion of which will further increase. However, replicase continues to make mistakes and produce new mutants. Some of these will replicate slower than mutant A, but possibly one will replicate faster. Call it mutant B. The proportion of B will increase at the expense of all the other species, including mutant A.

Spiegelman repeated this transfer from one tube to another seventy-four times. At each step there was a chance, but no guarantee, that at least one of the mutants would grow faster than any of its predecessors. The mutant that grew fastest did so at the expense of the others. At the end of the experiment, there was a dominant mutant that was being copied at about fifteen times the replication rate of the original RNA strands. It is reasonable to suppose that the shorter a mutant is, the quicker it is replicated, and in fact the mutant RNA that survived was only about 550 nucleotides long, compared with the original 4500. In the language of evolution, the RNA had "bred" for many generations, during which time a large number of mutations had occurred. The mutant that dominated at the end was that which survived because it bred the fastest in its (man-made) environment.

The Descent of Man

Man is descended from a hairy tailed quadruped probably arboreal in his habits.

—Charles Darwin

Copernicus and Galileo moved the Earth from the center of the solar system and invoked the wrath of the Church, because they had removed man from the center of the universe.[1] An earlier heliocentric hypothesis, put forward by Aristarchus in the third century B.C., was greeted by a demand that the author be indicted on a charge of impiety. Darwinism was worse; evolution not only denied the literal interpretation of Genesis but also questioned the uniqueness of man. The record shows that the main irritant was Darwin's attack on man's status, although the truth is that in the whole of *The Origin of Species* there is only one passing, although admittedly provocative, reference to man. On page 488 of the first edition the hope is expressed that "light will be thrown on the origin of man and his history." It was only in his *Descent of Man*, published twelve years later in 1871, that Darwin specifically stated, "He who is not content to look, like a savage, at the phenomena of nature as disconnected, cannot any longer believe that man is the work of a separate act of creation. . . . [M]an is descended from some lowly organized form." To rub it in he wrote, "My object in this chapter is to show that there is no fundamental difference between man and the higher animals in their mental faculties." The lighted match had been thrown into the petrol tank. No scientific text since the *Principia* has had such a widespread influence on thinkers, poets, writers, and society in general.

It did not help Darwin that he made no bones about his agnosticism. The ex-candidate for Holy Orders wrote that the Old Testament "was no more to be trusted than the sacred books of the Hindoos, or the beliefs of any barbarian." He was no kinder to the New Testament, and he was also not the first to have trouble with the problem of a God who stood by and watched suffering: "It revolts our understanding to suppose that his [God's] benevolence is not unbounded."

The agnostic Darwin was buried in Westminster Abbey on 26 April 1882. His *Origin of Species* remains the most widely talked about and controversial book in scientific history, and even if the Church would not burn him, it could not ignore him. On reading his book, the wife of the bishop of Worcester had appealed to heaven: "Let us hope it is not true, but if it is, let us pray that it does not become generally known!" The good lady's pious plea was not answered. The battles started within months of the publication of the book.

The first skirmish took place at Oxford on 30 June 1860 between the redoubtable

---

[1]In 1990, a third of the adults in England believed that the Sun went around the Earth (from a survey reported in the *Lancet*).

T. H. Huxley (who coined the term *agnostic* to describe himself) and the established Church, as represented by the bishop of Oxford, Samuel Wilberforce. Wilberforce, whose credentials were bolstered by the fact that he was a competent ornithologist, had taken advice from Richard Owen, a paleontologist and ex-friend of Darwin, who not only bitterly resented the attention that Darwin was getting but also had been challenged by Huxley in an argument over the difference between the brain of man and that of other animals. Owen had claimed that only in man was there a certain area of the brain called the hippocampus minor. Huxley's denial of this claim annoyed Owen intensely. Huxley, who was a loyal friend of Darwin, had been converted from his skepticism regarding natural selection by Darwin's book. At Oxford he won the debating duel with Wilberforce. At the same meeting a lengthy defense of Darwin was delivered by Joseph Dalton Hooker, a distinguished naturalist who, like Huxley, had initially been very critical of evolutionary ideas but had seen the light. Years later Bishop Wilberforce grudgingly conceded that there was some merit to Darwin's ideas, but the themes of the Oxford encounter remain to haunt all the subsequent confrontations between impassioned reason and impassioned faith. The antievolutionists return again and again to the affront to man's uniqueness and to God's omnipotence. Bishop Wilberforce's taunts as to the simian ancestry of Huxley have been echoed a thousand times in the Bible Belt. The attitude of the "antis" is that man is too important to have been a chance descendant of a one-celled ancestor—he could only have been personally created by the Creator. Consistent with this belief are those whose objections to evolution are predicated on the existence of an entity called the soul, which is presumed to be the unique possession of man. This problem was brought up ten years before Darwin's book, by Hugh Miller in *Footprints of the Creator* (1847). Miller pointed out that if the "development hypothesis" was accepted and *Homo sapiens* is supposed to have descended from lower forms of life, then it is arguable that the soul also evolved in these forms. What would be left of man's uniqueness if Bonzo had a soul? Miller concluded that man is not a continuation of the animal kingdom.

Science has advanced since Darwin's day. The evidence is now so strong that it would be almost impossible to find, even among the most rabid antievolutionists, someone who denies the existence of mutations or the mixing of genes due to sexual reproduction. And if these phenomena exist, then there is a mechanism for the creation of new species from old. A reasoned refutation of evolution has become so difficult that the opposition is now largely confined to fundamentalists, those who believe in the literal truth of the Bible, and specifically in Genesis 1:27: "And God created man in His image, in the image of God He created him; male and female He created them." This leaves little room for argument. Genesis is a powerful and brilliantly imaginative myth. As an account of the origin of the Earth and the creatures living on it, it is wildly improbable, but creationists believe that the theory of evolution is "simply the continuation of Satan's long war against God."

Fundamentalists have found ingenious ways to discount the scientific evidence for evolution. The classic example is the suggestion that fossils were deliberately concocted and placed in the Earth by Satan himself in order to deceive man, and that one can even detect a devilish whiff of sulfur at the sites of fossil remains. In fairness to the proponents of such views, it must be conceded that they cannot be disproved, either by reason or by observation. They are just staggeringly improbable, which is why I will ignore them.

Nevertheless, the argument over the *theological* validity of evolution is depressing. A world in which it took a Vatican committee thirteen years to grudgingly come to the conclusion (in 1992!) that Galileo had been wrongfully condemned by the Church in 1633 is a place where reason needs as many friends as it can get.

Darwin's Fan Club

It was predictable that part of the opposition to Darwin would come from organized religion. One of the few clergyman who bucked the trend was Charles Kingsley, whose best-known work, *The Water Babies,* contains references to the controversy over evolution and to the main antagonists. In particular, he refers to Huxley's spat with Owen over the hippocampus minor, writing that the victory in the great hippopotamus debate had gone to the well-known expert in necrobio-neopalaeonthydrochthonanthropopithekology, Professor Ptthmllnsprts. In the spirit of the Deists, Kingsley considered Darwin to have disposed of "an interfering God." He made no impression on his clerical colleagues, perhaps because he declared that in Heaven the saved indulged in continuous sexual intercourse. Which is one way of making converts.

It is often forgotten that, at the time, not everybody regarded Darwin as the spearhead of the threat to faith. A couple of months after *The Origin* came out, a collection of articles on religion appeared, entitled *Essays and Reviews* (1860). Written almost entirely by Anglican clergymen, this adopted a very liberal stance. Miracles were downplayed and reason lauded. The uproar was, if anything, more impassioned than the response to Darwin. The Church was being attacked from within. Unsuccessful attempts were made to have the authors prosecuted for heresy.

Likewise, despite the high-profile debate at Oxford, it is a mistake to suppose that the resistance to Darwin came solely from the Church. Darwin's own professor of geology at Cambridge declared that his former pupil was plunging humanity into "a lower grade of degradation than any yet recorded." Darwin, he said, was "deep in fallacy." Other devout scientists felt similarly. One of them was the fundamentalist marine biologist Philip Gosse, whose son Edmund describes his father's reaction to natural selection, in a classic and moving autobiography, *Father and Son* (1907).

Philip Gosse had already been shaken by Lyell's suggestion that the Earth had developed over a very long time period during which new species had appeared and others had become extinct. "Here was a dilemma!" wrote his son. "Geology certainly *seemed* to be true, but the Bible, which was God's word, *was* true." Philip Gosse spoke to Darwin in 1857 and was informed, in advance of the publication of Darwin's book, of the theory of natural selection. His son Edmund Gosse wrote,

> It was this discovery, that there are two theories of physical life, each of which was true, but the truth of each incompatible with the truth of the other, which shook the spirit of my Father with perturbation . . . he allowed the turbid volume of superstition to drown the delicate stream of reason. He took one step in the service of truth, and then he drew back in an agony, and accepted the servitude of error.

Philip Gosse went as far as publishing a book, *Omphalos* (1857), which claimed, "When the catastrophic act of creation took place, the world presented, instantly, the structural appearance of a planet on which life had long existed." The book was hammered by both atheists and Christians. Charles Kingsley wrote to the author that he could not believe "that God has written on the rocks one enormous and superfluous lie."

Darwin's dramatic vision of man's evolution from the primeval seas stimulated many of the writers and thinkers of the late nineteenth century. One who was swept off his feet by *The Origin of Species* was the author of *The Way of All Flesh,* Samuel Butler. Again the revolt against religion was lurking in the background, but this time Freud would have had got his notebook out. Butler did not get on with the authoritative figure of his father, a clergyman, a headmaster, and eventually the bishop of Lichfield. The son found in the theory of evolution a convenient crutch for his rejection of Christianity and God. But eventually Butler returned to a personal God of his own, rejected Darwin, and proceeded to write a number of books on evolution in which the theme was that Darwin was mistaken in his supposition that the variability in one generation of a species was just a question of chance. Darwin had removed will from nature; Butler wanted to restore it by postulating that living forms strive to overcome the difficulties in their environment and that because of this effort they developed better-adapted organs, or modes of action, which they passed on to their offspring. Butler succeeded in annoying both the Church and the followers of Darwin. For those of you who fear the domination of the computer, try reading *Darwin among the Machines*, in which Butler sees machines competing successfully with man, threatening to be the fittest to survive. Butler's conviction that evolution was driven by will had a deep influence on George Bernard Shaw, another one of those who liked to conjure up the life force—whatever that may be.

The poet Alfred Tennyson was an acquaintance of Darwin and was far more than averagely interested in science. Thomas Huxley said of him that he was "the first poet since Lucretius who has understood the drift of science."

Tennyson was strongly affected by Lyell's *Principles of Geology*, especially by the suggestion that the history of the Earth was far too long to be compatible with the traditional biblical account. Tennyson accepted Lyell's thesis and was a convert to the general idea of evolution before Darwin wrote his book. Tennyson was clearly concerned over the antireligious implications of evolution, and he was the prime instigator behind the formation in 1869 of the Metaphysical Society, a collection of eminent scientists, writers, and clerics whose aim was to attempt to "unite all shades of religious opinion against materialism." The group, which included William Gladstone, the duke of Argyll, two archbishops, two bishops, the dean of St. Paul's, the critic John Ruskin, the famous economist Walter Bagehot, Thomas Huxley, and others, discussed such topics as "What is matter?" "Is God knowable?" "What is death?" and "Has a frog a soul?" They had their lighter moments such as a proposal to test the efficacy of prayers—a "prayer-gauge"—by seeing how effective prayer was in a hospital ward. The society faded away in 1880. Tennyson whimsically explained: "Because after ten years of strenuous effort no one had succeeded in even defining the term 'Metaphysics.'" Tennyson saw man evolving into a more perfect being, the "crowning race" referred to in his great poem *In Memoriam*.[2] In

---

[2]It has been amply commented on that this poem contains passages that read like paraphrases from Chambers's *Vestiges of the Natural History of Creation* (1844), mentioned earlier.

the same poem it is also possible to find the influence of the geologist Lyell, for example, in these lines referring to the erosion of the Earth and the subsequent process of sedimentation:

> The moaning of the homeless sea,
> The sound of streams that swift or slow
> Drawn down the Aeonian hills, and sow
> The dust of continents to be.

Although Tennyson probably rejected the biblical account of the Creation, he finally convinced himself that science and religion were compatible. He firmly believed that science was a major factor in countering the superstition that he saw as darkening civilization.

There were those who were only too happy to have found in Darwinism an ally against religion and the Church. The notorious pagan, masochist poet Swinburne (known as Swineburne to his detractors) accepted evolution immediately. It spoke to a man who could write the line: "The supreme evil, God." The philosopher John Stuart Mill, who also saw Christianity as a failure but was loathe to say as much in public, was deeply affected by *The Origin of Species* and the controversy that followed its publication. His impression was that Darwin had weakened the image of religion to such an extent that the time was right for Mill to risk offering an alternative, in the form of the humanistic philosophy embodied in the thought of Jeremy Bentham. Mill's father had instilled in him the Benthamite doctrine that "the exclusive test of right and wrong [is] the tendency of actions to produce pleasure or pain." Directly influenced by what he saw as the beneficially disruptive effect of Darwin's work on Christianity, Mill wrote one of his major books for the general reader, *Utilitarianism* (1863). Utilitarianism is usually glibly summarized as "the greatest happiness for the greatest number," a phrase coined by the eighteenth-century English philosopher Francis Hutcheson. Mill did not see evolution as a direct basis for utilitarianism. Indeed, it is difficult to see how the survival of the fittest can in general be compatible with the greatest happiness for the greatest number. The point is that, for Mill, Darwin appeared to have engendered in England a social environment sympathetic to a *secular* approach to morality, which utilitarianism is. He miscalculated. Benthamite humanism did not sweep England.

Darwin's greatest champion was Thomas Huxley, who as a young man had also collected biological specimens, while spending three years as a medical officer in the South Seas on the HMS *Rattlesnake*. The famous educator, who believed in the abolishment of slavery and the education of women, deserves a book of his own. His attitude toward life is best summed up in his own words: "We live in a world which is full of misery and ignorance, and the plain duty of each and all of us is to try and make the little corner that he can influence somewhat less miserable and somewhat less ignorant than it was before he entered it." Huxley had written and spoken in support of evolution from the appearance of Darwin's book. When asked to give the prestigious Romanes Lecture at Oxford in 1893, he chose as his title "Evolution and Ethics." His theme was that natural selection does not automatically lead to *ethical* progress; that the fittest, as defined by Darwin, did not mean the best as judged on ethical grounds. This perhaps seems obvious, but there were weighty intellects, notably the philosopher Herbert Spencer, who thought otherwise. Spencer saw natural selection in the animal world as a model for the workings of human society, so that those who survived the human rat race were necessarily the fittest in *all* senses. Huxley, the great freethinker and liberal, was concerned to show that society should *not* be dominated by those who survive the rat race.

The critic and poet Matthew Arnold was another who was disturbed by the clash between science and religion. He realized that it was up to religion to make some concessions, or it would risk losing many of its followers. In particular, he pressed the opinion that it was akin to idolatry to take the Bible literally. How could one, after Darwin?

Darwin's fame soon spread to Europe and America. His prime disciple in Germany was the biologist Ernst Haeckel, who in his *Riddle of the Universe* (1899), went well beyond Darwin in applying the principle to "the motion of heavenly bodies," to the "growth of plants and the consciousness of man [which] obey one and the same great law of causation," producing "a vast, uniform, uninterrupted process of development." Moreover, he bestowed upon atoms a primitive consciousness or soul, accounting for the upward climb of matter to living forms. *Naturphilosophie* was alive and well. Of the scientifically active countries of western Europe only the French, except for their anthropologists, generally ignored or disparaged Darwin.

The amount of literature generated by the philosophical aspects of evolution was greater than that stimulated by any scientific theory since Newton. This was because, of all the revolutions caused by science, Darwin's struck most dangerously at the self-image of man. Apart from attacking the literal truth of Genesis, he had shown man to be an integral part of the animal kingdom. This historic turning point in our inner world is the factual justification for the increasing number of people who believe that we are not so much masters of this planet as part of it.

The Wrong Theory

One cannot help but be sorry for Jean-Baptiste de Lamarck. He clearly suggested the concept of evolution, half a century before Darwin. The trouble was that Lamarck's *mechanism* for evolution was completely wrong.

The man who invented the word *biology* apparently was the first person to sense the real significance of the similarities between fossils and the living species to which they bore most resemblance. Lamarck was a careful observer; he guessed that living forms had evolved from simpler organisms, and he tried to explain this in terms of an, at first sight, reasonable mechanism.

Lamarck believed in the *inheritance of acquired characteristics.* According to this theory, if you keep cutting off people's left hands for generation after generation, eventually a time will come when children will be born without left hands. Moreover, Lamarck held that, in nature, animals had an instinctive drive to improve themselves. The striving of individuals for improvement *was* the mechanism of evolution, when combined with the inheritance of those improvements. Consider giraffes. Lamarck would say that the evolution of the giraffe's long neck was due to the striving of each individual giraffe to reach the leaves on the trees. The efforts of a given giraffe led to a lengthening of its neck, and this *acquired* characteristic was passed on to its offspring. Darwin's explanation is completely different. He held that in every generation of giraffes there were those that had longer necks than the average because they were genetically favored in this respect. Since they had a better chance than the others to feed well on the trees, over the generations the longer-necked giraffes survived better—and passed on their favorable genes to their kids.

Condorcet, one of the outstanding figures of the Enlightenment, believed in the Lamarckian theory of the inheritance of acquired characteristics and argued that, apart from the value of education for the individual, the parents' education was, in some degree, inherited by their children so that there could be a cumulative educational effect through the generations. At the turn of the nineteenth century, a test of the inheritance of acquired characteristics was carried out by August Weismann (1834–1914), perhaps the first person to realize that the "germ plasm," what today we would call the genes, is not normally affected by damage to the whole organism. Weismann cut off the tails of several generations of mice—to no effect. None of the offspring of this line of martyrs was born tailless, or even with a significantly shortened tail. He might have spared the mice their suffering if he had just noticed that after hundreds of generations, Jewish boys are still born with foreskins. *Acquired characteristics are not inherited.* Contrary to Lamarck's theory, simple physical modifications of the parents are not inherited by their children. Damage to the parents' reproductive cell DNA will, of course, affect the nature of their offspring.

Lamarck's belief in the inheritance of acquired characteristics was anticipated by Charles Darwin's grandfather Erasmus, and has in fact persisted for centuries. It is part of the folklore of many societies and has been around too long to be killed by ugly facts. One example of the effect on their offspring of what their parents *see* is about 4000 years old. Genesis 30:37–39 reads: "Jacob then got fresh shoots of poplar, and of almond and plane, and peeled white stripes in them, laying bare the white of the shoots. The rods that he had peeled he set up, in front of the goats, in the troughs, the water receptacles, that the goats came to drink from. The mating occurred when they came to drink, and since the goats mated by the rods, the goats brought forth streaked, speckled, and spotted young."

The Greeks were equally inventive; they believed in *telegony*, a theory, supported by Aristotle, which says that if a woman has a child by her second husband it will have some of the characteristics of her first husband, if he consummated the marriage. It should be easy to discredit this idea by studying the children of film stars.

It is not often appreciated that Darwin himself accepted the idea of inherited characteristics. He was less concerned with the detailed mechanism of heredity than with what happened to the heterogeneous collection of individuals in a given generation. Neither he nor Lamarck could have known that the physical mutilation or modification of an organism (as against chemical or radiative damage) did not affect the DNA of the reproductive cells and that it is only through these cells and, more specifically, their DNA, that information is transferred from generation to generation.

The doctrine of the inheritance of acquired characteristics still has a few supporters. Its last great (if that is the word) promoter was the distinctly unpleasant Soviet charlatan Trofim Lysenko. This apparatchik declared that genes didn't exist and that it was the whole organism that contributed to heredity. By subjecting plants to a variety of environments, he proposed to modify them so that when they were bred they would have certain desired characteristics, for example, greatly improved resistance to extreme cold. Lysenko was supported by the political authorities, not for idealistic reasons but because they expected practical results in terms of improved yields. Only Lysenko's version of genetics was acceptable in the Soviet education system.

Since no genuine case of the inheritance of an acquired characteristic has ever been documented, Lamarckism should have died long ago. Its survival as an explanation for evolution is a tribute to the strength of folk mysticism.

The Perversion of Evolution

The Comte de Buffon, in the eighteenth century, declared—incorrectly—that chemical affinity was governed by a law similar to the law of universal gravitation. In the field of political science, Montesquieu saw the ideal king as being similar to the Sun, his gravitational force attracting to him the political institutions of his kingdom. Newtonian mechanics was dragged into totally inappropriate fields. It was thus predictable that there would be those who would see Darwinian evolution as a general principle, a means of classifying all human activities and explaining their development.

Everything that changes smoothly with time "evolves," but this is merely a case of verbal echoes. It is not apparent how Darwinian evolution can fruitfully be applied to art, literature, or philosophy. An early attempt of this kind was contained in a lecture by Huxley: "The struggle for existence holds as much in the intellectual as in the physical world. A theory is a species of thinking, and its right to exist is coextensive with its power of resisting extinction by its rivals." Karl Popper has expressed similar ideas, and the American pragmatist G. H. Mead declared that "the scientific method is, after all, only the Evolutionary process grown self-conscious," which is a neat thought. But the cold facts are that in the real, nonscientific, world, "fitter" theories do not necessarily eliminate less fit theories. In any case, the idea is sterile, predicting nothing of importance that could not have been deduced by common sense.

Darwin never used the phrase "the survival of the fittest." The man who did was Herbert Spencer. In the 1870s the name Spencer was synonymous with intellectual vision and power and profound philosophical insight into the workings of man and nature. But history can be cruel. Today, the man who started his professional life as a railway engineer and ended as the author of the massive multivolume *System of Synthetic Philosophy* is a footnote. Spencer's writings had a deep, if not lasting, influence on the way in which society viewed science. Whether or not one agreed with him, he brought science as a whole, and his concept of evolution in particular, into the intellectual world of Victorian England and post-Civil War America. He was an evolutionist at least a decade before Darwin, as shown by his first book, *Social Statics* (1850). In 1852 he published an article entitled "The Development Hypothesis," in which he strongly backed the concept of evolution as against the biblical account of special creation. The article attracted considerable attention and Spencer's imposition of evolution on the natural world was soon to be given a huge boost by Darwin and Wallace, who both admired Spencer greatly. Spencer frequently brought attention to the fact that he had preceded Darwin, so much so that someone remarked that he "seems to wish to take out a patent for the invention of the theory."

The more controversial part of Spencer's philosophy was his application of evolution to human history and institutions. Since, in his eyes, religion had lost its authority, ethics no longer had a spiritual basis. So Spencer melded science with social science and ethics, thus secularizing ethics and giving it, or so he thought, the authority of science. Spencer's ideas were well received in England and in Japan, but it was particularly in America that his writings and his visits boosted his reputation to guru status. His promotion of the concept of the survival of the fittest was welcome, if in practice superfluous, support for the rapidly growing class of millionaire entrepreneurs that was to typify American society from then on.

Spencer was a Lamarckian. In particular, he held that the mental exertions of parents were genetically transferred to the brains of their children. The reason for his belief in this totally false premise is clear. He saw *social* development as an evolutionary process in which each succeeding generation learned from the experiences of its predecessors. This process was not genetic, but it was Lamarckian; each generation inherited the acquired characteristics of its parents. Similarly, he saw *personal development* as a kind of Lamarckian evolutionary process, involving change through experience. It would have spoiled his grand thesis to admit that one component, namely, *biological evolution*, was not Lamarckian and therefore fell outside his scheme. Of course, he was right to suppose that social and cultural evolution are, in a broad sense, Lamarckian processes. We can improve on the legal systems of past generations by observing what was good and what was bad and rejecting the bad. But this "evolutionary" process has certainly nothing to with genetics and everything to do with conscious learning.

Almost no one reads Spencer anymore, unless they have to. Today the most interesting aspect of his life is perhaps his affair with the novelist George Eliot.

Spencer was not the last to use Darwin's name to push evolution into history. When, in 1883, Engels delivered the oration over Marx's grave in London, he said, "Just as Darwin discussed the law of Evolution in organic nature, so Marx discussed the law of Evolution in human history."[3] Another who used the language of evolution in dealing with society was Walter Bagehot, the English economist, in his book *Physics and Politics: Thoughts on the Application of the Principles of Natural Selection and Inheritance to Political Society* (1872).

Several misguided attempts to find evolution in unlikely places died at birth. Thus Sir William Crookes's reference to Mendeleev's periodic table as "inorganic Darwinism" is rubbish. The elements do not fight for survival, and none appear to have become extinct. However, things do become serious when an attempt is made to impose the concept of natural selection on sociology or history. There were those who found justification for war in the argument that war was just a human extension of the struggle for existence found in nature, and they could turn to Darwin himself, who wrote, "DeCandolle, in an eloquent passage, has declared that all nature is at war, one organism with another, or with external nature. Seeing the contented face of nature, this may at first be well doubted; but reflection will inevitably prove it is true."

It is also easy to see how Darwin could be used as a justification for an uncaring, competitive society. Shigalov, the most controversial character in Dostoyevsky's

---

[3]The widespread belief that Marx wanted to dedicate *Das Kapital* to Darwin has no foundation.

The Devil's, was based on the radical publicist Zaitsev, who used Darwin to defend the enslavement of blacks. One of the arguments against improving conditions in slums, used by nineteenth-century reactionaries, was that concern for the welfare of the unfit went against the principle of competition that underlay evolution. Man should not tamper with nature. On the other hand, there were those who wanted to help nature in the name of natural selection, and who held that the future of the race came above individual liberty. Disturbing ideas issued from quiet professorial rooms. Karl Pearson, the renowned statistician, declared that civilization resulted from "the struggle of race with race, and the survival of the physically and mentally fitter race." Francis Galton, a geneticist and a cousin of Darwin, coined the word *eugenics* for the policy of encouraging "good" human specimens to breed at the expense of the less "good." He suggested cash grants to encourage marriage and child production among the "fit" and sterilization of the "unfit."

Social Darwinism, the application to society of the principle of the survival of the fittest, attracted many Western intellectuals. A little-known example is H. G. Wells. Those who know Wells primarily through his science fiction might try reading his *Anticipations* (1901) and *A Modern Utopia* (1905). In both books racism, backed up by eugenics, is explicit. Wells speaks of "efficiency." Races that are not efficient will have to be kept under control by, among other measures, obligatory infanticide and abortion. Which races? That's an easy one: "the black . . . the yellow man . . . the Jew." Wells was an advocate of sterilization. One can hear Darwin in Wells's declaration that "the way of Nature has always been to slay the hindmost."

The Fabian Society, to which Wells belonged, was heavily tainted by social Darwinism. One of the society's outstanding spokesmen, Sidney Webb, warned of the national deterioration that would follow as a consequence of the presence in Great Britain of Irish Roman Catholics, and Polish, Russian, and German Jews.

We have sailed into dangerous waters. As applied to society, the "survival of the fittest" almost invariably means the survival of the thugs with the big sticks. Mental defectives and cripples were doomed in Nazi Germany, where the survival of the fittest meant the destruction of the weakest, where the definition of fitness was determined by the strong and the means of selection was the truncheon and the death camp. The final degradation of the human spirit was reached by Dr. Mengele standing at the head of a line of concentration camp inmates and, with a gesture of the hand, directing the fitter inmates to one side, to work, and the ill and old to the other side to die.

Natural selection is an *unconscious* process; the nonhuman world does not know that the better adapted will win out. Man can *consciously* help the weak, knowing that the survival and comparative happiness of our species depends on our control of our political and ecological environment, not on the elimination of biological weakness. On that path there is no logical end. Why not execute diabetics and dyslexics, migraine sufferers, and arthritics? Or the person who taught me so much of what I know about tolerance, my Down's syndrome son, Dan?

We have evolved from the living world, and one of the gifts that Darwin gave us was to place us in that world, not as conquerers with spoils but as inheritors with responsibilities.

24

The Gene Machine

One might compare it to the most perfect machine that God ever made.

—Diderot, on *the Stocking Machine*

In November 1895, Roentgen discovered X-rays. In 1912, von Laue showed that the diffraction of X-rays could be used to determine the arrangement of atoms in crystals. In 1952, Rosalind Franklin, working with the physicist Maurice Wilkins in King's College London, shone a beam of X-rays at a crystal of DNA and photographed the diffracted rays. She obtained beautiful but complex patterns. There are recognized and well-tried mathematical methods used to analyze the patterns in X-ray diffraction patterns and convert them into information on the spatial arrangement of atoms in the scattering crystal. Rosalind Franklin's tendency was to appeal to these methods, but it was clear that the huge number of atoms in the DNA molecule was going to make her task extremely formidable, and in her day powerful commercial computers were not available. James Watson, an American in his twenties, and Francis Crick, an English physicist, working together in the Cavendish Laboratory at Cambridge (yes again!) took a different approach to the hunt for the structure of DNA. They opted for intelligent guesswork based on what was known of the chemical composition of DNA. In practice they played with models, looking for a molecular conformation that was compatible with both the chemical and the X-ray evidence. They found the double helix.

The advent of modern molecular biology—the understanding of the molecular basis of life processes, and arguably the most important scientific development in the twentiethth century—dates from a one-page letter by Watson and Crick, published on 25 April 1953 in the British scientific journal *Nature*: "We wish to suggest a structure for the salt of deoxyribose nucleic acid (D.N.A). This structure has novel features which are of considerable scientific interest." To say the least. Their elucidation of the structure of DNA has had a profound influence on man's ability to understand and control the functioning of living cells and the mechanisms of heredity.

There is a certain injustice in the proportioning of fame for the DNA story. Many scientists have the feeling that at least one, and possibly two, of those who preceded the discovery of the double helix were worthy of more recognition than they got, both from the scientific establishment and from Watson, in his book *The Double Helix*.

Goethe once said, "After all it's pure idiocy to brag about priority, for it's simply unconscious conceit." Nevertheless, one can understand Watson and Crick's elation at having discovered the structure of the DNA molecule before the famous California-based chemist Linus Pauling, who was working on the same problem and breathing down their necks. It's very human to want to be first. It is not so easy to accept the oversight by which Watson neglects to mention the crucial observations of Fred Griffith, who in 1928 showed that genetic information could be transferred from one type of bacteria to another. Watson mentions but underplays the work of Oswald Avery, who in 1945 demonstrated that DNA was the material that transferred the genetic information. The book is also distinctly ungenerous to Rosalind Franklin. For example: "The thought could not be avoided that the best home for a feminist was in another person's laboratory." It was at a seminar that Franklin gave in 1952 that Watson saw her X-ray photographs and realized that they might indicate a helix. In a footnote to a 1954 paper, Crick and Watson say, "We wish to point out that without this data the formulation of our structure would have been most unlikely, if not impossible." They didn't say that they had obtained her photographs without her knowledge. Watson also claimed that Franklin rejected the double helix interpretation of her photographs, but this contradicts the evidence of her notebooks. In *The Double Helix,* Watson gives an unflattering picture of Franklin,[1] but others have felt that she was a victim of the "male club" syndrome. In 1962 Wilkins, Watson, and Crick shared the Nobel Prize. Rosalind Franklin died of cancer in 1958, at the age of thirty-seven. Many think that she deserved far more recognition than she got, and that Avery deserved a Nobel Prize.

DNA was discovered in 1869, ten years after the publication of *The Origin of Species.* There was no suspicion as to its real importance. It is doubtful if Darwin was ever aware of the molecule isolated by Friedrich Miescher, a Swiss biochemist, from pus sticking to discarded bandages. To give Miescher his due, he did venture the opinion that DNA had something to do with fertilization, but at the time no one followed up the idea. The true significance of DNA was revealed by Oswald Avery. The next step belonged to the chemists.

The Building Blocks of DNA

Your DNA contains the genes that control the structure and functioning of your body. Human DNA contains on the average something like a billion atoms, but its structure is simple. I say "on the average" because your DNA is not identical to my DNA. They are very similar, but if they were identical we would be indistinguishable, apart possibly from dueling scars.

DNA is built from only five kinds of atom: hydrogen, carbon, nitrogen, oxygen, and phosphorus. If the DNA molecule is gently broken up, we obtain three kinds of small molecules: phosphoric acid, sugar molecules (the sugar is ribose), and heterocyclic amines, which in this context are commonly dubbed "bases" (Figure 24.1a). There are four kinds of bases in DNA, and they hold the living world in their hands. Their structures and names are shown in the figure. They are usually referred to by their initials: A, T, G, and C. It is not necessary to remember the structure of the bases to grasp the way in which DNA works.

The General Structure of DNA

All living forms, except viruses, which are hardly living, contain DNA. The structure of human DNA illustrates the principles on which all DNA molecules are built.

---

[1]For a very different view of Franklin from that presented by Watson, see Anne Sayre, *Rosalind Franklin and DNA* (New York: Norton, 1975).

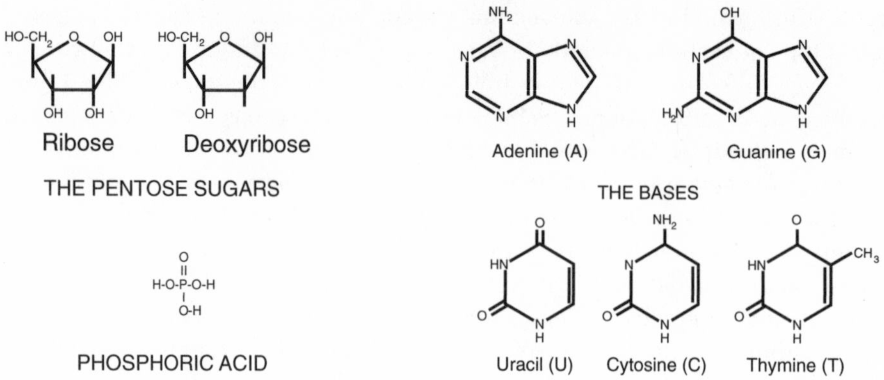

Ribose   Deoxyribose   Adenine (A)   Guanine (G)

THE PENTOSE SUGARS   THE BASES

PHOSPHORIC ACID   Uracil (U)   Cytosine (C)   Thymine (T)

Figure 24.1a. The molecular building blocks of DNA.

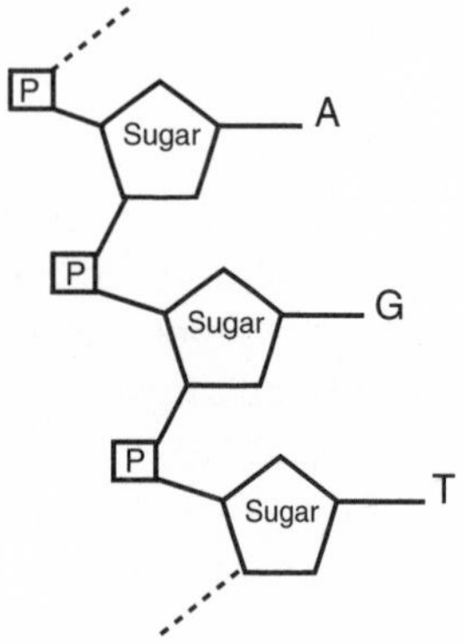

Figure 24.1b. Putting the blocks together. Phosphate groups are indicated by the letter P.

First link alternating ribose and phosphoric acid molecules to form a chain containing about 100 million of each. The DNA of all species contains this chain. Now take your stock of bases and start attaching them to the chain, each base being linked to a ribose unit (Figure 24.1b). In what order do I arrange the four kinds of base? The answer to that determines the difference between my generous brown eyes and my bank manager's inhumanly cold gray eyes. Human DNA contains about 100,000 genes, and each gene is essentially an ordered series of bases. Every gene has a different order of bases.

Where is the double helix? It's coming, but first we need a pivotal discovery made by Erwin Chargaff, who in the early 1950s published the results of chemical analysis on DNA from a number of species. Chargaff found that for every C (cytosine) molecule in a DNA molecule there was a G (guanine) molecule; if there were 74 million Cs there were 74 million Gs (within experimental accuracy). And for every A (adenine) molecule there was a T (thymine) molecule. The Nobel Prize was knocking at the window: Watson and Crick took Chargaff's finding and Rosalind Franklin's X-ray hint and proposed that the DNA molecule consisted of two strands lying side by side. Each strand had the same backbone of alternating ribose and phosphoric acid units. The two strands were held together by links between their bases, so that the total structure was reminiscent of a ladder (Figure 24.1c)—but a *twisted ladder*—a right-handed helix. And the rungs that held the ladder together were Chargaff's pairs of bases; every G held hands with a C, every A grasped a T. The pairs of bases are held together by gentle links: hydrogen bonds (see Chapter

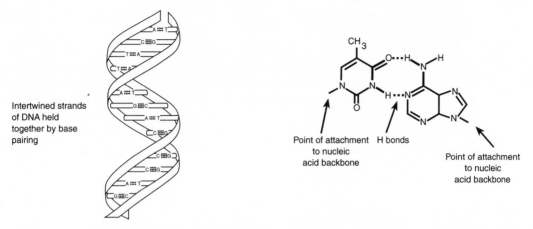

Figure 24.1c. A schematic sketch of the double helix. The way in which thymine pairs with adenine is shown on the right.

13). Note that because of the fidelity of G to C and of A to T, if you know the order of the bases on one strand you can immediately work out the order on the other.

So there it is. The structure of the most important molecule on Earth. Your DNA, unless you are an identical twin, is distinct from everyone else's in the whole Creation, the difference being, as we said, in the order of the bases. Because the DNA molecule is so large, it can be seen with sophisticated microscopes, appearing as a long wormlike structure.

DNA carries all your genetic information, a collection of genes that contain a set of instructions for building you. Your chromosomes contain your DNA, and since each chromosome carries a different set of genes, each chromosome contains its own kind of DNA, identical in *general* structure with all the DNA in the cell nucleus but differing from chromosome to chromosome in the order and number of the different bases that go to specify the different genes.

The Continuity of Science

Science builds upon itself. The discovery of the structure of genes would have been impossible without the discovery of X-rays, which allowed the general molecular structure of the DNA molecule to be unraveled. Also absolutely indispensable was the chemical knowledge that allowed the structure of the bases and the other components of the DNA molecule to be determined. Almost any scientist, if he looks at his research, can trace an *indispensable* line of discoveries through to the seventeenth or eighteenth century, if not farther back. The double helix owes much to Lavoisier.

How Does DNA Work?

The living cell builds most of the molecules that it contains. The exceptions are simple salts and a few small molecules that are absorbed from food and are used as building blocks for the construction of large molecules. In general the synthesis of all molecules in the cell depends on catalysts, the enzymes, which are protein molecules. *DNA, acting through the genes, is the master controller that is ultimately responsible for the synthesis of all proteins,* not only enzymes. It is worth looking at the general scheme by which proteins are synthesized, if only to marvel at the ingenuity of nature.

The secret is in the order of the bases. In human DNA there are about 6 billion bases, and it is this huge array that constitutes the collection of genes, the *genome*, that control the functioning and replication of all of us. (Don't get overwhelmed by how many bases we have; the humble lily has about 600 billion.) Although at one time it was believed that each gene was a continuous stretch of bases on the DNA molecule, in 1977 it became apparent that nearly all genes are split up into separate segments, the so-called *exons*, interspersed with other segments, the *introns*. If you need a mnemonic: *int*rons are *int*ruders. Thus a typical gene may consist of anything from two to fifty exons (the real stuff) separated by introns (the intruders). The introns can be very long, running up to about 100,000 bases in some cases. Why are they there? Functions have been uncovered for some introns, but many appear to be useless, as reflected in the nickname "junk." There is no firm answer at present. When we refer to a gene we will mean a series of bases, ignoring the introns. Most human genes contain 1000 to 3000 bases, without the introns. As we said, their task is to oversee the synthesis of proteins, and we will now see how they do it.

Building Proteins

Each gene is responsible for assembling a different protein. *The gene itself does not do the work;* it supplies the necessary information on the composition of the protein.

Proteins are built by assembling amino acids present in the cell, and the problem is to link the right amino acids together in the right order. The assembly of proteins is not done directly on the surface of the DNA molecule. The relevant gene is first copied and sent out of the nucleus into the extranuclear interior of the cell. This is how that happens:

1. First a portion of the DNA molecule is attacked by a collection of enzymes that "saw through" some of the rungs of the ladder so that the helix unwinds over the length of one gene, including introns (Figure 24.2a). Since the rungs are hydrogen bonds (Velcro), they are comparatively weak.
2. The exposed bases on a single strand of the DNA direct the formation of a new single-strand molecule—*not the protein*. The principle of the assembly line is simple: base pairing again. The newly formed molecule has the same backbone

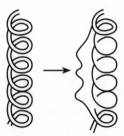

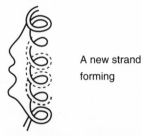

A new strand forming

Figure 24.2a. Unwinding a gene.

Figure 24.2b. The formation of ptRNA.

as the DNA, namely, a sugar-phosphate chain (the sugar this time being deoxyribose instead of ribose). *Along the new chain the order of bases is completely determined by the fact that every base must "match" that opposite it on the single DNA strand.* The only slight variation is that in the new strand the place of the base thymine (T) is taken by another base, uracil (U) (Figure 24.2b). Uracil pairs with adenine (A), and, as usual, G (guanine) pairs with C (cytosine). The new molecule, known as *primary transcript* RNA or ptRNA, can be regarded as a complement to the original strand, consisting of complementary gene and complementary junk (or complementary exons and complementary introns).

RNA stands for *ribonucleic acid,* and ptRNA is a shorthand symbol for a whole class of molecules. Its structure (the type, number, and order of the bases that it contains) is determined completely by the gene that is being copied. There are thus as many kinds of ptRNA as there are genes, but their general structure is the same.

Notice that when the DNA is sawn through it splits, for that part of its length, into two single strands. In principle both could be replicated, but only one, the so-called *sense strand*, carries genetic information, and it is the one that gives ptRNA. The replication of the other, *antisense strand*, would give nonsense, and in fact it is not replicated. It seems as though an enzyme involved in the replication process is involved in the decision not to replicate antisense strands.

3. DNA exists only inside the nucleus of the human cell. It is too large to get out through the membrane surrounding the nucleus, but ptRNA is a small enough molecule to escape. Once it is outside, in the cell cytoplasm, an extraordinary process occurs. Enzymes cut away the complementary junk and a smaller, smarter molecule is left: pure complementary gene (Figure 24.2c). This lean, mean molecule is known as *messenger* RNA, for short, mRNA.

It is on the surface of mRNA that the desired protein will be constructed, so the order of the bases on mRNA must contain the necessary information to ensure that the amino acids available in the cell are joined together in the right order to give that protein.

Now we know that there are in the cell something like twenty amino acids. We might assume that what we need is at least twenty assembly sites on the mRNA, each site having the correct shape to accommodate a different amino acid, so that

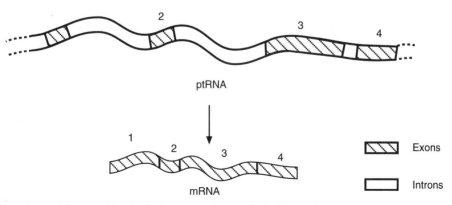

Figure 24.2c. Messenger RNA is obtained by cutting out the junk in ptRNA.

they can be lined up side by side before they link to each other to form the new protein molecule. Yet there are only four kinds of bases on mRNA. Our assumption that we need twenty sites thus runs up against a difficulty: if each base on mRNA was a site at which a specific amino acid could be held, this would, on a simple assumption, allow us to assemble proteins with only four different kinds of amino acids, each one somehow brought to the surface of its corresponding base, where it would eventually link hands with the neighboring amino acids to form a protein. This is not good enough; in this scenario the bases have provided us with a language of only four words. Such a language might not limit certain rock stars, but it is not enough to account for heredity.

It turned out, in 1961, that *every amino acid is associated with a site on mRNA consisting not of a single base but of a group of three adjacent bases.* Such a trio of bases is termed a *codon,* and every codon is an address, to which only one amino acid will somehow attach itself. As in a three-letter word the order of the letters—the bases—that compose the codon is important; different orders will give differently shaped surfaces, and it is the fitting together of compatible surfaces that controls the "docking" at the codon site. Thus in English "ODE," "DOE," "OED," "DEO," and "EDO" mean different things.[2] The codon CAU is the code for the amino acid histidine, but CUA is the code for leucine. There are more than enough (4 x 4 x 4 actually) different permutations of three letters chosen from C, G, A, and T to allow each amino acid to have its own codon. Actually, most amino acids have two to four codons. Thus glycine has the codons GGC, GGU, GGA, and GGG.

It is the order in which the codons are arranged on the mRNA that determines the order of the amino acids in the protein to be synthesized. Recall that the order of the codons in mRNA is completely determined by the DNA molecule that gave rise to it.

Now no amino acid is the right shape to fit snugly on any codon, like a lock and key. What happens is this:

4. Amino acids are brought to the codons on mRNA by our last type of molecule: *transfer RNA,* or tRNA for short (Figure 24.2d). This molecule has two "hands"; one hand grabs an amino acid, the other holds on to the corresponding codon on mRNA. It is this second hand that recognizes the codon. Thus, in the case of glycine, the tRNA holds on to glycine with one hand and links to one of the

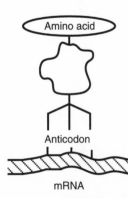

Figure 24.2d. tRNA has two hands: one holds onto an amino acid, the other is the anticodon to the amino acid's codon, which is on the mRNA.

[2]EDO: ancient capital of Japan; OED: *Oxford English Dictionary*. I haven't found a meaning for EOD, the sixth permutation.

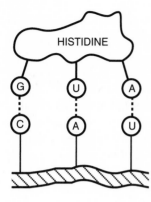

Figure 24.2e. How tRNA recognizes the site for histidine on mRNA.

codons for glycine on mRNA on the other. There are many kinds of tRNA; each one is so constructed that it is absolutely specific, binding only to one type of amino acid and to the corresponding codon. How does it know which codon to bind to? Easy! It base-pairs. Thus the tRNA for histidine has to bind to the codon CAU, and we already know that in general C binds to G, A binds to U, and U binds to A. So to bind to the codon CAU the tRNA has a "hand" consisting of the three adjacent bases: GUA, as you can see in Figure 24.2e. It is in this way that the twenty types of tRNA attach their specific amino acids to the correct codon sites. The order of the codons thus completely determines the order of the amino acids in the assembled protein.

The exact molecular square dance, on the surface of the mRNA, which eventually produces the linked amino acid chain which we call a protein, is merely hinted at in Figure 24.2f. Actually the process proceeds on an "assembly line," a cell structure called a ribosome that supports the mRNA. Also the protein is built up starting from one end of its amino acid chain and working along to the other end, and there is a codon at the end of the mRNA that is a "stop" sign, terminating the assembly process.

I have merely outlined the major players in the production and indicated their roles. Extraordinary as the mechanism is, all the known steps in protein synthesis are explicable in terms of the familiar physical and chemical interactions of molecules.

Samples of DNA isolated from fossils show exactly the same general structure as modern DNA, and the same codons have been identified. *The codons appear to be*

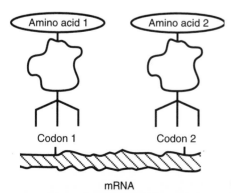

Figure 24.2f. Lining up amino acids. When the first two are in place, amino acid 1 transfers to amino acid 2 by a procedure that is not described here. The linked amino acids then transfer to the next amino acid brought to the surface of the mRNA.

universal in present-day species, from bacteria to man. Apart from viruses, whose genetic material is RNA, the DNA double helix is the unique genetic information carrier of the living world.

The Way in Which Cells Multiply

The fertilized egg develops into a fully formed organism by a process that is a long way from being completely understood. The essential fact is that fully developed organisms, except for the most primitive, are multicellular. As a new organism grows, after the fertilization of the egg, it splits into more and more cells, about 10 trillion in the case of an adult human. Almost all known types of cell have a nucleus, and every nucleus contains exactly the same DNA, exactly the same order of bases, as all the other cells in the organism. *The order of the bases in the DNA of the fertilized egg determines the order in all the other cells.* Your parents ensure that you have only *one* set of genes, one precious set from which all the trillions of others are derived. At each cell division the DNA of the parent cell appears in both daughter cells. Obviously there must be a mechanism for replication of the DNA molecule.

"It has not escaped our notice," wrote Watson and Crick, "that the specific pairing we have postulated immediately suggests a possible copying mechanism for the genetic material."

The mechanism by which DNA is replicated is simple: the DNA is split into its two component strands, each of which acts as a template for the assembly of a complementary strand (Figure 24.3). This is reminiscent of the more limited process, described earlier, for producing ptRNA from *one* gene. The new strands have their bases arranged in an order that is completely determined by the order of the bases on the single strands of the parent DNA. If, for the purpose of clarification, the two strands of the parent DNA are labeled 1 and 2, it can be seen that the new strand built around strand 1 is bound to be identical to strand 2, and the new strand built around strand 2 is just a replica of the original strand 1. Base pairing ensures that this is so. In this way each molecule of DNA generates two exact copies and, through billions of similar replications, the original DNA of the fertilized egg is passed on to every cell that has a nucleus. The DNA of the fertilized egg thus contains the information needed to control the activities of all the cells of the fully developed organism.

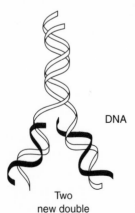

DNA

Two
new double
helices form

Figure 24.3. DNA replicates. After unwinding, the two separated strands act as templates for new strands.

On the pine tree outside my window there is a crow. She is a winged package of DNA, and her DNA will be given to her egg. The continuity of life depends on this flow of information.

Mutations

The molecule ethanol (ethyl alcohol, C_2H_5OH) is completely specified by its name. The molecule DNA is not completely specified by its name. Since, identical twins apart, everyone on earth has a different set of genes, one could say that there are several thousand million *different* molecules that are all called DNA. And that is without taking into account the rest of the living world, the DNA of ants and oak trees. *DNA* is a collective noun, like *mankind*. Furthermore, since man has evolved from hot soup, we see that DNA is an *evolving* molecule—it has the capacity to undergo change and yet, if that change is not too drastic, retain the ability to direct the development of a viable organism. The environment discourages the survival of those organisms whose DNA has given them characteristics which put them at a disadvantage compared to their colleagues. Thus, *evolution is the evolution of DNA.*

In Chapter 22 we presented mutants as one of the keys to evolution. From the viewpoint of this chapter, mutants are organisms whose DNA differs from that of their parents in that one or more genes have been modified; they have undergone *mutation*. Mutation means a change in the chemical structure of a gene; some chemical bonds in the molecule must be broken or new bonds formed. What agents can do this? The three most common are radiation, externally supplied chemicals, and cellular components.

Radiation has been recognized as a mutagenic agent since the 1920s, when a variety of organisms were irradiated with X-rays and subsequently produced offspring that differed macroscopically in some way from their parents. This is the origin of the concern about overexposure of human ovaries and testes to X-rays. Greater public concern has been expressed over the dangers of exposure to radiation emanating from radioactive elements, either in nuclear facilities, such as Three Mile Island or Chernobyl, or from nuclear weapons. Of the three types of radioactive radiation, γ-rays are the main danger from weapons since α- and β-rays have rather short ranges in air or matter. On the other hand, the ingestion of certain kinds of radioactive material can result in severe damage to cells due to the massive α-particles.

The damage incurred by the DNA of the *reproductive cells* can be so severe that the molecule is no longer capable of functioning. At lower levels the damage to DNA is often sufficient to result in an embryo that is not viable. At very low levels, damaged DNA can be, and often is, repaired by the cell itself. A genetic survey has been made of the children of parents who survived the two atomic bombs that were dropped on Japan. For parents who were near the explosions, their children showed only three mutations in a total of 700,000 genes examined. For parents who were well away from the explosions, 500,000 genes examined in their children also showed only three mutations. There is no statistically meaningful difference between these two figures.

The chief source of radiation hazard for the average citizen is natural radon gas originating beneath the surface of the Earth. Radon accounts for roughly half of the total natural radiation received by the body, but the level goes up to over 80% in some localities. Cosmic rays, which come from space and are thinned out by the air, normally account for 5% to 10% of the total radiation to which most people are exposed, but for those of us with frequent flyer cards who fly about 100 hours a year or more, the rays can constitute up to a quarter of the total.

External radiation is an extremely clumsy way of inducing mutations. It is nonselective, tearing through the cell like a crowd of hooligans, indiscriminately smashing anything to hand.

Many chemicals can cause mutations without completely inactivating the DNA. The first chemical mutagen to be studied was mustard gas, used as a weapon in World War I. Since then many substances have been indicted, among them cigarette smoke and oil of bergamot, the substance that gives certain teas their flavor. As in the case of any mutagen, chemical damage will not usually be confined to the reproductive cells, but only they can hand on the modified genetic information that produces mutant species.

The Enemy Within

DNA is a huge molecule that swims in an immensely complicated environment. Is your DNA stable, even in the absence of environmental threats? Is a mutagen needed for mutation?

The earliest evidence of the built-in instability of the DNA molecule, within the cell, came from the laboratory of Barbara McClintock, who in the 1940s came to the conclusion, on the basis of breeding experiments on maize plants, that some of the genes were mobile. They appeared to change positions on the chromosome in the transition from parent to offspring. Her unexpected findings were regarded with caution. When, in the 1950s and 1960s, the role and structure of DNA became clear, there still seemed no good reason to believe that the molecule was unstable. McClintock's results and her interpretation of them remained suspect, but her day was to come.

It has been possible for some time to determine the sequence of the genes in a DNA molecule—in principle. In the 1970s the sequencing of 1000 to 3000 bases, a typical number for *one* gene, would have taken ten to 30 years to complete. In the early 1990s the rate speeded up until roughly 10 million bases a year were being sequenced in the human genome, out of a total of about 3 billion. At this rate there would have been about another 300 years to go, but the joint effort of hundreds of scientists in dozens of laboratories so accelerated the sequencing that by 1995 it was clear that the project might be completed before the year 2005. Even the limited knowledge of gene sequences that has been available for the past couple of decades has been sufficient to show unequivocally that, in a variety of organisms, genes can change their positions on the DNA molecule, seemingly without any external prompting. The mobility of genes is an accepted fact. Barbara McClintock was awarded the Nobel Prize in 1983.

The term *transposon*, or more informally "jumping gene," is now used to describe any piece of DNA that moves to another location on the molecule. Thus a child may inherit a certain set of genes from its father, but the *order* of the genes on the DNA helix may be very slightly different in the child. Does this matter? It can.

Anything that alters the environment of a gene can potentially act as a *repressor*, preventing the gene from operating. Thus suppose that a certain transposon moves so that it is sitting next to a gene responsible for the production of a protein essential for the process of blood clotting. The gene may be so cramped by its new neighbor as to be put out of action, thus preventing the synthesis of blood-clotting factor. Such a case has been recorded. The transposon that did the damage was found in the parents of the unlucky child, but at a different location on their chromosomes. Transposons provide one mechanism for turning genes off or modifying their effectiveness.

A variation of the jumping gene trick has been found in certain bacteria. There are stretches of DNA, not necessarily genes or exons (bits of gene), typically several hundred bases in length, that are able to replicate themselves. Unlike the transposon, they themselves do not move, but the *copy* drifts away and forces its way into the DNA molecule at another point. If it squats in the middle of a gene, it can prevent that gene from performing its normal function. Alternatively, it could fall near a gene or on a gene regulator.

Despite the disturbing vision of pieces of DNA skittering around from one part of DNA to another, in general the standard account of inheritance, based on a fairly stable DNA molecule subject to occasional local mutations, still holds.

The Constant Codon

As far as we know, the codons are universal. Thus throughout that part of the animal kingdom for which data are available, the sequence CAU serves as the code for histidine. Research on the DNA of different species has revealed not only that our codons are identical but also that the *genes* of different species are very similar. In other words, the total base sequences of analogous genes are very similar. In this respect we are very close relatives of slime mold, radishes, and armadillos. Of course a bacteria, for example, hasn't got genes that control the color of its eyes; it has no eyes. But the synthesis of similar enzymes to ours is controlled by genes that, where their base sequences have been determined, are effectively the same as ours. Thus it has been found possible to cure yeast cells containing a mutant gene by bathing them in a solution containing human genes. Some of the yeast cells incorporate the corresponding "healthy" human gene into their own DNA and proceed to function normally.

Of course our DNA differs from that of other species, otherwise they wouldn't be other species; but the nearer a species is to us, the nearer its DNA tends to resemble ours. As annoying as it may be to some, our *set* of genes differ from those of the humble mouse by only about 2%.

You are the inheritor of the first DNA to have been created, about 4 billion years ago. That DNA molecule was much simpler than ours, but the DNA of the most primitive organisms on Earth is recognizably related to ours. As evolution proceeded, DNA evolved, a firm and continuous thread running through the whole of the living world.

Why Don't Your Eyes Grow Nails?

Consider the following fact: we have over 200 different types of cells in our bodies—muscle cells, nerve cells, leukocytes, and so forth. The DNA in all cells is *identical*, and therefore almost every cell contains *all* the genes responsible for the way we are. (Some cells have no DNA.) Now let us take one specific gene, that which controls the synthesis of insulin. It has been known for very many years that insulin is produced in the pancreas—and in no other part of the body. And yet the gene is distributed all over our bodies, in our skin cells, our muscle cells, our gum cells, and so on. Why don't all our cells make insulin? In fact, why don't all our cells do everything? Why don't our muscle cells grow hair, or our brain cells produce gastric juices?

Because there is a repression mechanism that closes down the operation of different genes in different cells. Usually an unwanted gene is literally covered up by a small molecule, a *repressor*. Thus the gene for insulin is inoperative in all cells except those in the pancreas. You may well ask, how does the cell know which genes to repress? We don't yet have a satisfying answer to that question.

There is more: a mechanism often exists for canceling the repression in time of need. For example, consider a bacteria that has a gene responsible for the synthesis of an enzyme that catalyzes the breakup of a certain sugar. If this sugar is not in the bacteria's environment, the gene is switched off, covered with a repressor. No enzyme is produced; if it were, it would have no role. However, if the sugar is fed to the bacterial colony, there is an advantage in having the enzyme since it could break up the sugar, providing the cell with energy. What happens is that the first molecules of the sugar entering the cell are chemically altered so as to produce a molecule, the *inducer*, which clamps on to the *repressor*, altering its shape so that it falls off the gene. The exposed gene now starts manufacturing the enzyme needed to break up the sugar. The process is referred to as *enzyme induction*; the enzyme is manufactured when it is needed instead of being pointlessly produced when its not wanted.

Wonderfully subtle. Evolution or Creation? Please put your money on evolution.

Why Sex?

We believe that we understand the broad outlines of heredity, but problems remain. We look at an important one.

Why are there two sexes? A case can be made for the thesis that sexual reproduction is a less than optimal means of ensuring the survival of a species. Consider this argument: Imagine an organism exists that has adapted itself exquisitely to its environment, by which we mean that a significant proportion of the population have a genome that could hardly be improved on from the point of view of the survival of that species. The best thing for the continued survival of the species would be for these "perfect" individuals to produce progeny who were exact copies of themselves. Now there are organisms that breed asexually; we say that they are *parthenogenetic*. Many insect species reproduce by parthenogenetic females laying eggs without the intervention, or existence, of a second sex. Reproduction by parthenogenesis, in the absence of mutations or copying errors, gives an exact copy of the parent's genome in the offspring. A "perfect" genome would be reproduced exactly—like mother, like daughter. Has this feminist process an advantage over sexual reproduction? Apparently yes, because during *sexual* reproduction the whole population is involved and the "perfect" genome of one part of the population can get spoiled by mixture with the less perfect genomes of another part. On the face of it, sexual reproduction is against the interests of the "perfect" individual—and the species—because the genome that is most favorable for survival is on the average degraded. In parthenogenesis it is preserved. Note that if, in a "perfect population," there are mutations producing less than perfect genomes, these individuals cannot, in parthenogenesis, "contaminate" the genome of others and will presumably be eliminated by the environment faster than their perfect colleagues.

If we believe that evolution is inseparable from the survival of the fittest, why has sexual reproduction, an apparently inferior method of maintaining a species, evolved and survived? Evolution is supposed to weed out those characteristics that mitigate against the survival of the species. It looks as though to be successful, species should breed parthenogenetically.

One fallacy in the argument is the implicit assumption that the environment is stable. A "perfect" genome is given this accolade because it is perfect for the environment in which it exists. Parthenogenetic species are all right as long as their environment is stable, but in relatively rapidly changing external conditions the very stability of the genome is a disadvantage because it prevents the genome from evolving so as to adapt to new stresses. Sexual reproduction is a far more efficient creator of new genomes because it mixes the parents' DNA. The greater the variety of genomes pitched against nature, the more likely it is that one or more of them will be better adapted to new conditions than the previous genomes. This may apply particularly to the resistance of a species to serious infection by microorganisms.

Part of the defense of the body against bacteria depends on genetic factors. Bacteria have, on their surfaces, molecules (car keys) that are recognized by specific receptors (car locks) in the defense systems of the body. This recognition process is the basis of the body's capacity to destroy the invaders. However, bacterial keys can change under the influence of mutation, which means that any given population of bacteria has a wide range of molecular "keys." Selection ensures that bacteria containing those car keys that are not immediately recognized by the body's receptors will flourish at the expense of less successful variations. One way that the species can try to defend itself against attack is to produce new receptors (change the car locks to match the new keys). The receptors in the body are also under genetic control, which means that the mixing of genomes that occurs in sexual reproduction will improve the chances that the offspring of a particular generation will produce receptors to which the "successful" bacteria are not suited. However, the bacteria will start to adapt to the changes, since once again natural selection will encourage the (minority) of bacteria that are not recognized by the new receptors. But sexual reproduction always aids the body to defend itself. The bacteria change keys, the body changes the locks, and neither side ever wins. This situation is a typical example of the Red Queen theory, which has gained popularity over the past few years and has been invoked in a number of contexts. The Red Queen in *Alice through the Looking Glass* runs faster and faster but stays in the same place.

By the way, there is some evidence to suggest that species which move completely to parthenogenesis do not last very long on the time scale of evolution. Extreme feminists take note.

The Wandering Gene

Genes can cross from one species to another. This happens, for example, in the cross-pollination of plants, but it appears that genes can be transferred between species by a number of other natural mechanisms. For example, a virus may have a transposon (a jumping gene) in its RNA that could be transferred not to another site on the RNA but to the DNA of the organism whose cells it invades. This is a specific example of what may be a more general phenomenon in which genetic material, or even complete organelles, from one species invades the cells of another. We saw in Chapter 22 that this may account for the origin of mitochondria. Here's another example.

It is believed by some biologists that very early in the history of life on Earth, relatively large complex cells first acquired the ability to carry out photosynthesis when one or more became fused with a photosynthetic bacterium. And that's how plants were born: the photosynthetic apparatus of the plant cell is primarily concentrated in the chloroplast, which originated as a bacteria.

In one of the most far-reaching developments in scientific history, man, emulating and surpassing nature, has also developed techniques for transferring genes between species. This is not a trivial accomplishment. There are dangers associated with this power, and ethical questions have already arisen. In the following chapter we will see how science meets morality in the shadow of heredity.

25

The Lords of Nature?

...and thus render ourselves the lords and possessors of nature.

—Renè Descartes, *Discours de la méthode*

Our rapidly developing ability to identify and manipulate genes is creating problems, both biological and ethical, that we have never had to face before.

Is Genetic Engineering Dangerous?

When the Whizen family of Los Angeles decided to donate the cost of a new biotechnology wing at my university, there was one condition: that there should be an annual series of conferences on "bioethics." Bruce Whizen was deeply concerned with the ethical implications of science. He was well aware that although biotechnology could be used for "the relief of man's estate," it also had the potential to be an ethical Pandora's box. He was thinking of that component of biotechnology that we call genetic engineering.

Genetic engineering is, broadly speaking, the art and science of removing genes from the cells of one species and introducing them into the DNA of a member of the same or a different species. The primary *commercial* purpose of genetic engineering is to induce another species to produce a substance which it would not normally produce, but which is useful to man. Remember, the role of genes is usually to direct the synthesis of specific molecules within the cell. Thus, in principle, if I put the gene for producing adrenaline into a bacteria, then it will produce this vitally important substance for me, a substance that bacteria never normally synthesize since it is not part of their biology.

The technique of taking DNA from, say, human cells and placing it in the cells of another organism has become commonplace. Let's take a classic example to see how this is done.

We need to be able to obtain large numbers of the desired gene (i.e., to *clone* the gene). This can be done by planting the gene in a helpless organism and waiting for that organism to multiply rapidly. For this purpose it is usually best to use bacteria as the host organism, since they breed so fast. Certain bacteria, including the ubiquitous *E. coli*, carry their DNA in circular form in their extranuclear protoplasm. There can be several of these rings, or *plasmids,* in one bacterial cell (Figure 25.1). Plasmids can be isolated and treated with *restriction enzymes*, which have the ability to cut open the plasmid ring. If the solution in which this is done contains genes isolated from another species, say man, it is possible for the exposed ends of the cut plasmids to link onto these foreign genes and reclose the ring, thus incorporating the visiting gene into the plasmid. This process is aided by another enzyme, *DNA ligase.* The resulting ring of hybrid DNA is known as *recombinant DNA*. At this

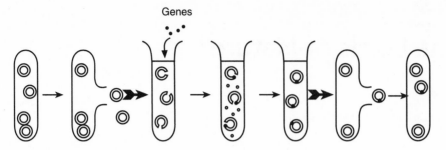

Figure 25.1. The grafting of a foreign gene into bacterial DNA. The ring-like plasmids are removed from the bacteria, opened, and mixed with a foreign gene. They close, incorporating the gene, and are returned to the bacteria.

stage the plasmid, feeling like a nest with a cuckoo's egg in it, can be induced to reenter bacteria present in the medium.

The exciting but, to many, vaguely menacing fact is that the doctored bacteria, containing recombinant DNA, reproduce, generating bacteria containing the identical recombinant DNA. A small number of bacteria can generate a huge number of offspring in a matter of days. Furthermore, and this is the point of the exercise, the doctored bacteria will synthesize not only their own proteins but also proteins whose synthesis is controlled by the foreign gene, a gene that could be of human origin. Here is an example.

The interferons are a class of small proteins that are produced in minute quantities in the human body and are capable of inhibiting the action of some viruses. The problem is that it is not feasible to obtain the amount of human blood that would be needed to satisfy the growing clinical need for interferon, and the cost of isolating useful quantities of the protein from blood would, in any case, be enormous. Furthermore, the isolation process is lengthy, on the order of months. A few miles south of Tel Aviv there is a company that has taken the human gene responsible for the synthesis of an interferon and placed it in bacteria. Interferon is now produced relatively cheaply by these bacteria, the rate of production being several hundred times as fast as the conventional method. Using the same general principle, growth hormones, blood-clotting factor, insulin, vaccines, and a growing variety of polypeptides used in medicine are being produced in several countries by genetic engineering of bacteria or yeast cells. The bacteria that produce human insulin do so so enthusiastically that they bulge with product.

The uses to which genetic engineering is being put are increasing yearly. Enzymes are proteins, which makes them very suitable for production by genetic engineering. Among those enzymes successfully produced is a lipase, an enzyme that breaks up fat molecules and is added to detergents. It was just necessary to isolate the gene for lipase and induce it to link up with bacterial plasmids. Genes placed in plants are producing, among other things, pharmaceutical peptides and melanin, the dark pigment of the skin, used in some sun lotions. Genes transferred to plants have conferred upon them heightened resistance to insects and disease.

There appears to be no reason why we should not be able to transfer *any* gene from a member of one species to the DNA of a member of the same, or any another, species. If this is so, then what was once unthinkable is becoming feasible: remedying genetic defects by replacing faulty genes with their healthy forms. The advocates of genetic engineering see a future in which such transfers will eat away at the ills to which man's flesh is heir, a future in which the blind cruelty of nature will find its match in the laboratories of the genetic engineer. Others are more wary.

Another Viewpoint

A doubt hovers in the background, a sense of the presence of deliberate or accidental evil. The ability to transfer genes has, as far as I know, been used only for harmless or beneficial purposes. Should such work be more closely scrutinized and perhaps controlled? It is now common practice to insert into mice, genes that are responsible for human diseases, such as the gene associated with Wilms' tumor, a malignant growth causing kidney cancer in children. Mice that carry the genes for human disorders can be of immense help in studying such disorders. This medical breakthrough is seen in a rather different light by those who support animal rights. Naturally, the transfer of genes to human beings has caused rather more concern.

Gene Therapy

Molecular biology has opened up the possibility of transferring genes to human beings, either from other human beings or from other species. HMS should be aware of this development and exactly what it implies biologically. The ethical implications are the subject of ongoing discussion.

To put things in perspective, it is worth noting that it is now possible to take a mouse and replace *any* gene that has been identified, by a mutated version of that gene. The replacement sits in the same location as the healthy gene. The purpose is usually to examine the role of a specific normal gene in the working of the organism, but that is not the point I wish to make, which is to emphasize the rapidly increasing degree of precision with which the genome can be analyzed and operated upon.

There are two quite distinct procedures in gene transfer: transfer of genes to the *somatic* cells and transfer to the *reproductive* cells. To clarify: the somatic (nonreproductive) cells, such as your muscle cells or your blood cells, do not hand on their DNA to the next generation. Only your reproductive cells do this. If the gene for producing comedians is integrated into the DNA of your liver cells, your children will not be the Marx Brothers. However, if there were such a gene and it was incorporated into your ovum or sperm, then you might run the risk of initiating generations of cigar-chomping, wisecracking babies.

The effects of *somatic* gene therapy are limited to the individual being treated; the invading gene does not appear in the sperm or ovum of the recipient. There are two possible motives for such treatment. One is to take an ostensibly normal human being and attempt to improve her. A contrasting motive is to repair genetic shortcomings. An example follows.

There are people who are born without the ability to synthesize an enzyme called adenosine deaminase (ADA). This failure arises from a defect in the gene responsible for the synthesis of the enzyme. Without ADA, there is a drastic weakening of the immune system, the body's defense mechanism against invaders. The resultant disease is known as severe combined immune deficiency (SCID). On 1 September 1990, W. French Anderson, then at the National Institute of Health, attempted to treat a small girl from Ohio who suffered from SCID. His approach was to give her the gene that was responsible for synthesizing the enzyme ADA, which in her case was defective. Normal ADA genes were transferred to a virus that had been rendered harmless. Cells were then taken from the girl's blood and exposed to the virus. The virus invaded the cells, to which they transferred their genetic material, including the ADA gene. The cells were then cultured so that they multiplied. They were then infused into the girl's bloodstream. They began producing ADA; they had the gene. A second child underwent the same treatment shortly afterward. Both children were partially cured. Every six months they return for an infusion, a small price to pay for a reasonably normal immune system.

The general technique of *somatic* gene therapy will almost certainly become a standard form of treatment for some diseases arising from genetic defects. Anderson's technique was suited to a disease of the blood—transfusion is a standard procedure. Replacing the faulty genes in an organ such as the kidney is far more difficult, in fact practically impossible at the moment, and the answer to most genetic diseases would seem eventually to lie in *reproductive cell* gene therapy. But this would help the children, not the parents.

Whereas I can see no scientific, legal, or ethical objection to somatic cell therapy, I can understand why there have been objections in the case of genetic manipulation of the *reproductive* cells. This is an area in which the issues should be very clear to HMS because, although the techniques are not used at present, the possibilities are legion.

Although reproductive cell gene therapy can be applied to animals other than man, and here one may have to consider ecological and other consequences, the ethical controversies in this area are naturally primarily centered on the application to man. In Europe, in 1988, the organization of Medical Research Councils put its stamp of approval on somatic gene therapy but objected on principle to reproductive cell gene therapy. In the same year, a similar position was taken at the Fifth Summit Conference on Bioethics in Toronto. The objection centered around the point that the child-to-be would have had no say in the genetic manipulation that gave her her genes, and that if the therapy had produced undesirable effects there would be no way back. If one wanted to be awkward, one could point out that at present there is no way back for almost any child that is born with a genetic defect, and that gene therapy performed on the ovum of a woman carrying a very undesirable gene could allow her to have normal children without the fear that they were going to inherit, say, cystic fibrosis or Tay-Sachs disease.

The British medical journal *Lancet* devoted an editorial to the subject of human gene transfer in January 1989, expressing the following opinion:

> What one needs is an educated public. They need to be sufficiently DNA-literate to appraise the advantages and disadvantages of gene therapy. The more people involved in the decision-making processes the better. Decisions should not be confined to a small group of experts, be they scientists or politicians, philosophers or doctors.

This neither condemns nor supports gene therapy. Understandably, when it comes to genetic manipulation of reproductive cells, religious bodies have a rather more disapproving point of view, as exemplified by the Catholic Church, which in its official policy statement in 1992 reflects its own but also many other people's, misgiving:

> It is immoral to produce human embryos destined to be exploited as though they were disposable biological matter. Certain attempts at engineering of chromosomatic or genetic matter are not therapeutic but are intended for the production of selected human beings according to sex or other pre-established criteria. This engineering goes against the personal dignity of the human being and his unique, unrepeatable identity.

For religious fundamentalists the answer is easy: one doesn't mess with God's creations. For the rest of us there is a real problem.

Understanding the techniques of reproductive cell gene therapy is less important than understanding the principles: we are talking about introducing foreign genes into the DNA of the reproductive cells of a potential parent who has a defective gene. The object is to avoid an inherited disease in the patient's offspring. One can foresee the day when it will be possible to take a sample of sperm from a man, remove a defective gene, and replace it with a normal gene. The sample could then be used to fertilize his partner's eggs. Once again, I can see no ethical objection to what would once have been regarded as a miracle. On the face of it, only medical good could come of such a manipulation. Of course it could go wrong, but so can any medical procedure. So why do the official bodies object, and a lot of other people worry?

On the purely emotional plane, there is perhaps a feeling that we are interfering with nature at the deepest level. Another predictable fear arises from the Frankenstein scenario: somewhere in a cellar the crooked (mad?) scientists, in the service of the Leader, are producing subservient, merciless monsters. But for this, one doesn't need a laboratory. Hitler produced the SS; Pol Pot produced the Khmer Rouge; Dr. Mengele and the Japanese performed medical experiments on human beings. Those who advise caution in the area of genetic engineering see such horrendous episodes as supplying them with legitimate backing, but on these grounds one would have to ban surgery, injections, and many of the accepted techniques of medicine.

It is my belief that those who fear genetic engineering have far more reason to be perturbed by the social doctrines of the kind promulgated by the nineteenth-century geneticist Francis Galton or the racial theories embraced by twentieth-century fascist regimes and their admirers. There is no need for genetic engineering if a government, or indeed an individual, wants to treat human beings like laboratory animals.[1]

Quite apart from ethical problems, there are biological dangers inherent in gene transfer. Those involved in genetic engineering are aware of them. In research institutes using recombinant DNA techniques, part of the work is carried out in laboratories that resemble inverted sterile rooms; nothing must be allowed to *leave* the room unless it is monitored. The fear is that genetically engineered bacteria could have unpredictable effects if they escaped and subsequently multiplied. This is a far more realistic concern than the cloned-superman fantasy. It is not that the bacteria could be pathogenic, although they might, but rather that they could have un-

---

[1]It will be interesting to see the long-term results of the use, in the United States, of a sperm bank confined to donors who are prominent scientists and intellectuals. Personally, I would prefer an average child of my own to an artificially inseminated child of a genius. But maybe I am revealing a vein of sentimentality or macho possessiveness.

predictable effects on the local ecological balance. Man has upset this balance before, by introducing foreign species into an ecosystem.

The biological consequences of genetic manipulation are not the only area of controversy associated with the human genome. The most controversial subject has been IQ, but a growing area of public concern is genetic testing. We glance at both.

Genes and Society I: Time for the Emperor to Get Dressed?

There used to be many Jewish boxers. My father reeled out their names, going back to the eighteenth century: Samuel Elias, known as Dutch Sam, the originator of the uppercut; the Sephardi Jew Daniel Mendoza, the first known scientific boxer.[2] Closer in time to me was the Lower East Side boy Benny Leonard, who retired in 1925, after eight brilliant years as undefeated lightweight champion of the world. When I was four, Max Baer held the heavyweight championship of the world. My father explained: "Jews are good at boxing, it's in our blood." Today there is hardly one prominent Jewish boxer.

Despite my father's claim, boxing ability is almost certainly not inherited. A more loaded question is whether IQ is inherited and, if so, whether some races are less intelligent and are doomed to remain so. Are there genes for intelligence—and where can we get some? Will we one day demand somatic or reproductive-cell gene transfer to improve our, or our children's intelligence?

Alan Turing, a genius who dealt with the general theory of computing, said that we were programmed by our genes. It is doubtful whether he believed that the environment had no place in controlling our abilities, both mental and physical, but this is what many IQ practitioners claim: that IQ is independent, or nearly independent, of the environment. Only the genes count.

Is the average IQ of the Scots higher than that of the English? It appears to be, from the published data. What does the result mean? A logical positivist would say that it proves that Scots obtain higher scores on IQ tests than Englishmen. Period. A Scot would be inclined to say that it proves that Scots are more intelligent, to which an Englishman might reply that IQ tests only test certain kinds of abilities, and are no more indicative of general intelligence than the ability to do crossword puzzles.

So far the argument has been only slightly barbed, but now translate the whole scenario to whites versus blacks and you end up with campus riots and visiting speakers being shouted down. Consider William Schockley, who shared the 1956 Nobel Prize for research into semiconductors and discovering the transistor effect. Schockley became obsessed with what he saw as the threat to the average IQ due to the higher birthrate among what he declared were low-IQ people. His spoken and written statements quickly developed strongly racial overtones. Thankfully, the vast majority of the scientific community repudiated his views, emphasizing that his excellence as a physicist lent no weight to his genetic ramblings, but Schockley remained convinced that the Negro, as blacks were called in his day, was less intel-

---

[2]Mendoza's lodgings can still be seen opposite Bethnal Green underground station, in East London. A natural middleweight, he was barefist heavyweight champion of England in the 1780s and gave boxing lessons to gentlemen. He wrote the first modern textbook on boxing, *The Art of Boxing* (1789).

ligent than the white man. This is not an argument that I want to get pulled into here, but there are some points that should be made.

In a world in which history has resulted in the vast majority of blacks being economically and educationally underprivileged, it is difficult to believe that the standard IQ tests, devised by a white middle-class society, can be applied equitably to the average black community. (There is at least one test, designed by blacks, in which blacks scored better than whites.) But there is a prior question. What does the IQ test measure anyway? Is it really testing a genetically controlled characteristic, and if so, what is that characteristic?

When I was a student at University College London, my first girlfriend was a psychology student, who put me through an IQ test. My score was quite high, and I asked what the number meant. "That you're the kind of person who does well at IQ tests." It was a glib joke, but I think it was near the mark.

Intelligence tests originated in the work of the French psychologist Alfred Binet in 1905. He was attempting to find a method of predicting which children were most likely to fail in grammar school. He and his colleagues found that if the tests were based on the kinds of problems the children faced at school, they gave a good prediction of the children's performance in school. Truly amazing.

The obsession with racial differences in intelligence goes back at least as far as Darwin's cousin Sir Francis Galton (1822–1911), the creator of eugenics. Convinced that intelligence was related to the achievement of eminence, Galton believed that the ancient Greeks had a much higher intelligence than present-day Europeans. On the same basis, because in his day there were more eminent Englishmen than Americans, he concluded that the Americans were mentally inferior. Galton, incidentally, wrote a book called *Hereditary Genius* (1869), in which he attempted to apply the theory of evolution to man. Darwin had effectively avoided this step, but, perhaps encouraged by Galton's book, he published *The Descent of Man* in 1871.

The high priest of IQ in the early decades of the twentieth century was an American, Lewis Terman. Something of the mind-set of this obnoxious man can be gathered from his statement that the higher birthrate of southern Italians as compared with Harvard graduates "threatens the very existence of civilization." He was a member of the Human Betterment Foundation, which recommended wholesale sterilization in California. His colleague Henry Goddard hung around Ellis Island, testing arriving immigrants. His results showed that 87% of the Russians, 83% of the Jews, 80% of the Hungarians and 79% of the Italians were "feeble-minded"! (And 100% of the testers?)

During the 1920s, the supposed irrelevance of environment was shown to be suspect, to put it mildly. In 1925, William Bagley discomforted the racist school of IQ by showing that in those areas where children stayed at school longer, the average IQ was higher. Worse still, whites in some southern states of America had a lower average IQ than blacks in some northern states. In order to answer the growing criticism that tests were ethnically biased, attempts were made to construct "culture-fair" tests, a peculiar development if IQ, as many of its practitioners had stated, is purely genetically controlled. But is it?

In his landmark book *The Science and Politics of I.Q.* (1974), Leon Kamin writes, "There exist no data which should lead a prudent man to accept the hypothesis that IQ test scores are in any degree heritable." You may feel that this is overstating the case for the environment, and Kamin himself, at the end of his book states, "To assert that there is *no* genetic determination of IQ would be a strong, and scientifically meaningless statement," and he puts the burden of proof upon those who wish to prove otherwise. Kamin states: "That burden is not lessened by the repeated assertions of the testers over the past 70 years." Little has happened since Kamin wrote those words to lessen that burden. But what does IQ measure anyway?

Intelligence is implicitly *defined*, for those who believe in IQ, by the IQ test. By calling these tests "intelligence" tests, the practitioners of IQ tests have tried to convince us for years that that is what they are measuring. But it surely cannot be that something as complex as what most (intelligent!) people call "intelligence," can be measured with a single number. In what meaningful sense is someone with an IQ of 134 superior to someone with an IQ of 125? Oh, sure, I forgot—one of them is better at IQ tests.

Worse still, the basic premise of IQ tests is that they are measuring something that is *independent of learning*. If it were, there might be a little more justification for supposing it to be genetically controlled, but surely what we normally think of as intelligence is affected by the stock of *knowledge* that we have. What we talk of as intelligence is in fact a function of an inextricable combination of our brain circuits and our learned knowledge. And our circuits may be affected by the learning process. I do not believe that IQ tests measure anything *quantitatively* meaningful, except perhaps that very low scores almost certainly indicate gross mental handicap, but that would be obvious without the intervention of "science." I am far from alone in the feeling that, as in the case of psychoanalysis, the emperor should get dressed. We have been led astray by the IQ gurus. The dangers of taking IQ too seriously are many. One can certainly sympathize with Noam Chomsky's opinion that there is no decent reason for attempting to find IQ differences between different races.

Let's try a different track. Are IQ test results an indicator of *creativity*? In 1953, Ann Roe measured the IQs of sixty-four prominent scientists. Some had high scores, but for others "none of the test material would give the slightest clue that the subject was a scientist of renown." Similar studies on the connection between IQ and creative achievement for a variety of professions in the humanities and sciences showed no correlation except among mathematicians, and this only weakly. The kind of problem-solving abilities that are tested by IQ are just not the kind that are relevant to most human activities. Common experience shows us that there are people with high IQs, and possibly impressive academic records, who are highly unimpressive when it comes to almost anything outside their professional sphere.

The field is wide open. David Perkins, at Harvard, has suggested, on the basis of studies on twins, that environmental effects are responsible for up to half of the observed variation in human IQ scores. Consistent with this viewpoint is the fact that the average IQ of a group of schoolchildren was raised by several points by as little as one year's special teaching. The same children also performed far better than a control group on more general tests of ability than the IQ test.

So what does IQ measure? Nothing very defined, and except perhaps for very high and very low scores, there appears to be no quantitative connection between IQ scores and what most people understand by intelligence. I have the feeling that the "inheritance of IQ" will soon be seen to be, if not a meaningless question, a very uninteresting one.

Genes and Society II: Genetic Screening

I was at one time the chairman of a committee for encouraging higher education among the Druze Arabs in Israel. One day a prominent sheik showed me a photograph of his family. There appeared to be no female members, although his children presumably had a mother and I knew that he had daughters. It was sons that mattered. In India many newly born female children are murdered. Should families preferring a son have the right to demand abortions if an embryo is identified as female?

The preference for male babies is the most obvious of a whole range of selective options that parents have. Thus a couple might decide to abandon a mentally defective child or a child with severe motor disabilities. The advent of molecular biology has now brought us to the stage where such choices can be based on a genetic analysis of an embryo, or its prospective parents. This is one of an ever-widening range of closely related problems of choice that arise directly from the ability to identify genes in embryos and adults.

It is now a common laboratory practice to isolate DNA from an organism and subject it to analysis, in order to detect the presence of specific genes. This kind of work has great potential for the improvement of human and animal health, but it also generates some very tricky legal and ethical problems.

DNA analysis carried out on an embryo might reveal that the child-to-be would suffer from a physical or mental disability. Is abortion justified if that disability is of an extremely serious nature, threatening the life or the quality of life of the child? If so, where would you draw the line? There is a research group that claims that shyness is genetically controlled. Another researcher believes he has found a gene that is responsible for baldness. Silly examples, but what about diabetes or a tendency to develop Alzheimer's disease?

Picture the day when DNA analysis is obligatory before you get accepted for a job. Improbable? Several American companies have already started genetic screening of their employees. Your analysis reveals that you have an above-normal chance of getting cancer or Parkinson's disease. Will you ever get employment? Or life insurance? Insurance companies are considering demanding a DNA scan before insurance coverage is granted. Supposing your girlfriend or boyfriend finds out that your DNA is suspect; will he or she reconsider the wisdom of marriage? Should your DNA be legally your business and no one else's? Is this, as I feel it is, eugenics creeping through the back door? These are issues that all of us should be aware of, and they will become increasingly pressing as the Human Genome Project nears completion. Although the complete *base* sequence may well take many years to complete—unless improved techniques become available—the identification of all the *genes* will require far less time. We can find out what a gene is responsible for, and its location, without knowing its complete base sequence.

The ability to tell a couple that they both carry the gene for a fatal disease is, it would seem to me, a legitimate tool in a doctor's black bag, but if the practice becomes widespread it should be accompanied by specialized counseling services to help those who receive bad news. Are we approaching the time when, as one commercial gene sequencer has suggested, we are all going to have our personal genome on a compact disc in the family doctor's records, or on a magnetic card, along with our credit cards? Brave New World?

The Human Genome

The Human Genome Project is a loosely coordinated attempt, being conducted in nearly forty countries, to completely sequence the 6 billion or so bases in human DNA, so that the codon sequence of every one of the approximately 100,000 genes will be known, as well as the order of bases in the introns. It is a huge task; there are international meetings on *one* chromosome, the investigators comparing progress on the elucidation of the spatial ordering of millions of bases. Undreamed of details of our makeup are being revealed. For example, preliminary estimates suggest that at least 550 different genes control the chemical and physical structure of the eye. The red blood cell, one of the simplest cells in the body is associated with only 8 genes; the white blood cell—far more complex in form and function—is represented by nearly 2200 genes, the salivary gland by 17, the prostate by about 1200, the ovary by some 500, the testis by over 1200 and the liver by nearly 2100. The brain is the work of close to 3200 genes. These figures will probably be subject to revision as research continues.

The potential benefits of a complete analysis of the gene sequence and base sequence of the human genome could be enormous. It is almost certain that improved techniques will eventually allow reliable pinpoint "surgery" on a patient's reproductive-cell DNA, allowing the almost routine deletion of faulty genes and their replacement by their correctly functioning versions. When one reflects that a defect in a single gene may be the cause of a life-threatening disease and that over 4000 single-gene diseases have been identified, it is clear that there is a rich field for the alleviation of human suffering.

Knowledge of the human genome will put increasingly powerful weapons into the hands of the practitioners of gene therapy—and into the hands of those who will screen people genetically for the government and for private companies. The dangers of this final invasion of biological privacy have prompted the drafting of a Genetic Privacy Act by an American professor, and a number of states in the United States have already passed laws partially aimed at preventing official bodies or commercial companies from making use of "private" genetic information.

Molecular biology has raised some strange legal and ethical questions—including the patenting of animals.

Patenting Animals

At Cold Spring Harbor Laboratory in 1982, there was a meeting of patent lawyers and biologists to discuss "The Patenting of Life Forms," specifically, the patenting of genetically engineered organisms. The first application for such a patent was filed with the U.S. Patent and Trademark Office on 7 June 1971, in respect of a microorganism cultured by Ananda Chakrabaty. The patent, U.S. Patent No. 4,259,444, was assigned on 31 March 1981. In April 1988 a mouse implanted with a human gene causing cancer—the "Harvard mouse"—was the first animal to be patented in the United States. Many plants have been patented, for example, a cotton plant genetically engineered to resist certain herbicides.

Should it be possible to patent animals? A number of companies are attempting to incorporate human DNA into goats, cows, pigs, and mice. The objectives vary, from overcoming the rejection of the human body to organs transplanted from other animals, to inducing human genetic disease in animals so as to be able to test drugs on them. There are ethical and legal issues here that become more complex in the case of gene implantation in humans. If I am implanted with a gene that guarantees that my son will be a musical genius, can the medical team who did it patent me or him? The question is absurd, but is it any more absurd than the wish of the American team that is working on the human genome to patent its findings and sell them? Incidentally, a French team working on the same problem has declared that its findings belong to humanity, an interesting difference in outlook. An acquaintance of mine in the United States, who headed a company that put a great deal of work into the partial mapping of the human genome, once stated, "We have fifty-four markers on chromosome 7 alone. We have mapped it in a way no chromosome has ever been mapped. We really own chromosome 7." He didn't mean it literally, but he clearly feels that an investment of several million dollars in research was not made so that he could freely distribute information to one and all. The company wishes to develop diagnostic tests for a variety of undesirable human genes. Is that any different from a drug company patenting its findings? His company is not unique; the techniques are now fairly common. What are the legal and ethical implications? Once again, HMS should be aware of what is being done and what it implies, both medically and ethically. When James Watson was put in charge of the American Human Genome Project he insisted that 3% of the budget should be set aside to finance the investigation of ethical problems, a praiseworthy demand that has borne little fruit.

The Future

The discovery of the genetic code and the construction of quantum mechanics are the most fateful events in the history of twentieth-century science. In the practical world, quantum mechanics has given birth to the information revolution, through such devices as the transistor, the microchip, and the semiconductor laser. The twenty-first century may be the century of the genome.

It is impossible to say where genetic engineering and genetic analysis will have taken us by the end of the next couple of decades. That it will pose more ethical problems is almost certain. For the science fiction fan one question, asked by the molecular biologist Lederberg, hovers in the background: "What stops us from making supermen?" To which he supplies the answer: "The main thing that stops us is that we don't know the biochemistry of the object that we are trying to produce."

The question as to whether we can create supermen is a sensational way of asking a more prosaic question: Can we dramatically improve the mental and physical characteristics of ourselves and our children by genetic manipulation? The answer is that in principle the means are, or one day might be, available. On which disquieting note I pass the discussion over to you.

Molecular biology, particularly its genetic aspects, has laid a rational infrastructure for Darwinian evolution and become an indispensable part of biological and medical science. The discovery of the double helix and the breaking of the genetic code have filled in a gap in the hierarchy of the sciences. The practical consequences may play a major part in shaping everyday life in the next century, a century in which, if we survive, we will increasingly manipulate the biological world, hopefully for our own good.

It would be misleading to suggest that a detailed knowledge of the human genome will rapidly lead to spectacular advances in the treatment of genetically controlled diseases. The identification of the location and nature of the normal gene responsible for a given biological characteristic does not necessarily mean that we can easily counter the effect of a faulty gene. The potential is there, but up to now the medical benefits of the analysis of the genome have been scant. Nevertheless, let's end on a hopeful note. While I am writing this, a fourteen year-old boy is dying in a London hospital because no donor has been found for a heart transplant operation. In another part of the world, scientists are rearing genetically engineered pigs with the object of producing internal organs that will not be rejected by the human body. Whether this project will be successful remains, at the time of writing, to be seen. If the approach succeeds, it could remove the great obstacle to widespread transplant operations: the difficulty of obtaining sufficient donors. The thought of carrying a pig's heart in your chest might worry a Romantic poet or an antivivisectionist—but not the parents of a sick child.

26 | Life: The Molecular Battle

"Dissolution, putrefaction," said Hans Castorp. "They are the same thing as combustion; combustion with oxygen—am I right?"

"To a T. Oxidation."

"And life?"

"Oxidation too ..."

—Thomas Mann, *The Magic Mountain*

In the last 350 years, the formulation of a small number of unifying concepts has allowed us to rationalize much of what we know of the living world. We have already seen examples, in Darwinian evolution, Mendelian genetics, and the genetic code. Other major concepts include the eighteenth-century system of classification of living forms and the nineteenth-century concepts of the cell and the germ theory of disease. The confluence of biology and chemistry in the twentieth century has brought a rich harvest and, while we can hardly be said to understand life, our knowledge of the workings and structure of cells has become increasingly detailed.

Life is a continual struggle at all levels. Darwinian evolution is predicated on the competition between *organisms*. Physiology, biochemistry, and molecular biology have shown that there is a continual *molecular* struggle within each organism, a struggle against dissolution, a struggle to maintain a constant internal environment, and a struggle to negate the deleterious consequences of invasion by both living and non-living intruders. The organism's ability to survive these battles is based largely on the huge variety of shapes that can be adopted by protein molecules. Shape is the key to most of the cell's activities.

But what is life? At what level of complexity does it begin? Certainly not at the elementary particle level, or the atomic level, but it was once supposed to start at the level of the single molecule.

What Is Life?

At one time, all known objects were very clearly either dead (e.g., the Pyramids) or alive (e.g., archaeologists). Both Hippocrates and Aristotle stated that an innate heat, originating in the heart, distinguished living organisms from dead matter. The sharp line separating dead from living was reinforced, in the eighteenth and early nineteenth centuries, by advances in chemistry that pointed to the conclusion that living matter was composed of special kinds of molecules. Substances were isolated from living material that were never found in the clearly inanimate world. A classic case was urea, separated from urine. It was believed that such substances could not be produced from any material derived from nonliving matter. Thus carbon dioxide (dead) could be obtained by heating chalk (dead), but there was no known way of making urea from inorganic (alias: dead) substances. And that was because living material is fundamentally different from nonliving, or so the story went. Living matter possessed a *vital force*. So organic chemistry was born, the chemistry of molecules that could supposedly be made only by living organisms. The cut between the living and nonliving world was firmly fixed at the molecular level, ergo life had been created separately from the creation of the inanimate world.

And then, in 1828, a German chemist, Friedrich Wöhler (1800–1882), attempted to make an inorganic substance, ammonium cyanate, by evaporating a solution containing ammonium and cyanate ions, both also inorganic substances. He kept on heating after the crystals appeared and isolated a white solid that on analysis turned out to be not the "dead" substance, ammonium cyanate, but the "live" molecule, urea.

Wöhler hesitated to publish, for two reasons: he didn't quite believe what he had done, and moreover it looked as though he was guilty of treachery. He had been a student of Berzelius, the originator of the vitalist theory that organic molecules could be made only by living organisms.[1] Wöhler's discovery should have put the vitalist theory down for a long count. An organic molecule had been created from inorganic molecules. But there was not a lot of attention paid to his feat at the time. What should have been the final decisive knockout came in 1845 when Adolph Kolbe (1818–1884), who introduced the word *synthesis* into chemistry, synthesized acetic acid from carbon disulfide (CS_2). Previously, acetic acid, the basis of vinegar, had been universally produced by fermentation of organic matter. One last effort to conserve the uniqueness of "living molecules" came, in 1898, from a certain Dr. Japp.

Build a molecule. Let it look at itself in a mirror. For very many molecules, what they see is another molecule, looking very like them, but differing from them in the way that your left hand differs from your right hand. When such molecules are synthesized in the laboratory, they usually appear as a mixture of the two mirror images—like gloves, or shoes, coming off a production line in pairs. This is not normally true of molecules synthesized by cells. The cell manufactures only one of the two molecular forms—gloves for your left hand only. Japp claimed that living organisms were the only systems that could produce optically active molecules (i.e., one glove without the other). He was proved wrong in 1917, when *asymmetrical syntheses* were first carried out in the lab.

There are no molecules that man has not made, or cannot potentially make, from inorganic precursors.

Today the term *organic chemistry* is misleading since it means nothing more than the chemistry of molecules containing carbon atoms. Not all organic molecules are associated with living material. Thus benzene is an organic molecule but is not normally found in living cells. Our conclusion is that *at the level of molecules, living and nonliving matter cannot be differentiated.*

Like Descartes, most scientists believe that living matter obeys the established laws of chemistry and physics.[2] Thomas Huxley said it: "I can find no intelligible ground for refusing to say that the properties of protoplasm result from the nature and disposition of its molecules." Life is a demonstration of the degree of complex-

---

[1] In his letter to Berzelius, Wöhler wrote, "I can no longer, as it were, hold back my chemical urine; and I have to let out that I can make urea without needing a kidney, whether of man or dog: the ammonium salt of cyanic acid is urea."

[2] It was because of this belief that Descartes thought he could find a way to hold back aging.

ity and organization that can be obtained from molecules, at which point the vitalists drop back to their final defensive line and desperately shout, "Yes, but what about thought? That separates us from stones, and there is no bridging that gap with chemistry and physics." I'm going to sidestep that one.

All of which leaves us without a definition of life and fails to answer the question: At what level of complexity can a system said to be living? The mistaken chemical differentiation between living and dead molecules has been removed, but at higher levels of organization there is clearly a difference between a spanner and a Spaniard.[3] What if we move one level up from molecules and instead consider viruses?

A drawing of the T4 virus is shown in Figure 26.1. It looks more like a space capsule than a pussycat. It has six spidery, jointed legs. Its head, which contains the virus's DNA, carries a long, hollow shaft. T4 is a bacteriophage; it attacks bacteria, landing on them and standing on its legs. Next it squats down and drives the end of the shaft through the bacterial cell wall. It injects its DNA into the bacteria. As in science fiction horror movies, this DNA takes over the cell, directing the production of alien RNA molecules and proteins from the molecules in the cell. With its DNA and the new RNA, the cell synthesizes enzymes that attack and destroy the bacterial DNA. The bacteria have now been converted into factories for the production of viruses. Is a virus alive? It can certainly reproduce, in its own perverted way. Is this the level at which life begins? Are we psychologically resistant to the idea that a mechanical-looking object is alive?

Let us resolve not to be worried by vague borders, and adopt the attitude of those working in the field of fuzzy logic: there are questions to which the answer is neither yes nor no. We will take a simple approach, asking what commonly observed characteristics differentiate undeniably living organisms from the nonliving world.

The Signs of Life

If we take a holistic approach, there are any number of properties of the whole organism that can be taken to be indicative of life, but three seem to be absolutely universal:

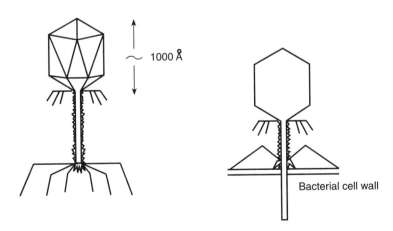

~ 1000 Å

Bacterial cell wall

Figure 26.1. A typical virus.

---

[3]With acknowledgments to John Lennon's *A Spaniard in the Works*.

- The need to take in a source of energy to maintain the organism's integrity
- The ability to reproduce
- The ability to respond to stimuli

One can quibble about all three. A starving man can survive for some time with no intake of energy. However, he is doing it at the expense of his internal sources of energy, molecularly autocannibalizing himself. A castrated man has no ability to reproduce, but he belongs to a living species. And a stone (certainly a pet stone) can respond to a stimulus—kick it and it moves, heat it and it cracks. However, it is difficult to think of an unquestionably nonliving object that displays *all* the preceding characteristics. The need for energy is perhaps more fundamental than the other criteria, for without energy intake the organism will, after a short time, not respond to stimuli or reproduce.

The Source of Energy

Without food all organisms die and disintegrate, breaking down to simpler molecules, most of which finally combine with the oxygen in the air. Plants take in energy directly from the Sun and also absorb air, water, and salts. They use the sunlight to synthesize sugars and other molecules. One of the first steps in this process is the breakup of a water molecule, giving a hydrogen atom that attacks carbon dioxide and thus starts the long chain of chemical reactions leading to glucose. Deprive plants of light for a long period, and their delicate and complex structure disintegrates. Animals cannot directly utilize the Sun's energy, and many animals rely on meat or fish as their primary source of energy, but ultimately the source of all flesh is plant material, so that animals indirectly use the Sun's energy. The average person in Western society may during his life eat 8 complete cows and around about 35 sheep and 750 chickens, but these animals live by eating plant material.

Because we ultimately depend on plant life for all our food, we also depend entirely on the Sun for our energy intake. *Photosynthesis maintains all life on this planet*, except for a few idiosyncratic microorganisms with exotic energy sources. And when you burn fossil fuels, such as coal or oil, you are releasing the imprisoned prehistoric heat of the Sun.

Life Is Cellular

At what level of organization, between quark and queen, does life appear? Are there any living molecules? We saw in Chapter 22 that RNA can reproduce itself in a test tube, but it needs a living creature (man) to put the right molecules into the system at the right time. This is not characteristic of the reproduction of animals and plants. And a molecule of RNA does not need energy to maintain its stability. It would seem that to ascend to the level of life, we need more than a molecule, however wondrous; we need a collection of different molecules, and it has become clear over the past century that we need a highly organized system of molecules. How highly?

The least complex living entities we know, down there among the amoeba, contain huge numbers of molecules coexisting in a flexible but mechanically stable package—*the cell*. Cells come in a variety of shapes, a few with well-defined geometries, like the red blood cells, but mostly roughly spherical or ellipsoidal. Some highly specialized cells, like nerve cells or sperm, have one or more thin, fiberlike extensions. All living systems are composed of one or more cells. A bacterium is single-celled; a human being contains something like 10 trillion cells. Repeating an earlier conclusion: the fact that the contents of the cell include very large (highly improbable) molecules, and that the cell is a highly ordered (highly improbable) structure, implies that living matter is in a state of comparatively *low entropy* as compared with the disorganized mess of small molecules into which it disintegrates when it dies.

Summarizing the facts, we take as our definition of a living form:

A low-entropy, dissipative system of molecules, organized into cells, responding to stimuli and capable of reproduction.

This reads like an interdepartmental memo, not a description of a living organism, so let's put some flesh on it. Living forms, and their cells, from the simplest to the most complex, share a number of characteristics, among which are the following:

- Cells arise by the division of existing cells.
- Cells differ widely in detailed appearance, depending on their function, but all have a membrane around them and a variety of internal "organelles" that are usually found to have identical or similar functions to those found in most other organisms.
- All living cells contain water as an essential component.
- Living material contains twenty or so of the chemical elements, of which carbon, hydrogen, oxygen, nitrogen, phosphorus, and sulfur are numerically the most predominant. The first four in the list make up about 99% of all living matter.
- All living material, from the amoeba upward, is composed primarily of about twenty amino acids, five heterocyclic nitrogenous bases (Figure 24.1a), a selection of lipids, and some sugars. Amino acids are the basic units from which proteins are constructed. The nitrogenous bases are essential components of the nucleic acids, DNA and RNA. Lipids include what we commonly call fats, and also the phospholipids that are basic components of all cell membranes. Simple sugars, like glucose, are the basic units of carbohydrate storage supermolecules like starch and glycogen.
- The immediate source of energy utilized by the cell is adenosine triphosphate (ATP). This molecule appears to be ubiquitous, having been identified in every kind of living organism.

What Is a Cell?

A cell is an organized, dissipative system of molecules, often capable of reproduction. It is possible to take small bits of tissue from plants and animals, and even single cells, and maintain them in a suitable culture medium, during which time they maintain their structure, respond to stimuli, absorb sources of energy and produce energy, and in many cases divide. Our definition of life says that they are alive. When they lose their structural integrity and stop importing and exporting energy, we say that they are dead. In practical terms it looks as though *the unit of life is the cell.*

The Cell as a Black Box

There is a continuous flow of matter into and out of living cells. Cells die without a supply of matter of the right kind, although some spores can survive for years in a state of dehydration and then come to life when water is supplied to them. Such systems are analogous to food in a deep-freeze—all chemical processes have been halted. Nevertheless, the *living* cell needs food, and one way of looking at the cell is as a chemical converter, a unit that transforms incoming molecules into material products. Many of the products are needed to replace worn-out structural components, but in multicellular organisms some products, such as adrenaline or the sex hormones, are produced in specialized cells and exported to other cells. The waste products of the cell's chemical activity include carbon dioxide, water, and a variety of other molecules that we excrete. The normal intake of the body cells includes oxygen, amino acids, glucose, lipids, vitamins, certain metal ions, and traces of other substances. Some of these are not essential, in that the body can manufacture them from other molecules. Thus the amino acid glycine can be manufactured from other amino acids, but tryptophan can't.

Cells are organized. They contain distinct structures, but the molecules of which these are composed are continually breaking down. The structures are maintained by the synthesis of new molecules. This is what Thomas Huxley meant when he said that "living protoplasm . . . is always dying, and, strange as the paradox may sound, could not live unless it died." *Synthesis is part of life's struggle to maintain itself.* The red blood cell has no nucleus and no DNA, so that it cannot synthesize proteins. In consequence, its lifetime is quite short, typically about 120 days. Moreover, in the absence of DNA it cannot reproduce, which explains why the body manufactures about 100 billion red blood cells every day to maintain the total number that flow through its 60,000 or so miles of blood vessels.

Cells vary in their size, shape, contents, and function. Muscle cells do not look like, or function like, brain cells. Most cells are too small to be seen with the naked eye. Conventional wisdom proclaims that the reason for this is that materials enter the cell at its border and make their way across the cell by diffusion, like the petrol vapor in Chapter 18. This is a very slow process on the macroscopic scale, so that a cell large enough to be visible to the unaided eye would be too large for material to find its way across the cell in a reasonable time. In some unicellular animals there appear to be ways of pushing material around the cell so that there is a little less limitation on their size. The largest cells are, in general, eggs, but they contain a built-in supply of food. Conventional wisdom about the size of bacteria was challenged by the discovery, in 1993, of *Epulopiscium fishelsoni*, a bacterium living in the intestine of the brown sturgeon of the Red Sea. This microscopic monster, shaped like a baguette, is a twenty-fifth of an inch long, a million times more massive than the best-known bacterium, *E. coli*, and visible to the naked eye.

Muscle cells are long and thin, as suits their function, which is to contract in a certain direction. Red blood cells have a large ratio of surface to volume, which is also consistent with their function, which is to absorb oxygen rapidly in the capillaries of the lungs and give it up rapidly within the rest of the body. Some cells have small threadlike projections that can be waved so as to propel the cell. Plant cells often have specialized organelles containing pigments that pick up the radiant energy of the Sun. The dominant example is the chloroplast, which contains chlorophyll. The catalogue of cell types is long, but in spite of the great diversity of structures and functions, certain features are common to the structure of almost all known cells.

Cell Components
The Cell Membrane

In 1855, Karl Nägeli showed that pigments penetrated damaged cells far faster than undamaged cells. He concluded that there was what he called a plasma membrane around the cell, a suggestion confirmed in 1897 by Wilhelm Pfeffer. This membrane needs to be stable enough to prevent the cell contents from escaping easily, yet permeable enough to allow food materials, oxygen, and other substances to come in, and the products of the cell to come out. Thus the cells of the pancreas need to ingest starting materials in order to synthesize insulin, and the insulin must be able to escape to perform its function in the body.

Cell membranes are highly complex structures. Different types of cells have different types of membranes; a nerve cell has a different membrane from a liver cell, because the cells perform different functions and part of their functioning depends on the structure of their membranes. All cell membranes are made up of several types of molecule, but nearly all membranes have something in common: they are based on a family of long, flexible molecules with heads containing a phosphorus atom and usually carrying an electric charge. It is the *phospholipid* molecules that give the membrane its mechanical stability and its flexibility. These molecules cling to each other, strongly enough to maintain the integrity of the membrane but not strongly enough to prevent distortion of the membrane, such as you will see if you examine the shape of an amoeba under the microscope. Many molecules that cannot penetrate the phospholipid membrane get in and out of the cell through specialized channels or are ferried across by tailor-made "boats." Both channels and boats are inherent components of the membrane, and the nature of their structure and functioning is an area of very active research.

Nerve cells have long extensions along which electrical impulses pass. These impulses are generated by a complex mechanism that involves the passage of sodium and potassium ions from one side of the membrane to the other. Specialized channels for this "ion transport" have been identified. They come with "gates," which are able to close under certain circumstances.

The membranes of all cells contain protein molecules, often with attached sugar-type molecules. One of the commonest roles of these components of the membrane is to act as *receptors* for specific molecules that arrive from other parts of the organism, or are parts of the external surface of invaders such as bacteria. Receptors have shapes that are complementary (lock and key) to one specific molecule or a small group of molecules of related structure. These receptors are essential to the functioning of most cells, particularly because one of the ways in which a multicellular organism controls its overall functioning is to send messages from one part of the organism to another. Some of these messages are electrical, via the nervous system, but very many are chemical. Substances are produced in one part of the body and travel to other locations, where they trigger specific responses. How does the messenger know which cells to go to? Why don't all the cells on its journey allow it to enter? The answer is the presence of specific receptors on the target cells. Only one lock suits the key. This theme will recur frequently in this chapter.

The Nucleus: The High Command

The nucleus is a distinct entity, often roughly spherical or ellipsoidal, having its own membrane and containing the cell's DNA. In some primitive cells there is no nucleus and the genetic material floats free in the cell. Cells with nuclei are termed *eukaryotic* cells, in distinction to *prokaryotic* cells. Eukaryotes encompass the whole of the living world except bacteria, blue-green algae, and a few other simple organisms. The first signs that we have of eukaryotic cells are from about 1.2 to 1.4 billion years ago, compared with the estimated emergence of life about 3.5 billion years ago.

The Mitochondria: The Oil Well

The cell synthesizes complex molecules like proteins from relatively small molecules like amino acids. In Chapter 24 we saw how DNA and RNA provide the surfaces on which proteins are assembled. However a production line is not enough; there has to be an input of energy.[4] In all known cells the immediately mobilizable source of energy used by the cell is adenosine triphosphate (ATP). This molecule is made in specialized organelles, small structures within the cell known as *mitochondria*.

It is the breakdown of the ATP molecule (by the loss of a phosphate group) that provides the energy for a wide variety of the cell's activities. The number of mitochondria in a cell is thus related to its energy requirements. Muscle cells need large amounts of energy when they contract, and they get this from the cellular stock of ATP. Once this is broken down, it has to be replaced, and this can be done more speedily the greater number of mitochondria that are present. In the muscles that drive the wings of insects or hummingbirds, there can be up to 1 million mitochondria in each cell.

The mechanisms by which ATP is synthesized and the way in which its energy is utilized by the cell are well understood but are too specialized to go into here. An average cell, on an average day, synthesizes about 1 million molecules of ATP per second, to replace those that break down to supply energy.

Problems
Development

All of us start our lives as single cells. Most adult organisms contain millions, billions, or trillions of cells, but a multicelled organism is not just a large collection of single cells, living independent lives. A healthy living organism is akin to a well-coordinated army, not to a disorganized rabble. This fact presents us with a series of related problems that are at the cutting edge of biological research. Two are of central importance: How does a single cell know how to develop into a coordinated mass of cells having different structures and functions? And how is a multicellular organism coordinated?

---

[4]For the chemist, strictly speaking, I should be talking about *free energy* here, but this would unnecessarily complicate the general lines of the argument.

The fertilized human egg, during its first few divisions, appears to be a cluster of *identical* cells, but soon the roughly spherical shape distorts and differentiation sets in. Primitive structures can be seen that develop into the organs of the baby-to-be. These structures contain cells that are typical of that structure. How do a small cluster of *identical* cells "know" how to develop along different structural and functional paths?

The developing human cell contains, in its DNA, the potential to become any of the over 200 types of cell in the formed adult. Every cell starts with identical DNA, yet at some stage different genes are turned on or off in different cells. One of our great areas of ignorance is the nature of this differentiation—how it is triggered; how a set of cells becomes a distinct, functioning organ; how a set of organs becomes a coordinated, functioning organism.

We know that in certain cases, at an early stage of development, if we remove a part of the embryo that is clearly destined to become a certain organ and graft it onto another location on the embryo, it will "choose" to develop like the cells in its new surroundings. Is the effect of the new environment due to the passage of chemicals passed into the intruder cells from their new environment? Presumably the role of such messengers would be to turn off certain genes and turn on others in the transplanted cells. Are receptors on the cell surfaces involved? And why doesn't the intruding tissue change the direction of the cells in its new environment? Is it outvoted?

We have no coherent understanding of the process of development, but it looks as though a modest breakthrough has been made in identifying the nature of the genetic control of development. A group of genes, the so-called Hox genes, have been found, which can apparently "switch on" cells to their destiny. All the cells in the embryo have the same set of Hox genes, but, by an as yet unknown process, only certain Hox genes are active in a given location. Thus the cells at one end of the embryo will develop into the head of the organism because the relevant Hox genes are the only ones working. In the cells farther back, another group of Hox genes turns on and directs the development of the cells into the organs appropriate to that location. At the time of writing, the mechanism by which different groups of genes are turned on is not known, although there has been some speculation that varying concentrations of certain molecules in different parts of the embryo are responsible. Thus in the fruit fly (Drosophila) egg, it has been shown that a chemical, acting at one end of the dividing egg, induces the cells at that end to switch on the genes that lead to development of the head. The story is complicated, but it is slowly unfolding.

Control

The variety of chemical reactions occurring in cells is enormous. All these reactions have to be coordinated *in* the cell and *between* cells, otherwise the organism would self-destruct. *Control over the organism's internal environment is part of the struggle to maintain life.*

We have more or less conscious control over some of our bodily movements, through the passage of electrical impulses from the brain to our muscles, although some muscles, such as those of the heart, are outside our conscious control. We have no conscious control over the reactions in our cells. The extraordinary thing is that within each cell, and within the complete organism, the multitude of reactions and messages is evidently under overall control. Thus under normal circumstances the amount of each of the thousands of molecules in the cell remains approximately constant. If an excess of, say, an amino acid is added to the cell, the speed of certain reactions increases and others slow down, the net result being that in a short time the cell restores itself automatically to its normal state, the concentration of the amino acid returning to its normal level. Without this and other intra- and intercellular control mechanisms, the vast collection of cells that constitute the animal body would be wildly unstable.

The problem of control is intimately connected with the mechanisms of communication within the body, which in turn depend heavily on the *shape* of certain molecules. Enzymes provide a good example.

Turning Enzymes On and Off

Enzymes are the chemical catalysts of the living cell. Without them the reactions within the cell could take place but would be very much slower, sometimes by a factor of a million or more. Life could not continue under these circumstances. But it is not enough that hundreds of these catalysts are found within the cell; they must be controlled. Thus there are enzymes that are responsible for the contraction of muscle cell; how do they know when to act and when not? Why don't they act all the time if they are sitting there?

The answers to these and thousands of other control problems lie in the beautiful range of ways in which enzymes respond to other molecules. Enzymes are all very large protein molecules, but only small areas on their surfaces are responsible for catalyzing the reaction between other molecules. Thus there are enzymes, the *lipases,* that speed up the rate at which water molecules attack and break open the fats that we eat. On the surface of these enzymes there are specialized regions that are so shaped as to accept lipid molecules. These active sites are the wrong shape to fit almost any other type of molecule with which they come in contact. When a lipid molecule sits on such a site, it is subtly affected by its surroundings so that one of its chemical bonds is weakened. This is the origin of the enhanced rate of reaction with water, a reaction that could go at a barely perceptible rate if the enzyme were not present. A molecule whose reactions are speeded up by sitting on the active site of an enzyme is termed the enzyme's *substrate.* An enzyme may have only one naturally occurring molecule as a substrate or it may act on a number of structurally related molecules.

We can often find, or design, a small molecule that will fit the active site of an enzyme. Such a molecule interferes with the action of the enzyme; it blocks the parking space, or, if you like, it puts a strange key in the lock. Such *inhibitors* might come from inside the cell or from another place in the body, or they might be ingested by the body. The strength with which inhibitors bind to an active site varies. A catastrophic extreme is when an inhibitor sits on an active site and refuses to leave. The molecules of certain nerve gases bind fiercely to the active site of enzymes connected with the transmission of nervous impulses, thus causing paralysis and death.

Inhibition provides one of the body's standard methods of controlling the concentrations of molecules in the cell. Often this control involves a *feedback* mechanism. A simple example occurs in the case of the synthesis of certain molecules. Suppose that a certain amino acid is synthesized by a chain of reactions, each catalyzed by an enzyme. The cell is supposed to have an ample stock of starting material and the synthesis proceeds, but at a certain stage it is advisable to tell the system to stop—to turn off the production line. The way that this is done could be that one of the enzymes in the production line has an active site that is a good fit to the *final product*, the amino acid. As the concentration of the amino acid rises, some of it binds to this enzyme, blocking its active site. When all the enzyme molecules are blocked by the rising tide of amino acid, the enzyme ceases to operate. Production stops. If the amino acid concentration subsequently drops, for any reason whatsoever, some of the amino acid molecules that are acting as inhibitors leave the enzyme, which can then start operating again, setting the production line rolling once more. Thus is the level of the amino acid concentration maintained by a feedback mechanism. The process depends on the fact that the amino acid is weakly bound to the enzyme. If it were too strongly bound, it would sit more or less permanently on the active site of the enzyme and permanently destroy the production line.[5]

Enzymes can be switched off, but they can also be switched on. There are many ways of doing this, most of which depend on the fact that an enzyme is a large, flexible molecule that can change its shape. Here is an example.

The 1966 World Cup

The contraction of muscle cells is a complex process that depends heavily on enzymes that act on certain components of the cell. We would have permanently contracted muscles if these enzymes acted all the time, and so they normally have shapes (conformations) that are inactive. When I was watching the 1966 soccer World Cup, my brain sent electrical impulses to the outer surface of my muscle cells. Each impulse resulted in the opening of a channel in the membrane of my muscle cells, a channel that allows calcium ions to pass. Now normally the concentration of calcium inside the cell is about 10,000 times less than it is outside the cell. When the channel opened in the membrane of my muscle cells, the excess calcium started to flow in. Inside the cell, it sat on two molecules called troponin and calmodulin. These are *effectors,* molecules capable of switching on one specific enzyme or a small group of enzymes, but they themselves are impotent without the calcium ion sitting on them. When calcium ion sat on, say, troponin in my muscle cells, it changed the molecule's conformation so that it fitted onto a site situated on one of the enzymes that is responsible for the contraction of my muscle cells. The effector changed the conformation of the enzyme so that it became active. My muscle cells contracted, and I clenched my fists. I also shouted, "Goal!!" when Geoff Hurst scored his first goal. When the electrical impulse stopped, the channel in the cell membrane closed and the excess calcium in the cell was pumped out. The effectors lost their calcium, and with it their shape, and fell off the enzymes. The active sites of the enzymes resumed their distorted shapes, the enzymes no longer affected the cell, and my fists relaxed—for a moment.

There are many variations on this game of enzymic control, and the cell has some pretty fancy ways of controlling the activity of its enzymes and thereby the concentration of the products that it produces or the material that it absorbs. All is

---

[5]Notice also that the feedback mechanism creates a *nonlinear process*; the rate of amino acid production is not proportional to the amount of starting material. Many of the processes within the cell are nonlinear, and this fact suggests that it may be profitable to examine them by the methods of chaos theory.

entirely explicable in terms of well-known chemical principles. This is true of every bodily function discovered in living organisms.

The Communications Network

The best-known communications system in the body is the nervous system, a branching system of fibers that spreads out from the brain and the spinal chord. The number of nerve cells, *neurons*, in the brain is about 100 billion, and each is linked to at least one other neuron and sometimes to as many as several hundred. Electrical impulses flow along the membranes that cover the fibrous extensions, the *axons*, that radiate from the central body of the neuron. At first sight the system looks as if it depends entirely on the passage of electrical currents. However, most axons are not very long, and what might appear to be a continuous nerve fiber consists of chains of neurons with the tips of their axons touching the surface of the next neuron in the chain—a bit like a daisy chain. Currents are incapable of passing between adjoining axons.[6] The tiny gap between the nearly touching neurons is termed the *synapse*, and it is a very poor conductor of electricity. The mechanism by which the current is maintained from axon to axon is chemical. The arrival of an electrical impulse at the tip of an axon stimulates the release of a chemical messenger, a *neurotransmitter*, which is a molecule that diffuses across the synaptic gap. When it reaches the other side it is absorbed on special receptors, situated on the outer surface of the second neuron's membrane, and so shaped as to fit the neurotransmitter molecule. (For the thousandth time we see the significance of molecular shape.) Once absorbed, the neurotransmitter results in the formation of a new electrical pulse, which passes along the axon of the second neuron, and so on. The neurotransmitter does not survive very long after it has been released. If it did, it would carry on stimulating the second neuron, and even in the absence of a stimulus from the first neuron, currents would never cease, which could mean that long after someone had stuck a pin in your finger, you would keep feeling the point going in. Thus, a common neurotransmitter, acetylcholine, is destroyed, by a purpose-built enzyme, within two thousandths of a second after secretion. There are a wide variety of neurotransmitters, the best-known being dopamine, which acts in the synapses of one part of the brain, and acetylcholine, which acts in some of the synapses of the *autonomic nervous system*, that part of the nervous system that controls our internal workings and over which we have no conscious control.

The venom of some snakes contains substances that fit into those receptors on muscle cells which adjoin the tips of nerve axons. A bite from such a snake can result in paralysis, which can be reversed by a similar substance to that in the venom, one which is shaped so that it can compete for the same receptor but is not so strongly bound that it cannot fall off afterward.

Molecular Messengers

The nervous system is only one of the two main types of internal communication system used by the body. We also use a very wide range of chemicals that carry messages to specific organs or cells. Neurotransmitters do this over the extremely small distances between adjacent axon tips.

---

[6]The truth is that these are not conventional currents, that is, movements of charge along the nerve. This is not important for our present tale.

The human body contains a number of *endocrine glands,* and other organs, which can secrete *hormones.* Hormones are a major means of communication between the organs of the body. Among the better-known endocrine glands are the pancreas, the ovary, the testes, the digestive system, the thyroid gland, and the pituitary gland. The pancreas, for example, produces insulin and glucagon, which, as already mentioned, control glucose levels in the blood.

Hormones are released in response to specific or general stimuli. Thus the secretion of the hormone prolactin, which stimulates milk production, is the result of a specific message to the anterior pituitary gland. On the other hand, epinephrine (better known as adrenaline), which is produced by the adrenal medulla, raises both the heartbeat and the concentration of glucose in the blood, and can be secreted in response to a variety of stimuli, including fear, the new boy on the block, or listening to the muscle-tensing, driving syncopation in Beethoven's Seventh. All hormones act initially by being recognized and adsorbed by specialized receptor sites. The biologists talk of target cells. Thus, a cell that has no receptors for thyrotropic hormone will be completely blind to this molecule, which is produced in the pituitary and acts on the cells of the thyroid gland.

Sometimes the receptor is within the cell, not on its surface. The steroid-type hormones, those related to cholesterol, can usually move straight through the outer membrane of the cell and find their receptor site either in or near the nucleus of the cell, influencing the working of the cell's DNA. Other hormones, for example, proteinlike molecules, cannot penetrate the membrane of the cell, and their receptors are on the surface. If such hormones are to affect the internal working of the cell, they must presumably do more than sit on the doorstep. One way in which externally absorbed substances can act is exemplified by the neurotransmitters, which, when they are adsorbed by their receptors, induce neighboring ion channels to open so that ions can flow across the cell membrane. It is this flow of ions that produces electric currents in nerve cell membranes. Another mechanism for externally absorbed molecules to influence the cell involves the presence of a protein molecule, sometimes called a G-protein, which sits in the membrane, next to the receptor. Adsorption of the hormone by the receptor changes the latter's conformation, which in turn affects the G-protein. This protein is situated so that part of its surface is in contact with the contents of the cell, and changes in its conformation can trigger enzymes that are situated next to it, in the interior of the membrane. Thus is a series of reactions initiated, which one hopes will be of benefit to the organism.

We are only just beginning to understand the remarkable control mechanisms of the cell and of the whole organism. The focus repeatedly returns to the role of shape.

Defense Matters

It is not enough for the organism to control its internal environment when times are good. It has to be able to defend itself when times are tough.

Living organisms have a whole range of weapons with which to fight invaders. I concentrate on human defense. There are some simple protective systems, capable of repelling microorganisms before they reach our internal systems, although there are bacteria living in our digestive systems, with which we are happy to coexist. The skin provides a mechanical barrier to microorganisms, which are also discouraged by a variety of secretions, but these are certainly not enough. Evolution has provided us with an immune system, a collective term for a highly complex, amazingly versatile collection of cells and proteins that do battle with those invaders, dead or alive, that penetrate our outer defenses. *All the mechanisms of the immune system depend on shape recognition.*

Phagocytes

The basis of the immune system is a set of cells, all of which are manufactured in the bone marrow. Some of these cells, the *phagocytes*, literally eat invading bacteria, engulfing them. Phagocytes, which account for over half the blood's white cells, circulate throughout the body's tissues. Over 100 billion are produced daily by the blood marrow, so that in cases where the bone marrow is dysfunctional a vital part of the body's defenses is absent or weakened.

How do phagocytes know which cells to attack? Why don't they eat red blood cells? By now you know that shape is the answer. Bacterial cell walls contain a variety of "markers," molecules that fit into receptors on the cell surface of the phagocyte and trigger its activities. The fact that animal phagocytes carry such receptors is clearly the result of a long evolutionary process.

The phagocytes are backed up by a group of protein molecules in the blood, collectively known as *complement.* They have several modes of action, one of which is to attach themselves to alien microorganisms, thus providing markers for phagocytes to recognize. Complement molecules can also attach themselves to large invaders, such as parasitic worms. These organisms are much too large to be engulfed by phagocytes, but once the complement has labeled them with a marker, they are recognized by a group of specialized cells, the *mast cells, basophils*, and *eosinophils.* The last named can now bind to the invader and destroy it by a variety of mechanisms, including the use of an enzyme to chew up the invader's body wall. Mast cells and basophils do not challenge invaders head-on. One of the ways in which they act is by releasing molecules that help the body's tissues to fight invasion. A well-known example of such a molecule is histamine, which widens the capillaries in the neighborhood of the invasion bridgehead, resulting in what is known as *inflammation*, a phenomenon that provides a better blood supply to the invasion point.

Lymphocytes

Apart from the system of blood vessels, vertebrates have a *lymphatic system,* a branched network of tubes that carries a practically colorless liquid, the lymph. This liquid, which is found in the tissues of the body, has its origin in the serum of the blood that filters through the walls of the capillaries, and out of the blood vessels. The lymphatic system provides a means of channeling the liquid back to the bloodstream. Part of the system is a series of *lymph nodes,* clumps of tissue through which the lymph flows and which, when swollen, you can often detect under the skin. The lymph nodes are one of the main locations for *lymphocytes*, a collective noun for a group of three kinds of cells, the B cells, the T cells, and a third group that contains so-called natural killer (NK) cells. The important thing about B cells is that they produce *antibodies.*

The antibody defense mechanism shows the principle of shape recognition at its most powerful. The B cell produces receptor molecules that sit on its surface. Now these receptors, known as *immunoglobulins,* are essentially chains of amino acids (polypeptides), and their synthesis is controlled by the genes of the cell. Evolution has ensured that B cells have a variety of genetic makeups, which means that there is a tremendous range of receptors on the surface of the body's B cells. Receptors for what? For almost anything! In contrast, phagocytes are limited; they have a number of receptors that fit markers on most bacteria, but they cannot cope with invaders which do not have markers that suit the phagocytes' range of receptors. The B cells have so many types of receptor that one of them is almost certain to fit onto some region of the invader, be it molecule or virus. An invader that binds to a receptor is known as an *antigen*—think of this as short for *anti*body *gen*erator. Once an antigen is recognized by a B cell, a flurry of activity results (Figure 26.2). The cell divides into *plasma cells*, which are just production lines for more of the successful receptors, the immunoglobulins, on the original B cell. The plasma cells produce very large numbers of the receptors, which are released into the lymph and at this stage are renamed. They are the famous antibodies, which by hanging on to the invaders either render them inactive or provide markers for the receptors of complement (see above) which allows phagocytes to recognize the joint complement-microorganism complex.

The tremendous versatility of the B cells is both an advantage and a disadvantage. In the Paris flea market there used to be a stall that had boxes full of keys. If you had lost the key to a lock, you could bring the lock with you and spend an hour or two rummaging through the boxes hoping to find a replacement. The chances were that you would find something because there was such an enormous choice. This is what happens with an invading antigen—molecule or virus—but the problem is that there is usually only a very small number of B cells that carry a suitable antibody. The other keys are useless. It is true that as soon as the antigen binds to the correct B cells they start turning out antibodies, but they cannot always do this fast enough to cope with a massive invasion. The versatility of the B cells comes at the expense of their overall effectiveness.

One way that man has found to help the B cells is by vaccination. This depends on the fact that even when a harmful microorganism has been "killed" by, say, heat treatment, its surface may still carry unaffected antigens. These can induce the formation of large amounts of antibody when they meet the right B cell. The antibodies remain in the body and, until their numbers decrease drastically by the normal wear and tear of life, they are ready to immediately attack the active form of the microorganism if it appears. There are people who regard this procedure as an offense to God.

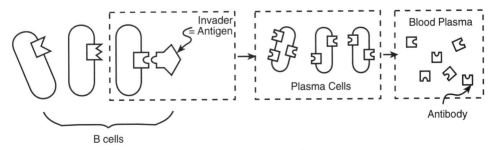

Figure 26.2. B cells react with a suitable antigen to produce plasma cells, which act as factories for the production of antibodies.

HIV

There are other types of lymphocytes, some of them involved in the division of the B cell into plasma cells once it recognizes a specific invader (Figure 26.3). Tc cells attack and destroy body cells that have been invaded by viruses, thus preventing the virus from reproducing itself. Both B cells and Tc cells, which are in the forefront of the body's defenses, are stimulated by other cells, the Th (T helper) cells, and inhibited by Ts cells. The HIV virus has found the Achilles' heel of the immune system. It attacks the Th cells and gradually stops them from helping the B and Tc cells, preventing the former from producing antibodies and the latter from attacking virus-infected cells.

I have only touched on some of the aspects of the body's defenses against invasion. The immune system is responsible for the rejection of molecules or cells that differ from those it meets in the healthy body. The downside of this ability is manifested by the phenomenon of rejection, which is a major problem in organ transplants and can be countered to some extent by using drugs, the best known of which is cyclosporin, to suppress the immune system.

The markers which each of us has on our own body cells are, of course, genetically determined. Six genes control the synthesis of markers, and they lie close to each other on chromosome 6. These genes exist in over a hundred mutations, and different people have different combinations of these variants. The chance of two strangers having the same set of mutants, and therefore the same set of markers on their cells, is tiny. Thus the immune system of one person will "see" foreign, unfamiliar markers on the cells of a transplanted organ and will produce antibodies to these cells, resulting in their destruction or interfering with their functioning. That's rejection. However, because the genes are close to each other on the chromosome, they are generally inherited as a complete group, rarely being broken up by crossing-over (see Chapter 21) in passing from parents to offspring. This means that the inheritance of these genes is fairly straightforward and a child has a roughly one-in-four chance of having the same six genes as its brother or sister. This is why

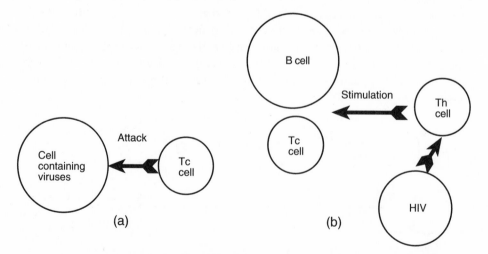

Figure 26.3. (a) Tc cells attack normal cells that have been invaded by viruses. (b) Both B cells and Tc cells are stimulated by the Th cells, but the HIV virus attacks Th cells.

transplants between matching siblings are often more successful than between strangers, or between siblings who have a different set of genes.

Life Is Inseparable from Defense

The elucidation of the nature of the defense mechanisms of the living cell and of the whole organism is one of the keys to combating some of the cruelest diseases plaguing man—and beast. I have limited the discussion here to a cursory glance at some central subjects. In discussing life I have avoided whole areas, preferring to illustrate some of the general characteristics of the cell and of living material. In particular I wished to convey the central importance of molecular shape in the normal functioning and the defense of the body.

Life or Death

Why do cells die? Why shouldn't a cell live forever?

One of the most surprising findings, which came early in the history of embryology, was that the development of the embryo is accompanied by the apparently natural death of many of its cells. One can see why this could happen in, for example, the development of the foot, where one mechanism for the formation of toes from a simple block of dividing cells would be for cells situated between the future toes to die, leaving "bays." Subsequently, it was found that in many developing eggs, cell division was invariably accompanied by a selective pattern of cell death. Reproducible phenomena have a reason, and in cells it is natural to ask if cell death is gene-controlled. The idea arose that there might be a "death gene" in normal cells. Such a gene has been found in a wormlike creature, about 1 millimeter long, called *Caenorhabditus elegans*. This organism has a gene known as *ced 3*, which appears to dictate the suicide of the cell. In mutant organisms in which *ced 3 is* inactive, cells seem to be immortal. There are many questions to be asked, the first of which is: why do we need a death gene? The answer that springs to mind is that in the absence of such a gene, there would be nothing to stop cells from dividing and proliferating indefinitely, as they try to do in cancerous tumors. But if there is a gene that programs cell suicide, how do we stay alive?

A gene dubbed *ced-9* has been found in *C. elegans*, and if *ced-3* is a death gene, then *ced-9* can be termed a *life gene*. When we examine mutant cells in which *ced-9* is inactive, we find that they quickly die. At about the same time as this work was going on, it was found that there is a gene in humans, named *bcl-2*, that appears to do the same job as *ced-9*.

This research raises a multitude of important questions. Is there a genetic clock that determines our lifetimes? What is the role of these genes in cancer? In aging? There is a long way to go, but we have the first signs of a genetically controlled mechanism that decides whether cells should live or die. That this mechanism is far from simple is suggested by the finding that the longevity of cells is apparently dependent on whether or not a specific cell is adjoined by neighbors. In healthy pieces of cell tissue, the cells can be induced to commit suicide by simply separating them from one another. What kind of communication is there between cells that counteracts the death gene? We know that these two genes are not the end of the life-death story.

There is a gene, *p53*, which seems to be able to detect damage to the cell, specifically to its DNA. There are repair mechanisms within the cell that are often capable of repairing minor damage to DNA. It seems as if *p53* prevents cells with damaged DNA from dividing and producing more defective cells. It lifts this sanction once the cell is repaired. However, in over three-quarters of the cases of human cancer in which it has been investigated, the *p53* gene has been found to be mutated and therefore not operational. Defective cells can multiply. The implications of findings of this sort for the treatment of disease are revolutionary. Cancer cells have seemingly lost the knack of dying, and most conventional methods of therapy are based on destroying tumor cells. Maybe we should be attacking these cells at a subtler level, inducing them to commit suicide or preventing them from proliferating.

The mystery of life is also the mystery of death. (No, Goethe didn't say that.) Understanding the workings of the simplest living one-celled organism, let alone the coordinated activities of the billions or trillions of cells in vertebrates, is still one of the great scientific challenges. Living matter, and living organisms, make the most complex human-built systems look clumsy and primitive. But the computer boys are coming.

The Day of the Golem

Today, Capek's robots,[7] the Golem,[8] and Frankenstein's monster are conceivably attainable artifacts, at least in the dreams of scientists working in the area of artificial life. At the simplest level robots move around the laboratory floor, avoiding some objects, approaching others, hopefully manipulating objects that it "likes" and possibly assembling them. These tasks require sensing capabilities but also a "brain" designed to manipulate the incoming signals and generate an "intelligent" response to them. For nonliving systems only the computer can fulfil such a task at or above any but the most trivial level. Since computers are getting more powerful, and programming more subtle, the performance level of robots is rising; although they are hardly threatening at this stage, the future seems almost open-ended.

A somewhat different activity in the artificial life field is typified by computer programs that give ever-changing and apparently random patterns on the computer screen. The hope is that occasionally the pattern will clearly become nonrandom, performing what appear to be simple tricks, like constructing a series of regular steps crossing the screen or endlessly reproducing some feature of the pattern. With time, and increasing computer power, the programs have become more complex and patterns have appeared that are complex enough to suggest (with a bit of goodwill) that they could only be the product of a living brain. The researchers feel justified in calling their subject artificial life. A closely related subject is artificial intelligence, which has tended to concentrate on the construction of computer programs that are dedicated to the solution of a narrow range of problems.

One has to be very careful with language here. There is no more in a program than has been put into it, and no more can be got out of it. The fact that unexpected order emerges from a program is merely an indication that the human brain is not always capable of predicting all the consequences of a long series of operations.

---

[7]The play *R.U.R.* (1920) described robots constructed from man-made protoplasm.
[8]The Golem, who was the subject of Jewish folklore, was a computer-like perfect servant. He was given written instructions, usually put into his mouth. His main fault was that he took his orders completely literally.

The achievements of these programs are indeed sometimes surprising, but hardly staggering. And they *are* all products of the human brain. There is no doubt that the programs will produce more and more surprising behavior, and if this book were being written in 500 years' time, its contents might be chosen by OMNIWRITE, who (who?) would also edit it, while a titanium alloy AUTHORSOOTHESS would bring me a whisky and stroke my noble brow.

It would be foolish to attempt to predict how far these experiments will take us. Some practitioners see the present as the era in which artificial life originated in a warm pool, but this time a pool of program writers. They foresee two species on the Earth, *Homo sapiens* and *Homo even more sapiens*. There will be soft living matter and hard living matter. Whether we will get that far no one knows, but it would be rash to suppose that in 5000 years' time there will not be entities with a large range of capabilities, including possibly the ability to reproduce by building new entities and to improve their own programs using the techniques at present being investigated by the artificial intelligence experts. Is this what we really want to do? I admit that a large part of my objection to such a course is based on irrational dislike of the idea of humanoids, and on a rational fear that such humanoids would reach a stage at which they were a threat to the existence of soft life. It would, to my taste, be a nightmare world in which there was no one left to feel a pain in her chest when she read George Seferis's lines:

> Man frays easily in wars;
> man is soft, a sheaf of grass,
> lips and fingers that hunger
> for a white breast
> eyes that half-close in the
> radiance of day . . .

> From *The Last Stop* (translated by
> Edmund Keeley and Philip Sherrard)

But perhaps, God help us, there will be soft humanoids.

A Designing Hand?

The sheer complexity of life, the astonishing way in which living organisms are adapted to their environment, the emergence of life itself, the fact that it is almost incredible that the universe should have developed so as to allow the formation of living matter—all this has led many thinkers to suppose that the emergence of life and of man and woman cannot have been accidental. This is an old argument, famously presented in 1802 by the theologian William Paley (1743–1805) in his tale of a man finding a watch. Such an intricate construction, reasoned Paley, could not conceivably have come about by chance; the watch must have been the result of a designing hand. This is true but irrelevant. Watches are indeed designed by an intelligent being, but man is not; he *evolved*, and the obvious counter to Paley is that machines cannot reproduce and evolve. Living forms can, and this is what they did. Evolution disposes of teleology. Paley is always quoted in this context, but his thesis was far more vigorously pressed by the eighth earl of Bridgewater, who in his will left £8000 to finance a project in which eight authors would write texts illustrating "the Power, Wisdom, and Goodness of God" as manifested in the design apparent in the animate and inanimate world. Eight volumes were duly published, the last in 1836.

There is perhaps more to the design hypothesis than Paley's naive analogy, because it has been revived in a rather different form by a number of modern physicists, in general those with an interest in cosmology, which is not surprising since an important part of the new evidence that they marshal in support of their thesis comes from a consideration of the development of the universe, in particular the synthesis of the chemical elements. We will glance at their ideas in Chapter 35, where we look at the creation of the universe.

A somewhat different area of controversy is the relevance of biology to morality, a question with very long roots indeed, one of which is Aristotle's belief that every organ in a living organism is created by divine intervention and is specifically designed to fulfil a definite function, acting in cooperation with every other organ. This is a teleological doctrine—the assumption that activities and structures are adapted to a predetermined end. Clearly, teleology has much in common with the design hypothesis. Both Aristotle and William Paley would say that the gods knew that they wanted you to see, so they designed eyes. Teleology can, and has been, extended to suppose that man is designed not only so as to function biologically in a certain way but also to function ethically in a manner related to his biological nature. This is different from the attempts to apply evolution to ethics that we ran across in Chapter 25. Here it is the design and existence of the single individual that are supposed to have ethical implications, quite independently of the existence of evolution. It is surprising how persistent this belief is.

It is possible that some readers will sense a connection between biology and ethics. It's a good topic for a warm summer evening, on a flower-covered balcony.

27

The Origin of Life? Take Your Choice

To believe that physical and chemical forces could by themselves bring
about an organism is not merely mistaken but, as already remarked, stupid.

—Schopenhauer, *Parerga und Paralipomena*

I do not know the origin of life. Those of us who hold, like I do, that life emerged
spontaneously from inanimate matter are, we must admit, at a distinctly embarrassing disadvantage: we have not yet come up with a convincing mechanism for
abiogenesis. In his presidential address to the 1871 meeting of the British Association for the Advancement of Science, Lord Kelvin stated, "Dead matter cannot become living without coming under the influence of matter previously alive. This
seems to be as sure a teaching of science as the law of gravitation." If he is right,
which I doubt, then life must have been present in the universe for all past time. Either that or we must turn to the finger of God in the Sistine Chapel, and indeed,
after reading this chapter you may well conclude that that is our only hope.

During the nineteenth century, abiogenesis was given a boost by the successful
synthesis of organic molecules from inorganic precursors, but the fact that we can
synthesize amino acids and nucleic acids from inorganic starting materials does
not explain how life started. We are intelligent beings who can purposefully bring
together chosen chemicals under carefully controlled conditions. This is very different from accounting for the *spontaneous* formation of living systems in an inanimate world empty of all intelligence. And we have come nowhere near creating
life in the laboratory.

There have always been those who have held that life is a property that cannot
possibly arise out of inanimate matter, not because they can't conceive of the chemical pathway but because it offends their view of the universe. This is the "Life-is-
something-special" school of thought, for whom the uniqueness of life is threatened by mean little scientists in scruffy lab coats trying to prove that a proto-Bach
originated in a mixture of gases that was struck by lightning:

> Science! . . .
> Why preyest thou thus upon the poet's heart.
> Vulture, whose wings are dull realities.
>
> Edgar Allen Poe, "To Science"

Those who prey upon the poet's heart know that, in tackling the problem of the
origin of life, there are two basic, and interdependent, questions that have to be answered:

1. Living systems are based on extremely large and complex carbon-containing,

molecules. As Sir James Jeans remarked, "Life exists in the universe only because the carbon atom possesses certain exceptional properties." In contrast, the nonliving world from which life is presumed to have arisen is composed of small, very simple molecules—except for silicates, which can form enormous chains of atoms but are not a basis for life. *How did the large, complex, carbon-containing molecules of life originate in the first place?*

2. Living systems are very highly organized. The living cell is a very spatially structured entity, operating through a sophisticated system of catalysis, transport, and feedback mechanisms. It is a highly improbable structure, or, in other words, it has low entropy. *How did the molecules of life get organized?*

One can believe that a complex system like the living cell is capable of manufacturing large, complex molecules from small, simple precursors, but the original manufacturing mechanism has to come from somewhere. Factories making CD players do not assemble themselves. Believers in abiogenesis are forced to accept that life is a self-organized invention.

To construct any convincing theory of abiogenesis, we must take into account the condition of the Earth about 4 billion years ago.

The Violent Childhood of the Earth

The Earth is about 4.5 billion years old. Fossil remains of microorganisms have been found dating back about 3.5 billion years. Water is essential for life, and the Earth was certainly cool enough for oceans to exist at that time, although they could have been warmer than they are now. We can get some hint of conditions on the young Earth by looking at the Moon. The Moon has been around for about the same time as the Earth. Because of the lack of water, or a significant atmosphere, its surface has not been shaped and smoothed in the way that the Earth's has. That is why the craters on the Moon's surface, created by meteorites, asteroids, and assorted interplanetary debris, have remained largely unchanged from the time of impact. The wealth of craters on the Moon gives us good reason to suppose that the Earth, too, suffered from regular and massive cosmic bombardment. After the emergence of life, gigantic collisions could have wiped it out, over most or all of the Earth. Thus it has been estimated that the arrival, probably in the Caribbean area, of an asteroid measuring about 6 to 7 miles across might have resulted in the destruction of the dinosaurs and the elimination of over half the living creatures on Earth about 65 million years ago. It is possible that there was a shower of objects around this time. Very large falling objects could have generated enough heat on landing to boil the oceans, and if an object fell on land, huge clouds of dust could obscure the Sun for long enough to threaten the continuance of plant life. There is evidence, from fossil remains, that there have been at least four periods in the last half billion years in which up to 80% of marine species have been wiped out by climatic changes, by changes in sea level, and by periods of enormous volcanic activity. In fact, about 250 million years ago it looks as though cataclysmic volcanic eruptions in Siberia eliminated all but about 5% of the living forms, plant and animal, on Earth. The belief is that Siberia was effectively drowned in lava and that monstrous dust clouds hid the Sun. The resulting changes in climate initiated an ice age and global downpours of acid rain.

I have set this violent scene at the outset to underline the forces ranged against the survival of life once it did emerge. How on Earth (sorry) did a fragile scattering of life maintain its hold, whatever its origin? Or was it wiped out several times? And did it arrive here several times? Over 20,000 meteorites a year arrive here, most of them not much larger than an egg. Do some carry living material?

The "They-Came-from-Outer-Space" Hypothesis

The supposition that life came from space answers the question Where did life on Earth come from? but not the question Where did life come from?

Comets and asteroids usually contain organic matter, and some scientists have said this is how life reached Earth. One problem is that the collision between a large object and the Earth would easily generate enough heat (microscopic kinetic energy) to incinerate any organic material in the projectile to carbon, carbon dioxide, oxides of nitrogen, and other small inorganic molecules.

A variation of the life-from-space theory envisages the arrival of complete organisms from space, which is 1950s B-film territory. A prominent advocate of this view was the English astronomer Fred Hoyle, who has suggested that many human ailments, including cancer and AIDS, originated in "space-bacteria." The book *Lifecloud* (1978), by Hoyle and his collaborator Chandra Wickramasinghe, was described by the biologist Lynn Margulis as "wanton, amusing, promiscuous fiction." It was also not entirely original; the Swedish chemist Svante Arrhenius had, about eighty years previously, also proposed that life had arrived here in spores propelled through space by the pressure of light.[1] Considering the ability of spores to survive in adverse conditions for many years, and to germinate in favorable conditions, one cannot sweep aside these ideas, but there is not even one supporting piece of evidence. Conclusion: first theory—very shaky, and very unsatisfactory, since it does not explain the origin of life.

Next in line, but basically the same as the meteorite theory, is the IDP hypothesis, designed to provide a source of large organic molecules that would be a construction kit for life. This focuses on *I*nterplanetary *D*ust *P*articles, which are supposed to be, and could well be, continually settling on Earth. Since they are presumed to be small, with diameters considerably less than a millimeter, they will float gently through the atmosphere, bringing with them not life but complex organic matter. The heat created by their impact on landing would be negligible. If such particles are still snowing down on us, then those that have fallen on the iced-over areas, such as the regions around the poles, must be trapped in the ice, which, if melted, should reveal the presence of organic material. The experiments have been done. The results are not encouraging; traces of organic matter have been found, but negligible amounts of amino acids, and a cynic might challenge their extraterritorial origin.

Between 1977 and 1981, Hoyle and Wickramasinghe came out with a series of strange articles on the composition of interstellar dust clouds. These papers could not be ignored because of Hoyle's prestige (as an astrophysicist, not a biologist), but they have tarnished his image forever. First, the authors kept changing their minds

---

[1]Light *does* have a pressure. Maxwell proved this theoretically, and it has since been demonstrated experimentally.

about the chemicals that they claim to have identified in interstellar dust clouds; second, their science is unacceptable. What they did was to examine the infrared (IR) light coming from these clouds. We have seen that the radiation emitted or absorbed by different materials can be used for identification purposes. However, convincing identification depends absolutely on the presence of well-defined frequencies in the spectrum (see Figure 15.6). This is very far from the case in the IR spectra of dust clouds. The gross uncertainty of the authors' methods is tellingly illustrated by the history of their publications. They started off by "identifying" graphite and silicates in the cosmic dust, but about five years later they changed their minds, now "identifying" an organic polymer (see Chapter 13) called polyoxymethylene, a molecule that has no biological significance. This was not the end; two years later they claimed that they had identified a mixture of eighteen different organic polymers. Within about a year they had identified a different suspect altogether. This time it was cellulose that had been seen at the site of the crime. I won't go into too much detail, but their comparison of the dust spectrum and that of cellulose was, how shall I put it—imaginative. Incredibly, this was still not the end of the shameless psuedoscience. One bright morning the authors woke up to the realization that the dust spectrum was really (wait for it, folks) that of a mixture of bacteria and algae that had gone through a process of freeze-drying in space! One almost regrets that the series of papers did not continue, finally identifying the spectrum as belonging to a mixture of Maxwell House and freeze-dried ET. In any case, the theory suffers from all "they-came-from-outer-space" theories: even if true, it doesn't explain how organic material (carbon-containing molecules) or life originated "out there." Let's get back to this planet.

Terrestrial Abiogenesis
Creative Lightning

In 1924 the Russian scientist Aleksandr Oparin made the first clear statement of a modern theory of abiogenesis. Oparin claimed that Engels had influenced his theory, and he was one of the few Soviet biologists who sided publicly and vociferously with the charlatan Lysenko (see Chapter 23).[2] Oparin proposed that when life emerged it was on an Earth with a very different atmosphere from today's. The "air" then was composed mainly of hydrogen, methane (CH_4), ammonia, and water vapor. There was no oxygen. Under the influence of the Sun's light, of lightning, and of heat from volcanic eruptions, these gases reacted to give organic compounds. The source of carbon was obviously the methane.

Over a period to be measured in hundreds of millions of years, the concentration of those organic molecules that ended up in the oceans increased enormously. Those that fell on Earth did not interest Oparin since the emergence of life, as everyone agrees, needs an aqueous environment. The next stage is a huge handwave: "Life developed from this marine solution of molecules."

A very similar theory was advanced in 1928 by the English scientist J. B. S. Haldane, who was unaware of Oparin's work. Both Oparin and Haldane based their

---

[2]He was also a fan of Olga Lepeshinskaia, who was awarded the Stalin Prize for 1950, for supposedly generating living cells from solutions containing egg white, among other things. She branded Pasteur as a reactionary and an idealist, presumably because he had denied the existence of spontaneous generation. She should have collaborated with Hoyle.

hypotheses on the proposition that the atmosphere that prevailed when life first emerged was effectively free of oxygen. Earth is the only planet in the solar system with a significant percentage of oxygen in its atmosphere, but it is supposed by some scientists that the source of that oxygen is almost entirely due to the process of *photosynthesis* in plants. In this process the green leaf uses sunlight to build sugars from water and carbon dioxide. One of the products of photosynthesis is oxygen. If photosynthesis is the sole origin of atmospheric oxygen, then there could indeed have been no oxygen in the atmosphere of the pre-life Earth, since there were no plants.

The Oparin-Haldane hypothesis remained on paper until 1953, when Stanley Miller, working with the eminent chemist Harold Urey at the University of Chicago, performed an experimental test of the theory. The idea was simple: to attempt to produce organic molecules by taking a mixture of small molecules and subjecting them to conditions presumed to simulate those on Earth, 4 billion years ago. They guessed at the composition of the Earth's atmosphere, as it was before life emerged. Into a flask they put water, hydrogen gas, ammonia, and methane (CH_4). Water is not controversial. The other gases have the virtue of being simple, and although methane is technically an organic molecule, it is certainly some way, in terms of chemistry, from say amino acids. Ammonia is there to provide a source of nitrogen, essential for amino acids. Then they added lightning, or at least its equivalent, electric sparks, which they passed through the mixture of gases and water vapor. The flask was heated, stirred, and sparked for several days, during which time the contents turned into a reddish brown mess. The experiment was carried out many times, with differing percentages of the reactant molecules and different intensities of electrical discharge. The results can be summed up simply: the methane disappeared from the gas, and the main product of the reaction was tar, which is the name that organic chemists give to a thick dark, gluey organic reside that sometimes appears in reactions between organic molecules. Analysis of the residue in Miller and Urey's flask showed the presence of about a dozen smallish organic molecules plus traces of many other substances. Of the major products, two were of special interest, the amino acids glycine and alanine. These are two of the twenty amino acids occurring in the proteins of living organisms.[3]

This looked like a stupendous breakthrough, especially as other mixtures of simple gases produced additional biologically important molecules such as sugars. Furthermore, Sidney Fox showed that if a solution of amino acids is repeatedly heated, dried, and redissolved—a cycle that could happen in nature—small, roughly spherical clumps of proteins begin to be formed.

The Miller experiment is the most quoted work in the field. No one had previously come anywhere near producing such biologically vital molecules as amino acids from simple molecules, in conditions that might conceivably have approached those of 4 billion years ago.[4] Some scientists thought the search for the origin of life had almost ended. Nucleotides, needed for the synthesis of RNA and

---

[3]To simulate the effect of sunshine, some mixtures were bathed in ultraviolet light. They also yielded organic molecules.
[4]A competent organic chemist can synthesize amino acids from simple molecules, but she will be using suitable apparatus, and many other molecules or inorganic compounds, some of which act as catalysts. These were not the conditions prevailing 4 billion years ago.

DNA, would probably turn out to be obtainable under similar conditions, and then all that was necessary was to fill in the steps leading from the amino acid mixture to the first primitive cell. No one thought this would be easy, but the Miller experiment appeared to have broken down the main gate of the castle. Perhaps, after all, life started in this way, in what Darwin called "a warm little pond." So what is the bad news? There is plenty.

The simplest criticism of the Miller experiment is that it does *not* account for the production of *nucleotides*, the essential components of RNA and DNA. Life depends on nucleotides; without DNA (or RNA) the cell as we know it could neither synthesize its components nor reproduce. Now all nucleotides contain phosphorus atoms, and it is certain that inorganic phosphorus compounds existed "in the beginning"—we know that from geological studies. Nevertheless, it has not been possible to produce nucleotides in a Miller-type experiment. One way around this problem is to suppose that when life started it was based only on proteins. We will return to this hypothesis shortly.

Another serious criticism of the Miller experiment concerns the role of methane. Miller and Urey chose methane for two reasons. First, they needed a source of carbon atoms; otherwise, no life. Second, they wanted what chemists call a *reducing atmosphere.* Without getting technical, you can appreciate that doing the opposite, using an *oxidizing atmosphere*, would be doomed to failure. Thus supposing you added oxygen to the mixture to be sparked. Sparking and heating would result in *burning* the other components, where possible. The hydrogen would end up as water, the ammonia as a mixture of water and oxides of nitrogen. In general the reaction between oxygen and small organic molecules gives smaller inorganic molecules such as water and carbon dioxide. There is no building up of more complex molecules like amino acids. That is why Miller and Urey chose to exclude oxygen. But did the atmosphere of the Earth ever contain large amounts of methane?

It is true that methane is one of the components of "natural gas," which is trapped in huge underground cavities in some parts of the world and is used as a fuel. But this methane is produced by the effect of millions of years of pressure and heat acting on prehistoric *plant* material. There had to be life before this natural methane could be formed. Could there nevertheless have been methane in the atmosphere of the early Earth, before life? We can get hints by looking at other planets in the solar system. Mars, and even more Venus, are not drastically different from the primordial Earth. Their atmospheres are very rich in carbon dioxide, not methane. This is a problem.

Since Miller's first experiments, there has been a change of opinion about the early Earth's atmosphere. One thing is certain: it could not have contained significant amounts of both hydrogen *and* oxygen. The first flash of lightning would have initiated the mother of all explosions. There is at present no evidence that the atmosphere was reducing (methane and hydrogen), and the prevalent opinion at the moment is that the Earth's atmosphere, at the time that life emerged, was mainly carbon dioxide and nitrogen, with small amounts of hydrogen. In the absence of methane, Miller-type experiments with carbon dioxide, hydrogen, ammonia, and water produce only the amino acid glycine, in very small amounts. It is possible to get other amino acids if the ratio of hydrogen to carbon dioxide is greater than one, but as Miller himself admits, it is highly unlikely that there was enough hydrogen in the atmosphere of the Earth to maintain such a ratio. The gravitational pull of the

Earth on hydrogen molecules is rather weak, and they also travel very fast (see Chapter 1), and thus tend to escape from the atmosphere.

There are other doubts about the Miller approach. One criticism, which could be aimed at almost any theory of abiogenesis, is that even if a mixture of amino acids were produced, it would not be particularly stable. Among the processes tending to destroy the "primitive soup" is adsorption of organic molecules on rocks, although there are of course those who would see this as part of a new theory, in which life forms on the surface of underwater minerals. We will look at two theories of this kind shortly.

An interesting aspect of Miller's experiments is the relative amounts of amino acids that he obtained. An average over many experiments shows ratios of different amino acids that are not too different from those obtained by the analysis of the organic matter found in some carbon-containing meteorites. The similarity strongly suggests that there have been natural Miller-type experiments in space. If this is so, and *if* (big if) life started by Miller's mechanism, the obvious conclusion is that there is a reasonable chance that life started elsewhere in the universe. It would still have to be water-based, to be life of the kind that we know. In 1996, evidence was presented that strongly suggests that there are fossil microorganisms in rocks recovered from Mars.

The Haldane-Oparin-Miller hypothesis is out of fashion. Of the forty or so simple molecules that would be needed to form a primitive cell, the experiment produces two. It is worth bearing in mind that glycine contains only ten atoms and alanine, thirteen. The simplest nucleotide contains thirty atoms. The probability that a given large molecule will be produced by chance from small molecules, by sparks, falls drastically as the molecular size increases.

It has to be realized that even if heat, radiation, and lightning, on the young Earth, had produced all the amino acids and nucleotides needed for present forms of life, the gap between an aqueous solution of these molecules and a living cell is stupendous. It's a question of organization: in the absence of a guiding intelligence, how did the components get together? Even *with* a guiding intelligence, present-day scientists are not doing very well. For the moment, let's show the Miller experiment to the side door and see who is next in line in the waiting room.

The Call of the Depths

In the 1970s, exploration of the sea bottom revealed that there were tiny volcanoes blasting molten matter into the ocean bed. These hydrothermal vents can heat the water to well over $100°C$, and there are scientists who believe that life began in this saunalike atmosphere. It's hot down there, but there are bacteria that live in hot springs in near-boiling water and there are many microorganisms that have adapted to the eerie environment near the vents, using sulfur-containing molecules as a source of energy. How could life *start* down there? Well, if you take a solution of simple molecules, put them in the scientific equivalent of a pressure cooker, and leave them to cook, you can produce an impressive yield of amino acids. What simple molecules? Ah, there's the rub. Those used by the experimenters included formaldehyde, cyanide ion, and carbon dioxide. No one will quarrel over the carbon dioxide, but the other two molecules have been handpicked, with malice aforethought. They are both very chemically reactive, and it is understandable that cyanide was chosen since it contains both carbon and nitrogen. Is there good reason to suppose that cyanide and formaldehyde are, or were, that common down there, or indeed anywhere except in chemical supply companies? Not at present. And where are the nucleotides? This is an aquatic variation of the Oparin-Haldane hypothesis, and it suffers from one of the major defects of the Miller experiment: the starting molecules have been chosen with little or no evidence that they were really present in nature. Next theory, please.

Convenient Surfaces

There are at least two theories based on the premise that certain inorganic minerals have surfaces which, because of their shape, are capable of acting as catalysts for the formation of key biological molecules. The appropriate small molecules, drifting randomly through the primeval soup, are trapped when they strike the particular part of the surface that fits them. During the time that they are immobilized, another type of molecule may chance on the trapped species and link up with it. The surface eases the linkup process by holding the reactants in suitable positions and inducing shifts of their electron clouds that facilitate the formation of new bonds. The new molecule then sits around waiting for the next component. There are many examples of solids acting as catalysts for synthetic chemical reactions, and their action clearly depends on one or more of the reacting molecules being adsorbed on the surface of the solid. Of course the geometry of the surfaces, on the atomic scale, is crucial to the biogenesis story, as it is for any solid-state catalyst.

One variation of the theory has been presented by the Glasgow-based scientist Graham Cairns-Smith. The materials he focuses on are clays, which you may say haven't got a well-defined surface, but at the microscopic level they have a definite and regularly repeating structure because they are silicates of the kind illustrated in Figure 13.6. Cairns-Smith envisages the chance production of a huge variety of surfaces, some of which are so shaped as to act as templates for vital organic molecules. I will not go into detail, but the theory is certainly ingenious, original, and no less convincing than the other theories. Unfortunately, it is also no more convincing. And it will take a hell of a job to prove.

Another theorizer, Gunther Wachtershauer, a patent lawyer with a doctoral degree in organic chemistry, also suggests that inorganic crystals were the substratum on which life developed. He favors a substance known as iron pyrites, a compound of iron and sulfur that forms beautiful iridescent crystals that gave rise to the common name "fool's gold." It has an advantage over clays in that it is naturally reducing, being prepared to give up electrons to the right acceptors and thus play an active part in the reactions presumed to occur on its surface. Wachtershauer suggests that nucleotides could form on such a surface, and that the first form of life might have been a grain of pyrites surrounded by a membrane.

Stanley Miller refers to the pyrites theory as "paper chemistry." He has a point. The fact is that neither of the proposers of the clay and pyrites theories have gone into the laboratory one morning with a mixture of clay and some really simple molecules, and come out sometime later with a complex organic molecule. Thus it has been possible to build polypeptides from amino acids in the presence of clay, but the process also needed the helping hand of a molecule called adenosine 5'-phosphate, a molecule that is extremely unlikely to have been around 4 billion years ago. And the real problem remains: even if there were polypeptides, it's a huge jump to get to a living, reproducing cell.

Which Came First?

Where are we? Amino acids have emerged from Miller-type experiments, but not nucleotides. This brings us up against a fundamental problem. We have seen that DNA is responsible for the replication of the cell. *If a primitive cell contained only proteins, it would have no future.* Proteins cannot replicate themselves. Such a cell would eventually age and die without progeny. We are not aware of immortal cells.

On the other hand, imagine a primitive cell with *only* nucleotides. We know that DNA can direct duplication of the cell, initially by duplicating itself. *But that duplication needs certain enzymes,* and in a cell with no proteins, DNA could not duplicate—remember that all enzymes are proteins. Neither could such a cell direct the synthesis of proteins—a process that itself requires enzymes (see Chapter 24). In any case, it is rather difficult to see how such an enzymeless cell would carry out the thousands of functions that are typical of a living cell and which depend absolutely on the presence of enzymes.

We see that nucleotides (DNA and RNA) and proteins are an interdependent system as far as the replication of cells is concerned, and without replication the glorious and miraculous assembly of the first living cells would have led nowhere. Life would occasionally spring up here and there only to die every time, in sterile loneliness. Like the twinkling of glowworms in the dark.

It stretches even the credulity of a materialistic abiogenesis fanatic to believe that proteins and nucleotides persistently emerged simultaneously, and at the same point in space, from the primeval soup. We are in trouble enough without adding events of an astronomical improbability.[5] And if simultaneity is not feasible, which came first? Nucleotides or proteins? And how did the primitive life-form survive if it was deprived of either of them?

Glimmers of Hope
The Naked Gene

It was accepted for many years that there was a clear division of labor between proteins and nucleic acid: proteins were the catalysts of the living cell but could not be replicated *without DNA,* while DNA did not appear to act as a catalyst but could replicate *with the help of proteins (enzymes).* In other words, there were two complementary systems—genes and enzymes, both dependent on the other if they were to maintain the living process but neither able to do the work of the other. New avenues of thought have been opened up by the remarkable work of Sidney Altman and Thomas Cech, which earned them the Nobel Prize for chemistry in 1989. They showed that under the right circumstances RNA can act as a catalyst, thus *possibly* dispensing, at least initially, with the need for a protein in a primitive life system. In other words, they opened up the possibility *that only one type of molecule was necessary for the creation of life.* They showed that RNA had properties that were reminiscent of those of enzymes.

A second advance, in the laboratory of Leslie Orgel in the Salk Institute in California, was the demonstration that, in a solution containing other simple mole-

---

[5]Of course you could say that the primeval oceans were so stuffed with amino acids that *anywhere* that nucleotides appeared they landed in the middle of amino acids. No one has suggested this up to now, but it would require staggering amounts of carbon.

cules, mononucleotides (the units from which nucleic acids are constructed, see Chapter 24) can be induced to replicate themselves without the help of proteins. Furthermore, we know from Spiegelman's work (see Chapter 22) that RNA, in the right circumstances, can evolve. Since amino acids might conceivably have been produced in the "soup," RNA may have evolved into a form capable of interacting intelligently with them, starting the trail to polypeptide synthesis.

Once again, much is paper chemistry, but nevertheless, all this is very suggestive, since it opens the way for a theory of the origin of life based on RNA, or at least on a self-replicating molecule, needing no protein to help it in that process. The suggestion is that life began not with a primitive cell but with a structure more akin to a gene, and that the gene evolved a structure to look after itself—that structure being the cell. Those who have read Richard Dawkins's provocative book *The Selfish Gene* (1976) will be strongly reminded of his picture of the gene ensuring its survival by "living" within the body of its host.

If RNA did appear at an early stage, and we take into account its catalytic activity, then there might have been far more complex behavior in the primeval soup than had, until recently, been envisaged. We need that complexity in order to produce even the simplest entity that displays the characteristics that we demand of life: the ability to reproduce and the need to take in and produce energy. RNA can carry information and could conceivably replicate without the need for enzymes. The "nucleotides first" advocates have been the first to admit that a massive stumbling block stands in their path: the synthesis of RNA *by chance* is a highly improbable process, and as yet no one has presented a mechanism by which it might have occurred. It's tough enough doing it in the laboratory, and no one has come anywhere near producing RNA by mixing *simple* molecules and subjecting them to the imagined conditions prevailing when the Earth was young. Even when you do have RNA, the process of self-replication in the laboratory is not at all straightforward, and it requires considerable intervention on the part of the experimenter. The RNA trail is fascinating for the light it has thrown on the behavior of this molecule, but we are still a long way from explaining its genesis, let alone that of life.

The Naked Protein

In complete contrast to the previous scenario, in which RNA is the only complex molecule needed to originate life, there have been suggestions that life started with *self-replicating* proteins: it is the proteins that are the only type of molecule needed to fill the roles of both replicators and catalysts. This avoids the difficulty of explaining the simultaneous abiogenesis of the nucleotides, which are far more complicated molecules. The late arrival of the nucleotides on the scene would then herald a whole new level of sophistication in the cell's ability to evolve, which could account for the fact that after the appearance of single-cell organisms about 3 to 4 billion years ago, evolution seemed to stand still for a couple of billion years before multicellular organisms appeared.[6] It is not beyond the bounds of possibility that under the right circumstances a protein could be self-replicating. There is nothing in the laws of physics or chemistry that forbids this. It could be objected that all the

---

[6]It appears to me that another explanation of the huge stretch of time for which unicellular organisms failed to evolve is that bombardment by meteorites, comets, asteroids, and other interplanetary junk could have wiped out life several times in the first couple of billion years after its first appearance.

proteins that have complex functions in the cell (e.g., enzymes) are composed of a dozen or more (up to twenty) different amino acids, and that it is highly improbable that such a range was available in the "soup." This is not an insurmountable objection. We know that the *shape* of enzyme molecules plays a central role in their functioning, and it is possible that a small number of amino acids would be sufficient to confer the required flexibility of shape, and therefore function, in building primitive proteins. In this case, Miller's two amino acids (glycine and alanine) are more significant than we supposed. Again I must sing the party pooper's refrain, "This is paper chemistry." It is; the experiments are waiting to be done.

The Birth of the Cell?

Life exists in units enclosed by a sack—the cell membrane. Such an enveloping structure is needed to maintain the physical integrity of the cell, as well as to perform other vital functions. A universal component of all cell membranes is the phospholipid molecule. If you take a dollop of most types of phospholipid, put it in a nearly neutral aqueous solution, and then gently agitate the solution, the greasy lump will break up and it can be shown that the liberated molecules will spontaneously aggregate into closed, microscopic, approximately spherical envelopes—or *vesicles*, as they are called in the jargon of the lab. It is not beyond the bounds of reasonable possibility that phospholipids could be formed in the soup. The problem is that no one has done this in Miller-type experiments. Another class of molecules that form roughly spherical, closed structures in water are detergents, which are considerably simpler than phospholipids. The tendency of the greasy ends of detergent molecules to clump together and avoid water results in the formation of closed structures having their greasy side facing inward. The structures formed by detergent molecules are far less stable than those formed by phospholipids, but if we are doing paper chemistry, then they could form a cell-like capsule in which the first simple cell developed. I'm not at all sure that I believe that, but some do.

It is significant for all theories concerning the origin of life that the phenomenon of self-organization, as exemplified by phospholipid and detergent molecules, is now a hot subject in chemistry. If you want to follow this fascinating trail, look for "supramolecular chemistry" in the scientific literature.

Back to Miracles?

None of the present theories of the origin of life is strongly favored by the scientific community. In fact, I would not blame you if, by this stage, doubt is creeping into your mind, or is perhaps already firmly installed in the throne room. Francis Crick has written, "The origin of life appears to be almost a miracle, so many are the conditions which would have had to be satisfied to get it going." The finger of God is certainly a tempting way out. The probability of a crowd of small molecules forming the needed large molecules to start the long, complex path to a single cell seems to be almost zero.

Going back to our 200 dollars' worth of chemicals again: no one would deny that attempting to create Kim Bassinger/Mel Gibson by heating and shaking the mixture, although worth trying, might take forever. But that doesn't prove that life has supernatural origins. Up to now, in attempting to prove that it hasn't, we've probably just made the wrong starting assumptions. The problem is that we don't know what the right ones are.

One thing is absolutely certain: we get no help from spontaneous generation, the emergence of completely formed complex organisms from inanimate material. This nonexistent phenomenon had a long history and should have dropped dead when Pasteur showed that a broth that had been boiled and then left in a sterile atmosphere did not show signs of bacterial growth for many days.[7] The little museum in the Institut Pasteur in Paris has some of the vessels in which Pasteur sealed boiled broth. They were, when I last saw them, apparently free of all bacteria. Later experiments of this kind purported to show that other nonliving systems could generate life. Probably the last serious attempt was by H. C. Bastian, who in 1872 described his results in his huge book *The Beginnings of Life.* His experiments resembled those of Pasteur. He placed infusions of hay and a variety of other organic material in hermetically sealed flasks and boiled them for many hours, long enough for all life to have been destroyed—in his opinion. The flasks were sealed and left to cool at room temperature, and within a day or two it was obvious that there was bacterial growth. The experiments were repeated by John Tyndall. By meticulously excluding bacteria, Tyndall showed that the contents of the flasks remained sterile, stone-dead. Spontaneous generation has not reared its head again.

Paradoxically, Tyndall would have wished it otherwise. He was a notorious agnostic, some said atheist, and in general it was scientists of his opinions who looked to spontaneous generation, or abiogenesis, as a phenomenon that supported their inclination to dispense with the need for supernatural intervention. Tyndall's famous presidential address to the 1874 meeting of the British Association for the Advancement of Science, in fact, contained a defense of the continuity of nature, the inseparability of the inanimate and the animate. This was too controversial for his time, and the address was to create a public furor.[8] It is ironic that the very scientist who needed spontaneous generation to dispense with miracles should be the one to finally kill it.

Self-Organization

The basic problem facing anyone who is looking for the origin of life is to account for the formation of a complex, *very highly organized, self-sustaining and self-replicating* system out of a mixture of chemicals that, certainly in the early days of the soup, displayed none of these characteristics. As Jacques Monod, the eminent French biologist wrote in *Le Hasard et la nécessité* (1970), "The universe was not

---

[7]Of course, in a way, abiogenesis *is* spontaneous generation, if by that phrase we mean the appearance of life from nonlife. However, the proponents of spontaneous generation were talking about the appearance of *fully formed living forms*, anything from mice down to maggots and bacteria, not the gradual evolution of living matter from simple molecules over a long period of time.

[8]Most Victorians could not understand that the man they saw as a deluded materialist, could, in his address say: "The world embraces not only a Newton, but a Shakespeare—not only a Boyle, but a Raphael—not only a Kant, but a Beethoven—not only a Darwin, but a Carlyle. . . . They are not opposed, but supplementary—not mutually exclusive, but reconcilable." He sounds sympathetic.

pregnant with life nor the biosphere with man. Our number came up in the Monte Carlo game."

We asked two questions near the beginning of this chapter, but we have discussed only one of them: the origin of the molecules of life. We found no convincing answers. Regarding the second question—the production of a stable system with lower entropy (less probability) than its normal equilibrium state—is not only possible, it has been observed (see Chapter 18).

In considering the possibility that a system can spontaneously organize (move to lower entropy), let us remind ourselves once more that the second law of thermodynamics only says that the entropy of an *isolated* system increases due to any changes. (An isolated system is one in which neither matter nor energy can cross its borders.) There is nothing in the law that says that entropy cannot decrease or hold steady for a system in which either energy, or matter and energy, can cross its borders. Such a system is the living cell.

This dispensation, granted to *closed* or *open* systems, to maintain or decrease their entropy completely changes the rules of the game. It removes one of the great conceptual barriers that at one time threatened to overthrow any theory of abiogenesis, namely, the mistaken "principle" that systems could not lose entropy and move from comparative disorder to comparative order.

But how could primitive cells maintain a highly organized (low-entropy) state, even if they somehow achieve such a state?

How Long Can We Keep the Fire Alive? Life versus Entropy

In the nonliving world it is almost universal to find that natural chemical and physical processes are accompanied by an *increase* in entropy, that is, a loss of order. There are no systems more ordered than the living cell and its components. We expect matter that is in a highly organized form, such as an enzyme, a DNA molecule, or a living cell, to be unstable with respect to breaking up into smaller components. That's the way the universe goes. Living systems continually undergo chemical and physical change. Why do they not invariably disintegrate into smaller, more probable, molecules, thereby increasing the entropy of the cosmos, in obedience to the second law?

They do. When a living organism dies it decomposes, which just means that the fantastically complex molecules of which it is made break down into far simpler components. The spontaneous process of decomposition goes the way that the second law predicts: the complex, highly ordered system breaks up into a highly disordered mess. The structure of the cell collapses (becomes disordered), and the large, entropy-poor (highly organized) molecules that it contains disintegrate into small, entropy-rich molecules, destroying the miraculous order and leaving the fragments of the cell to wander about in a disordered manner. The entropy (alias disorder, alias probability) of the universe increases. This happens spontaneously, as the second law predicts. The problem is that while it is alive an organism maintains its highly ordered, low-entropy state. It fights the second law.

Living matter would much rather be dead. Living matter is a curiosity. Rephrasing our question, we ask:

Granted that life exists, how can a living organism be at low-entropy *equilibrium* (and they usually are) if the obvious equilibrium state is a decomposed corpse, the high-entropy end point of spontaneous processes?

The answer is contained in our previous consideration of systems at *dynamic* equilibrium. Living forms are in dynamic equilibrium, like the heated tube of helium in Chapter 18. The tube was kept far from its normal equilibrium state by a flow of energy through the system. How does the body maintain the complex structural integrity of its cells and of the molecules within the cells, in the face of the destructive influence of the spontaneous tendency to raise its entropy by decomposing? It eats. Without food input we die and then drift rapidly away from equilibrium. What is the nature of this ingested material?

Plants take in very simple, high-entropy molecules such as water, carbon dioxide, and nitrogen-containing salts, and from these build up complex, low-entropy molecules, especially carbohydrates. They manage to produce low-entropy molecules from high-entropy precursors because the building up process is driven by the energy in the Sun's light.[9] *Plants are largely synthesized from high-entropy molecules.* Animals, on the other hand, cannot live on carbon dioxide, water, and salts. We eat plant and animal material that has large, *low-entropy* molecules in it. These molecules, in the end, are derived from photosynthesis, the great creator of order on this planet. Thus animals take in relatively complex molecules through their intestinal membranes—molecules such as amino acids and simple sugars— and build them up into even more complex materials, such as proteins. Animals are highly ordered systems that, in contrast to most plants, are *largely synthesized from highly ordered (low-entropy) molecules.* In our bodies we continually synthesize complex molecules to replace those that naturally break down, and we are helped in this struggle against entropy increase by taking in comparatively highly organized, low-entropy molecules. Thus we can look at the stable equilibrium of living matter as a dynamic state in which low-entropy living matter is defended against the irreversible march to higher entropy by means of an intake of suitable material. We are removing low entropy from the external world and using it to keep down our own.

It might be thought that we could directly utilize very large low-entropy molecules in our food, such as DNA and proteins. This is not what we do, and cannibalism does not replace the cannibal's DNA and enzymes with those of his victim. We do take in very large molecules, but we partially break them down into smaller components, such as amino acids and simple sugars, which are still comparatively low-entropy molecules, and use these to build up our own cell constituents. The Austrian physicist Erwin Schrödinger (1887–1961) said that living organisms were eating low entropy from their surroundings to maintain their own entropy at a low level. Of course this is not the only role of food; we need various elements to maintain our health, and we also need a source of energy. Note that the metabolism of food in our cells results in the formation of heat, which we give up to our surroundings. Heat is the form of energy with the greatest entropy (due to the disorder of the moving molecules), so we are taking in a source of low entropy and exporting

---

[9]This energy is in a highly available form; it induces photosynthesis whatever the temperature of the plant. It is not a degraded form of energy like heat, and in fact there is justification for saying that it is in a state of low entropy.

high entropy. That's how we fight the battle against our own natural tendency to drift to higher entropy.

Postscript

The enigma of the origin of life is fascinating and of little practical significance. As far as solving the problem is concerned, we may in the end have to be satisfied with a small selection of credible possibilities.

To those who see no way out but divine intervention, I would point out that you are asking more from scientists than you have a reasonable right to expect at this stage. You are asking them to account for the creation of the most highly organized mechanisms in the universe, in terms of a world that vanished nearly 4 billion years ago, leaving very little indication of what it was like except in rather general terms. The journey between Mendel and the genetic code would have seemed utterly impossible to an early-nineteenth-century scientist; they didn't even know the structure of atoms. But we made it.

I had one long conversation with J. B. S. Haldane, which started off with politics and ended up with science. When I questioned him about evolution, one of his remarks sparked my interest, and sent me to the library that evening: "Evolution's not the problem. Life is." Then he said, "Oparin and I once had an idea about that, but we'll never know the real answer."

An Apology

Much has been left unsaid. I have limited myself to pointing out a few central themes in the scientist's way of dealing with living organisms. The biological sciences are far larger in scope, and far richer in detailed knowledge, than I have suggested. But we are also vastly ignorant, especially of two related areas: consciousness and the interaction of mind and body. I promised in the introduction to keep away from consciousness, and I will keep my promise. Regarding mind and body I will also pass. There are those who state firmly that what we call the mind is merely a manifestation of the body. This may well be true, but I'm not going to be drawn in on that one. And if you think that the nature of the mind is elusive, just wait until we get to the next chapter.

The Dissolution of Reality?

Mount then, my thoughts, here is for thee no dwelling
Since Truth and Falsehood live like twins together.
Believe not sense, eyes, ears, touch, taste, or smelling.

—John Dowland, "Tell Me, True Love"

The twentieth century has seen a second scientific revolution that
boasts two of the most powerful conceptions of the human mind—the
general theory of relativity and the quantum theory. One has changed
our image of space and time, and has shown us that matter and energy
are two names for the same shadowy entity; the other has allowed us to
understand and predict the nature of matter and radiation with a
degree of detail that would have seemed miraculous to the scientists of
previous centuries. It has also put in question our ability to grasp
reality.

28

The Inexplicable Quantum

I think I can safely say that nobody understands quantum mechanics.

—Richard Feynman

The quantum theory is not explicable in commonsense terms. Its axioms, and some of its predictions, go against all our built-in preconceptions as to the nature of reality. In the 1930s Niels Bohr, one of the theory's fathers, was reported to have said that if you weren't confused by quantum mechanics you didn't really understand it.

Perhaps the most powerful theory that science has constructed grew out of the need to explain aspects of nature that were utterly inexplicable in terms of Newton's mechanics and Maxwell's laws. It is, for example, impossible to say much of importance concerning the behavior of nuclei, atoms, and molecules without using quantum theory.

When Leibniz read Descartes's method of stripping away all doubtful beliefs and then building up a true philosophy, he mocked its generality. The method boiled down, he said, to "take what you need and do what you should, and you will get what you want." This could better be said of the way that quantum mechanics is almost universally used. We put in some physical facts, follow the rules for obtaining the needed results, and almost always get what we want. The comparison between the theoretical results and experiment is rarely disappointing.

It can be said of quantum mechanics that we are thankful, impressed, but, in the still of the night, profoundly uneasy, because the quantum theory presents us with paradoxes more puzzling than the questions that it solves. Einstein always saw the quantum theory as imperfect and transitory. One can understand his discomfort. Quantum mechanics implies that to many of the questions that we ask of physical systems we will, *in principle*, only obtain answers couched in the language of probability. We will never be able to say, "If you do this to a molecule, then that will happen," but only, "If you do this, there is such and such a probability of that happening." Some physicists believe that there will be some nontrivial changes in quantum theory before very long. Others claim that any major change is impossible; the theory as it stands is just too successful to stand much alteration.

We will go back to the beginning of the century, when the great breakthrough occurred, and see why and how it happened. If at the end you find the theory unsatisfactory, you are in excellent company, but you might bear in mind Bohr's reproof of Einstein: we must not tell God how to run the world.

The Grainy Texture of Nature

As a child you probably swung on a swing at some time, perhaps pushed by one of your parents. If he or she pushed steadily, you swung backward and forward repeatedly, through the same arc. My father used to do this for me, until I was banned from that park for shooting arrows at people. (I was dumbfounded—the arrows were blunt.) We will call the child and the swing, together, an *oscillator*. My father could control the height I reached at the end of the arc by altering the force with which he pushed me. *This means that he was controlling the total energy of the oscillator.* The higher the end of the arc, the greater my potential energy at that point—or, if you like, the more work my father had to do to get me there. At the extreme of the arc I was stationary, and all my energy was potential energy (see Chapter 17). Suppose that at the extreme point of one particular arc I had a total energy of exactly 600 joules. Could I be given an energy of 599 joules? Obviously yes. All that would be necessary would be for my father to push me a little less forcefully, so that I rose a little less higher at the end of the arc. Could he have given me an energy of 599.9 joules? Of course. And of 599.9683885214 joules? In principle yes, although the accuracy of the measurements needed to verify such an energy would test the experimenter.

Common sense says that if the oscillator can be given two different energies, say zero (don't push the swing) and 600 joules, we can in principle give it *any* energy between 0 and 600. In the same way, common sense says that if a car can travel at a maximum speed of 120mph, then it can travel at *any* speed between zero and 120 mph, and thus it can have any predetermined kinetic energy up to the maximum value. Until December 1900, it was universally believed that we could choose the energy of an oscillator to be any value that we wished, between zero and some physically determined maximum. It would have been said that the allowable energies of an oscillator formed a *continuum*.

There are a number of basic quantities besides energy that appear to have this kind of continuity. Length is the simplest. We intuitively believe that we can, in principle, separate two bodies by any distance that we choose. Similarly, common sense—that most insidious of brainwashers—tells us that time is continuous.

Freud's *Psychopathology of Everyday Life* appeared in 1900, opening a new era in the way that people thought about their internal world. It is doubtful if any of those who took voluble sides with or against Freud realized that a paper published by a German physicist on 14 December of that same year was to create an even more thorough revision of the way we see the external world. Max Planck's suggestion seemed innocuous enough; it was that the total energy of a collection of oscillators can assume only *discrete* values.[1] It is as though Planck (1858–1947) had said that in the playground in the park, the *total* energy of all the children on the several swings could only have certain discrete values, just as the total number of swings can be only an integral number. A collection of oscillators sounds a bit remote from daily life, but it isn't. As we saw in Chapter 16, the atoms and molecules of any solid body, for example, are ceaselessly, and unstoppably, vibrating. You, and much of your surroundings, are collections of oscillators. Planck was saying that the energy of such systems of oscillators comes in little packets, *quanta*. If you have 37 quanta you can't go up to 37 and a half, the next stop is 38.

---

[1]For the scientific reader it should be pointed out that, contrary to the impression left by most textbooks, Planck did not originally equate the energy of a *single* resonator with an integral multiple of $h\nu$, although he did repeatedly writes expressions like $U(N)=Ph\nu$ with P an integer, where $U(N)$ is the *total* energy of the N resonators.

Planck had a reason for plucking his highly curious assumption out of the air. He was trying to explain Figure 15.8b. Was it possible to predict what fraction of the total radiant energy of a blackbody was associated with a particular frequency? Planck's solution fitted the facts exactly. As others had done before him, he modeled the blackbody as a collection of oscillators giving off radiation. The novelty that he introduced into the calculations was his assumption of the *discreteness* of the total energy of the oscillators. It worked: if you gave him the temperature of the heated body, he could predict accurately what fraction of the total energy of the radiation would be associated with any particular range of frequencies. This success did not necessarily validate Planck's assumption, but it certainly made people think. If it was true, what could it mean—that a system could have only certain values for its total energy? It was like saying that the sum of the heights of a group of people could be only an exact number of inches. As though my daughter and I together could total 134 inches in height but not 134.5. The next allowed figure would be 135. This is obviously not true, so why should the energy of oscillators behave in this strange way?

Planck said of his epoch-making assumption: "It was an act of desperation. . . . I had to obtain a positive result under any circumstances and at whatever cost." The justification was that it worked, it solved the black body problem.[2] But, although he told his son that he thought he had done something as important as Newton's work, much later, in 1922, he said that the quantum had called forth "a break with classical physics far more radical than I had initially dreamt of."

Planck was worried. For at least a decade he struggled to find a way of solving the blackbody problem *without* assuming "quantization of energy." He was tied to the past. One can understand his reservations; no one had ever dreamed that energy was anything but a continuous entity, to be poured out like water, not counted like sheep.

Enter Einstein

Einstein realized that Planck's hypothesis could be taken to imply that the *individual oscillators* that appear in nature, say vibrating atoms or molecules, are each characterized by a certain discontinuous set of energies. They can vibrate more or less violently, but they cannot have *any* arbitrary energy. Their energies can be increased only in finite steps, by packets of energy, *quanta,* just as American money cannot be counted out in units less than cents. The energy of a packet would be proportional to the frequency of the oscillator. If something vibrated very fast, then its quanta would be larger than that of a slower oscillator. The oscillator can only have a total vibrational energy equal to an integral number of quanta. The amount of money you have in your pocket cannot be twenty-three and a half cents.

The size of the packet of energy, the energetic cent which we call a *quantum,* was formally given by:

$$\varepsilon = h\nu$$

where ε is the energy of the quantum, ν is the natural frequency of the oscillator,

---

[2]In fact Planck made another revolutionary assumption in his paper, but it is not amenable to simple explanation. If there are any specialists reading this: Planck anticipated Bose-Einstein statistics.

and h is the famous *Planck's constant*. The value of ε is thus proportional to the natural vibration frequency of the oscillator.

Push a swing and let it go. It has a natural frequency at which it oscillates. This is the frequency that we put into the preceding equation. Tuning forks have natural frequencies, and so do atoms and molecules, although you can't see or hear them. The equation says that if *any* oscillator has a certain natural frequency, ν, then it can only have an amount of energy that is equal to one, two, three, or more quanta—packets of energy—each packet equal to h times the frequency of the oscillator.

For the swing in the park, the size of each quantum, in round numbers, is about 3×10^{-34} joule, as can be seen from the footnote,[3] where it is also shown that to add one quantum of energy, my father would have had to increase my height at the end of each arc by about 10^{-36} meters. The differences in both energy and height are far too minute to be detected by present-day techniques. What this says is that the packets of energy of the swing are so incredibly small that to all intents and purposes the energy of the swing is a continuum. It is as though you were climbing a staircase in which the height of each step was a billionth of an inch. It would look and feel like a continuous slope.

So what do I need quanta for if they are so small that I can't detect them? It's rather like the post–World War I German mark. Money only came in multiples of marks, but one mark was so worthless that it had no practical significance. In 1900 the only use for quanta seemed to be in the solution of the blackbody problem. But not for long.

The Photoelectric Effect: Bullets of Light?

The idea of energy "quantization" was not greeted with wild acclaim. Why couldn't an oscillator have any energy that it wanted to?

Einstein, never afraid of the dark, or the light, pressed on. What Planck had done was to suppose that the blackbody oscillators, the atoms in the walls of the cavity, *gave up or accepted* energy only in finite packets—quanta. This does not really mean that energy is quantized. Einstein likened it to serving beer in tankards, which didn't imply that beer *existed* only in tankard quantities. Einstein went much further than Planck. He quantized light. He was saying that beer *did* exist only in units of a tankard. Next, he ran a bulldozer through one of science's most sacred beliefs, declaring light to be composed not of waves but of discrete packets, tiny immaterial pellets, each carrying one quantum of energy. By doing this he solved another outstanding problem: the photoelectric effect. He won himself a Nobel Prize.

The explanation of the photoelectric effect is a turning point in science. It is the counterpoint to Young's two-slit experiment. One man showed that light was

---

[3] Let's see how this applies to the swing. The frequency, ν, of the swing can be taken to be about one cycle (trip back and forth) in 2 seconds, or 0.5 cycles per second. The value of h is 6.6×10^{-34} joule seconds. So the size of the packets of allowable energy, the quanta, that can be given to the swing is $ε = hν = 6.63 \times 10^{-34} \times 0.5$ joule or $\approx 3.3 \times 10^{-34}$ joule. This is a tiny quantity. To see how small, suppose that my father wishes me to have more vibrational energy and that he wants to increase my energy by *one* quantum, the minimum allowable increment. To do this, he has to do the same amount of work. That means increasing my height at the end of the arc. The work done in increasing my height is mgH, where H is the extra height by which I am raised (see Chapter 17). If my mass was 30 kilo, $mgH \approx 300H$. The value of H that makes this equal to the energy of one quantum is 1.1×10^{-36} meters.

wavelike, the other that it was particlelike. Einstein's article effectively created quantum mechanics.[4] He published his ideas in 1905, in the *Annalen der Physik*. In the same year and in the same journal, he published two papers on special relativity, which opened a new era in physics, and a paper on Brownian motion, which alone would have made his name. There has never been such a concentrated exhibition of genius since 1666, the year (and a half) in which Newton formulated for himself, but did not yet publish, the law of universal gravitation, the laws of motion, and the elements of the calculus—apart from splitting light with a prism. He was twenty-four. Einstein was twenty-six.

In 1887, Hertz found that when he created a voltage difference between two metal plates he could make sparks jump between them. When he gradually separated the plates, he reached a distance at which sparks no longer appeared; we would say that electrons were no longer jumping from one plate to the other. However, when he turned on an electric arc lamp near the plates, the light, shining on the surfaces, brought back the sparks. The following year Hallwachs showed that if an uncharged metal surface was illuminated with UV light it became positively charged. The explanation of these observations had to wait until J. J. Thomson discovered the electron. It was Thomson who correctly surmised that UV light was ejecting the negatively charged electrons from the metal surface and leaving an excess of positive charge. It was these electrons that were jumping across the gap between the UV-irradiated plates in Hertz's lab. This is the photoelectric effect, the basis of most devices that open doors or operate toilet flushes when you break a light beam.

So far so good, nothing particularly revolutionary. But in 1902, Philipp Lenard (1862–1947), a racist[5] Nobel Prize winner, studied the effect of varying the intensity and frequency of the UV light falling on the metal. If the intensity of the light was increased, the number of electrons per second leaving the surface increased but, completely unexpectedly, they still had the *same velocity*. Why didn't a more intense beam give the electrons a bigger kick? A larger wave in the ocean certainly creates more havoc because it carries more energy. The peculiar aspect of the experiments was that the velocity at which the electrons left the surface *did* increase when the *frequency* of the UV light increased.

It is worth looking closely at Einstein's controversial explanation of all this. Einstein proposed that light was not continuous but consisted of packets, light quanta, each having a fixed energy, completely determined by the *frequency* of the light. A beam of light behaved like a stream of bullets, not a wave. Moreover, he proposed that the energy of each of these packets, which are now called *photons,* is given by h times the frequency of the light, $E = hv$, just like the packets of energy that characterized Planck's oscillators. Thus UV light having a frequency of 10^{16} Hz consists of a stream of photons each having 100 times the energy of photons of IR light having

---

[4]Other scientists developed quantum mechanics into a form that Einstein could not accept. He acknowledged that the theory was a valid practical tool for accurately predicting the behavior of light and matter, but he could not accept a central axiom of quantum mechanics, namely, that the outcome of any given measurement could be stated only as a probability, not a certainty. We will come back to this matter in the next chapter.

[5]Lenard, whom we will meet again, published a treatise called *Deutsche Physik* as an antidote to "Jewish science." He once remarked, "Science, like every other human product, is racial and conditioned by blood."

a frequency of 10^{14} Hz, which is why UV light can cause skin cancer and IR can't. Both photons travel at the same speed, the speed of light. Increasing the *intensity* of a light source increases the *rate* (but not the speed) at which photons leave the source. If their *frequency* has not been changed, neither has the energy of the individual photons. A mechanical analogy is that firing a pistol faster does not change the velocity, and therefore the kinetic energy, at which the individual bullets leave the muzzle, it merely changes the rate at which bullets strike the target.

What are photons? In a letter that he wrote in 1951, Einstein admitted: "All the fifty years of brooding have not brought me any nearer to an answer to the question, 'What are light quanta?'" Newton must have been smiling, for he had stated that light was a stream of "corpuscles."

Einstein used the idea of photons to quantitatively explain the photoelectric effect. The electrons in a metal are obviously held to the metal by some force; otherwise they would leave spontaneously. Picture such an electron in a beam of light as being bombarded by a stream of light bullets. If the blows are too weak the electrons will not be pushed away from the metal, no matter how intense the stream of bullets. If you can't throw a ball hard enough to dislodge a coconut, it doesn't matter if you increase the rate at which you are throwing balls. The only way to dislodge the coconuts is to throw the balls harder, to increase their kinetic energy. This is how Einstein saw things, and he declared the energy of a photon to be determined by the *frequency* of the light, through the relationship $E = h\nu$, not by the *intensity* of the light beam, which only determines the *number* of photons hitting the target in a given time. The higher the frequency the more energetic the photon. If the frequency of the light is too low, the energy of the photon will not be enough to eject the electron. Increasing the intensity (the rate of impact) of low-frequency light will leave the electrons in the metal unimpressed. However, as the energy (frequency) of the photons is raised, the electrons will feel harder and harder blows until at some value of the energy (that is, at some value of the frequency of the light) the electrons will be knocked free because they are struck by photons with sufficient energy to dislodge them. This is exactly what is observed: below a certain frequency, which is characteristic of each metal, no electrons are ejected, no matter how intense the light.

In 1901 Planck obtained a value of 6.55 x 10^{-34} joule.second for his own constant. In 1916 Robert Millikan analyzed his experiments on the photoelectric effect, using Einstein's theory, and found a value of 6.57 x 10^{-34} joule.second. Despite the apparent verification of Einstein's ideas, Millikan, in 1915, commented: "Einstein's photoelectric equation . . . appears in every case to predict exactly the observed results. . . . Yet the semicorpuscular theory by which Einstein arrived at his equation seems at present wholly untenable." In other words, whatever happened to waves? Lord Rutherford was equally puzzled: "There is at present no physical explanation possible of this remarkable connection between energy and frequency." He still thought in terms of waves. The energy of a wave *is* associated with its amplitude, that is, with its *intensity*. Bigger waves carried more energy. This is why Rutherford expected that if the intensity of a light beam was increased there was more chance that it would blast electrons out of the metal surface. But, as we said, if you can't throw the balls hard enough, you won't knock off the coconuts, no matter at what rate you throw them.

For over a decade Einstein was practically alone in believing in light quanta—

photons. Just as Thomas Young had been ridiculed when he declared light to be wavelike, so Einstein was now smiled upon pityingly when he declared light to be corpuscule-like. The wheel had turned full circle.

Today no physicist denies the existence of photons. They can be detected. When light shines on a screen covered with a suitable scintillator, the screen shines. If the intensity of the light is reduced, the intensity of light from the screen falls with it, but at a certain stage, when the light is very weak, the screen ceases to radiate equally from all parts of its surface. Spots of light begin to flicker against a dark background. These are the places that photons are striking. Similarly, photoelectric cells, which respond to radiation by producing an electric current, will give an apparently continuous output of current in a strong light, but when the light is very weak, the steady current begins to falter and finally becomes a random series of pulses. Photons are at work. Light has to be very weak for these phenomena to be detected, as you can see from the fact that an average 100-watt lightbulb gives off about 3×10^{20} photons every second, a photonic Gatling gun giving a steady rain, rather than the isolated shots of a Wyatt Earp.

A New View of Nature

Planck's concept of quantization, Einstein's photon, and the explanation of the photoelectric effect put a heretical question mark next to both Maxwell's wave theory and the belief that energy could be reeled out like curtain material and cut into any length that the customer required.

During the quarter of a century following Planck's inspired guess, it became clear that when we move down to the scale of the atom, nature behaves in ways not explicable in terms of the classical laws that govern Newton's and Maxwell's world. Classical physics was not demolished; it was shown to have very deep limitations.

We are going to tell two quantum stories. One will be about the atom, because that is the basis of our life on Earth. The quantum theory accounts completely for the behavior of atoms and molecules.

Our other story will be about quantum mechanics itself. The first story is an attempt to convince you that quantum mechanics works and is useful. The second story is far stranger, it is the scientific equivalent of a psychedelic "trip."

The Atom

The discovery of the electron and the nucleus left classical physics in an embarrassing situation: how did you put them together to make an atom? Rutherford had shown that the atom was much bigger than the nucleus, which implied that the electron was careering around in the nuclear sky. How did it resist the attraction of the nucleus? Why didn't it "fall to Earth"? Easy. It rushes around in a circle, like the Earth round the Sun, which would have solved the problem had Maxwell never lived. His equations predict that an electron behaving like a planet to the nucleus's sun should, because of its motion, emit electromagnetic waves. For the classical picture of the atom this emission of radiation has a suicidal consequence: since the emitted waves contain energy and it must come from somewhere, it can only come by a continuous depletion of the electron's kinetic energy. The electron will slow down, and as it does so it will spiral inward, ending up by crash-landing on the nucleus. It obviously doesn't do this.

In 1913 Niels Bohr invented a new theory of the atom. He used one of the great principles of science: if the theory denies the fact, reject the theory. Future historians may well ask why this principle seems to be so neglected in other areas of human life.

The first principle of Bohr's atom was that an electron *does* move in a stable path, circuiting the nucleus, rather as a moon circles its planet. His second assumption was that, in contradiction to classical physics, an electron moving around a nucleus does *not* radiate waves and lose energy. Question: Why not? Answer: Don't know; but, like Planck, let's keep going and see what turns up. The third assumption concerns what is known as the *angular momentum* of the electron. This is a simple concept. For a particle of mass m, going around in a circle of radius r, with a velocity of v, the angular momentum is just mvr. Now in the sensible, staid world of classical physics, for a given body at the end of a given string, say the iron ball whirling around a hammer thrower's head, the only way to alter the angular momentum is to change v, the velocity, and obviously v can be given any value that is desired. In short, in the classical world, angular momentum, like energy or length or time, is a continuous entity; it can have any value under the Sun. You can guess what Bohr did. He quantized angular momentum, declaring that it could exist only in multiples of a certain basic unit.

Bohr considered the hydrogen atom, which has only one electron. The electron, said Bohr, could circle around the nucleus, provided that it had an integral number of packets of angular momentum. The result of this limitation is that there are restrictions on the energy of the electron, so that the electron in the atom was allowed to have only certain discrete energies (Figure 28.1a). This restriction on the energy of the electron was accompanied by a restriction on the paths that were allowed to an electron. It could not choose any path it liked, as the Earth can in Newton's world. Bohr thus envisaged an atom in which the electron had a *limited choice* of paths around the nucleus, and in any given path the electron had an energy associated with that path. *An electron could not choose a path, or an energy, other than one of those allowed.* In this, it differs radically from the planets in the solar system. In Newton's world the paths of the planets are completely arbitrary, fixed only by an accident of birth—the way in which the original disk of matter that formed the solar system broke up. There are an infinite number of possible paths for the Earth to travel around the Sun, and they form a continuum in the sense that I can

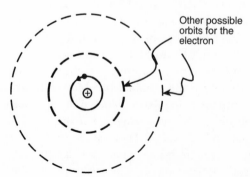

Figure 28.1a. The atom that Niels Bohr made. The circles represent some of the only possible paths allowed to the electron.

find a path with *any* particular energy that I choose. Bohr's electron was riding on a carousel with an infinite number of circles of horses, but these circles were associated with a discrete scale of energies. An electron could sit in any circle that it liked, but not between the circles, and each circle had a different energy. Bohr's conception of a nucleus surrounded by electrons, each in one of a set of allowed paths, or *orbits,* is the model behind the conventional but highly misleading drawing of an atom (Figure 28.1a).

Each of Bohr's orbits had a different distance from the nucleus. An electron that had a path farther from the nucleus would be easier to remove from the atom than one nearer the nucleus, because the electrostatic attraction between nucleus and electron falls off with distance. In the normal hydrogen atom the electron would try to get as near to the nucleus as it could, because of their mutual electrostatic attraction. It would therefore normally occupy the nearest allowed orbit, which is usually called the *lowest* orbit, because it has the lowest energy. You have to do work to "lift" it into an orbit that is farther from the nucleus. You would be fighting the pull of the nucleus, just as it takes work to lift a boulder higher above the Earth. This work, in moving between any two orbits, was one of the things that could be calculated from Bohr's theory. We can thus calculate the *energy difference* between two orbits. This allowed a connection to be made between Bohr's invisible, somewhat peculiar, and definitely hypothetical atom and the real world.

Suppose, for some reason, that an electron found itself in one of the higher orbits. Its natural tendency, due to the attraction of the nucleus, would be to fall into a lower orbit. In doing so it would lose energy, like Joule's weight falling and giving up energy to the water. Where would this energy appear?

Bohr turned to Einstein's concept of photons. The energy would appear as a photon, a "particle" of radiation, having an energy equal to that which the electron had lost. But Bohr could calculate the energy lost by the electron in "falling" from a higher to a lower orbit. It is simply equal to the energy difference between the orbits. Now the energy of the photon is connected to its frequency by Planck's relationship, $E = h\nu$. Bohr worked out the energies of the different orbits and deduced the energies, and therefore the frequencies, of the photons that would be emitted if the electron jumped between any two of them. He was predicting the frequencies of the light given off by a heated hydrogen atom (Figure 28.1b).

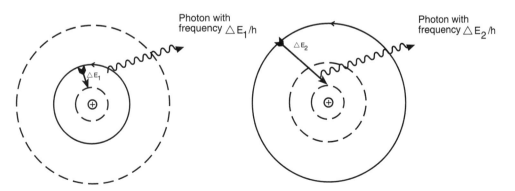

Figure 28.1b. Two of the notes that the hydrogen atom sings when an electron drops from one orbit to another.

Years before all this, several diligent researchers had observed the light emitted by the hydrogen atom when it was at high temperatures. Many frequencies were observed, but there seemed to be no pattern connecting them. And then history amused herself. A math teacher in a Swiss girl's school, Jakob Balmer (1825–1898), who was obsessed with numbers, was given the first four frequencies (or wavelengths) in the hydrogen spectrum, in response to his claim that he could find a formula to fit any four numbers. In 1885 he succeeded in fitting the wavelengths, λ, to a simple expression that was later rewritten in the following form:

$$(1/\lambda) = R[1/2^2 - 1/n^2]$$

where, if n is put equal to 3, 4, 5, and 6 the observed four wavelengths are obtained. In the expression, R is a constant that had to be given the value 109,702 cm^{-1} to make things fit. Thus if you put $n = 3$, the value obtained for λ is 6.56 x 10^{-5} cm, and this is one of the frequencies emitted by heated hydrogen atoms. No one had the faintest clue how to give a physical rationale for this infuriatingly simple expression.

The expression that Bohr obtained *theoretically*, for his model of the hydrogen atom, was:

$$(1/\lambda) = R[1/n_1^2 - 1/n_2^2]$$

which includes Balmer's expression as a particular case (when $n_1 = 2$) and allows the calculation of lines that Balmer's formula did not cover.[6] Bohr calculated R from his theory. He found a value of 109,677.656 cm^{-1}.

Whether one does or does not understand the meaning of quantized energy and angular momentum, whether or not one believes in photons, it cannot be denied that Bohr's achievement was remarkable. There must surely be some truth in a theory that can accurately predict the wavelengths of scores of lines in the spectrum of an atom. This was no trivial feat.

Bohr's Atom

The significance of Bohr's achievement is twofold:

1. Although his theory was soon replaced by a more basic approach, he had shown that the introduction of *quantization*, coupled with the courage to ignore some of the hallowed tenets of classical physics, opened up the intimate structure of matter to quantitative explanation and prediction. Today the consequences of Bohr's breakthrough are everywhere, from solid-state electronics to fiber-optics, from basic chemistry and molecular biology to the pharmaceutical industry, from the understanding of superconductivity to the behavior of white dwarfs and the forces between elementary particles.

2. Bohr's theory, by its success, showed that in attempting to explain the behavior of matter, the principle of quantization, a nonclassical idea, probably had to be extended to most of the basic properties of matter, and not only to radiation. Planck quantized the vibrational energy of oscillators, Einstein quantized radiation, and Bohr quantized the angular momentum of the electron; but it has turned out, for example, that we also have to quantize translational energies, the kinetic energy of motion. Does this mean that a molecule in a box is allowed to

---

[6]For example, putting $n_1 = 3$ and $n_2 = 4$, a wavelength of 1.876 x 10^{-4} cm is obtained. This line appears in the observed spectrum of hydrogen atoms.

travel only at certain discrete speeds? In principle yes, but the quanta of translational energy (the amounts by which you can increase or decrease an object's kinetic energy) are generally too small to be detectable.

Let's temper our enthusiasm, and clear the ground a little:

- The Bohr model is part of what physicists came to call the "old quantum theory." It gave the right answers to a small number of questions, but it was based on a number of seriously false assumptions.
- The picture of electrons revolving, somewhat like planets, in well-defined paths is false. This is not what they do at all, although they do have well-defined energies. The occasional parallel drawn between the solar system and atomic structure is ill founded. The laws controlling planetary movement and the behavior of electrons are utterly different.

In the years that followed the Bohr atom, more surprises were to come; matter was far curiouser than anyone had imagined. In the next two chapters we take a walk through the quantum Disneyland, but here we jump forward from 1913 to today and see what practical doors quantum mechanics has opened.

The Elements Revealed

The chemical properties of matter depend overwhelmingly on the properties of the electron. Quantum mechanics reveals what electrons are doing inside atoms, metals, proteins, what have you. It is a straightforward mathematical exercise to use quantum mechanics to calculate the properties of the hydrogen atom—it has only one electron. The calculation of the properties of atoms other than hydrogen is more difficult, but such calculations are getting better and better and have for long been sufficient to allow a very good understanding of the way that atoms of the chemical elements behave. Take one example, the sodium atom.

The atomic number of sodium is eleven, which means that the neutral atom has eleven protons and therefore eleven electrons. Quantum mechanical calculation shows that two of these must be quite near the nucleus on the average. It takes a lot of (calculated) energy to pull them out of the atom. Another eight electrons are, on the average, farther from the nucleus and take less energy to dislodge. Finally, there is one electron that, again on the average, is clearly farthest from the nucleus and easiest to take from the atom. We can say that the electrons are grouped in three "shells," two in the lowest shell, eight in the next, and one in the "highest" shell. We begin to understand why it is that the sodium atom has a tendency to lose one electron (see Chapter 13) and almost always appears as the sodium ion, having one positive charge. If you go to the next atom in the periodic table, magnesium, you will find that it has an atomic number of twelve and therefore has twelve electrons. These are arranged with two in the lowest shell and eight in the next, as with sodium, but there are two electrons in the highest shell, not one, as in sodium. When magnesium metal reacts, it almost always gives up two electrons, obviously those in the highest shell, which are the farthest from the nucleus. The periodic table suddenly starts to make sense. Quantum mechanics has provided an explanation of the general features of the electronic structure of all the atoms known to man. This is one of the pillars supporting modern chemistry and biochemistry, and many areas of physics.

When a molecule forms, there is a rearrangement of the electrons of the constituent atoms (see Figure13.4a). By using the equations of quantum mechanics, we can calculate, to greater or lesser degrees of reliability, the energies and the spatial distribution of all the electrons giving us a picture of the size and shape of the molecule. This knowledge allows a rational explanation of most of the properties of molecules, including their willingness to react with other molecules, and the expected results of such reactions.

The electronic structure of metals and semimetals is of cardinal importance in explaining their properties and designing new materials and devices. The study of the *quantized* electronic energy levels in metals, and particularly in semiconductors, is a central part of the theory of these materials and the explanation of the properties of transistors, semiconductor lasers, and the other sophisticated and fascinating paraphernalia of modern micro- and optoelectronics. There is a whole discipline called quantum electronics.

The Music of Matter

The light given off or absorbed by atoms and molecules is a mixture of many frequencies, a mixture that is characteristic of each atom or molecule (see Chapter 15). Bohr had shown why. For atoms it is the falling of electrons from one orbit to another that is the origin of emitted light. Similarly, when an atom absorbs light it does so because, when a photon is swallowed, an electron jumps from a lower to a higher orbit. The energy of an electron depends on which orbit it is in, and the light that it gives off, or absorbs, when it jumps from one orbit to another has a frequency that depends on the difference in energies of the two orbits. Now different atoms will of course have different sets of energies for their orbits, each atom being distinctive in this respect. Consequently, each atom will be characterized by a different set of frequencies for the light that it will absorb or emit, and this is in fact also true of molecules. *Each atom, and each molecule, sings a different tune.* Which is why, when we examine the light from the stars, we hear familiar melodies; we can recognize the chemical elements, even though we may never even reach the nearest star. Atoms or molecules for which the energy difference between certain orbits is of the appropriate magnitude can emit or absorb *visible* light when an electron moves from one of those orbits to another. This is the origin of color, and it is possible to calculate whether a molecule, or atom, will be colored.

The Greenhouse Effect

The Earth appears to be warming up. The blame has been laid at the door of certain gases, both manufactured and natural. The basic mechanism of the greenhouse effect is simply explained by quantum mechanics.

In discussing the greenhouse effect, it is necessary to take into account the vibrations and rotations of molecules. As you would expect, a molecule cannot vibrate or rotate at any old frequency. Just as for a simple oscillator, like a swing or a pendulum, there are a set of allowed vibrational frequencies for each different molecule and, as in the case of the swing, if you want to make a molecule vibrate more vigorously (with a greater amplitude), you have to give it a packet of energy corresponding exactly to the difference in energy of two given vibrations. The same applies to rotations. A molecule cannot rotate in any way that it chooses. It has certain allowed rotational energies; the carousel cannot rotate at any speed that it feels like. If it is rotating at one speed and jumps to another higher speed, it must be given the energy difference between the two "rotational levels." But you know that—from pushing playground merry-go-rounds. Thus when radiation induces a change in either the vibration or rotation of a molecule, the process involves the taking in or giving out of photons at well-defined energies, which means well-defined frequencies.[7] Molecules that have absorbed suitable photons will vibrate and rotate faster, and we can say that they are hotter. Indeed, if we put a thermometer into a sample of a gas that is absorbing light, we would find that the temperature was rising, which is not very surprising since the molecules are absorbing energy. This is the basis of the greenhouse effect, as we will now see.

The frequency distribution of the Sun's radiation, which is a typical blackbody curve, is shown in Figure 28.2. The Earth's atmosphere is effectively transparent to the Sun's rays, which means there are no molecules in the atmosphere that strongly absorb light at the frequencies emitted by the Sun. All the visible light comes through, and so do almost all the other frequencies. This radiation warms the Earth, but it does not directly warm the atmosphere because it is not absorbed. Now since the temperature of the Earth over the centuries has not risen continuously (until recently), most of the energy we get from the Sun must be sent back into space. The Earth does give off radiation, but certainly not visible light, otherwise we should have full daylight in the night. The distribution of frequencies given off by the Earth is also shown in Figure 28.2. The curve resembles that for a blackbody at about 0°C. *The frequencies sent out are generally much lower than those in sunlight and are concentrated in the IR range.* This is because the Earth is a much cooler body than the Sun. When this radiation travels upward through the atmosphere, some of it *is* absorbed by molecules in the air because some of the frequencies now match the vibrational and rotational absorption frequencies of a number of gases—look at Figure 28.2. The gases involved are primarily carbon dioxide, methane, water vapor, and

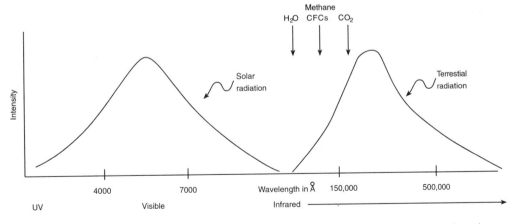

Figure 28.2 The radiation of the Sun is hardly absorbed by the gases in Earth's atmosphere. However, the radiation emitted by Earth is absorbed by a number of molecules, at the wavelengths indicated.

---

[7]Molecules can also give out or absorb radiation when their electrons jump from one molecular orbit to another, but this a process that is not relevant to our discussion of the greenhouse effect.

CFCs, the last being the (now unpopular) gases in some spray canisters.

The gas molecules that have absorbed energy from the Earth's radiation will vibrate or rotate faster; they are hotter. The atmosphere warms up. That is the greenhouse effect. The finger of blame is usually directed at carbon dioxide, but there is a growing danger from methane. This molecule is produced in a number of ways, including coal mining—which liberates underground pockets of gas—garbage dumps, rice paddies, and the thoroughly unsocial personal habits of cows. A methane molecule can trap about thirty times as much energy as a carbon dioxide molecule. The concentration of both of these molecules in the atmosphere has been growing at a rate of about 1% a year for some time.

Some scientists believe that the greenhouse effect has been exaggerated, that atmospheric warming is small enough to be a temporary glitch within the bounds of normal statistical variation. On this issue, HMS should listen to both sides but err on the side of pessimism.

Quantum mechanics has been applied to the nature of the most basic particles from which matter is constructed. It has spread into biology, into electronic technology, into cosmology, into drug design, and is even being brought into discussions on the operation of the brain. If the reader stops here, he will have come away with the impression that we have acquired a highly useful means of systematizing the behavior of matter and a reliable predictive technique. Quantization is perhaps a little strange, but you've probably heard taller stories.

This impression is correct, but it doesn't tell the whole story. Quantum mechanics has forced us to reconsider our whole picture of the universe—at the deepest level.

29

New Ways of Thinking

I have therefore found it necessary to deny *knowledge* in order to make room for *faith*.

—Immanuel Kant, *Critique of Pure Reason*

The Magic Formula

The way that Bohr treated the hydrogen atom was a mixture of concepts taken from classical physics, mixed with Planck's idea of quantization and Einstein's concept of photons. The theory did not survive; a far more serious break with classical physics was soon to come.

Before we start, the reader should glance at the above quotation from Kant. The bases of quantum mechanics are not observable or derivable. At the present stage of scientific history, all that we can say is that most scientists accept them—on faith.

The most important words in the language of those who use quantum mechanics in order to deal with matter is *wave function.*

It is a definite faux pas to ask your hostess where an electron is. You ask what its wave functions are. Every system has a *set* of wave functions, from the smallest particle, whatever that turns out to be, to the whole universe. You do too.

A wave function is a mathematical expression; its form depends on what system you are concentrating on. The set of wave functions for an atom, a molecule, or any physical system contains *everything* that can be known about the system, by which I mean that if you subject the wave functions to the right mathematical manipulations you can find out the value of any physical property: the allowed energies of the body, its magnetic properties, its electrical properties, its color, and so on. It is wave functions that give a size and shape to our molecules-in-a-box. If this is so, it sounds as if the mystery of matter has been solved. *In principle* this is true, just as for macroscopic bodies traveling at speeds much less than that of light, Newton's laws, also *in principle*, provide the means for predicting motion. In both cases the complexity of the system may result in practical limitations to our predictive power.

Finding the wave functions for many systems is a computational task that is beyond the capabilities of our largest computers, so we usually calculate approximate wave functions of large systems because that is the best that we can do. In fact, the wave functions for individual atoms and molecules containing up to twenty atoms or so can be found to a very high degree of accuracy.[1] The wave functions for larger systems have to be crudely approximated, but it is surprising how much information can be obtained by cunning. Thus it is far beyond us to obtain the wave functions for a living cell, but in dealing, say, with the molecular mechanism of human

---

[1]Even then we have to be content with wave functions that are in numerical form—that is, they are composed of tables of numbers, or combinations of simple functions, rather than compact expressions. Our inability to find analytical solutions to problems involving the motion of more than two bodies still hounds us in the corridors of quantum theory.

vision, it is helpful to know details of the structure and reaction to light of the visual pigments in the retina, and we know enough about the wave functions of these molecules to understand why they do what they do.

At this stage the reader may be asking, "Yes, wonderful—but what *are* wave functions?" That question will be sidestepped to make way for another one: "In practice, where do wave functions come from?"

Answer: To construct wave functions we have to know:

- which physical particles constitute our system,
- what their basic properties are, and
- what forces operate between them.

Thus to find the wave function for the hydrogen atom, we need to take into account the fact that there is one proton and one electron, each with its own mass and charge. We also need to know that Coulomb's law (see Chapter 8) operates between the two particles. At this point, if we were working in the eighteenth century, we would proceed to use Newton's laws of motion to calculate the paths of the electron. We must abandon this way of thinking. Instead, we put the facts into an entirely new universal equation, the *Schrödinger wave equation,* after the poetry-writing Austrian physicist Erwin Schrödinger (1887–1961).[2] This equation, published in 1926, generates wave functions.

Schrödinger constructed his equation intuitively. The equation, which in its form is very closely related to Maxwell's equations, can be regarded as a quantum mechanical equivalent of Newton's laws of motion, but its output is very different.

The first problem that Schrödinger fed into his equation was that of the hydrogen atom. He obtained exactly the energies that Bohr had; in other words, he could duplicate the observed frequencies of the hydrogen spectrum. But his equation was general; it applied to any system under, over, or including the Sun, while Bohr had constructed a theory for one system—an atom with one electron. What Schrödinger said was, tell me what elementary particles there are in your system, tell me the forces acting on them, I'll put them in my equation and, in principle, I will give you a set of mathematical expressions (wave functions) that will describe your system completely. If I can't, it's because my calculator is too small. Although the energies derived for the hydrogen atom by Bohr and Schrödinger were identical, the electronic "paths" obtained by Schrödinger were entirely different.

The solutions to the Schrödinger equation, for any system, always come as a set of wave functions, each of which is linked to an energy.

Each wave function can be used to give a picture of the whereabouts of the electron when it has one of the allowed energies. Thus for the hydrogen atom we get a series of "paths," each of which has a definite energy.

The wave functions thus appear to be the equivalent of Bohr's orbits, and so far there does not appear to be such a startling contrast between the classical and quantum world; all that we have been asked to believe is that the energy of a system is quantized, it cannot assume any value that it chooses. But there are basic differ-

---

[2]Although not Jewish, Schrödinger moved to England when the Nazis started dismissing Jews and later spent many happy years in Ireland, where he was given great support by Prime Minister de Valera.

ences between the way that quantum mechanics claims that electrons behave and the paths of the planets or the nice circular orbits of Bohr's atom. It is here that we begin to feel that we are drifting far away from the classical view of the cosmos.

The wave function for an electron differs fundamentally from the classical equations that give the path of a stone. For the stone we have expressions that link the time with the exact position of the stone. If we put a value of the time into the expression, we can calculate the position and the velocity of the stone at that time, and we can experimentally check our conclusions. The wave function doesn't give us a path for the electron, and we cannot predict with certainty that an electron in an atom will be at any one place in particular. We can calculate the *probabilities* of finding an electron at different points in space, but we can say nothing about the electron's path between those points. The electron is not a train; it may leave one place and turn up at another, but there is no point in looking for the railroad track. The smile of the Cheshire cat may appear up one tree, fade away, and then appear on the roof, but don't ask where it was in the meantime. So what does it mean when I wrote earlier that the hydrogen wave functions give the "whereabouts" or "paths" of the electron in the hydrogen atom?

Pick one corner of the room that you are in, where three walls meet at the bottom of the room. This is to be your zero point, to which all measurements of position are referred. The three edges of the room sprouting from this point will be labeled x, y, and z; the vertical edge is labeled z. Now any point in the room can be labeled by three numbers, the values of x, y, and z corresponding to that point. For instance, one possible position for the tip of your nose might be defined by x = 2.5, y = 4 and z = 1.6 (we are using meters, and I assume that you are tall enough to have your nose at 1.6 meters from the floor). Now take a simple mathematical expression—a function—in only two variables: xy, that is x times y. Plot it, within the confines of the room. This we do by finding out the value of the function at a large number of points in the room (Figure 29.1a). All of the points will be on the floor, since z does not appear in our function. A complete map of these numbers gives the value of the function at all possible points on the floor, which is not particularly useful.

The wave functions for the hydrogen atom are mathematical expressions in *three* variables, which we can choose to be x, y, and z, although there are other choices.[3] In an atomic or molecular system, any three values given to these vari-

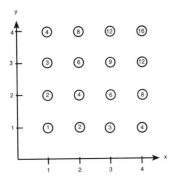

Figure 29.1a. Values of the function xy for a few points in the x-y plane.

---

[3]I have ignored a part of the wave function that deals with the so-called spin of the electron. This will not concern us.

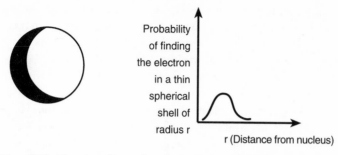

Figure 29.1b. The whereabouts of an electron in an average hydrogen atom. The imaginary sphere, which is centered on the nucleus, includes the places that account for about 99% of the probability of finding the electron. The graph shows the probability of finding the electron in a thin spherical shell centered on, and at an arbitrary distance from, the nucleus.

ables pick out a place in the atom or molecule, just as any point on the floor of our room corresponds to definite values for x and y. Given a certain wavefunction, we can choose many trios of numbers and work out the value of the wave function at many points in space. In other words, we can assign a number to any point. We use one of the wave functions that Schrödinger found for the hydrogen atom. A hugely simplified impression of the resulting map is shown in Figure 29.1b, and is described more fully in the caption. We call such maps *orbitals*. The map is the quantum mechanical analogue of one of Bohr's orbits, but it is very different in nature.

Nothing Is Certain

Max Born (1882–1970), another refugee from Hitler's Germany,[4] suggested how to interpret wave functions. Born said: Choose a very small volume in space, find the value of the wave function at a point in that volume, square it,[5] and the resultant number is proportional to the *probability* of the electron being in that small volume.[6] If we were inside the spherical orbital in Figure 29.1b. we would experience a cloud the density of which would be proportional to finding the electron at a given point. Such clouds are called atomic *orbitals*—not atomic orbits. *No path is detectable in an orbital.* Another way of presenting the probability of finding the electron in the spherical orbital is shown in Figure 29.1b.

What Born was saying was that we can only talk in terms of *probabilities* of finding the electron at a certain place, and that those probabilities are contained in the wave function. Schrödinger never accepted this interpretation. Like Einstein, he refused to believe that we can describe reality only in terms of probabilities.

Born's interpretation says that you cannot describe the flight of an electron in an atom. The electron might be at a certain point and it might not. If the value of the

---

[4]The two great theories of modern physics, relativity and quantum mechanics, are largely the product of German-speaking scientists. In many of their biographies you will find the phrase "left Germany (or Austria) in 1933."

[5]If it is a complex number, then calculate its magnitude.

[6]The exact probability is equal to the magnitude of the wave function times the volume of the box. If you always choose the box to have the same volume, then the magnitude of the wave function is proportional to the probability of finding the electron in any given box.

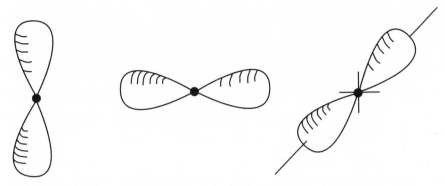

Figure 29.2. Some other possibilities for the whereabouts of an electron in the hydrogen atom. As in the previous figure, the imaginary surfaces enclose a "probability cloud." Each of the three orbitals is oriented at 90° to the other two.

wave function in a tiny box around that point is squared and the answer is 0.18, and at another point it is 0.09, then there is twice the chance of finding the electron at the first point. But there is absolutely no guarantee that it is there. It is not like a planet for which, if we know its position at a certain time, we can predict its exact position at a later time.

The Schrödinger equation automatically provides us with an energy for each wave function. The normal state of the hydrogen atom, the state with the lowest allowable energy, is that in which an electron is somewhere in the orbital in Figure 29.1b. We can say no more about its whereabouts. Thus normally, a hydrogen atom is best visualized as a spherical cloud representing the blurred image of the electron as it moves unpredictably about the tiny nucleus.

As in the Bohr atom, the electron can be induced to jump to other orbitals, having higher energies. These orbitals are based on different wave functions, just as Bohr's various orbits described different trajectories in the nuclear sky. A few of the higher orbitals of the hydrogen atom, based on the solution of the Schrödinger equation, are shown in Figure 29.2. Note that the three hourglass-shaped orbitals have identical shapes, differing only in orientation. This is a hint of the role of symmetry in shaping nature, a role that was fully revealed only in the mid–twentieth century.

Once we have a set of wave functions for a system, we can in principle manipulate it mathematically to derive not only the different possible physical distributions of the electrons but also the values of *any* physical property of the system that we want. The reliability of our results depends on how accurately we have calculated the wave function, just as, in the world of classical physics, a prediction about the position of a comet depends on how accurately we have calculated its path.

To summarize the quantum approach to the structure of matter:

- In the realm of atomic and molecular quantum mechanics, the most commonly used equation is the Schrödinger wave equation. It is the quantum replacement for the classical laws of motion.
- To solve the equation for a given system requires a prior knowledge of the types of particles that constitute the system, their mass and charge, and the forces acting between them. This is no different from the knowledge needed

to treat similar problems for macroscopic bodies, using classical laws.

- The solution to the Schrödinger equation gives the allowed, discrete energies of the system. With each energy come one or more wave functions.
- Only certain spatial distributions are allowed for electrons in a system, and those distributions are derived from the squares of the wave functions.
- There is no mention of a path for the electron. We are not told how it moves.
- We can only expect probabilities when we ask for the whereabouts of an electron.

It is particularly in the last two points that the break with classical physics is so striking, and has given rise to much speculation about the breakdown of the deterministic world picture.

Werner Heisenberg said that we should not even attempt to create a visual picture of the atom. Influenced by the positivistic philosopher-physicist Mach, he believed that all we can know are measurables, such as the frequency of spectral lines. In 1925 he produced a formalism based on measurables only. His methods, which were slightly prior to, and apparently differed radically from, those of Schrödinger, gave the same results. It was a remarkable intellectual feat, based on an area of mathematics called matrix theory. Later the two approaches were shown to be completely equivalent, merely differing in their mathematical structure.

Quantum theory has revolutionized our understanding of the interaction of matter with radiation. We can use a knowledge of the energy levels of atoms or molecules to understand the absorption and emission of radiation. We can design lasers, light sources producing synchronized photons. Using quantum mechanics, the initial steps in photosynthesis and in vision have been clarified. We know exactly why certain substances are colored and how to design colored molecules. We understand how radiation can sometimes change the shape of a molecule or even rupture its bonds.

Quantum theory is not earthbound. Thus the concept of energy levels has been used, for example, to give an explanation of the behavior of the stars known as *white dwarfs* and of the nuclear processes in the Sun that give us energy.

Since relativity is a universally applicable theory, it too should be taken into account in any theory of the behavior of matter, but it does not make an appearance in Schrödinger's equation. This defect was first remedied in 1928 by Paul Adrien Maurice Dirac (1902–1984), the English son of a Swiss father. In conformation with our approach, we will not write down Dirac's equation for an electron, but we cannot ignore one of its consequences.

The solution of Schrödinger's equation gives a series of energies to go with the wave functions. The solution of Dirac's equation gives *pairs* of energies, one positive, one negative. A lesser scientist than Dirac may well have been tempted to ignore, as meaningless, the negative energies that came out of his equation, but Dirac had a highly original mind, and the consequence that came out of his musings was that there should exist a particle, with the mass of the electron but with a *positive* charge. The positron was discovered by Carl Anderson in 1932. Dirac shared the 1933 Nobel Prize with Schrödinger. His equation also correctly predicts that the electron, and also the positron, is magnetic—that it has "spin," a property that we mentioned in Chapter 8.

The positron was the first antiparticle to be discovered, but Dirac predicted that every particle that is charged has an antiparticle, with the same mass but opposite charge. The antiproton was discovered in 1955, but the neutron, being electrically neutral, has no antiparticle. We can classify all these antiparticles as *antimatter*. The justification for the label antimatter is that, for example, a meeting between an electron and a positron results in mutual annihilation, the destruction of their charge and mass, and the formation of two photons. This phenomenon has been the basis of the fear that there might be a universe, out there somewhere, composed completely of antimatter. If there is, we can only pray that ne'er the twain shall meet.

Science never fossilizes, at least not in this century. Bohr's atom was soon overtaken by deeper theories; the interpretation of Dirac's ideas has been modified. And the quantum theory itself, the most fascinating theoretical framework that man has yet devised, remains a minefield of paradoxes. It is to those paradoxes that you may wish to turn next, but before we do there is one fact that deserves a short mention.

The Sameness of Matter

No two stars are exactly the same. No two systems of a planet and its moons are identical. But all electrons are the same. All protons are the same, as are all neutrons and all the other known subatomic particles. Thus, as far as we know, every electron has exactly the same mass, the same charge, and the same magnetic moment. "They all look the same to me." We don't know why matter comes in the form of absolutely identical particles; it is one of the great mysteries.

30

The Land of Paradox

We live in her midst and know her not. She is incessantly speaking to us, but betrays not her secret.

—Goethe

This chapter is an attempt to indicate the enchanting strangeness of quantum mechanics. We concentrate on ideas, not mathematics. We have Richard Feynman's assurance that quantum mechanics is not the most understandable of subjects, an opinion with which the reader may well concur by the time he is halfway through the chapter.[1]

Quantum theory is most concisely expressed in the language of mathematics, but since mathematics is one of the more resistible temptations in life, I have rejected the elegant formalism and contented myself with a tiny selection of straightforward equations. I am bolstered by Leo Szilard's well-known remark that elegance is for tailors.

Einstein, to the end of his life, regarded the quantum theory as being fundamentally defective. A number of modern physicists concur. One can sympathize: quantum mechanics has spawned some of the most peculiar concepts that science has ever produced.

If you took quantum mechanics at college, you have probably become blasé about some aspects of the theory that at one time were guaranteed eyebrow raisers. Maybe energy does come in little packets; so what's so strange about that? Shouldn't we accept nature for what it is? After all, the discreteness of energy may be a strange phenomenon, but it is not unbelievable. The first European to see a giraffe was probably astounded. But the giraffe was merely unexpected; it was not incomprehensible. There are some aspects of quantum mechanics that not only defy common experience, they defy reason.

Paradox One: Oh! A Life on the Ocean Wave

Prince Louis de Broglie (1892–1987), an aristocrat who used his brain, served in the French navy as a young man, where he installed the first radio on a French ship. He was thus doubly suited to make one of the most startling, and basic, advances in quantum mechanics—the suggestion that *matter behaves like waves*.

To understand De Broglie's ideas we need to know what momentum is. Momentum is simply the product of mass times velocity. If, like the sprinter Linford Christie, your mass is about 100 kilo and you are running at about 11 meters per

---

[1]Except for the material contained in the sections entitled "Paradox Two: Measurement" and "Paradox Four: Uncertainty," this chapter is not a preliminary to any other material in this book.

second, then your momentum is 100 x 11 = 1100 kilo.meter/sec. Momentum gives an indication of the impact a body has when it hits something. A double-decker bus has far more momentum than a bicycle traveling at the same speed.

In 1924 de Broglie presented his doctoral thesis to the examiners of the Sorbonne. His proposition was that if a particle had a momentum of p, then it also had a *wavelength*, λ, given by:

$$\lambda = h/p$$

where h is Planck's constant. What he was saying was that *any* moving body, say a bicycle, has a wavelength. Very strange.

The examining committee for de Broglie's doctorate didn't want to make fools of themselves by accepting the thesis if it was sheer rubbish. They sent a copy to Einstein, who gave it his stamp of approval. De Broglie was awarded his doctorate, but apparently no one but Einstein took him seriously. After all, what did it mean to say that moving matter had a wavelength? Have *you* got a wavelength when you walk along the street?

Let's put in some numbers. Linford Christie's wavelength as he crossed the line to win the 1992 Olympic 100 meters was given by λ = $h/1100$ meters, which is roughly 6×10^{-37} meters. This is about 1 million trillion times smaller than the radius of the proton. It is a fairly remote possibility that waves of this vanishingly small wavelength will result in observable everyday phenomena. And, in any case, how do we interpret the finding? In what sense was Christie a wave?

Forward. Take a very small body, an electron (mass = 9.1×10^{-31} kg), and accelerate it until it is moving at a speed of 13 million kilometers an hour. This just requires a suitable voltage difference in an evacuated container.[2] If you put these figures into de Broglie's equation, you find that the wavelength of such an electron is about 2×10^{-10} meters. This is about the kind of distance that is found between adjacent nuclei in many crystals.

Enter Clinton Davisson (1881–1958) and L. H. Germer. They directed a beam of electrons onto a crystal and found that *the beam behaved as though it was a ray of light striking a regularly patterned surface.* The electron beam bounced off the crystal in a variety of very well defined directions; in other words, it was *diffracted.* A bell rang. Davisson was aware of the then new field of X-ray crystallography. In Chapter 15 you saw that when X-rays, with a wavelength similar to the distances between nuclei in crystals, are passed into those crystals, they are diffracted. But diffraction is a wave phenomenon. How could electrons, which are particles, be diffracted?

Davisson realized that his electron beam was being diffracted just as though it were an electromagnetic wave having exactly the wavelength predicted by de Broglie. The moving electrons behaved like a wave! He published his work in January 1927. *De Broglie's matter waves were real.* J. J. Thomson's son, G. P. Thomson, carried out similar experiments, which also showed the wave nature of the electron, and he and Davisson were awarded the Nobel Prize for physics in 1937.[3] Incidentally, you can duplicate Young's two-slit experiment (see Chapter 15) with elec-

---

[2]The electrons in your TV cathode tube are moving at about 70 million km/hr.

[3]It is standard practice in freshman physics courses to mention that J. J. Thomson proved the electron to be a particle and his son showed it to be a wave. Another Nobel Prize–winning father and son were Neils Bohr, who won his prize for his work on the (electronic) structure of atoms,

trons, and get interference, just as for conventional waves.

It was de Broglie's thesis that set Erwin Schrödinger thinking about an equation to describe matter. His equation has similarities to the well-known equations for describing the movement of waves.

The atmosphere in the scientific world of the 1920s was a mixture of intense excitement and complete bewilderment. Einstein had shown that the photoelectric effect was explicable if light waves are particles; Davisson and Thomson had shown that a beam of electrons could behave like a light wave, as de Broglie had claimed. My advice to HMS is to adopt the following attitude:

1. Accept the experiments. They cannot be questioned. They have been repeated for decades in lab after lab.
2. If someone sidles up to you in Jimmy's bar and asks you whether matter is made of waves or particles, tell her or him that in some experiments it *behaves* like particles and in some it *behaves* like waves.
3. If the bartender in Harry's bar asks you if light is composed of waves or photons, tell him that it is made of photons that in their *behavior* often mimic the behavior of waves.

In both bars you will probably be refused further alcohol. At present it is accepted that both matter and radiation demonstrate what has sometimes been termed *duality*; they have properties associated with both waves and particles. I have been careful in phrasing the previous sentence. I have not said that matter *consists* of particles, or radiation of waves.

If de Broglie's relationship is correct, it should be possible to detect wavelike behavior in other particles besides the electron. Since, according to the relationship, large bodies, like Olympic sprinters, have undetectably small wavelengths, we can only expect to easily detect waves for rather small particles. This has been done for protons, neutrons, the helium atom, and the hydrogen molecule, H_2. These extraordinary findings appear to completely blur the distinction between matter and radiation, between particles and waves.

It usually does no harm to think of electrons, or protons, as if they *are* particles. Experimental evidence for the "wavelike nature" of electrons can be taken to indicate that the movement of electrons is determined by *laws* that have a wavelike nature. A cork bobbing in the sea is not a wave, but its motion shows wavelike characteristics. The analogy is extremely dangerous: it suggests that the electron is "really" a particle. We don't know that. If the thumb-screws were applied, most physicists would admit that they have no proof that the elementary particles *are* either particles or waves.

Paradox Two: Measurement

I am going to give you a faulty analogy. Toss a coin in the air, so that it revolves. This is your "system." You are now going to make an "observation" on the coin, on the system. You are going to see whether it is "heads" or "tails." To do this you have a measuring apparatus—your hands, assisted by your eyes. As the coin is in flight

and Aage Bohr, who won his in 1975 for theories of the structure of the nucleus. A unique mother and daughter pair were Madame Curie, who won the Nobel Prize in chemistry in 1911, and Iréne Joliot-Curie, who won the prize in 1935.

you trap it, so that it lies horizontally on the back of your lower hand. You look at the coin and note that it is tails. You have made a "measurement" on the system. There are only two allowed results for a measurement—heads or tails. "Heads" means a state in which the coin is *horizontal* with head upward. While it is in flight, the coin is in general neither heads nor tails, since it is rarely horizontal. If you want to define what its "state" is while it is turning, you can say that it is a continually changing mixture of heads and tails. Thus, you can imagine the state of the coin (the system) before measurement as being a mixture of the two states that you are able to detect by measurement. For example, when it is vertical, you could say that it was half heads and half tails. If you "measured" it in that state, by using your apparatus (your hands), you would find a head in about 50% of the measurements. Remember, "heads" means that the coin is *horizontal* when you look at it.

Now imagine that you could not see the coin while it was in flight. It is clear that you can say nothing about its state before you make a measurement. At a given time it is a mixture of the two *basic states,* heads and tails, but you cannot know what mixture. Second, when you do take a measurement, you force the coin into one of the basic states, *destroying its state just before the measurement.* This means that you are not really measuring the state of the system in the same way that you measure the length of a piece of wood. You know that a measurement of 2 feet means that the length not only *is* 2 feet; it *was* 2 feet before you measured it.

In the world of small particles, what you measure is the state that the system is in *after interaction with the measuring apparatus.* You usually destroy the state that the system was in before the measurement, and in doing so you lose all hope of knowing what its state was.

A system, in quantum mechanics, is usually in an unknown state before we take a measurement on it, just as the turning coin is. You can object that there are simple ways of knowing the state of the coin at any time—high-speed photography is the obvious answer. You are right, and you have found one of the faults of the analogy. In quantum mechanics there is rarely any way of completely characterizing a state before you take a measurement, and taking a measurement destroys the previous state, so you won't learn about it anyway. To clarify this, let us take a famous example, which will be of use to us later on. Figure 30.1 will help.

In Chapter 15 we spoke of polarized light, light in which the vibrations of the electric field were in one plane. We saw that there were crystals that have a special plane (see Figure 15.5), reminiscent of a slot, such that light polarized in that plane will pass through the crystal while light polarized perpendicular to the plane is stopped. We also saw that light polarized in an intermediate plane will partially pass through the crystal. We explained this by saying that we can regard such light

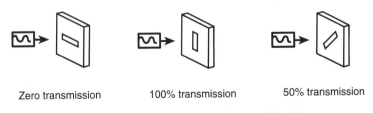

Zero transmission   100% transmission   50% transmission

Figure 30.1. The behavior of polarized photons on meeting "slits." Photons polarized at 90° to a slit will not pass. Of those oriented at 45°, half will pass, and they will emerge polarized in the plane of the slit.

as being made up from two components, polarized at right angles to each other. Only the component in the appropriate plane passes through the crystal. Thus part of the light fails to penetrate, and the intensity of the emergent beam is reduced. This explanation is all right for waves, but what about photons? Photons are indivisible. We cannot account for the reduced intensity of the light emerging from the crystal by saying that only a part of each photon goes through.

To explain the observations in terms of photons, we suppose that some of the photons go through and some don't (Figure 30.1). Those that do must emerge polarized in the plane of the crystal, since that is what happens to light that goes through such a crystal. The photons that fail to go through must end up polarized perpendicular to the crystal plane. The coin must be either heads or tails, *no matter how it fell on your hand.* The photons that interact with the crystal end up in only one of two conditions.

The probability of a given photon going through the crystal is determined by the angle of the plane of polarization. If the plane is in the optical plane of the crystal the probability is one; every photon of this type goes through. If it is perpendicular to that plane, the probability is zero (Figure 30.1). What happens when the plane of polarization of the photons is at an arbitrary angle to the polarization plane (the "slot") in the crystal? In the language of quantum mechanics we must ask this question differently: What is the wave function of the photons? We are not going to write down explicit mathematical expressions; we don't need to. Let us call a photon polarized in the plane of the crystal, A, and a photon polarized perpendicular to that plane, B, and call their respective wave functions, $\psi(A)$, and $\psi(B)$. If the plane of polarization of the light is at 45° to the crystal plane, we can suppose that the wave function of the photon is a *mixture* of equal contributions from $\psi(A)$ and $\psi(B)$, or part heads, part tails. The wave function, Ψ, of a photon polarized at 45° is given by:

$$\Psi = (1/\sqrt{2})\psi(A) + (1/\sqrt{2})\psi(B) \qquad (30.1)$$

Why $1/\sqrt{2}$? Why not 1/2? Because Born said that if you *square* the coefficient of $\psi(A)$, you get the probability that the photon will be found in state A. The square of $1/2$ is one-half, which is the probability of finding the photon in state A after the measurement. Similarly, the probability of finding the photon in state B is also one-half. The sum of these probabilities is one, which it must be since the photon has to emerge as either A or B; there are no other choices.

Wave functions often appear as combinations of the above sort, where the square of the coefficient before each component gives the probability of finding the system in that state. Thus for a different plane of polarization, I might deduce that the wave function for the photon was $0.5\psi(A) + 0.866\psi(B)$. The probability of finding the emergent photon in the state A is $(0.5)^2$, which equals 0.25, one-quarter. The probability that the photon will emerge in the B state is $(0.866)^2$, which is 0.75, three-quarters. Notice that the sum of the probabilities again equals unity, which it must do. The photon must either go through or not; there are no other probabilities.

Notice again—because it has very far-reaching implications—what happens when a photon strikes the crystal (the measuring apparatus). Either it fails completely to pass or it passes in its entirety. *The measuring apparatus forces it into one or the other of the basic components*—heads or tails. If it comes through the crystal, it will have the pure wave function, $\psi(A)$. Thus, if the plane of polarization is at 45° to the crystal plane and a certain photon goes through, we can say that it has "turned into a photon polarized in the plane of the crystal." A physicist would say that its *wave function collapses* into that of a photon polarized in the plane of the crystal. The coin has been forced to be either heads or tails. For that specific photon, the other component of the wave function, the $\psi(B)$ part, vanishes instantaneously. On the other hand, if a photon fails to get through, then that must be because its wave function has become pure $\psi(B)$. Its wave function has also collapsed. The part containing $\psi(A)$ has vanished.

In the classical picture, we said that the intensity of 45° polarized light is reduced by one-half in passing through the crystal, and we accepted intuitively that "half the light gets through." What does *not* happen in the quantum mechanical picture is that part of a photon goes through. All of it does. But only half the photons make it. Which photons go through and which don't? *We can never know beforehand.* The only way is to test them. Those that do, do. Those that don't, don't. But if the photons are all identical, what decides that one photon goes through and another doesn't? We can say nothing about this whatsoever, a statement guaranteed to irritate Einstein, who believed that one day we would know how to sort out the A and B photons before they pass through the crystal.

Our inability to predict which photon goes through the crystal is, according to quantum mechanics, not a question of yet-to-be-revealed knowledge. It is not a situation paralleling our past ignorance of the cause of malaria. That ignorance was overcome when it was discovered that malaria was carried by mosquitoes and the responsible microorganism was identified. The effect was known, and the cause was found. Einstein, and a small number of other physicists, held or hold the view that one day we will be able to know which photon will go through and which not, because there are undiscovered mosquitoes in the atomic world—they are termed "hidden variables"—which when uncovered will provide the means by which we will return to a straightforward cause-and-effect explanation of quantum phenomena. This is definitely a minority view among physicists.

Nineteenth-century science dealt in certainties. The behavior of any system could in principle be calculated on the basis of *deterministic laws*. This confidence was part of Newton's inheritance. But the times were changing, and, as in the case of entropy, quantum mechanics nearly always speaks not in certainties but in probabilities.

In quantum systems, those on the atomic or molecular scale, a measurement involves a gross interference with the state of the system, revealing only its state *after* interaction with the measuring apparatus. We believe that we can know only what we measure. This attitude is heavily dependent on the great authority of Niels Bohr in the early years of the quantum theory. As for the observed probabilities that emerge from experiment, Bohr saw them as the deepest knowledge that we could have. Some regard this as an arid view of reality, but there are no convincing signs at present that quantum theory, and in particular the probabilistic interpretation of the wave function, is going to be replaced in the near future.

Bohr and Einstein had a prolonged and famous correspondence on this issue, Einstein periodically constructing thought experiments that refuted Bohr's probabilistic interpretation of quantum mechanics, and Bohr invariably refuting the refutations. Today there is a feeling that, although Einstein was probably wrong in his attempt to bring complete determinism into quantum mechanics, the final word on quantum mechanics has not been spoken.

We saw that when a photon with an arbitrary plane of polarization passes through a suitable crystal, the wave function of the emerging photon consists of one component only of the wave function of the entering photon. The "collapse of the wave function" that eliminates one part of it is a predicted consequence of the measurement process. But the theorists have a problem.

We have suggested that the wave function of a "quantum system" collapses when it interacts with a "measuring device." Now it is a basic axiom of quantum mechanics that if a quantum system is left *undisturbed* its wave function changes with time in a completely deterministic manner. There is no question of probability jumps, or of the wave function collapsing. *The coin keeps on turning until it interacts with something,* and whether I use quantum mechanics or Newton's mechanics it will behave in a completely deterministic manner. Only an observation will break this deterministic evolution and introduce probability. With our photons we have brought in a measuring device, a crystal "pair of hands" to collapse our wave functions, but we have been criminally inconsistent. The measuring device is also subject to the rules of quantum mechanics; it should be included in a larger wave function that includes both the device and the system—the photons. If we do this we have a larger quantum system—the crystal and the photons—that will evolve deterministically and will *not* give a definite reading on the measurement device: a coin cannot stop turning in midair and become heads or tails. This would mean that the total wave function had collapsed and eliminated the part that corresponded to other possible readings. This, according to the axiom, cannot happen: a quantum system cannot get rid of part of its own wave function; the system has to be observed for its wave function to collapse. There has to be a measuring device—like my hands—*that is not part of the quantum world.*

The conclusion is that in measuring a quantum system I can never collapse the wave function if I am part of the system. The only way to get a definite reading on my measuring device, *which is part of the quantum system,* is to use another, *external*, measuring device to collapse the enlarged wave function. This means setting a device to look at the device, which doesn't help, since now, to be consistent, I have to include the new device in the total system. After all, why shouldn't it be governed by quantum mechanics as well?

Where does it end? If I believe that quantum mechanics is universally applicable; it must apply to everything, to the whole universe. In other words, the whole universe is a spinning coin. How can we ever get measurements if a quantum system cannot, on its own, collapse its wave function?

Since devices do take measurements when I observe them, then perhaps it needs the presence of a conscious mind, which is presumably not subject to quantum mechanics. The mind is the "pair of hands" that catches the coin. But why isn't the mind part of the quantum system? One way to escape this paradox is to say that quantum mechanics itself is faulty, but the problem is that when it has to answer questions, it does it so well.

We are in difficulties, but there is worse to come. The concept of collapsing wave functions can have incredible consequences. I use "incredible" in its literal sense.

Paradox Three: The Phantom Pussycat

By 1924 science had been shaken to its foundations. But there were barbarians

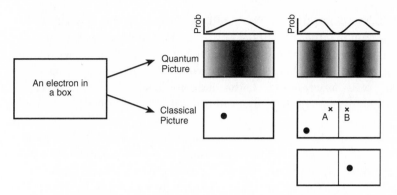

Figure 30.2. An electron in a very flat box, according to classical and quantum mechanics. In either theory the probability of finding the electron at the symmetrically placed points A and B is the same. Quantum mechanics gives us a wave function from which we can calculate the probability of finding the electron at any given point.

coming who would not leave one stone standing upon another. One of these was Erwin Schrödinger. He presented a paradox that is famous in the corridors and mousetraps of science departments: the paradox of Schrödinger's cat, which is based on the collapse of a wave function. Let us initially take a simpler example of collapse so as to concentrate on the essentials.

An electron is in a brick-shaped box (Figure 30.2). The walls of the box are opaque. Like all physical systems it has a wave function, a mathematical expression that we can use, among other things, to estimate the probability that the electron is at any given place in the box. Logic tells us that an *imaginary* partition placed straight across the center of the box would divide the wave function into two halves that, when squared, would look identical.[4] We believe this because intuition and experiment tell us that the probability of finding the electron at two symmetrically related points (A and B in Figure 30.2) must be identical. This implies that the squared wave function is similarly symmetrical, because the probabilities are related to the square of the wave function at any point.

Now slip a real, extremely thin, partition across the box. Since we cannot see inside the box, we cannot know in which half the electron is trapped. What is the wave function of the electron? Presuming that the electron cannot be found inside the partition, the wave function must go to zero there. However, whatever the form of the wave function, its square remains symmetrical about the partition—the wave function continues to exist, in *identical* form (when squared), in both halves of the box. Why identical, if the electron can be only in one half of the box? We might expect the wave function to vanish on the empty side of the box, because the probability of finding it there is zero and this implies that the square of the wave function, and thus the wave function itself, must be zero everywhere in that side of the box.

We have forgotten Born. We can make no assumption about the whereabouts of the electron if we cannot detect it.*We have not yet looked inside the box.* A gam-

---

[4]The two halves could differ in sign, one being positive and the other negative, but after squaring to get probabilities this difference vanishes.

bler, without a doctorate in physics, would tell you, correctly, that the probability of finding the electron in either side of the box is one half. What quantum mechanics says is: if you believe that probabilities are given by the values of the wave function, you cannot, *without knowing that the electron is definitely in one half of the box,* wipe out the wave function in the other half. To do so would be to unjustifiably imply that there was no probability of finding the electron in that side of the box.

Now, we determine in which half the electron actually is. What is the wave function now? Whatever its shape, it is zero everywhere in the empty side of the box, because the probability of finding the electron there is now zero—you *know* that it isn't there. Before the observer observed, the wave function covered the whole box; afterward it was confined to one side. The wave function has collapsed into one side of the box—*because of an act of observation.* There is no need to open the box to observe the electron; the box could be made of glass with a cloth draped over it.

If the preceding theory is correct you are daily responsible for the collapse of wave functions. Every time that an act of observation (a measurement) is made on a system that can, prior to an observation, give more than one answer to our question, the part of the wave function representing the answer that we don't get, collapses.

You fear that Fingers Mulloy has stolen your American Express card? You don't know? First construct a wave function for your card, a combination of that for your card when it is in your pocket, and that for the card when it is in Mulloy's pocket. Now look. It's in your pocket. The component of the wave function in Mulloy's pocket collapses. That will do nicely.

Is it necessary for the observer to be human? Couldn't I train a dog to detect electrons in boxes? Would the wave function also collapse?

Let's try a deliberately provocative version of the preceding experiment. Construct a wooden box that can be smoothly partitioned into two identical and separable parts. Put my cat Schwarz in the box. Now, slide the partitions in place, separate the two boxes, and send one to the farthest galaxy known to man. Because we don't know which box contains the hapless Schwarz, the wave function of the cat must be divided *equally* between the two boxes. What does the wave function in each box look like? Like that for half a cat? Certainly not, for the wave function in both boxes must be the same. There is a chance of finding a completely formed Schwarz in either box, not a mutilated Schwarz. Now open the box on Earth. Thank God! Schwarz is in this half. This means that *instantaneously* the wave function on the edge of the universe must go to zero—there is no possibility of the cat being there. But how can that be? How can the information have reached the second box immediately? The fastest means of communication is light, and light travels at a known finite speed. It would take millions of years for the information to arrive. During that time, presumably, the second wave function has not collapsed, which means that there is a 50% probability of finding Schwarz at the edge of the universe when I know that he is on Earth.

At present there is no answer to this question. It may be that the speed of light is irrelevant, that the "connection" between the two boxes relies on something else entirely. But we don't know what that something is.

Schrödinger was no less cruel than I. He placed his famous feline in an opaque box in which there was an atom that was radioactive. Suppose the atom to have a 50% chance of throwing out an α-particle in the next five minutes. Also suppose that when the α-particle does emerge it triggers a mechanism that kills the cat. Schrödinger made it a hammer smashing a vial of cyanide. What is the wave function of this legendary beast, five minutes into the experiment? Well, radioactive decay is completely unpredictable. It can be discussed only in terms of probabilities, which means that at five minutes there is a probability of one-half that the cat is alive and a probability of one-half that it is dead. Its wave function must contain both these probabilities, and we can do this by constructing a wave function that is half that for a dead cat and half that for a live cat:

$$\Psi = (1/\sqrt{2})\,\psi(\text{dead cat}) + (1/\sqrt{2})\,\psi(\text{live cat})$$

As Schrödinger might have said, wass is das? The living dead? A cat in the Twilight Zone? This wave function is certainly nothing to do with classical science. Of course we can look in the box, but then the wave function collapses to that of a cat or that of an ex-cat. The following question has been asked: If the presence of the observer collapses the wave function, then is not the cat an observer? And if a cat can induce wave functions to collapse, what about mice, disgusting beetles, curious ants? Surely a mouse would be able to tell the difference between a live and a dead cat. And is the presence of, say, a mouse sufficient, or does the mouse have to *know* that the cat is dead or alive? Einstein was skeptical about the possibility of a mouse changing the wave function of the universe. The great Hungarian scientist Eugene Wigner held that if a machine made a measurement, for example, by detecting in which half box an electron was, the wave function would not collapse until a conscious mind observed the machine. This brings us back to the previous section. If Wigner is right, and he was a little cautious about this at a later stage, it means, for example, that during the whole history of the universe before conscious life evolved, there was no means of collapsing a wave function. This is a real problem. Supposing that the creation of life depended on a certain molecule reacting with another molecule and that there was a 1% probability of them reacting and a 99% probability of them not reacting. In the absence of a conscious mind to collapse the wave function containing these two probabilities, the creation of life would be in the position of Schrödinger's cat—neither here nor there. *We* know that the reaction must have taken place, so have we collapsed it retroactively?

If all this sounds like Ionesco or the *Monty Python Show*, and you don't like this idea of collapsing wave functions, take a breath and see how the following argument grabs you.

Hall of Mirrors

Everything we have said about collapsing wave function stems from the basic principle that, in quantum mechanics, any system is associated with a mathematical expression that includes within it the possible result of *any* measurement on the system. If I can get either of two possible answers, then that information must already be in the wave function. When I observe the system, I can get only one answer. After the observation the system must therefore be in such a state as to give me the answer that I observed; the alternative answer is no longer possible, and the part of the wave function associated with it must have disappeared. Until an observation is made, all possible answers remain a possibility, which means that the wave function has not collapsed. Thus the basic assumption is that a wave function will not collapse unless it interacts with an "outside" system. But what is an outside system? Why, in the experiments with Schwarz, don't I take myself to be part of a quantum system that includes me, the box, and the cat? And then we could not record an observation (collapse the wave function) unless an "outside" observer observed us. Which just puts the problem one step back. Why not take the second observer to be part of a quantum system that includes . . . (There was an old lady who swallowed a fly . . .)

The logical end of the tale is that we have to take the whole universe to be one large quantum system. The mighty wave function includes all possible answers to all possible questions. How can it change? How can the spinning coin stop spinning if I, the observer, am actually part of the system? There is a very strange way out.

Consider the cat-in-a-box experiment. There are two possible outcomes: either the cat is dead and you are conscious of a dead cat, or the cat is alive and you are conscious of a live cat. Note that you have two possible states of consciousness, one associated with the dead cat, the other with the live cat. The complete wave function for the cat *and you* should include both your states of consciousness. Now suppose that the wave function cannot collapse; there is nothing outside the quantum universe that can "stop it spinning." But something does happen—when we open the box, the cat is either dead or alive. And here comes the crazy solution. Because there are two possibilities as to what will happen, *both things happen*. The universe instantaneously splits into two universes—one contains a live cat, the other a dead cat. The wave function does *not* collapse; both its components survive, each in a *different* universe. *And there is a copy of you in both universes.* Why? Because before the "observation" there was a probability of having the image of a dead cat in your mind, and also a probability of having the image of a live cat in your mind. You were a part of the wave function; you can't be collapsed.

In this interpretation, wave functions don't collapse, they survive—in this case one in one universe, one in the other. One of you is looking at a live cat, while in the other world one of you is looking at a dead cat. There is no communication between these two worlds. The general suggestion is that every time an event can occur in one of several ways, it occurs in all those ways, each possible event occurring in a different universe. Thus, in the experiment in which photons go through a crystal, every time that a photon makes a choice and emerges from the crystal, another world splits off in which the other choice was taken. The wave function doesn't collapse; part of it survives in each universe. See how neatly this solves the problem of Schwarz and the farthermost galaxy? The universe splits in two, and the "other" universe contains a cat in a box in a distant galaxy. No need for messages traveling far faster than the speed of light.

Or are you saying: "If that's a neat solution I'm a fax machine."

The theory of multiple universes, first suggested in 1957 by Hugh Everett III in his doctoral thesis at Princeton, is as bizarre as anything that the rather prosaic practitioners of spiritualism and satanism have thought up—and perhaps less believable. Where are all these universes?

Has anyone yet communicated with any other universe? Not in my family, but the idea is a boon to spiritualists. I presume that I exist in a huge number of copies. I am dead in some of them, because I have many times been in situations where there was a reasonable probability that I could have been killed, and my wave function at the time of these incidents, must have been:[5]

$$\Psi = c_1\Psi(\text{live Brian}) + c_2\Psi(\text{defunct Brian})$$

After each hair-raising incident, the universe split. In this universe, I made it. In the other, I was buried, perhaps along with Schwarz. It gives a whole new meaning to the expression "Let's split."

Among the very few people who take the many-universe theory seriously are those attempting to apply quantum mechanics to cosmology, the study of the large-scale structure and dynamics of the universe. But the theory has not provided an explanation of any observations for which conventional quantum mechanics has failed.

Richard Feynman advises us to "avoid being confused by things such as the 're-duction of the wave packet' and similar magic," but quantum mechanics provides the most damning evidence that science is a marvelously effective battleship floating on a deep, ever-changing sea of illusions.

Paradox Four: Uncertainty

> Something unknown is doing we don't know what.
>
> —Sir Arthur Eddington on the uncertainty principle

Descriptions of large lumps of matter specify their *position* and *velocities.* The accuracy with which these variables can be measured is limited only by human and technical shortcomings. In principle we can go on improving this accuracy to undreamed-of heights.

It will help us later in this discussion if we agree that, knowing the velocity of a body, we can, provided we know its mass, immediately work out its momentum, since this is defined as mass times velocity, $m \times v$, or just mv. In classical physics, we believe that a body, say you or me, always possess a *definite* momentum and position—how could it be otherwise?

One major tenet of quantum mechanics, enshrined in the uncertainty principle, is the proposition that a body *cannot* have both an accurately known position and an accurately known momentum. That, *in principle,* we cannot simultaneously measure these two things to any degree of accuracy that we choose is now accepted by the vast majority of physicists. We are edging into strange terrain here, but let's press on through the trees and listen to the language of the natives as they describe the motion of an electron.

The uncertainty principle was stated in 1927 by Werner Heisenberg (1901–1976), who ended up in charge of Hitler's effort to build an atomic bomb and later claimed that he wasn't really trying. As my grandfather remarked when my father denied smoking, "I believe you, but don't do it again."

Heisenberg attacked the most basic problem in experimental science—the nature of measurement. If you measure any property of a system, say the mass or velocity of a particle, there is a certain error involved. The size of that error is primar-

---

[5]This is not the complete wave function. I should include a part that represents the rest of both universes, except me.

ily dependent on the instrument used and the care of the observer. Suppose that we are measuring the position of a particle along a line that we label x. We call the uncertainty in measuring the position of the particle, Δx (delta x). Using the naked eye, Δx has a minimum value of about one-fifth of a millimeter. It is difficult to distinguish two points separated by a smaller distance. Now suppose that we simultaneously measure the momentum along the same line. Again there will inevitably be an uncertainty that we will label Δp. Classical physics puts no *theoretical* limitations on our ability to reduce these uncertainties as much as is technically possible. Heisenberg thought otherwise. The consequence of his mathematical formulation of quantum mechanics is that the *product* of the two uncertainties, Δx times Δp, or $\Delta x \Delta p$, has a lower limit; it is proof against any attempts whatsoever to reduce it. This lower limit is $h/2\pi$, where, as usual, h is Planck's constant. If we had perfect instruments, we could reach this limit. We could write:

$$\Delta x \Delta p = h/2\pi \qquad \text{(1)}$$

In general we have to write $\Delta x \Delta p > h/2\pi$, where the greater-than sign, $>$, indicates that the product can be larger than $h/2\pi$, due to human failings. What Heisenberg's uncertainty principle says is that you cannot simultaneously measure the position and momentum of a particle to any degree of accuracy that you want, *no matter how sophisticated your apparatus is*. This is a property of nature. We take a very important example.

The uncertainty principle precludes the existence of a particle that is stationary. For such a particle there can be no uncertainty in its position, a fact that can be expressed as $\Delta x = 0$. Put this into (1) and you find that $\Delta p = $ infinity. What does this mean? It means that if you try to measure the momentum of this supposedly stationary particle it can have *any* momentum. For if there was a value of the momentum above which there was no possibility of finding the particle, then the uncertainty in its momentum would not be infinite, it would be finite. But if there is a chance of finding *any* value of the momentum, then we have the absurd situation that we have a stationary particle that is moving. For to have momentum it must move. The conclusion is that there are no completely stationary particles, which is what I asked you to believe in Chapter 16.

If you cool down a chunk of metal, the vibrations of the atoms become less violent, but you will reach a point at which the amplitude of the vibrations refuses to get any smaller. You can do nothing after this; you can't stop the atoms dead. The jargon for this minimal energy is *zero-point energy*. The fact that it exists, and has been shown to exist experimentally, is another extraordinary achievement of scientific theory. But who in his right mind would say that it is impossible to stop something vibrating?

A Different Point of View

Look back at Figure 15.7 and recall the conclusion that we came to concerning wave packets. If you know the position of a packet fairly accurately (Figure 15.7c), then the uncertainty in its position, Δx, is small. However, a strongly localized packet cannot be given a well-defined frequency because the packet contains such a wide spread of frequencies. Now suppose, in the spirit of de Broglie, that the packet was really a particle and that it had a momentum *determined by its wavelength*. Its wavelength is not well defined because its frequency is not well-defined.

Thus there is a big uncertainty in its wavelength, and as follows from de Broglie's equation, a correspondingly big uncertainty, Δp, in its momentum. Δx is very small and Δp is very large. Just as Heisenberg says, if one uncertainty, in this case Δx, is very small, then the other, in this case Δp, must be big.

If you look at the packet in Figure 15.7b, you will see that in this case the uncertainty, Δx, in its position is quite large. However, since it is composed of waves with very similar frequencies, the uncertainty in its wavelength, and therefore in its momentum, Δp, is quite small. Again, raise your hats to Heisenberg.

Heisenberg gave a practical illustration of the impossibility of measuring velocity and position, showing that measurements on a particle inevitably involved interaction with the instrument. For example, in a microscope the specimen under observation is bombarded by photons. For a large object this makes little difference, but on the atomic scale the photon will disturb the object, changing its position or velocity, or both, and thus destroying any chance of measuring both variables accurately. This has been described as the newspaper reporter making the news. Analysis of the experiment gives equation (1).

The probabilistic interpretation of quantum mechanics, and the added fuzziness due to the uncertainty principle, have had an appeal to those searching for free will. They see an unpredictability about nature, a looseness that opens up the steel chain of cause and effect. This is a very provocative subject, but one that would take too much room to discuss properly, for this book.

Paradox Five: Nothing Up My Sleeves

The uncertainty principle's most remarkable trick is to produce something out of nothing. The claim of the uncertainty principle is that it can conjure matter out of a vacuum. This is better than the ten little fishes and the five loaves of bread.

To perform this miracle we use an alternative form of the uncertainty principle. The form that we need uses energy and time:

$$\Delta E \text{ times } \Delta t \gtrsim h \qquad (2)$$

What does Δt mean? For our purposes, the best way to interpret the meaning is to regard it as the lifetime of a system having an uncertainty, ΔE, in its measured energy. A simple example is a particle that has an infinite lifetime, a stable particle. Then Δt is infinite and ΔE is zero. In other words, the energy can be measured exactly. Now consider a hydrogen atom in which the electron has jumped from the lowest orbital, the stable state, to a higher orbital. The experimental fact is that the electron very rapidly falls back again. We can say that the *excited state* is very short-lived. It is also an experimental fact that the energy of such a state is "blurred"; when we attempt to measure its energy, we find a range of values that is wider, the shorter the lifetime of the state. *Very short-lived states have very ill-defined energies.*

Now, what we are asked to believe is that particles can suddenly materialize out of a vacuum. By analogy with the zero-point energy of vibration, it can be supposed that electromagnetic fields also have a kind of zero-point energy. Then we can imagine that energy is "borrowed" from the field for a fraction of a second and a particle is created. We are assuming the equivalence of mass and energy, demonstrated by Einstein (see Chapter 32). After a short time, the particle sinks back into the sea, but a short lifetime implies a large uncertainty in the measured energy of the particle. In other words, it can have a large energy. We will call such phantoms *virtual particles*. Notice again that the shorter-lived the particle, the smaller is Δt and the larger is ΔE.

Now Einstein tells us that $E = mc^2$. This means that a particle with a lot of energy can, in many circumstances, be regarded as a particle with a lot of mass. Putting everything together, it appears as if virtual particles that are very heavy have very short lives. This will be of relevance to us in the next chapter.

The only excuse for this far-fetched story is that it explains certain facts. See how you feel about it after this.

Imagine a *real* electron sitting in the vacuum. All around it there are pairs of *virtual* electrons and positrons continually rising from, and returning to, the fishy deep. The real electron will attract the positively charged virtual positrons and repel the virtual electrons. This means that a real electron will maintain a permanent coat of virtual positrons around it. Most electrons in nature have such a coat; they are said to be "dressed." This has an observable effect. Imagine that you are another electron approaching our dressed friend. You will feel the repulsion of her negative charge, but that will be slightly reduced by the attractive force due to the coat of positively charged virtual positrons. The other electron will feel the same about you, because you too will be dressed. The normal coulombic interaction is thus measured between dressed electrons. However, when two speeding electrons in an accelerator approach each other, they can get to distances of approach that are far smaller than those expected in the everyday world. When the distance between them is very small, they wildly throw off their coats and become "naked" electrons. This increases the repulsive force between them to a value that is larger than we might have calculated from the coulombic repulsion at larger distances. This effect has been observed experimentally.

A very important virtual particle is the virtual photon. Some people envisage the virtual photon as emerging from the electron, buzzing around for a very short time and then being reabsorbed. It is as though the electron was surrounded by a permanent cloud of virtual photons, being born and dying almost as soon as they emerge. Although virtual photons don't last that long and generally return to their maker, they occasionally die by interacting with another electron. *This is the way that quantum mechanics sees the coulombic interaction.* This is an important turning point for us, and we pause to chew things over.

The idea of force at a distance, as Newton found, came under attack from a variety of scientists and philosophers. Now we have an entirely different interpretation of force. Electromagnetic force is pictured as the exchange of virtual photons. A variety of analogies are used to give students a feeling for this explanation of force. The standard ones involve two people skating parallel to each other and throwing objects, back and forth to each other. The thrower moves backward as he throws, for the same reason that cats move backward when they stand on ice and sneeze. The catcher moves backward because of the momentum of the object caught. The skaters appear to repel each other. The analogy works only for a repulsive force. An attractive force needs a rather strained analogy, usually involving boomerangs!

This radically new way of looking at force will be employed when we come to the elementary particles. It is a classic example of the way in which science reinterprets reality. The previous picture, of a force acting through space, obeying the coulomb law, works when we use it for most problems. Down to a distance of about 10^{-11} cm, the coulomb equation works perfectly, and calculations of the force between charged bodies are not reliant on the supposition that virtual photons are bopping around. The main justification of the new picture lies in other areas, particularly in the combination of quantum mechanics, Maxwell's equations, and relativity that is primarily associated with the name of Richard Feynman and is known as QED—quantum electrodynamics. This theory gives an unbelievably accurate account of all phenomenona involving the electromagnetic force. QED's party piece, rolled out whenever its credentials are under scrutiny, is the agreement between the calculated and experimental value of the g-factor of the electron. The meaning of the g-factor is irrelevant;[6] it is the agreement between experiment and theory that is supposed to astound you (it astounds me):

Experimental: 2.0023193048(8)
Calculated by QED: 2.0023193048(4),

where the numbers in parenthesis are the uncertainties in the last figure. Can anyone doubt that all three components of QED—Maxwell, relativity, and quantum mechanics—have some truth in them?

My conscience drives me to add a footnote to the last sentence. The theory relies on a trick: dividing both sides of an equation by infinity. This is frowned on in the best mathematical circles. Moreover, the theory is not drawn entirely from thin air; it is necessary to put in the *measured* mass of the electron before you can get the right calculated values of other physical quantities.

Do virtual particles exist? Perhaps they are just a convenient way of explaining a number of phenomena, but no more than that.

Paradox Six: They Seek Him Here . . .

We have not dared to choose between Newton and Einstein on one hand and Thomas Young and Maxwell on the other. One set of Giants spoke out for corpuscular light—Newton using intuition, Einstein showing that photons explained the photoelectric effect. On the other hand, Young's experiment is so convincing that it is difficult to see how light can be anything other than waves. And then along comes Maxwell with a theory that is preposterously successful in predicting the speed of light and the production of all types of electromagnetic waves. And, as we asked previously, how can interference be explained by particles? Can bullets annihilate each other? Let us wander farther into the enchanted wood.

Here is a very strange, eerie experiment. We are going to repeat Young's classic experiment (see Figure 15.4a), sending light through two slits and observing the resultant alternating dark and light lines. We thought we knew why the lines appeared: because the waves from the two slits could constructively or destructively interfere, arriving in phase or out of phase. But how can we possibly explain the pattern in terms of photons?

We will take advantage of the availability of extraordinarily sensitive detectors. We are going to use instrumentation that can detect *one* photon.[7] This allows us to use extremely weak light.

---

[6]It gives the relationship between the spin of the electron and its magnetic moment.
[7]Nature doesn't do too badly; there is evidence that some of the visual rods of the human eye can be triggered to give an electrical impulse to the brain, by *one* photon.

When we perform the two-slit experiment with extremely weak light, we find that the detectors pick up *individual* hits of the photons as they strike the second screen. There can be no doubt that photons are arriving, not continuous waves. You can detect the localized flashes on the screen. It is as though rain were gently spattering the marble paving stones of a previously dry piazza. This looks good for photon fans. But not for long. As the hits are recorded, they start to build up a pattern and gradually the pattern emerges: alternate lines! Of course this is what we should have expected; that's what the result of the two-slit experiment always is. The problem is that now the dark lines in Young's experiment seem not to be produced by interference but simply by the fact that photons never arrive at the locations of the dark lines. We never record a hit at the center of the dark stripes. This means that darkness does not come about by photons arriving together and annihilating each other. They just never arrive.

Photons spreading out from a particular slit apparently "know" that they must not travel in the direction of the dark stripes. There is something distinctly weird going on here. Imagine that you are a photon passing through a slit. How on earth do you know not to go toward the location of a dark line? How does a raindrop avoid certain places? Maybe it is due to some kind of interaction between the slit and the photon?

In the scientific community there is a well-known phrase: "One experiment too many." We are going to risk another experiment.

If a photon "knows" that it can't go to a dark line *because it interacts with the slit as it goes through,* then shutting the other slit shouldn't make any difference. How can a photon possibly "know" that the other slit is closed or open? We should get the interference pattern with one slit open! Try it, do one experiment too many.

The interference pattern disappears. Of course it does. You wouldn't expect anything else; shining a light through a single slot should give a simple, slightly blurry-edged image of the slot. No dark lines, no interference pattern. Young could have told us that—the trouble is that his explanation is not the one we want. He would have said that interference is impossible with only one slit because the part of the wave going through one slit no longer has anything to interfere with. The photon model just leaves us with a headache: a photon going through one slit when the other is closed loses the ability to avoid the dark areas. The photons just spread out over the whole (previously patterned) area. If you open the second slit, the pattern reappears. The path of each photon appears to be influenced by the *presence* of the other slit!

Since it seems unlikely that a photon going through one slit can "know" about the state of a second slit, let's try another hypothesis. Maybe the photons from two slits cooperate in some way, while they are in flight, giving each other instructions? Crazy? Yes, but anything seems possible in the quantum world, so let's eliminate that possibility. We will lower the intensity of the light so much that on the average there is only *one* photon in flight at any one time. It's no use—the striped pattern builds up! Incidentally, this experiment also completely rules out the possibility that, at higher flows of photons, there are photons arriving in the dark spots together but annihilating each other.

At this stage our position is that the pattern is produced by single photons traveling completely independently of each other. This implies that a given photon going through a certain slit knows where to go and where not to go (dark or light stripe), without interacting with other photons. (OK, you didn't expect interaction, but the experiment had to be done). How does the photon know where not to go? This is very difficult to understand. So difficult that no one has explained it.

How does a photon, or an electron, know that the other slit has been closed? And why does opening the other slit ensure that an individual photon going through a slit knows that it must avoid the location of the dark stripes? These findings, checked and rechecked, defy common sense and classical physics. One way out of the marsh is to suppose that a photon simultaneously travels along several paths. I won't elaborate here, except to comment that we are in very strange country.

The results of the two-slit experiments cut away at the very basis of our understanding of the way the universe is. Richard Feynman used to say that the two-slit experiment was *the* problem in quantum mechanics. Some waffling noises have been made by a number of theoreticians, but there is clearly something very basic about the stuff of the universe that either is waiting to be revealed, as Einstein firmly believed, or is simply outside the capabilities of our brains to grasp.

Paradox Seven: Do We Really Understand What's Going On?

In 1935 Einstein, with two younger men, Boris Podolsky and Nathan Rosen, wrote a paper that became so famous that it is now known simply by the initials of its authors: EPR. Until recently I frequently saw Professor Rosen walking along the paths of the university. He was a small, courteous, soft-spoken gentleman, the last person one would have expected to cause so much trouble.

Heisenberg's conclusion that it was *in principle* impossible to simultaneously measure certain pairs of properties, for example, a particle's position and its momentum, gave Einstein no rest. He was perturbed, not so much by the supposed impossibility of simultaneously measuring the position and the momentum of a body to any accuracy but by the deeper statement that a body could not simultaneously *have* a definite position and a definite momentum. What could such a statement possibly mean in the real everyday world? Einstein searched for an experiment that would demonstrate that the position and momentum of a particle could be determined exactly, and reproducibly, and that the properties of a system could be predicted before they were measured, thus showing that they really did exist.

EPR is an attempt to show that there are experiments that can sidestep the Heisenberg uncertainty principle, in the particular case of position and momentum, and also allow the correct prediction of a particle's properties before measurement. The experiment was not performed by the authors; it was a thought experiment.

The idea of EPR is simple. It relies on the principle of the conservation of momentum, a principle of Newtonian mechanics that quantum mechanics does not question.

If a system has a certain amount of momentum and no external forces act on it, the total momentum of the system remains constant, forever. The easiest case is a body, say a spaceship, moving through space, with no forces acting on it. The first law says that its velocity remains unchanged. But if this is true, then the product of its velocity times its mass, which is its momentum, also remains unchanged. Another example: imagine two identical pieces of chewed chewing gum, each having the same mass and the same velocity, sliding toward each other on a nonsticky surface. In this case the velocities come with a sign; if the velocity of one piece is v

The Ascent of Science

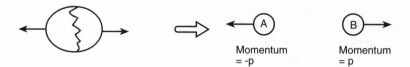

Figure 30.3a. The EPR thought experiment deals with two identical particles with identical, but oppositely opposed, momenta.

then we must write the other one as -*v*, because it is in the opposite direction. Thus if the momentum of one piece is *mv* the momentum of the other must be -*mv*. This means that the sum of the momenta (+*mv* + -*mv*) is zero. When they hit each other, they will stick together and both stop moving. The total momentum remains *unchanged*, because no external forces acted on the system. It is still zero.

Now suppose that an initially motionless particle disintegrates into two *identical* fragments that fly apart, the two fragments moving along the same straight line (Figure 30.3a). According to Heisenberg, we cannot get an exact measurement of the position *and* momentum of either particle. The notoriety of EPR rests on the claim put forward in the paper that there is a simple experiment that overcomes this quantum mechanical ban, and allows both the position and the momentum of either of the two particles to be measured to any degree of experimental accuracy. We follow the argument of EPR.

Keep in mind that we are going to measure the position and momentum of particle B. Since initially the single, pre-explosion, particle has no momentum, and there are no forces acting on it from without, the total momentum of the system (the two identical, moving fragments) must remain zero after the disintegration. This means that if one fragment has a momentum of *p*, the other *must* have a momentum of -*p* in the other direction (Figure 30.3a). Thus if we measure the momentum of fragment A very accurately, we can immediately *deduce* the *momentum* of B, without observing it directly. This is the first part of the EPR experiment, and note that we have made no direct observation on B.

Now there is nothing to stop us from accurately measuring the *position* of B; the uncertainty principle does not limit the accuracy of *one* measurement on a system. Thus we can measure the position of B (directly) and its momentum (by inference), to any degree of accuracy that we like. Heisenberg says that you can't do this.

But there is another point here. Heisenberg says that a particle cannot simultaneously *have* a defined momentum and position. EPR denies this. Consider particle B. After measuring the momentum of A, we can predict, with absolute accuracy, the momentum of B, *before* we measure it. A central tenet of the conventional theory of quantum mechanics, associated with Born and particularly Bohr, is the doctrine that you cannot know anything about a particle until you measure it. Only when you trap the spinning coin can you say something definite. Einstein says, vis-à-vis EPR, "If, without in any way disturbing a system, we can predict with certainty . . . the value of a physical quantity, then there exists an element of physical reality corresponding to this physical quantity." The particle B has a real position and a real momentum, real because predictable, since either can be predicted from the values measured for A. Again Einstein took the commonsense view of the world. There are real objects out there with real properties *that do not depend on the observer.* Heisenberg and the other founding fathers say no.

Bohr was adamant. There had to be a flaw in EPR—particles did not have prop-

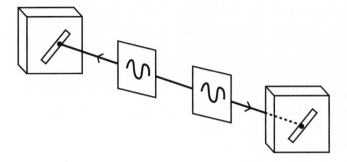

Figure 30.3b. The two identically polarized photons both have a 50% chance of passing through their slot. The "choice" of whether a given photon does or does not might be expected to be completely independent of the behavior of the other photon. It isn't.

erties before they were observed. There is one very unlikely way around the problem, a supposition that will save Heisenberg. We suppose that fragments that once were together are always connected, in the sense that measurements on one fragment, contrary to EPR, *do* affect the other. In this "spooky" scenario (the adjective is Einstein's), a measurement of momentum on fragment A would induce an uncertainty not only in the measurement of the position of the same fragment but also in that of fragment B. Some kind of blood brotherhood unites the fragments forever. Rubbish? How can it matter to one fragment what is being done to the other? It has since been shown, *experimentally,* that it does. And we have no idea how. EPR doesn't work. Thought experiments, even by one of the two greatest scientists in history, need checking experimentally.

Notice that the argument of EPR depends on the assumption that measurements on one of the particles *cannot* affect what is measured, or what happens in any way, to the other particle. This sounds straightforward; why should there be an effect, unless we are extremely careless? This is called the *assumption of locality*, and it seems so obvious as to be taken for granted in classical physics.

A crucial event in the history of EPR, and of quantum mechanics in general, was the publication of a paper by John Bell in 1964, in the journal *Physics.* By simple arguments Bell proved that if the two parts of the system, say the two fragments above, cannot affect each other, then it is impossible to obtain the results observed in the actual experiments. The experimental results prove that locality doesn't apply—particles A and B are "talking to each other."

The first real experiment showing that the thought experiment of EPR was contrary to experience was performed by Alain Aspect in 1982. It was not exactly the EPR experiment, although that has now been carried out, but an equivalent experiment on photons. The purpose of the experiments was to show that two photons are in some way in communication with each other, so that measurements made on one affect the measurements made on the other. Instead of going through Aspect's experiment, we take a particularly simple case that illustrates the principle involved.

Two photons fly apart (Figure 30.3b). They both have a perpendicular plane of polarization, and they both strike crystals with polarization planes ("slots") at 45° to the vertical. As we saw, such a photon has an equal probability of going through or being stopped. Now, common sense says that these two photons make their choices independently; whether one of them goes through or not can have no effect on the choice of the other one. In fact, experiment shows that the "choice" of one photon, whether to go through or not, is conveyed to the other proton and affects its "choice." The incredible thing is that the experiment can be performed in such a way that at the time of the measurement on one photon the other photon is too far away to receive a light signal before it too undergoes a measurement. Reason says that it cannot be "told" what to do by the other particle before it itself is measured. And yet it seems to know what the result of the other measurement was, and to react accordingly. The particles not only speak to each other but also appear to communicate faster than light.

There have been some very clever variations of these experiments, but they all point to two inescapable facts:

- The photons or particles involved are not "localized"; something connects them across space. There is no reason to suppose, on the basis of the experiments, that they could not affect each other *instantaneously* across astronomical distances.
- Where there is a choice of results for the measurements on one fragment, the observed result, whatever it is, affects the result of a measurement on the second. It cannot therefore be said that the second particle possessed certain fixed values for its properties *before* they were measured.

The first property, *nonlocality,* is perhaps the strangest finding of quantum mechanics to date. *Spooky* was the right word. Much has been made of this weird togetherness. The universe has been envisaged as a collection of particles bound to each other by past associations but unable ever to break those ties, no matter how far apart they drift. A holistic cosmos, but bound by what? And what is it that allows particles to communicate faster than light, the supposed fastest means in the universe?

Are we capable of understanding reality? Is the scientific approach an approach to reality? Is there a way out of the maze?

Familiarity Breeds Consent

As a schoolboy I often met aggressive anti-Semitism in the playground, sometimes from considerably older and bigger boys. My father's advice was simple: "You can fight better than any of them. Single out the leader. There's always one. Go up to him and punch him really hard on the nose." "And what about the others?" "They won't do a thing." They didn't. It worked in three schools. The mindless bullying stopped immediately. I was dangerous, even grudgingly admired. To me the method seemed completely senseless. I was always outnumbered, but my father's apparently irrational recipe worked. That's how I feel about quantum mechanics. Our scientific fathers, those whom we revere—Schrödinger and Heisenberg, Born and Dirac, Bohr and de Broglie—have given us a recipe for success. It works every time, but it flouts common sense. As I grew up I began to appreciate the rationale of my father's advice. It really did make sense. Will it be the same with quantum mechanics? The augurs are not too favorable.

How can thousands of scientists use a theory that has irrational features to it? Because where theoretical results can be compared with experimental observations, one can only pray that all theories were as reliable. No scientist would dream of stopping using quantum mechanics because he doesn't understand its foundations. We keep hitting the leader of the pack on the nose.

Among scientists there is no uniformity in response to this unsatisfactory situation. The predominant opinion by far is: no opinion. But many theoreticians are unhappy with the assumption that quantum mechanics, as conventionally formulated, is a correct, final theory. The fact that, to deal with quite straightforward questions, like Schrödinger's cat, we might have to stray into multiple and continually splitting universes, makes many scientists ask whether quantum mechanics is missing something important.

Personally I feel that the possibility that we are too limited should not be discarded. As Lords of the Universe, it does not flatter us to be told that there are things beyond not only our ken, but even our capacity to ken. Since Lucretius there has been an implicit belief among scientists that "in the end" Nature will be comprehensible. This may not be true, if to "comprehend" means to reduce all explanations to concepts that are acceptable in terms of logic and current common sense. It is more likely that we will get used to using concepts that we don't really understand, content in the fact that they allow us to "explain" what we see as objective reality and to reliably predict the outcome of experiments. In the end we may be no more capable of understanding the big questions about the universe than Schwarz is able to write haiku.

The paradoxes of quantum mechanics have provoked many thinkers to pick up their pens. Unfortunately for the mental health of HMS, in addition to serious, reasoned reflections on this subject, there is a steady stream of popularly available dubious thinking.

Quantum Sociology

We should suspect an intention to reduce God to a system of
differential equations.

—Sir Arthur Eddington

There sometimes seems more danger of us doing the opposite. Quantum mechanics has been a fertile breeding ground for those who see in its ambiguity a stepping-stone to social philosophy or religion. Thus one recent writer on the supposed social relevance of quantum mechanics claims, "Through the precise and testable imagery of physics we may learn a new language for describing other, related domains of our daily experience." We may, but I hope we don't. Our present language suffices, if used accurately. All we can borrow from science, in this respect, is a greater awareness of the places in life where reason should be paramount.[8]

Lately there has been an attempt to use the wave-particle duality as a basis for building a new view of society. It is claimed that we too display duality—we are individuals, but we are also part of society. This parallel is facile in the extreme and can be repeated ad infinitum: a note is a note, but it is also part of a symphony; a grain of salt is a grain of salt, but it is also part of a sausage. *And* a particle can look like a wave! And . . . therefore? And therefore nothing. The fact that we are both individuals and members of society is extremely important, but it has nothing to do with quantum mechanics and the parallel is sterile. It is essentially a misuse of lan-

---

[8]Incidentally, as William Empson showed many years ago in *Seven Types of Ambiguity* (1930), it is the very ambiguity of language that gives literature much of its subtlety and power.

guage—the use of the word *duality* to create the impression that there is something sociologically significant in quantum duality. There isn't. There is no consequence of our social duality that owes anything to quantum mechanics. It could be a good theme for a poem, but HMS should take care; although sociology can use mathematical modeling to study the dynamics of social intercourse, it can learn almost nothing from physics.

Einstein was much concerned with questions of morality and meaning. When asked what effect relativity had on religion, he replied, "None. Relativity is a purely scientific theory, and has nothing to do with religion." Nevertheless, there is a long history of fruitless attempts to relate the specifics of mathematics or physics to man's religious and moral life. This is the wacky side of the Pythagorean heritage. The most general lesson that man can learn from science is the need to apply the highest standards of reason *to those problems to which they can be applied.*

31

The Elementary Particles

What are little girls made of?
What are little girls made of?
Sugar and spice and all things nice,
That's what little girls are made of.

Little girls are made of quarks and leptons. Likewise little boys, and everything else in the universe.

The reign of the triumvirate of the proton, neutron, and electron was short-lived. In 1932, the year that Chadwick discovered the neutron, Carl Anderson detected a fourth particle, with the mass of the electron but positively charged. The *positron,* which Dirac had predicted, is stable if left on its own but has a permanent suicide pact with the electron, both particles vanishing and leaving two γ-rays. The positron was discovered by observing the particles generated by the collision of *cosmic rays* with matter.

Cosmic Bombardment

In 1910 Father Theodor Wulf, a Jesuit priest, ascended the Eiffel Tower and detected more radiation than he had expected. Intrigued by this observation, Victor Hess made ten ascents in a balloon, between 1911 and 1912, some to over 5000 feet, and in 1936 he received the Nobel Prize for discovering cosmic rays.

Cosmic rays fall on us from outer space. They are composed of particles traveling at very high speeds—mostly protons, but with small percentages of the nuclei of several elements. They have come a long way, probably originating in the stars within our galaxy, and they are very powerful missiles. Nuclei hit by cosmic rays can readily split into fragments, not only into protons and neutrons but also into other particles. Cosmic rays can also produce fragments by the conversion of some of their enormous kinetic energy into matter. This is relativity showing its fingerprints; the conversion of energy into mass can occur when extremely energetic particles are stopped in their tracks by collision.

Because cosmic rays can be lost before they reach the ground, because of collisions with the molecules in the air, many early experiments were done on mountaintops or by sending apparatus up in a balloon. In 1937, using cosmic rays, Anderson, together with S. Neddermeyer, struck gold again, discovering another two particles: the *muon* and the *antimuon.* These two particles, one positively charged, the other negatively, had about one-fifth of the mass of an electron and spontaneously disintegrated in a matter of a millionth of a second. The hunt for new particles intensified.

The Accelerators

It seemed that if you hit elementary matter hard enough you could fragment it, but cosmic rays were not an ideal projectile, and the need became evident for a man-made source of bombarding particles, allowing a choice in the control, type, and energy of the projectile. Thus was born, in the 1950s and 1960s, the age of the accelerator. An accelerator is simply an extraordinarily expensive evacuated tube, surrounded by powerful magnets. A variety of projectiles can be accelerated to enormous speeds in these tubes. There are accelerators that can produce beams of electrons traveling at 0.999 999 999 86 of the speed of light. The huge and growing expense of accelerators is a consequence of the need, or more accurately the desire, to attain higher velocities. With the use of very high-speed protons, neutrons, electrons, and other projectiles, hundreds of particles have been characterized, and at one time it seemed that every new issue of *Physical Review Letters* or *Nature*, included a report of a new particle.

The speed of the colliding particles in an accelerator is so great that we have to take into account the theory of relativity (see Chapter 32), one result of which is that bodies moving near the speed of light increase significantly in mass. This means that, on collision, two particles can have a joint mass far larger than that which they have when they are at rest. It is thus not surprising that comparatively heavy particles can be formed as a result of such collisions.

By the 1990s there were about 400 "elementary" particles, which suggested that the Creator picked up almost anything to hand as he built the cosmos. This wealth of new toys was exciting but puzzling, but if it gives the impression that particle physics is an expensive kind of beachcombing, it is thoroughly misleading.

Although a few of the new particles had been predicted on theoretical grounds, at one time there seemed to be no unifying principle behind the variety of masses and charges. Nevertheless, over the past few decades, the theorists have performed miracles of interpretation and have succeeded in bringing a degree of rational order into the welter of disparate data on elementary particles. The story is too long and complex, the theory too advanced, and the number of particles too large for the layman to appreciate the finer details—and I include among laymen those scientists, like myself, who are not in the field. However, the general lines are clear and the picture is fascinating. Our basic approach here is to give a brief overall view of the field, not following the trail of detection but going to the back of the book to see who done it.

It is fair to say that if you want to know how the world works at the level of atoms and molecules, you will lose almost nothing by forgetting about several hundred particles and contenting yourself with the properties of the electron, the proton, and the neutron. On the other hand, you will be missing out on a very strange tale and whole lot of colorful characters.

The Particles and Their Antiparticles

The elementary entities of which the universe is constructed can be classified into two groups (Figure 31.1) with names reminiscent of tribes of science fiction aliens:

- The *quarks*. Quarks are never found alone. They exist in various combina-

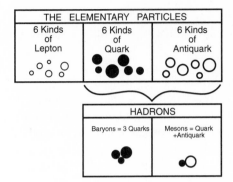

Figure 31.1 The classification of the elementary particles.

tions that form a large group of particles known as the *hadrons*. Several hundred hadrons have been identified, of which by far the best known are the proton and the neutron.

- The *leptons.* There are only six known leptons, of which the electron is the most familiar.

Both quarks and leptons have *antiparticles* having the same mass but *opposite charge*. The first antiparticle to be found was the positron, the electron's antiparticle, which can be classified as an antilepton. Just as quarks are never found in isolation, so antiquarks are never found alone, but the various combinations of these particles, the hadrons, do have detectable antiparticles. Thus there is an antiproton. The neutron has no antiparticle, since it is electrically neutral. Almost all the particles, hadrons or leptons, disintegrate spontaneously and rapidly and can only be studied immediately after they are formed in a collision experiment. The average lifetimes of particles and antiparticles are identical, except in one famous case, relevant to the Creation, which we will discuss later.

The only other known objects in the physical universe are the entities that mediate the physical forces, of which we have so far met only the photon.

Quarks and Leptons

Leptons appear to be genuinely elementary particles, having no smaller components. Thus, *as far as we know, quarks and leptons, and their antiparticles, are the irreducible components of matter.* No one has ever observed either a single quark or a combination of more than three quarks. Combinations of three quarks are called *baryons*, and include the proton and the neutron. Those composed of a quark and an antiquark are called *mesons*. Baryons and mesons together form the hadrons.

The quark was postulated in 1964 by Murray Gell-Mann and Georg Zweig. Early in Zweig's career, his appointment to a post in a Swiss university was blocked, on the grounds that he was a charlatan. His first paper on quarks was rejected; it was obviously nonsense. He is now a professor in California, and the concept of quarks is a cornerstone of particle physics.[1]

---

[1] The name quark comes from the following lines in James Joyce's Ullyses: "Three quarks for Muster Mark! / Sure he hasn't got much of a bark / and sure any he has it's all beside the mark." Quark is also a kind of cheese. Quark and apricot flan is partic-ularly good.

Just as the periodic table suddenly made sense when the electron, proton, and neutron were discovered, so the quark has brought a wonderful order into the world of the elementary particles.

Color

There is a science fiction story in which it turns out that there have been, for 10,000 years, two human species occupying this planet, each invisible to the other, and with no channels of communication.[2] This is how electromagnetic and gravitational forces are. They don't appear to speak to each other in any way. Electric fields have a counterpoint that we call charge. Gravity has a counterpoint that we call mass. Electric fields no more act on mass than gravity acts on charge. Each to his own. A priori there is no reason there should not be other forces in the universe, each acting on some property, *other than mass or charge*. Particle physics has shown that this is so.

Until recently, we have been in the position of a man who has had a clothespin put on his nose, been blindfolded, and then asked to sort out a bunch of assorted flowers. He could only go on shape and size. Remove the peg and the blindfold, and he finds that flowers have other properties—color and smell. These are revealed by the interaction with his sense organs. There are properties of the elementary particles that have revealed themselves in their interactions with each other, and that are completely different from mass and electric charge. These properties help us to classify the particles. The best known are *color*, which we will talk about, and *strangeness*, which we won't.

Two forces were known at the beginning of this century, the electromagnetic force and gravity, but it became evident that the forces between quarks were far too strong to be due to either of these. What holds quarks together inside hadrons, such as the proton? What is the nature of this new force, which is called the *strong force?* And on what does the force act?

For quarks, particle physicists invented a new property, analogous to mass and charge. A new property was needed because if there was a new type of force between quarks there had to be something on which it could act. They called it *color charge* because it has nothing whatsoever to do with color. Ordinary charge has two variations—positive and negative. Color charge has three variations—red, green, and blue—which have nothing to do with red, green, and blue, and might as well have been Harpo, Groucho, and Chico. Thus, in addition to mass and to charge, quarks have another property, *color charge*, that acts as an anchor for the strong force. For short we will call it *color*.

The Roll-call of the Quarks

How do you get several hundred different hadrons out of combinations of two or three quarks? By postulating that:

There are *six* kinds of quark, each having a choice of *three* colors, and each having a fractional electric charge (Figure 31.2).

Five of these quarks have been characterized on the basis of experiments in accel-

---

[2]When I mentioned this story to a feminist friend of mine, she declared that the two species were men and women.

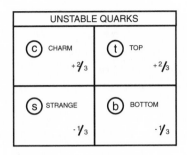

Figure 31.2. The classification of quarks.

erators. The sixth, although not yet found, is confidently predicted to exist. The names of these quarks, and their symbols, are: u, for the *up* quark; d, for the *down* quark; *these are the only stable quarks*. In addition there are four unstable quarks which appear momentarily in accelerators and were probably around at the creation of the universe: s, for the *strange* quark; c for *charm*; b, for *bottom* (although the more genteel call this quark *beauty*), and t, for *top*, which some call *truth*. Truth has yet to be found, or if you prefer it, we have not reached the top. Three of the quarks (u, c, and t) have a charge of +2/3 (on a scale in which the charge on the electron is -1). The other three (d, s, and b) have a charge of -1/3. Every quark has an antiquark with the opposite charge. Thus the antiquark of u is denoted $\bar{u}$ and has a charge of -2/3. The belief that there are six kinds of quark and six kinds of leptons is part of what is known among theoreticians as the standard model of the elementary particles and the forces that act between them. Are the quarks and leptons themselves built from simpler particles, like Russian nested wooden dolls? We don't know.

In 1995 it was claimed that experimental results supported the existence of an elementary particle with more mass than most atoms. This could be the shy top quark but stronger evidence would be welcome.

The letters u, d, s, and so on, are referred to as the *flavor* of the quark. Thus a quark has both *flavor* (which indicates, among other things, the magnitudes of its mass and its electrical charge) and *color charge* (which indicates how it behaves with respect to the specific strong force that quarks have between them.)

Protons and neutrons, the familiar bricks of matter, contain only up and down quarks. The proton is built of three quarks: uud, giving a total charge of +1. The neutron is ddu, with zero charge. The π^+ meson is $\bar{d}$u, which also gives a total charge of +1. All mesons have one quark and one antiquark.

Given that there are 6 quarks and 6 antiquarks, each having a choice of three colors, and that they can be combined two at a time or three at a time, there are easily enough different possible combinations to account for the few hundred known "elementary" particles. Moreover, although the masses of the quarks are not accurately known, they cover a wide range of values, which would account for the great variety of masses among the hadrons. The quark story, as described here, is schematically summarized in Figure 31.2.

Back to the Trinity

Unless you are interested in particles that generally last less than one millionth of a second, modern particle physics leads to the conclusion that, after finding and rationalizing several hundred particles, the everyday world can still get by with only three. Once we had the proton, the neutron, and the electron. Today we can manage nicely with the up quark, the down quark, and the electron. Both the proton and the neutron are built of combinations of the two different quarks, and the electron has remained inviolate since J. J. Thomson found it. That's it. With these three you can build the periodic table—every element in the cosmos.

A positive electric charge and a negative charge cancel each other in the sense that they produce a body that is not acted upon by an electric field. Analogously, a combination of three quarks, each having a different color, is "neutral" toward the strong force. This neutrality holds for both the proton and the neutron, each of which has one red, one blue, and one green quark. Thus, at first sight, such "color-neutral" hadrons should not have any "color" interaction with other hadrons. If both the proton and the neutron are neutral in this sense, the question is: After all this theorizing, what holds the nucleus together? What force prevents two protons from rushing apart? To answer that I refer you back to Figure 13.4b and Chapter 13, where we discussed the (gentle) van der Waal's force between neutral atoms. There we saw that even though the *net* charge on each of two atoms was zero, the individual electrons and nuclei could attract and repel in such a way that the net result was an attractive force. This was because the two kinds of charged particle have a certain amount of freedom in how they distribute themselves in space. An analogous situation holds for the interaction of color-neutral hadrons. Just as the van der Waals force between atoms is much weaker than the internal force holding the electrons of each atom to their nuclei, so the force between two hadrons is much weaker than the force between their constituents, the quarks. The force between color-neutral hadrons is a weak offshoot of the strong force. As the number of protons in the nucleus increases, the electrostatic repulsion between them eventually overcomes the forces holding the neutrons and protons together, which is why it is becoming more and more difficult to produce new elements of greater atomic number than those found or synthesized to date.

The Shy Quark

Quarks have never been observed outside the nucleus. When nuclei or single particles are bombarded with high-velocity particles, collisions are invariably followed by the appearance of a wealth of short-lived particles that often disintegrate to give other species. The tracks revealed in bubble chambers are by now a familiar icon of the late twentieth century (Figure 31.3). Quarks never appear; they only give signs of their presence. The problem of "quark confinement" is one that has given rise to much speculation but at the time of writing has not received a universally acceptable solution. But we know that they are in there.

There is a parallel between Rutherford's discovery of the nucleus and the experiments that revealed the inner structure of protons. Rutherford's α-particles generally went straight through the metal foil, except when they had a head-on, or nearly head-on, collision with a nucleus, in which case they were deflected through very large angles. If a beam of electrons or neutrinos is fired at protons, most of them go straight through, but some are violently internally deflected—presumably by the quarks. A popular picture of a hadron is of a bag containing three quarks. But what keeps quarks together?

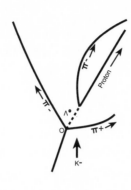

Figure 31.3. Charged particles leave tracks in a bubble chamber. A negatively charged meson comes in from the bottom of the picture and hits a proton at point O. The result is a negatively charged meson and a particle that immediately disintegrates into a neutral Λ° particle (which leaves no track) and a positively charged meson. The lambda particle moves a short distance before it disintegrates into a proton and another meson. (Photo from the Lawrence Radiation Laboratory, Berkeley, California.)

Contact Glue for Quarks

If we accept the picture of forces being associated with special particles—the electromagnetic force with the photon, and gravity with the (as yet undetected) graviton—then we can expect that the strong force, which holds quark to quark, will also have its own messenger. It has been demonstrated experimentally that the strong force is mediated by *eight* entities. The *gluons* (from glue) are stable but have no mass and no charge. One may well ask what they do have, but that question arises in our minds because we grew up in a world that had mass and charge as the only fundamental properties. Remember that photons have no mass or charge, but they exist.

There is no point in going into great detail about gluons. The force that they are responsible for is enormous. The energy required to separate two quarks can be as much as 10,000 joules, or the work required to lift a 50 kilo woman to a height of 20 meters. Gluons can interact with each other, considerably complicating the picture of what goes on in the nucleus. Photons, in contrast, do not interact with each other. The theory that deals with the interaction of quarks is called *quantum chromodynamics* (*chromo* for color), or QCD, and it is less developed than its counterpart, QED, that deals with electromagnetic interactions.

The Leptons

The leptons are simple compared with the quarks. There are six leptons (one of which is the electron), each with its antiparticle. They show no signs of having an internal structure.

One of the strangest individuals in the Great Particle Show is the lepton known as the electron neutrino, commonly just called the neutrino. The main natural source of neutrinos in our neck of the woods is the Sun. Every second, thousands of trillions of neutrinos fall to Earth; they are passing through you now. They can also be produced in proton accelerators. Neutrinos are probably the most penetrating particle in the cosmos. There are good reasons for this: the neutrino contains no quarks, and it is therefore unaffected by the strong force. But it is also electrically neutral and therefore blind to the electromagnetic interaction. In addition, the neutrino has so little mass that there is doubt as to whether it has any mass at all. All this would suggest that it had no interactions of any kind, but it can interact with quarks, by the mediation of a particle known as the Z boson. Nevertheless, since this action is very weak, the neutrino is an extremely introspective character, barely interacting with its surroundings. A neutrino could pass back and forth through the Earth a billion times and stand a fair chance of not interacting with anything. As John Updike wrote:

> Like tall and painless guillotines, they fall
> Down through our heads into the grass.
> At night, they enter at Nepal
> And pierce the lover and his lass
> From underneath the bed—you call
> It wonderful; I call it crass.

The Weak Force

The fourth of the known forces of nature to have been characterized is the weak force. It operates within the nucleus. The relative strength of the four forces is indicated by the approximate force in newtons between two protons placed at a distance of 10^{-15} meters from each other. They have to be that close, for the very short-ranged weak and (indirect) strong forces to have a chance of showing their strength: Gravity: 10^{-34}; weak force: 10^{-10}; electromagnetic force: 10; strong force: 10^3. Note that the strong force between two protons is the "van der Waals'" shadow of the enormous strong force between quarks.

How does the weak force manifest itself? Between what is it a force? You can look upon the force as one that is able to change the *flavor* of quarks, for example, from a down quark to an up quark. It was first invoked to account for the phenomenon of β-decay, the radioactive ejection of an electron and a neutrino from an atomic nucleus.

It is impossible to go into the brilliant theoretical analysis, mainly from work by Sheldon Glashow, Abdus Salam, and Steven Weinberg, that has been applied to the weak force, but it has been shown that just as electric and magnetic forces are really two aspects of the same physical phenomenon, so the weak force and the electromagnetic force are sisters under the skin, manifestations of a single phenomenon labeled the *electro-weak force*. It is not possible to really grasp the meaning of this without going through the physics—which we will not do. Nevertheless, it should be realized that this is a major step in the physicist's desire to construct a unified theory of all the forces in the universe, a theory which would show that all forces are different aspects of one force, as Kant and the *Naturphilosophers* believed. It still remains a belief, a reflection of the impulse to create order, and of the, perhaps mistaken, faith in the simplicity of nature.

Theory and Experiment

Why do we think there are six quarks if only five have been found? Recall the periodic table. Before they were found, certain "missing" chemical elements were predicted to exist because there were obvious holes in the table. It often happens in science that there are very good reasons for supposing the existence of a particle, molecule, or process, without our having observed it.

In recent times, the prediction of the existence of unrevealed particles and, more generally, the understanding of the known elementary particles, has depended crucially on considerations of symmetry.

The study of symmetry, as a branch of mathematics, began long before it became obvious in the twentieth century that symmetry was one of the most powerful tools that the physicist and chemist have in discussing the nature of matter.

Symmetry puts restrictions on the possible properties of matter. If I tell you that an animal must be symmetrical with respect to reflection in a mirror running through its body, I am immediately putting limitations on its form. If it has one leg to the left of the mirror plane it must have one, identical but mirror image, leg to the right, and so on.

The branch of mathematics that deals with symmetry is called *group theory*. When applied to physical objects, the basic question that group theory asks is: What changes leave an object looking the same? (The mathematicians would say: under what set of operations is the object invariant?)

Many objects have some geometrical symmetry, but there are subtler symmetries. Newton's laws are "symmetrical under time-inversion," which is a posh way of saying that they are unchanged when time runs backward. They are unchanged by taking the equations and changing the time t, into $-t$. When we get to relativity, we will see that all physical laws are left unchanged by certain changes. The fact that a law remains unchanged by reversing time, or by performing experiments in a moving train, are, like the conservation rules that we have noted, aids to understanding the properties of physical systems. If a physical system has a certain symmetry with respect to space or time, that immediately puts restrictions on the way it can behave.

We cannot go into group theory here. All I want to stress is that group theory has been used to classify the elementary particles into related families and that the theory helps us predict when there are particles—members of the family—that are missing. After all, if someone told you that he was holding an article that had the symmetry of a square, and you could see only three corners, you could predict that there was another identical corner hidden by his hand. When particle physicists talk about "SU(3)" and the "eightfold" way, they are talking about symmetry, which has its roots very deep in the theory of quantum mechanics and relativity, but is not one of the alluring sirens that we will be dropping in on to sip wine with on this particular trip.

The Final Force?

The British physicist Peter Higgs has suggested that there is another undiscovered particle, now called the *Higgs boson*, whose properties are such as to give particles the *appearance* of having mass. If true, this could explain the origin of inertia. The search for the famous Higgs boson is one of the justifications put forward for the proposed giant Superconducting Supercollider—the SCSC or SCS.

Another idea in the particle world, which has been the cause of many a seminar and symposium, is an entity on which great expectations were placed a few years ago: the "string," which has been followed by an improved model, the "superstring." I would not dream of attempting to go into detail about the superstring, but I bring it up to indicate the strange landscapes through which the theoreticians are wandering. There are variations of the theory, but all postulate a threadlike beast.

Particles, in the standard model, are supposed to have no size; they are point particles. As you might imagine, a particle with zero size can give funny answers to some calculations. Thus calculating the energy of such a particle often came up with the awkward answer that the energy was infinite. Infinities are the curse of theoretical physics, so why not assume that the really basic particles are not points but are threadlike.

It would take about 10^{20} of the hypothesized "superstrings" to stretch from one side of a proton to the other. These loops of something are supposed to be able to vibrate, and each of the modes of vibration corresponds to a different elementary particle. The large number of possible modes of vibration could provide a basis for the hundreds of different known elementary particles, and might even give a rational explanation for their masses, which up to now have given no hint whatsoever of a pattern. Just as Pythagoras found that certain mathematical relationships were connected with certain natural musical tones, so the allowed vibrations of the superstring might, said the theoreticians, give the masses of the particles. This hope has not materialized.

The excitement created by the theory is, according to my theoretical colleagues, dying down somewhat. Mathematically it is complex; one variation claims that the strings exist in ten dimensions, another that they occupy twenty-six dimensions! The real problem is that so far the theory has not come up with any new predictions that have been experimentally verified.

Rashomon

Our ignorance is vast.

—Karl Popper

We have divided the cosmos into particles and fields, but what justification have we for preaching this doctrine? There is no doubt that it fits in with our visual experience of the world—there are things that we can see and that take up space, and there are things that we cannot see but whose presence is evidenced by the behavior of falling bodies, compass needles, and transformers. But the wave-particle story teaches us to beware of simple conclusions, however obvious they seem.

Consider the photon. We tend to regard it as a particle because we can detect it when it hits a suitable device. A flash of light on a screen would seem to indicate that a shell has fallen, not a wave. And yet nothing in practice depends on our belief that the photon is a "body," something that Descartes would say "had extension." In fact, the theoreticians prefer to view the photon as a manifestation of the electromagnetic field, a kind of bunching up of the field or, if you prefer, a wave packet in the field. This approach abolishes the distinction between particles and fields. This is not too daunting in the case of photons; they are not supposed to have mass anyway. But the same idea has been extended to all "material" particles.

The theory was first propounded by Heisenberg and Pauli in 1929. Theoreticians now treat all particles as though they were manifestations of a field. Thus a proton is associated with a field and can be regarded as a gathering together of this field. According to modern *quantum field theory*, every particle, whether it be electron, quark, muon, or whatever, can be regarded as being a bunching up of a field, on which surreal note we take our leave of matter—in this book anyway.

The theory of particles and their interactions has taken us far from the simple ideas that guided the construction of the atom in the first quarter of the twentieth century, when three particles and the coulomb force seemed to suffice. Modern particle theory is a soup containing quantum mechanics, relativity, and group theory, and it will probably never, at any time, be fully digestible by more than a few thousand people. What we have done here is merely to give a slight indication of its richness, fascination, and complexity. It is one of the great intellectual achievements of our time, even if its practical fruits for the last forty years have been negligible.

Relevance

Many scientists feel that science should be useful if possible. I am one of them, but the spirit of this book is, I believe, consistent with those scientists, of whom I am also one, who regard science as a search that needs no practical justification. Particle physics has undoubtedly produced some of the most imaginative and subtle ideas in science, and the spectacle of the elementary particles being "tamed" is a magnificent example of physical intuition, mathematical awareness, and great leaps of the imagination. It is also a very expensive hobby.

The need to attain higher and higher energies, to liberate, for a fleeting moment, a new particle has resulted in the design of monster accelerators costing $10 billion or more. A few years ago, those with their hands on America's purse strings began to ask questions, and the particle physicists had to go on the defensive. Congressional committees, conscious of the national debt and the demands of society for better social services, are not too friendly toward the cry of "science for science's sake." Perhaps this accounts for the decidedly strained efforts of certain particle physicists to justify research in particles as being much more socially useful than it is. The truth is that to date its practical applications have been meager. Neutrons have been used in cancer therapy, and positrons are revealing increasingly detailed information on the working of the brain. But it has been several decades since a new particle has affected the lives of anyone but the discoverer's professional colleagues. One point should be clear to HMS, *if we had discovered only protons, electrons, and neutrons (p,e,n), and never found one more particle, it would have made little difference to HMS.* The p,e,n trio has underpinned our understanding of atomic and molecular structure and dynamics, and has formed the basis for much of the technology and medical advances that have changed all our lives. If you want to be a fundamentalist, you can say that the everyday universe is built of up quarks, down quarks, and electrons. If you throw in the neutrino to account for β-decay, you have enough components to account for most of the history of the universe. The rest of the exotic menagerie of particles have affected no one except those who write articles-on-particles. The likelihood that the next particle that needs a multimillion dollar accelerator to produce (and then survives for only a quadrillionth of a second) is going to be widely useful is remote. Too remote, perhaps, to justify the enormous expenditure that could, at least in principle, be diverted to a large variety of other no less interesting, and possible more useful, scientific projects.

In October 1993 the United States House of Representatives voted to withdraw support for the construction in Texas of the huge Superconducting Supercollider, which was to be the most powerful accelerator yet built. This giant was to have a circular tunnel, 54 miles in circumference. As one Representative said, "It's good science, it's simply not affordable science."

The prospect of discovering all the fundamental particles and unifying all forces has drawn physicists into believing in the attainability of a theory that accounts for *everything*. These theories (for there will surely be several!) are named theories of everything, or TOE for short.

HMS should not be taken in by the nomenclature. Does anyone seriously believe that within the next millennium TOE, which will be basically theories about particles and forces, will be a more effective means of studying medical, sociological, chemical, political, biological, economic, psychological, geological, and other problems than an attack made within the framework of these disciplines and related disciplines? The use of the word everything in this context has always struck me not as grandly all-inclusive but as insufferably parochial.

What kind of priority should HMS give to the funding of particle physics? That is her own choice, but it is as well to be aware of the propaganda. The proponents of the new accelerators and TOE tend to write evangelical prose. They often imply, but hold back at the edge of actually saying, that the discovery of the top quark, or the Higgs boson, will be the high point of man's history. Why do the rest of us have our doubts? Leon Lederman, the distinguished director of the Fermi National Laboratory (Fermilab), has drawn a parallel between the superaccelerators and Gothic cathedrals: "Both provide spiritual uplift, transcendence, revelation." Embarrassing. A cathedral was the spiritual center of the *whole* community. The teaching of the priest, right or wrong, directly affected the daily life of his flock. Accelerators express the faith of no one except a few thousand scientists. The results of their activities will almost certainly have little or no practical result, and will be truly understood by a negligible minority of mankind. Steven Hawking has declared that if we can discover the Higgs boson we will know the "mind of God." A scientist can take this as a piece of poetic licence, but HMS may have no guidelines to help him weigh such statements. If taken at face value, and I'm not so sure that Hawking didn't mean it that way, it shows a complete lack of proportion. If the Higgs boson represents the mind of God, then I suggest that we are in for a pretty dull party. The particle will probably be found, because the theorists have an impressive record of prediction. The populace will not rush into the streets waving flags. Many of them will spend the evening listening to Mozart or Sonny Rollins—two of the infinite voices of God.

The Croatian Visionary

Rudjer Joseph Boskovic, born in Dubrovnic in 1711, was a Jesuit priest, which is an occupation conducive to the exercise of reason. He traveled all over Europe as a diplomat and a scientist. He was a remarkably active person, reminiscent of his acquaintance Benjamin Franklin, but he was a far better mathematician and a deeper thinker than Franklin. He advised the Vatican on the threat posed by the cracks in the cupola of Saint Peter's in Rome, made a plan for the draining of the Pontine marshes, and designed a new harbor for Rimini. At the age of 24 he read Newton's *Principia* and *Optiks*, both of which were decisive in determining his scientific activities. The latter book triggered his practical work on the construction of telescopes, but it was the *Principia* that was his secular Bible. Boskovic published many papers and books, but his magnum opus was *Philosophiae naturalis theoria redacta ad unicam virium in natura existentium*, published in 1758 and, after a number of subsequent editions, printed in a revised form in Venice in 1763. In this book Boskovic showed a remarkably imaginative approach to physics, and though his ideas on atoms and forces had no experimental backing, they were an influence on a number of scientists. It is not easy to show a direct line from Boskovic to Faraday's concept of fields, but Faraday was aware of Boskovic's book, as was Lord Kelvin. Faraday's belief that electricity was a force, not a substance, was in line with Boskovic's view of things.

Boskovic's picture of atoms not as material entities but as centers of force was highly influential in the nineteenth century. Michael Faraday used it, acknowledging Boskovic's influence at the British Association meeting in 1837 at which he joined in an attack on the material atom. Boskovic's attempt to construct a mathematical framework that would account for the behavior of matter in terms of one universal force, a central theme in modern physics, is an underappreciated landmark in science. His conception of atoms as being a kind of gathering together of a field, rather than something called matter that was separate from the field, is very close to some modern ideas. It may turn out that what we call matter is indeed nothing but a knot in the void. Or is this all a fantasy dreamed up by professors? They continue to churn out psychedelic ideas, some of which live not much longer than muons.

Is Science for Real?

> And God-appointed Berkeley that proved all things a dream.
>
> —W. B. Yeats

We have drifted far away from the picture of little bouncy dumbbells that we started off with in Chapter 1, the "molecules" that rushed around blindly in a box. In the early twentieth century our dumbbells were replaced by three elementary particles—protons, neutrons, and electrons—but the newborn quantum theory still left us with the two elements of reality that we had loved and known since we were this high: matter and energy. Then the holy trinity of particles were replaced by usurping quarks and leptons, and we were told that forces are mediated by particles, mass is interconvertible with energy, the vacuum can produce particles. And now particles are merely manifestations of fields.

At this point reality is rapidly running through our fingers. What are these fields? Are they more than mathematical constructs that have been put together so that when suitably manipulated they give some of the answers?

What is left? A universe that is a maelstrom of interacting fields? In Chapter 12

I suggested that you look at your hand and then reflect that it was mainly empty space, but I left you with the illusion that there were particles. Look again. Maybe you are looking at nothing but disembodied lumps in fields. Is that what we are, cowpats in space, thrombosis in the cosmic bloodstream? Is this the last model? How real is all this?

Science in the end is a construction of the human mind. Eddington saw this as a guarantee that it was rational. Our theories get better *as judged by how much they can consistently explain.* The ability of the quantum theory to explain the periodic table makes it a better theory than those that it displaced. If there is a reality out there, our ability to predict its behavior is growing spectacularly. For many of us this is enough; we don't care how the medicine cures the horse, as long as it does. We can always hide behind the positivist philosophy that says that the meaning of a proposition is its method of verification. The experiment has been defined in all its details, the dials have said what they had to say. That is the meaning. This is how, at least at the beginning of his career, Werner Heisenberg saw quantum mechanics.

Are we approaching the ultimate answers? One might think so from some of the disciples of TOE. They remind me of a remark Max Born made in the late 1920s, to the effect that "physics as we know it, will be over in six months." To which Einstein might well have answered, "By whose clock?" For not all clocks seem to run at the same speed.

viii

In the following chapters we leave behind the Newtonian universe where space and time are separate entities, and no one asked awkward questions like "How did it all start?" and "Where is it going?" Here we will travel through the space-time that Einstein built, and we will see how science has tried to answer what may well be unanswerable questions.

32

Relativity

Light is the principal person in the picture.

—Édouard Manet

Einstein fathered two theories of relativity. The first, which came out in 1905, is the special theory of relativity, which generated much of the popular imagery of relativity: shrinking spears, lethargic clocks, bodies accumulating mass, youthful space travelers, and so on. The theory essentially concerns itself with what an observer sees when he looks at a system—a spacecraft, an aeroplane, a car, a comet, an electron, and so forth—that is moving uniformly with respect to him. If Einstein had not come out with the theory, someone else would have done so; the mathematics was standard, and many of the ideas were sitting around waiting to be drawn together. The second theory came out in 1916. The general theory of relativity is a theory of gravity, supplanting that of Newton and transforming the way that cosmologists dealt with the universe. This is the origin of the warped space so beloved by science fiction writers. Einstein, in the general theory, completely rebuilt the way in which we mentally construct space and time, a transformation that he had begun with the special theory and that owed a debt to the work of the Russian-German mathematician Hermann Minkowski.[1] The general theory of relativity is one of the great monuments of the scientific imagination.

Of what use is the theory of relativity? Very little to HMS. Its most notorious consequence, the equation $E = mc^2$, is the theoretical backing for the atomic bomb, but Einstein was no more responsible for the atomic bomb than Newton was responsible for intercontinental ballistic missiles. The theory was essential to the Dirac equation for the electron, and for quantum electrodynamics (QED). Neither of these subjects has disturbed the sleep of HMS.

Relativity affects the design of particle accelerators because particles traveling at speeds close to that of light gain in mass, which affects their trajectories. Small effects of relativity can be detected in the electronic structure of heavy atoms, but this is a very limited, esoteric subject. Otherwise relativity mainly appears in cosmology.

Why all the fuss, then? Because relativity has made us think again about where we live, which is in time and space. Because Einstein pulled aside the curtains and revealed that, in describing nature, space and time are woven together, and matter and energy are one.

---

[1]There is a story by Anatole France in which the aged King Herod is asked about Jesus but has difficulty remembering him, in differentiating him from the other Judaic rebels. Einstein was a pupil of Minkowski in Zurich, but the teacher barely recalled his student.

Absolute Space and Time?

> But all the clocks in the city
> Began to whirr and chime:
> "O let not Time deceive you,
> You cannot conquer time."
>
> —W. H. Auden, "As I Walked Out One Evening"

Time and space have tortured scientists, philosophers, and poets since the Greeks. Saint Augustine, in his *Confessions*, answers: *Si non rogas intelligo* in reply to the question: What is time? Locke expansively translates this as: "The more I set myself to think of it, the less I understand it." Bishop Berkeley was more confident: "Time is the train of ideas succeeding each other." (But can you have the idea of "succession" if you don't assume the idea of time?) Is time absolute, coming off God's production line in a steady stream? Newton thought so: "Absolute, true, and mathematical time, of itself, and from its own nature, flows equably without regard to anything external, and by another name is called duration."

Both Newton and Saint Augustine held that time began at the moment of creation, a view subscribed to by one of the main medieval Islamic schools of philosophy and also by many modern cosmologists. Newton also believed that space was absolute, existing out there—a ghostly framework, independent of man: "Absolute space, in its own nature, without regard to anything external, remains always similar and immovable." Newton's concept of space, God's dwelling place as he called it, was clearly spelled out by John Keill in a lecture in 1700:

> We conceive Space to be that, wherein all Bodies are placed . . . that is altogether penetratable, receiving all Bodies into itself, and refusing Ingress to nothing whatsoever; that it is immovably fixed, capable of no Action, Form or Quality; whose Parts it is impossible to separate from each other, by any Force however great; but the Space itself remaining immovable, receives the Successions of things in motion, determines the Velocities of their Motions, and measures Distances of the things themselves.

For Newton and Galileo time and space were absolute. And if an angel at the edge of the universe glanced at his Rolex at half past two in the afternoon, and then at three o'clock, the same half hour would pass in Cambridge and in Padua. Intervals of time, and distance, were everywhere what they were on Earth. Common sense tells us that this is the nature of time and space; we don't expect to become distorted when we run, or to find that when we return home after a drive, our watch has lost an hour. The world makes sense.

But there was a feeling of unease. Newton realized that there was a problem in locating this absolute framework. If a body moves with respect to another body, how do we know which is moving and which is stationary? Maybe they are both moving. Unless we have God's chalk marks on the stage, we cannot know: "It is indeed a matter of great difficulty to discover, and effectually to distinguish, the true motions of particular bodies from the apparent; because the parts of that immovable space, in which these motions are performed, do by no means come under the observation of our senses."

Indeed they don't. Consider an astronaut in a spacecraft, seeing another space-craft overtake his own. He can give an accurate estimate of their *relative* motion, but he cannot state anything about the absolute motion of either craft, simply from his observation. The other craft might be stationary, and he might be going back-ward. He has no way of knowing. He could of course ask Houston to give him the speeds of both craft, as estimated from Earth, but that will not give him his *absolute* speed, only what he asked for: his speed relative to Earth. He could find out the Earth's motion with respect to the Sun and then work out his speed with respect to the Sun. But the Sun is moving within the Milky Way . . . And so on. There is no way that he can find *the* absolute framework. I cannot know if I am moving with re-spect to God's dwelling place because no one has seen it.

Bishop Berkeley, Newton's faithful adversary, who might have made a better physicist than a cleric, dismissed absolute time, space, and motion as fictions of the mind. Leibniz, also consistently anti-Newtonian, rejected absolute space, on the grounds that it is unobservable and can have no observable effects. One can only say where something is in relation to other bodies. So it was with time. Time might stretch out like an endless line, but there were no fixed, sacred milestones on it. All we had, to give us some kind of arbitrary scale, were recurrent natural events and man-made timepieces. There seemed to be an internal time, our consciousness of a sequence of events, one coming "after" another, but why suppose that this had any connection with the time of the physicists?

Space arises as a concept in two rather different but related ways. It may first have emerged with the need to specify the *location* of objects. The question is, does location have any meaning in the absence of all material objects? How can I say where some point is if I have nothing to refer it to? If you were alone in an other-wise empty universe, how could you describe where you were? (Except in hell). Another concept of space arises in the context of a beer bottle. There is something called a "space" within the bottle. This something is capable of being filled. The bigger the bottle, the bigger the space. But the space is defined by the bottle. Once again, if we remove all material objects, what meaning is left to this space? Try as you may, any attempt to define it relies on the presence of a reference system. As Einstein said, "There is then no 'empty' space."

Every definition of time comes down to a statement of coincidence. When I say that something happened at three in the morning, I mean that it coincided with a certain position of the hands on my watch, and that they could be correlated with a radio signal from Greenwich, and so on. Time eludes us. We don't understand why it flows, or even what that phrase really means. But there is a class of phenomena that suggests that Newton was right about absolute space, that there really is a fixed framework out there.

Centrifugal Force

Suppose that you are an astronaut and you look out of your window and see two unpleasantly purple little aliens, each holding on to opposite ends of a rope, and apparently circling about the center of the rope. How do you know that it is they who are circling, and not you who are circling them? Well, one thing is certain: if two children hold the ends of a rope and circle each other in a playground, they feel a *centrifugal force*. The rope between them is taut and under tension. Are they undergoing *absolute* rotation? You can't get rid of the tension in the rope by suggesting that the children, or the aliens, are stationary and you are going around. Rotation, in contrast to linear motion, appears to be with respect to some fixed *absolute* frame.

Again Berkeley disagreed. Anyone rotating saw the so-called fixed stars circling the other way. It was, he said, with respect to the fixed stars that we judged rotation, not to some more fundamental spatial frame. Berkeley's ideas on rotation were picked up, at the end of the nineteenth century, by Ernst Mach, who had a deep influence on Einstein. Mach was a no-nonsense positivist: all we could know was what we could observe, and since the universe is filled with matter, it would never be possible to observe a body rotating in a universe empty of matter and there was no point in hypothesizing about it. If centrifugal force existed in the real universe, it was, said Mach, due to the influence of the fixed stars, which were themselves not necessarily fixed. Mach, in fact, credited the same stars with inducing the property of inertia in matter, in contradiction to Newton, who saw inertia as an inherent property of all matter. Mach's principle, as Einstein called it, has attracted renewed attention in recent years but has never been proved.[2]

Apart from rotation, the other phenomenon that gave pause for thought was Foucault's pendulum. Jean-Bernard-Léon Foucault (1819–1868) hung a 28-kilogram iron ball from the dome of the Pantheon in Paris. The cable was 67 meters long, and the suspension allowed the plane of oscillation to change—which it did, slowly rotating. The change is such that it can be readily interpreted as a change in the orientation of the Earth, due to rotation. The pendulum keeps swinging in the *same* plane. One might regard this as evidence that the pendulum "feels" absolute space and remains oriented in the same position with respect to this framework. But is the truth that it feels, not a fixed absolute space, but the stars? We don't know. You can see copies of the pendulum in the United Nations building, the Science Museum in London, and other places.

Because Einstein had a tendency to go for the big questions, he was bound to get involved with the puzzle of space and time. He came to it through thinking about light.

The Problem of Light

When Einstein was eight years old, Michelson and Morley's experiment failed to detect an ether (see Chapter 15). The physicists were in trouble. On the one hand, they wanted a medium for light waves to wave in; on the other hand, they couldn't detect it.[3]

There has been much discussion about whether Einstein was aware of, or took much notice of, the Michelson-Morley experiment. He himself is ambiguous on this, but it is likely that he would have produced his theories of relativity with or without the famous null experiment.

---

[2]Einstein pointed out something odd about centrifugal force. Every other force seems to be an inherent property of all, or certain, kinds of matter. Gravitation is associated with mass, electromagnetism with charge, the strong and weak forces with the properties of elementary particles. But centrifugal force is apparently a product only of movement.

[3]Faraday's intuition once again put him on the side of the angels: "The view which I am so bold to put forth considers radiation as a high species of vibration in the lines of force which are known to connect particles, and also masses of matter together. It endeavors to *dismiss the ether* but not the vibrations" (my italics).

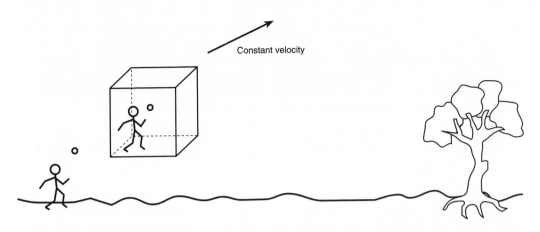

Figure 32.1. The two frames are described as *inertial* because they are moving at a constant velocity with respect to each other. The two men see exactly the same behavior from the ball they throw.

Einstein thought about light for ten years before he produced the special theory of relativity, but the real origin of the theory lay in Newton's *Principia* and Galileo's *Dialogue Concerning the Two Chief World Systems.* Einstein did not accept the absolute space and time of Newtonian physics, but, fatefully, he did accept another basic axiom, conceived by Galileo and accepted by Newton: the concept of *inertial frames.* This concept is absolutely central to the theory of relativity.

Both Newton and Galileo believed that if you are in a box (a "frame") moving with *constant velocity* with respect to the Earth, the laws of motion that hold on Earth hold in the box (Figure 32.1). Maybe you are reading this in a steadily moving car or commuter train that is traveling in a straight line. In your box, a body will obey the first law of motion: it will move with constant velocity unless acted on by a net force. Leave a cup on your table, and it will not move with *respect to you* because *in the frame* (the train) there is no horizontal force acting on it. Now drop your briefcase and it falls *vertically* to the floor, accelerating with the acceleration due to gravity, obeying the second law of motion, just as it would if the train were stationary. (You will be repeating the seventeenth-century experiments of Gassendi, who dropped objects from the masts of moving ships.) Throw a ball, inside the train, and its path *with respect to the train* will be identical with that which it would have followed, *with respect to the Earth,* had you thrown it outside the train. Newton's laws hold.

We say that the box and the Earth are "inertial frames." The room that you are sitting in is an inertial frame. The Earth and the train,[4] provided it is moving at constant velocity, are inertial frames. Any two frames moving at *constant velocity* with respect to each other are inertial frames. A simple visual way of grasping the concept of inertial frames is to realize that if you are in one you can never prove, solely from the behavior of bodies within the frame, that you are moving. If the train carriage didn't vibrate and the blinds were drawn, you might think you were standing in the station, and there is no experiment, *confined to the train,* that you can do to show that you are moving with respect to the Earth.

---

[4]You will find that trains regularly steam through discussions of relativity. This stems from Einstein's own original explanatory illustrations of his theory.

Physicists use a number of jargon-laden phrases to convey the preceeding idea. Some talk of Galilean Invariance. All they are saying is that if two observers are moving at a constant velocity with respect to one another, the way in which the world behaves, *within either of their respective frames,* is controlled by identical physical laws. Note that to define inertial frames, we have no need to know the absolute speed of the two frames, only their *relative* motion.

In inertial frames, there is no experiment that can be done, within the frame, that will reveal that it is in motion with respect to another frame. All physical laws are identical within the two frames.

The definition of inertial frames is based on the fact that each frame is moving at a *constant* speed with respect to the other. If the train is not moving at a fixed velocity in a *straight* line, it is not an inertial frame with respect to the earth. In an accelerating frame, all you have to do is put a glass on your table to see that the train is moving; the glass starts to move across the table seemingly of its own accord, in the absence of an external force. This apparently contradicts Newton's first law. I say apparently because, in the absence of friction between glass and table, someone standing on the platform will see a stationary glass, obeying the first law *with respect to the earth.* Each observer must keep his eyes focused within his own frame if he wants to decide whether a body in that frame is obeying a certain physical law with respect to the frame. Similarly, in an accelerating train, bodies drop not straight down to the floor but diagonally, even though the force of gravity is acting vertically. Either of these two simple observations, *within the frame,* is enough to reveal that the frame is moving with respect to the earth, thus showing that the train and the earth are not inertial frames.

A spacecraft moving at a constant velocity with respect to the Earth, and the Earth itself, form a pair of inertial frames. The laws in the spacecraft should therefore be the same as those on Earth. It is true that astronauts float freely in their smoothly moving spacecraft—behavior that can hardly be said to be earthlike. It isn't, but their world is controlled by the same basic *laws,* and their weightlessness doesn't show that the spacecraft is moving with respect to the Earth. Newton's laws still hold in the capsule. They are just too far away from the Earth for the force of gravity to be effective. If someone jumps on the Moon, or throws a rock, the motion in both cases is described completely by the usual laws of motion. The law of universal gravitation still holds, as it does in the spacecraft—it's just that, in the equation (equation (1), Chapter 5), we have to replace the mass of the Earth with the mass of the Moon.

There is no experiment that you can do *within the spacecraft* which will give the game away. Only if you look *outside* can you tell that you are moving with respect to the Earth, and that counts as an experiment outside the frame. This inability to perform an experiment within the box that would show that you are moving is a pivot of Einstein's derivation of special relativity.

The belief in the unchanging nature of physical laws in inertial frames is very deeply ingrained in physicists, but Maxwell's proof that light traveled at a fixed and finite velocity threw a spanner in the works.

The Image of the Undead

Remember the pond skater in Chapter 15, the cowardly insect that rode along on the crest of a wave in the water? If you had asked it what the velocity of the wave was *with respect to him,* he would answer "zero." Notice that the wave never passes the pond skater. Einstein was troubled by a similar problem: What would happen if he were traveling at the speed of light and attempted to look at himself in a mirror?

Imagine Einstein to be holding a torch that shone on his face. Light is reflected off his face. But, if he was traveling at the speed of light, the light could never move ahead of him. It would never reach the mirror, and like Dracula and the undead, he would not see his face (Figure 32.2). This is the kind of question that geniuses worry about while the rest of us watch TV.

Why did this problem worry Einstein? Note that if he is traveling at the speed of light with respect to the Earth, then he, and the mirror, are both in an inertial frame. But if he really couldn't see himself, then he had a way of knowing that he is in motion. For him, the observed speed of light would be zero, in contrast to the value observed on Earth. An experiment within the frame has shown that he is in motion with respect to another inertial frame. This means the collapse of the principle of Galilean invariance, which says that such an experiment cannot be constructed.

Einstein realized that the problem of Galilean invariance could be solved if he assumed that light traveled at its normal speed *in his flying frame.* If this were so, then he would see himself, the speed of light would seem normal to him, and he would once again have no way of knowing that he was moving. But there was a very serious problem with this solution.

If light were to travel at its normal speed in Einstein's frame, then an observer on Earth would presumably see the light from Einstein's flashlight traveling at *twice* the speed of light. If Robin Hood's arrows travel at 100 mph with respect to his bow, and he shoots an arrow forward from the top of an express train traveling at 50 mph, he will still see the arrow traveling at 100 mph with respect to himself, but a spectator by the track will see the arrow moving at 150 mph. Now that's OK for arrows, but it doesn't work for light. The reason is contained in our discussion in Chapter 15. We saw there that *the observed speed of light waves is independent of the motion of the source.* This can be shown to be true for electromagnetic waves by looking at Maxwell's equations. In other words, to an observer on Earth, light could not travel faster than its speed on Earth; and if Einstein was traveling at the speed of light with respect to the Earth, then the light could not leave Einstein's face; he would be riding on the crest of the wave and could not see his face. Ergo, Galilean invariance was a myth, there *was* an experiment that could reveal that you

Static frame
S

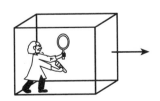

Moving with the speed of light
with respect to frame S

Figure 32.2. In the static frame Einstein sees himself because the light from the torch is reflected, first off his face, then from the mirror, finally reaching his eye. In the frame moving at the speed of light the light reaching his face can never leave it—he keeps catching up with it. Like Dracula, he will not see himself. At least that's what common sense says.

were moving, even if you were in an inertial frame: just measure the speed of light.

This problem was to bring down Newton's absolute space and time, so let's state it concisely:

- For a flashlight moving at the speed of light, Maxwell's equations show that, *to an observer on Earth,* the speed of the light from the source is equal to that on Earth.
- If the preceding is true, then for Einstein, who is moving at the speed of light, the speed of a forward directed light beam is zero, and he would know that he was moving with respect to the Earth. This contradicts the principle of Galilean invariance, since Einstein and the Earth are in inertial frames.

Which was Einstein to throw out? He could adjust the speed of light for a moving observer, by modifying Maxwell's laws. Or he could reject Galilean invariance.

The principle of Galilean invariance was so deeply installed in Einstein that he was not prepared to give it up. This meant that he had to leave all physical laws unchanged in the moving frame, and that included Maxwell's laws! In other words, if Maxwell's laws were applied in the moving frame, they would predict the speed of light to be quite normal, not zero. Einstein would not be riding on the crest of a wave. But now another dilemma emerged!

If the speed of light was normal in the moving frame, then the same beam that Einstein saw to be traveling at the speed of light would be traveling at *twice* the speed of light to someone on Earth (think of Robin Hood). Which, for the earthbound observer, would contradict Maxwell's laws. Remember, the speed of a wave is independent of the speed of the source.

Could a way be found to ensure that a beam of light traveled at the same speed for the man in the frame (thus rescuing Galilean invariance), but that the *same* beam as viewed from Earth also appeared to be traveling at the normal speed of light, as demanded by Maxwell's law? This sounds impossible, as though, standing by the track, we saw Robin Hood's arrow traveling at 100 mph with respect to the Earth, and yet he also saw it traveling at 100 mph with respect to the moving train.

With the boldness of genius that comes along only every few centuries, Einstein stood by his belief in Galilean invariance and declared that, whatever reason said:

The observed speed of light is the same for any observer.

If Einstein was traveling at the speed of light, and carried a flashlamp, he would see the light leaving the flashlamp at 300,000 km/sec and (this is the crazy part) so would an observer on Earth.

There are two pillars to the special theory of relativity: the concept of Galilean invariance and the belief that the velocity of light is the same for all observers.

Actually the second statement is included in the first if you believe absolutely in the principle of Galilean invariance, which states that you can't tell if you are moving in an inertial frame.

The Special Theory of Relativity Is Born

The need to ensure that light traveled at the same speed for any observer created the theory of special relativity. The theory ensures the invariance of the speed of light in all inertial frames. Einstein would see his face, but from the Earth the speed of light from his flashlight would still appear to be 3 x 10^8 meters per second. How?

Speed is estimated by measuring how long it takes something to travel a certain distance. The measurement of speed thus requires a measurement of length and of time. Now both Einstein and the observer on Earth are looking at the *same* beam of light. For them to measure the same speed relative to themselves, either Einstein's ruler or clock, or both of them, must have different scales from those of his earthbound assistant.

> Give identical rulers to two observers, in two inertial frames, and they will look exactly the same provided each observer looks at his own ruler. But if an observer looks at his own ruler and at the ruler in the other frame, they will appear to be different. Likewise, if you look, from the vantage of your frame, at a clock in another inertial frame, it will appear to be running at a different rate from yours.

If the differences in scale are chosen properly it *is* possible for both Einstein and the earthbound observer to see the beam from his torch traveling at the same speed. This conclusion is at the core of the special theory of relativity. *Distance and time intervals in inertial frames depend on which frame the observer is standing in.* Einstein had demolished the Newtonian-Galilean concept of an absolute time and an absolute space.

We have no intention of getting into the formalism of relativity theory, but at least it is worth seeing that the famous and central subject of "time dilation" is very easy to understand. Let us use Einstein's railway coach again (he did).

In the moving train we will place a clock of a kind thought up by Richard Feynman (Figure 32.3b). The lamp sends out a regular stream of light pulses, each of which is reflected off a mirror and falls onto a receiver that emits an audible beep and simultaneously triggers the next pulse. The time it takes for one cycle of the mechanism will be our unit of time. The man on the platform has an identical clock (Figure 32.3a), but let's see how the clock on the train appears to him. *In the eyes of the stationary observer (SO)*, the light pulses takes a longer path between lamp and receiver, due to the motion of the train. However, according to Einstein, the veloci-

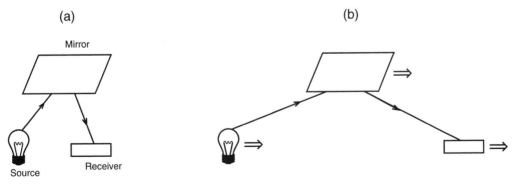

(a)   (b)

Mirror

Source   Receiver

Figure 32.3. Feynman's clock. The regularly emitted photons are reflected off the mirror and reach the receiver. To an observer on Earth the time taken for the journey in the moving frame (b) is longer than that in the static frame (a) because the path is longer and the speed of light is independent of the motion of the observer.

ty of light is the same for all observers so that *for SO,* the light pulse in the moving frame takes longer to cover the path from lamp to receiver. The passenger does not see this extended path; he sees an identical path to that seen, for his own clock, by the man on the platform. The man on the platform looks at his clock and notes that with respect to that clock, the clock on the train is running slow; it's not beeping fast enough. In fact, if the train is running at three-quarters of the speed of light, its time intervals will appear to him to be about one and a half times longer than those of the stationary clock. To the man on the train, life seems perfectly normal. If the train is traveling at 99% of the speed of light, the factor rises to over 50. Note once again that these are identical clocks and the path of the light *in his own clock* looks identical to both observers. To the stationary observer, time has passed more slowly on the train. The whole argument depends on common sense plus one totally un-common assumption: that the speed of light on the train appears to be the same to the man on the platform as it does to the man on the train.

If you want to work out the difference between the clocks, the formula for the in-terval of time between beeps is:

$$t = \frac{t'}{\sqrt{[1 - v^2/c^2]}} \qquad (1)$$

where t is the interval observed on the stationary clock and t' that, say 10 seconds for example, observed on the moving clock. v is the velocity of the moving clock with respect to the earthbound observer and c is the velocity of light. Note that when v approaches the speed of light, (v^2/c^2) approaches unity and t approaches in-finity. In other words, a clock traveling away from Earth at the speed of light would appear, *to an earthbound observer,* to have stopped. When v equals zero, and the two frames are at rest with respect to each other, $t = t'$, and we are back to the nor-mal Newtonian world.

This so-called time dilation is the origin of the famous twin paradox, in which one twin journeys into space at great speeds, returning to Earth to find that he has crossed far less days off the calendar than his brother. Which clock was right? Nei-ther! They were both right, because there is no Master Clock, guarded by old Father Time and counting off the *real* seconds. To emphasize this, put yourself in the train. The man on the platform appears to be moving away from you, and you will see his clock going *slower* than yours, the reverse of what he experiences. *There is no absolute time.* We have not measured something called time, we have counted off events, comparing two pieces of apparatus. That, according to the positivist philosophers, is about all we can say about time. And, by the way, although we haven't shown it, there is no sacred length; that too depends on the observer.

So Fast and No Faster

Einstein went a step further. He emphasized not only that light cannot travel in-stantaneously between two bodies, but neither can force.

Thus, suppose that you could suddenly switch off the charge on a proton. Would an electron, say a meter away, instantly feel that the electric field around it had gone out? No. The field would die out in the time that it takes for light to travel between the two particles. Admittedly, this is a very small interval of time, but it is not zero. Why should the force be connected to the velocity of light? The easiest way to see this is to recall that the electromagnetic force is mediated by photons, and they certainly travel at the speed of light. This is why Einstein saw the velocity of light not only as being an upper limit to the velocity of any body but also as a limitation on the "velocity" of forces. No force can be transmitted between two bodies faster than the speed of light. For example, if our fiendish friend Dr. Moriarty decided to operate his antigravity gadget on the Sun, we would not know about it for at least eight minutes, which is roughly the time it takes for light to reach us from the Sun. This assumes that gravity is propagated as we believe electromagnetic fields are propagated, taking time to cross space. No other type of signal can reach us faster than the speed of light—until you ask what is happening in Alain Aspect's experiment (see Chapter 30).

Fact, Not Fiction

Time dilation has been observed in several experiments. A particularly dramatic example is based on the spontaneous breakup of elementary particles, or of atomic nuclei (see Chapter 12), a process that proceeds at a pace unaffected by man. Each kind of elementary particle, when in isolation, has its own half-life. In 1937 Anderson and Neddermeyer detected a short-lived particle in cosmic rays. Muons have a half-life of 2 millionths of a second, which means that after this time half the muons in any given sample will have disintegrated. The elementary particles can be thought of as having an internal coin thrower, different types of particle having different rates of throw. Heads, I disintegrate; tails, I live a bit longer. The thrower knows when to throw because he has a clock. Pressure and temperature changes have no effect on the rate of this internal clock. In experiments at CERN, a beam of muons has been induced to rush round a circular tube at speeds of 99.94% of the speed of light. If you put this figure into equation (1), you will find that 1 second as observed on the coin thrower's watch (inside the muon) is seen as about 29 seconds by the stationary observer watching his own clock. To the coin thrower in the muon, disintegration goes on at the usual rate. To the experimenter, the lifetime of the muons should increase by 29-fold, from 2 to nearly 60 millionths of a second. The actual observed difference between experiment and Einstein's prediction was in fact less than 0.2%. "Time" had slowed down, if you want to put it that way. Mach would say that we have no right to bring time into it; all that had happened was that, as seen from the stationary observers' viewpoint, one clock ran twenty-nine times slower than the other. Either way, Einstein rules, OK.

Mass

One of the more notorious consequences of special relativity is that the observed mass of a body depends on its speed with respect to the observer. We will not derive this fact, but the mathematical treatment shows that as the body accelerates its mass increases. If it travels at the speed of light, its mass becomes infinite. This is the origin of the statement that nothing can travel faster than the speed of light.

Notice that the increase in mass is in the eye of an observer *in a frame with respect to which the body is moving.* If you travel with the body, it is motionless with respect to you, and the additional mass observed by the stationary observer is not observed by you. You will see what is called the *rest mass* of the body. Again we see that what we observe depends on the relationship between our frame and that of the object that we are observing.

The mass increase at high speeds, predicted by theory, has been confirmed experimentally. As mentioned earlier, it has to be taken into account in designing particle accelerators, in which velocities close to that of light are reached. The electrons in your TV cathode tube travel at roughly 7% of the speed of light, which means that they have about 7% more mass than their rest mass. Electrons in a linear accelerator can reach speeds within one part in a billion of that of the speed of light, and they behave, *experimentally*, as though their observed masses are over 10^{13} as large as their rest masses. This is as though an ant increased its mass to 1 million tons.

New Laws for Old

The second law of motion ($\mathbf{a} = \mathbf{F}/m$,) assumes that m, the mass of the body, is constant. We have seen that from an observer's point of view the mass depends on the relative speed of the body. If m increases, then, for constant force $\mathbf{F}$, the acceleration $\mathbf{a}$ decreases. At the speed of light the mass is infinite and the acceleration is zero. The force appears to have no effect at all, with respect to the stationary observer. Two hundred and twenty years after the *Principia*, Einstein rewrote Newton's second law and unknowingly gave the explanation, but not the impetus to the atomic bomb:[5]

$$a = \frac{F[1 - v^2/c^2]^{3/2}}{m} \qquad (2)$$

The superscript 3/2 means take the expression in the brackets and multiply it by its square root. If you don't like this reminder of high school, it doesn't matter; the two important things to note are that:

- When v, the velocity of the body, is equal to c, the quantity in the brackets vanishes, which means that the acceleration is zero. We can regard this as equivalent to the body having infinite mass.
- If v is zero, the equation reduces to Newton's $\mathbf{a} = \mathbf{F}/m$. Furthermore, even for quite large velocities, the equation is so close to Newton's as to make no practical difference.

As an example, for a body moving at a velocity of one-tenth of the speed of light (i.e., 3000 km/sec, which is enormous by any normal standards), the equation becomes a = 0.985F/m, which means that the acceleration for a given force is only about 1.5% less than Newton predicts. This illustrates a general conclusion: when we are dealing with velocities well below the speed of light, the equations of special relativity reduce to those of Newton. Thus we see that Newton is not "wrong"; his equations are very good approximations if we deal with bodies traveling at low speed. At low velocities all the mechanical equations of relativity reduce to those of Newton. It could be said that we never noticed that anything was wrong with Newton because we never traveled fast enough, and we never knew that we needed quantum theory because we aren't small enough.

---

[5]Einstein's theories had almost nothing to do with the trails of research that led to nuclear weapons. Such weapons undoubtedly would have been developed in any case.

The Unification of Mass and Energy

If we exert force on a body, then its observed mass increases when it accelerates with respect to us, and a weird and wonderful consequence can be derived. The apparent increase in the mass is due to our having done work on the body. (Recall the definition of work, as force times distance.) The work has appeared as both an increase in the kinetic energy of the body, because its velocity is increasing, *and* an increase in mass. It is apparent that the work has been converted into energy *and* *mass*. Now work is a form of energy and can only be converted into other forms of energy. This implies that the increase in the *mass* of the body can also be looked on as an increase in its *energy*. The conclusion is that mass and energy are two aspects of the same thing. The mathematics gives $E = mc^2$, where m is the body's mass and c is the speed of light. We are so used to this equation that we are no longer amazed by its wonderful simplicity. If there is a scientific analogue to minimalist art, then surely this equation, and Newton's $F = ma$, would take joint first prize at the Venice Bienalle.

Consider a practical example. A piece of coal has three kinds of energy: potential, chemical, and "mass" energy. One kilo of coal when burned yields about 30 million joules. This is its chemical energy and is very roughly the amount of potential energy that the coal would have if it were at a height of 3000 kilometers. Now let's turn the coal into energy by destroying its mass. This is *not* a chemical change like burning. Mass is genuinely disappearing. You can do the calculation immediately; you get $mc^2 = 1$ kilo x $(3 \times 10^8$ meter/sec$)^2 = 9 \times 10^{16}$ joule, or about 3 billion times as much as the chemical energy. It is clear why great efforts are being made to duplicate, on Earth, the reaction in the Sun, in which part of the mass of hydrogen atoms is turned into energy by nuclear fusion (see Chapter 34). The amounts of energy obtained for a given quantity of hydrogen are so large that the available hydrogen in the oceans looks infinite in this context. Incidentally, every second the Sun converts 5 million tons of its mass into energy. In a billion years, if the energy conversion proceeded at a constant rate, about 1.6×10^{23} tons would "burn" away. This amount is less than one ten millionth of the Sun's mass.

In nuclear reactions in the Sun or in nuclear reactors, the conversion of mass to energy is not complete. What happens is that two entities meet and produce products that have less total mass than that of the reacting particles. The *difference* in mass then appears as the kinetic energy of the particles produced. It is this kinetic energy that is the source of the power that we get from nuclear reactors.

A New Conservation Law

What are we to do about Lavoisier's law of conservation of mass and Clausius's first law—the conservation of energy? Neither mass nor energy, in the conventional sense of these terms, is conserved in the nuclear reactions mentioned previously. Let's take a more extreme example: an electron and a positron meet, annihilate each other, and turn into a pair of photons, which are definitely radiation; no mass survives. We cannot deny that mass has disappeared, and if we believe in the equivalence of mass and energy we must assume that the mass has been *entirely* transformed into the energy of the two photons. We can take the input of "energy" as being roughly the sum of the masses of the electron and the positron. We ignore any contribution from the kinetic energy of the colliding particles, which will usually be negligible. If we use $E = mc^2$, we find that the total energy equivalent of the colliding particles is about 1.537×10^{-13} joule. This must be equal to the energy of the two photons, which, assuming that they are identical, is given by $2hv$. Using the value for Planck's constant, we find that the frequency of the photons is about 1.2×10^{20}Hz, which corresponds to γ rays, agreeing with the observation that electron-positron annihilation produces these rays.

We see that what is conserved in nature is a mixture of mass and energy. (A physicist might tell you that the quantity conserved is the *energy-momentum four vector*, but don't be cowed.) Normally, in the everyday world, there is no significant interconversion of mass and energy, and we can keep both Lavoisier's and Clausius's laws, but nevertheless, mass and energy are two aspects of the same phenomenon.

The Defect and the Remedy

The defect in the special theory of relativity, which only becomes apparent in the general theory, is that its laws begin to break down seriously in the presence of large masses. Although the equations of special relativity predict strange effects on mass, the *basic* equations that Einstein constructed did not contain mass. They predict changes in the apparent mass of a body at high speeds, and they led to the equivalence of mass and energy, but the equations themselves, such as (1), include only space, time, the relative velocities of frames, and the speed of light. In this sense they have an affinity to classical physics—there is an arena of space and time into which we put matter. Time and space become inextricably mixed in the equations, but not because of the presence of matter. Matter does not affect the architecture or the clock of the stadium; only the relative motion of inertial frames does that. But Einstein had more rabbits in his top hat. The general theory shows that matter, by its presence, distorts "space-time." The general theory provides a very different account of the working of gravity from that of Newton, and it was thinking about gravity that led Einstein to formulate the theory.

General Relativity

The professor came down in his dressing gown as usual for breakfast but he hardly touched a thing. I thought something was wrong, so I asked what was troubling him. "Darling," he said, "I have a wonderful idea." And after drinking his coffee, he went to the piano and started playing. Now and again he would stop, making a few notes then repeat, "I've got a wonderful idea, a marvelous idea!"

—Mrs. Einstein

The physics and mathematics of the theory of general relativity, which was published in 1916, are far more complex than those for special relativity. We will simply indicate what the theory is about and what its practical consequences are. I start by comparing two startling predictions of the theory with experimental fact. Subsequently, I will use a tiny handful of equations, but at that point, those of you who are violently allergic to math may choose to close the book and play golf instead.

The Proven Facts, Part I

Einstein's theory predicted that the path of light would be bent near massive bodies. Newton had also believed that his light corpuscles would be attracted by gravity, and Faraday looked unsuccessfully for the effect of gravity on light. Einstein's prediction was tested in 1919, during an eclipse of the Sun. The man who organized the experiment, which involved sending observers to Brazil and Príncipe, an island in the Gulf of Guinea, was the English cosmologist Sir Arthur Eddington, who was an early convert to relativity. Telescopes, directed at the rim of the eclipsed Sun, detected light from stars that were known to be slightly behind the Sun. Because the starlight was bent inward toward the Sun as it passed, the image of these stars appeared to be just on the edge of the Sun (Figure 32.4). The measurements were made during an eclipse so as to blot out most of the light of the Sun, which would otherwise have made the observation of the stars impossible. The angle through which the light had turned was derived from their apparent position and the known position of the stars, as recorded on previous photographs. The angles through which the light was deflected were small, but enough to convince Eddington that the effect was real.

The experiment had been on a grand scale. An account of the observations was presented at a joint meeting of the Royal Society and the Royal Astronomical Society, in London, and the announcement of the results by the astronomer royal, Sir Frank Watson Dyson, was a dramatic high point in the history of science. The deviations recorded by the two expeditions were 1.61 and 1.98 seconds of arc, although the experimental error was rather large. Einstein had predicted a deviation of 1.74 seconds of arc. The chairman of the meeting, J. J. Thomson, proclaimed, "This is the most important result obtained in connection with the theory of gravitation since Newton's day [and] one of the highest achievements of human thought." That was on 6 November 1919. The following day there was an announcement in the London *Times*, and within days Einstein's name became known to more people at one time than that of anyone else in the history of science.

The effect is now so frequently observed as to have become commonplace among astronomers. This is partly because of the use of radio waves, the detection of which is quite unaffected by the presence of visible light. One of the sources of radio waves are a group of objects called *quasars* (quasi-stellar objects). In 1978 the British astronomer Dennis Walsh detected a double quasar, two closely spaced sources of light. Their spectra were suspiciously similar, and the suggestion arose that they were two images of the same object. This has been verified, and there are

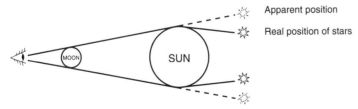

Figure 32.4. The light from the stars bends as it passes the Sun, allowing us to see stars that are behind the Sun. I have grossly exaggerated the bending angle.

now thousands of related cases in which part or all of the light from a distant object has been deflected by the presence of a large mass intervening between the object and the Earth. It is as though these large masses acted as imperfect lenses. The phenomenon of "lensing" was predicted in 1937 by the Swiss-American astronomer Fritz Zwicky, who based his suggestion on Einstein's theory. He saw that if two galaxies were aligned such that, seen from the Earth, one was behind the other, then the gravitational field of the nearer one could distort the light from the farther. In the best traditions of science, he was ignored. Dennis Walsh's observations restored Zwicky's good name.

The Proven Facts, Part II

Another prediction of general relativity is based on the prediction that a clock will run slower if it is moved from a weaker to a stronger gravitational field. This is quite distinct from the slowing down due to relative motion that we encountered in discussing special relativity. Perhaps it would be better to say that the clock slows down as it moves into increasingly distorted regions of space-time, but I stick to the old-fashioned way of talking. Thus consider a clock at ground level and one high above the Earth. The theory says that an observer on Earth should find that the airborne clock is running faster, because the gravitational field of the Earth weakens with altitude.

The first really convincing experimental proof of this prediction was carried out in 1976 by sending a clock up to a height of about 6000 miles in a rocket. At the high point of its flight, the clock was running at nearly a billionth of a second per second faster than those on Earth, as Einstein's theory predicted.

If time is slowed down by an increase in gravity, then the frequency of light emitted by an atom is also reduced. You can think of it as a slowing down of everything connected with time. Slow the video and the tenor becomes a baritone.[6] Thus a light source in a strong gravitational field (sorry, near a large mass) should exhibit a redshift, a lowering of frequency. This is not a Doppler effect such as that due to the relative movement of source and observer; it occurs even for a source and observer at a fixed distance. The effect has been observed in the light emitted by the Sun, and far more markedly in the light emitted by the gravitationally lavishly endowed white dwarfs, where clocks appear to be running slow by over an hour a year with respect to us.

There are no alternative explanations—as yet—for the observed phenomena predicted by the general theory of relativity. The effects are small, but they can be measured accurately, and they agree with the theory. What has all this to do with everyday life? Nothing, except that there is something remarkable in man's ability to create such subtle explanations for the way the planets and the stars stretch time and play with light. The one area where the theory is absolutely essential is in cosmology; the conditions at the time of the Creation were such that Einstein cannot be ignored.

Space-time

Of all the terms that relativity brings with it, the *fourth dimension* has been the

---

[6]Why don't the *colors* of the video film change?

most quoted, the most misunderstood, and in a way the simplest. It has been latched onto in particular by some nonscientists, and used to give a false aura of scientific authorization to anything irrational. As Eddington remarked, in the early years of the notoriety surrounding relativity: "In those days one had to become an expert in dodging persons who were persuaded that the fourth dimension was the door to spiritualism."

Every event in history needs four dimensions to describe it. At the Battle of Hastings, King Harold was hit in the eye by a Norman arrow. We can use the exact map reference to locate the spot on the ground above which Harold's eye was at the time. This requires two map coordinates. A third number is necessary to determine the height of the eye above the ground. A scientist would probably label these three numbers as x, y, and z. That takes care of the three dimensions that we need in ordinary space. However, we need a fourth number, the time that the arrow struck. That is the fourth dimension. Simple. All *events* take place in four dimensions— three of space and one of time. So why wasn't a fuss made about it until relativity came on the stage? To answer that, we have to go back to the Greeks.

Our scientific, as compared with our intuitive, concepts of space and time grew out of the way that the Greeks constructed geometry. The master was Euclid, whose axioms dominated geometry for well over 2000 years. Euclidean geometry is a self-consistent system, by which I mean that if you accept the axioms, then all the rest follows logically. Man lives in a world in which Euclidean geometry is useful. Any measurement that involves a plane, or a body constructed from planes, whether that body be a math lecturer's concept or a real everyday object, can be handled by geometry, where the qualifier "Euclidean" is rarely used. Newton's *Principia* is heavily dependent on Euclidean geometry for its proofs.

Euclidean geometry was subtly identified with the geometry of space. Man had looked around and seen that he lived on what appeared to be a flat plane. When he built a temple, he laid out lines on the ground. They were straight lines. When he drew a triangle, it was on a flat surface; the angles always added up to 180°. Euclidean geometry is built on man's experience of his immediate neighborhood, and man extended his straight-line universe out into space, imagining that space had the same Euclidean geometry. Newton's universe is based on Euclidean geometry.

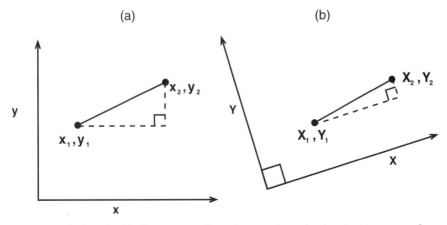

Figure 32.5. The length of the line in a given frame does not depend on how I orientate my reference axes; in this case x and y, but in general x, y, and z.

The laws of motion spoke of bodies that, in the absence of forces, moved in straight lines. Gravity, according to Newton, acted in a straight line, and he certainly saw the framework of absolute space as built of straight lines. The point that is relevant to us is that in that universe the shortest distance between two points is a straight line. Let us look at that a little closer.

The length of a straight line can be calculated if you know where its ends are. The formal way to do this is to first set up three axes, as in Figure 32.5a. In a room it would be convenient to use the edge where two of the walls meet and the two edges where the same walls meet the floor. Now mark off scales along these "coordinate axes." To simplify the explanation, I consider a straight line marked on a wall. The coordinates (x,y) of the two end points are determined as shown in the figure. The information is enough to allow the length, l, to be determined from the (Pythagorean) expression:

$$l^2 = (x_2 - x_1)^2 + (y_2 - y_1)^2$$

or, in a condensed notation:

$$l^2 = (\Delta x)^2 + (\Delta y)^2 \qquad (3)$$

The generalization to three dimensions is:

$$l^2 = (\Delta x)^2 + (\Delta y)^2 + (\Delta z)^2 \qquad (4)$$

The first thing to note is that these formulas do not work for a line drawn on the surface of a sphere or in general on any curved surface. In fact, we could turn our argument around and say that if (4) does give the length of a line having the shortest distance between two points, then we are dealing with "Euclidean space."

Now the length of a line drawn on a wall cannot possibly depend on how we choose our coordinate axes. Figure 32.5b shows the same line enclosed by two different sets of axes. The length as referred to the (a) set of axes is given by:

$$l^2 = (\Delta x)^2 + (\Delta y)^2$$

In the (b) set of axes is given by:

$$l^2 = (\Delta X)^2 + (\Delta Y)^2$$

We see that in both sets of coordinate axes the *expression* for the length is exactly the same. Only the *numbers* on the right-hand sides of the equation will alter, but the final length must be the same. Again generalizing to three dimensions, we can say that expression (4) is an *invariant in Euclidean space.* In other words, we can choose any frame we like, and the Euclidean expression (4) will give the length of the line. This length is not dependent on what frame we choose. This is true as long as all the frames are static with respect to the line.

The form of the invariant (4) underlies the Newtonian concept of space. In Newton's universe the length of a line had to be same whichever observer measured it, and the measure of its length was given by the Euclidean expression (4), *even if the frames were moving with respect to one another.* According to Newton, a line drawn on the side of a train would appear to have the same length whether or not the train was moving with respect to the observer.

The importance of the invariant can be appreciated by considering the problem of calculating the path of a discus. In Newton's universe the path of the discus and the distance it travels cannot possibly depend on the observer, static or moving. The method of calculation is based on that belief, even if you don't realize it. The method is founded on my conviction that the way I measure distance is invariant to any changes I may make in my frame, in other words, any changes in either my co-ordinate system (where I put the x, y, and z axes) or my velocity. The independence of velocity is reflected in the fact that the expression for the length of a line doesn't include time. The invariant is my guarantee that I don't need a new system of equations every time I change my frame.

When we get to special relativity, a problem arises. For two inertial frames traveling with respect to each other, the length of a line as seen in the two frames is different. As you have probably guessed, the line is shorter as seen from the frame not containing the line. Thus the length of the line, as defined by expression (4) is *not* an invariant in the theory of special relativity. The value of the expression, which gives the observed length of the line, alters depending on which frame you are in. What this really says is that the nature of space and time is not reflected by the invariant—it is, in fact, not an invariant in the space and time described by special relativity. So what is an invariant in special relativity? We can make a guess that if we are bringing motion into the picture we will have to bring time in somewhere.

Instead of considering two *points*, let us look at two *events* occurring at different places and times, one at the point (x_1, y_1, z_1) at time t_1 and the other at point (x_2, y_2, z_2) and at time t_2 (Figure 32.6). Now, by analogy with the expression for the length of a line, define a quantity s, by:

$$s^2 = (\Delta x)^2 + (\Delta y)^2 + (\Delta z)^2 - c(\Delta t)^2 \quad (5)$$

where Δt is $t_2 - t_1$, c is the speed of light, $\Delta x = x_2 - x_1$ and so on. It can be shown that this quantity has the same value no matter what inertial frame the observer is in. The individual values of the coordinates of the points and the times of the events, as seen from your frame, may be different from those seen by someone in another inertial frame, but the *sum* , s^2, will always end up the same. Explicitly: if, according to an observer in a certain framework, an event took place in that framework at the point (x_1, y_1, z_1) and the time t_1, and another event occurred at the point $(x_2, y_2,$

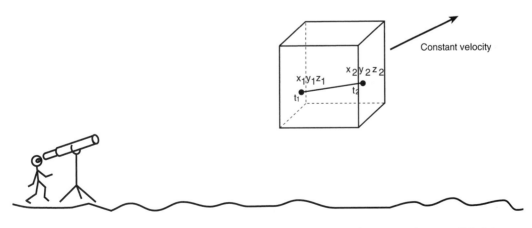

Figure 32.6. For frames moving with a constant velocity with respect to each other, a given observer will find the same value for the quantity s^2 separating two *events* and defined in equation (5).

z_2) and t_2; then for a second observer, traveling in another frame at a constant velocity with respect to the first, the expression (5) would give exactly the same number s^2. In general, two such observers would not see the same length of line. Neither would they see the same difference in time between the two events.

Expression (5) is Einstein's improvement on Newton's invariant; it is the invariant of special relativity. It is called not a length but an *interval*. It includes three space dimensions and the "dimension" of time. Now you see where the "fourth dimension" comes from, and why we speak of space-time. It has nothing to do with spiritualism. All that has happened is that we need to mix space and time together in one expression in order to obtain a quantity that looks the same to all observers.[7] Incidentally, although we are not in a Euclidean space, we are still in what can be called a "flat" space, a space where lines can get shorter depending on the observer but will not become curved. The shortest distance between two points is still a conventional straight line; we have not yet "warped space-time."

Here endeth all the mathematics we are going to encounter in *special* relativity. There is nothing in the least mysterious here, nor particularly complex. The expression for s is used in calculations, and the results are quite independent of any visual significance you may attach to s.

Neither HMS nor most scientists are likely to have to use the invariant of special relativity. The fact is that two observers moving at much less than the speed of light with respect to each other will see the same distance between two points and the same time interval between two events. They can go back to expression (4) and forget about the infinitesimally small "mixing" of space and time that becomes important only as the relative speed of the observers approaches that of light.

Flexible Space-time

For centuries it was assumed that there was only one kind of geometry. This is not true. Had man lived on a planet that was only 100 meters in diameter, he might have grown up thinking in curved lines. Stand on the North Pole of such a planet and draw two "straight" lines. They would be what we call lines of longitude. From where we stand we see two straight lines diverging. Euclid tells us that two straight line diverging from a point will diverge forever. But if you follow these two lines, they first diverge and then converge to meet at the South Pole. Two straight lines can contain an area. Now you will say that this is cheating; these are not straight lines in the normal sense. What you mean is that they are not *Euclidean* straight lines. I agree, but I would point out two things. First, a straight line can be defined as a line on which the shortest distance between any two points on it only runs through points on the line. This is true of the traditional straight line, but it is also true of the "curved" lines of longitude on a sphere if you accept that you cannot leave the surface of the sphere. Second, if Euclid had been born on our miniplanet, he probably would have constructed an altogether different system of geometry, based on different axioms. The theorems of this geometry would be at variance with the familiar and much-hated theorems we learned at school. To start with, he would have drawn triangles, squares, and other "straight line" figures, and found

---

[7]For the interested reader: the laws of physics are second-order differential equations with respect to time (Newton) or interval (Einstein).

that on a sphere, in contrast to "Euclid," the sum of their angles depended on the size of the figure. Euclidean geometry is convenient for flat surfaces, but clearly we need other geometries for other surfaces.

Other geometries were constructed in the nineteenth century by a number of mathematicians, the outstanding trio being the Russian Nicholas Lobachevsky (1792–1856), the Hungarian János Bolyai (1802–1860), and the German Bernhard Riemann (1826–1866). They had been preceded by Gauss, but he never published anything systematic on non-Euclidean geometry.

Non-Euclidean geometry was largely ignored by other nineteenth-century mathematicians, let alone physicists, and it was only when Einstein was looking for a geometry with properties that suited general relativity that a friend introduced him to the work of Riemann. When he was twenty-eight years old, Riemann published a paper entitled "On the Hypotheses Which Lie at the Basis of Geometry." The ideas therein were to be the basis for a great deal of modern mathematics and for our present conception of space-time. To give a hint of the difference from Euclid, consider Riemann's axiom that all lines are finite in length but endless. Worried? You wouldn't be if you lived on a sphere. Walk along any line, and you will find that you will return to the point of origin (the line is finite) but that your line has no ends. Now let us see why the general theory of relativity needs to forsake Euclidean geometry. Fasten your safety belts, space-time is about to be warped.

Comment on the General Theory

If asked to sum up the general theory in two sentences, I might say something like this: "In the inertial frames of classical physics and of special relativity, Newton's laws held in any frame, but they do not hold for an *accelerating* frame, say an aircraft taking off. One of Einstein's achievements in the theory of general relativity, was to construct physical equations that hold in any frame, even one that is accelerating." That's one way the physicists see it, but we don't have to look at it through their eyes.

The theory is sometimes said to be a new theory of gravity, and it does reveal gravity in a brilliant new light. But above all it is a revolution in the way that we need to view the relationship between space-time and matter. In the Newtonian universe, space and time are separate, absolute concepts. In special relativity, space and time get mixed up, but the extent of the mixing depends only on the relative speeds of the inertial frames—it does not depend on the mass of bodies.

In the theory of general relativity, matter distorts space-time.

The effect can be ignored for small masses, which means that special relativity, or even Newtonian physics, usually suffices for these cases.

There are at least two possible responses to the statement that matter distorts space-time. One is to attempt to form a visual impression of space-time and then ask what the presence of mass does to it. The other is to say that the human mind is not adapted to thinking in terms of space-time or its distortions—Einstein presumably gave us some trustworthy equations, so, let's see what they can do. In the end we are going to use them to explain observed facts or predict new ones; their success has nothing to do with questionable geometrical images. The second approach is undoubtedly easier; a theory stands or falls on its explanatory and predictive successes. However, it is tempting to try to form visual images to go with the theory, so I will attempt to give a short account of the simplest visual interpretation of general relativity. It is this interpretation that contains the psychedelic features of the theory, the warping of space and the bending of light.

The origin of the general theory is contained in an observation we made in Chapter 5, namely, the fact that someone in a falling elevator would not feel the force of gravity. This set Einstein thinking about the connection between gravity and frames of reference. Newton had seen gravity as a force, but Einstein realized that the effects of gravity could apparently be reproduced by the acceleration of the frame of reference. Acceleration is embedded deep in Newton's laws of motion, but Newton was limited to Euclidean space. What happens to the laws of motion when we go to other geometries?

When I fly from London to New York, the plane usually passes over Newfoundland and New England. On the map in the airline's brochure, this looks like the long way around, a very curved path. Of course, the reason is that the map is a distorted representation of the surface of a sphere. The flight path on a globe is also a curve but only because a plane cannot take the shortest straight line path through the Earth. It flies along a great circle, which is the shortest path permitted by the shape of the globe. We say that the *geodesics* of a sphere are great circles. In general, for any space, geodesics are the shortest lines between two points, taking into account the limitations imposed by the geometry of the space. For example, the shortest path between two points on the surface of a cylinder is, in general, part of an ellipse, as you can guess from stretching a piece of string between two points at different heights on a tennis ball canister.

In the geometry of Euclid, we learn that the shortest distance between two points is a straight line. We could turn this statement on its head and say that the fact that the shortest distance between two points is a straight line means that we are living in a "flat" world, that what we call "space" is the familiar space in which Euclid's geometry works. If you like, you can picture this familiar space as containing an invisible rectilinear scaffolding. In such a Euclidean space, the geodesics are what we are used to calling straight lines. We saw that before Einstein and the physicists came along, there were mathematicians playing with other spaces. More correctly, they had constructed geometries based on modifications of Euclid's set of axioms. Thus they threw out the famous axiom about parallel lines, which implies that parallel lines never meet. If you stand on the equator of a sphere and draw two closely spaced parallel lines of longitude, they will indeed be parallel, but if you follow them to the poles, they meet. Again, don't be tempted to say, "But they are not parallel lines"; they are, *on a spherical surface*. Relax your mind, take a puff, and journey into distorted geometries. Imagine that you were born on a sphere that had a diameter of 20 feet. The shortest distance between two points would always be a curved line, a geodesic, and resist the temptation to say, "Yes, but I could tunnel through the sphere and make a straight line." Not permitted! Now, throw away the sphere but leave the invisible geodesics. When you go for a "straight line" walk with your girlfriend, you will find yourself back where you started after walking "straight" for about 63 feet. You have been controlled by the geodesics.

Now go back to Newton's first law of motion. It says that a body moves in a straight line if no force acts on it. You could rephrase this to say that if a force is not acting, the body moves along a *geodesic in Euclidean space*. Now take this space and distort it, bending the geodesics. One can envisage a new first law, which says that when there is no force acting on a body it travels not in a straight line but along a geodesic which, if space is warped, need not be straight in the Euclidean sense. This puts a new light on the paths of the heavenly bodies. As an example, consider the case of a comet passing objects in its flight. It would effectively ignore a small asteroid, but the Sun has a dramatic effect, turning the comet about it and sending it off in an entirely different direction. Newton would attribute the effect to gravity, Einstein would say that both the asteroid and the Sun have distorted space-time and its geodesics, but the degree of the distortion depends on the mass of the distorter, and the Sun has a far greater effect.

The warping of space-time is, according to Einstein, a result of the presence of mass. Note again that the special theory says nothing of this effect. The most famous example of the warping of space-time is the bending of light as it passes the Sun. Newton had predicted that the Sun's gravitational force would deflect the "corpuscular" rays of light coming from the stars. Einstein saw this differently. The Sun distorted the space-time around it, "curving" it, so that light followed a curved path.

Why do we talk about the distortion of space-time and not of space? There are two ways of facing this question. They are not really different, but they give different insights. The first is to go back to the subject of invariants.

Invariants allow us to calculate what an observer sees when he is in the same or a different frame from what he is observing. They also say something fundamental about the physics of the space and time that we live in. Newton's is the comfortable, familiar invariant that treats space and time separately because they don't get mixed up in our everyday life. Newton's universe was Euclidean, and we found that the invariant in that universe was the length of a straight line. Einstein came along with the theory of special relativity, accepted Euclidean space, but showed that the Newtonian length was not invariable. It depended on the relative motion of the line and the observer. There is, moreover, no expression that is invariant for inertial frames unless it includes time. Instead of taking the distance between two points, Einstein took the *interval* between two events. Special relativity weaves time and space together when we want to account for what we see. General relativity broadens our outlook to account for the effects of mass and acceleration. Expression (5), for the interval, contains no reference to mass. Einstein remedied this in the theory of general relativity. He looked for a way of expressing what an observer would see not only if he were moving with respect to another frame but also if there were masses involved. As with the universes of Newton and of special relativity, there is an invariant in general relativity, and its form tells us a great deal. We will not use it, but it is worth writing down, just to make some nonmathematical points. Here it is in its simplest form:[8]

$$s^2 = Ac(\Delta t)^2 + B(\Delta x)^2 + C(\Delta y)^2 + D(\Delta z)^2$$

$$+ Ec(\Delta t)(\Delta x) + Fc(\Delta t)(\Delta y) + Gc(\Delta t)(\Delta z)$$

$$+ H(\Delta x)(\Delta y) + J(\Delta x)(\Delta z) + K(\Delta y)(\Delta z) \tag{6}$$

---

[8]I have simplified the notation. The 10 coefficients, A,B,...K are written g^{00}, g^{11}, g^{22}, g^{33}, g^{01}, g^{02}, ...g^{23}. Together they are known as the *metric tensor*.

(c is the speed of light). The essential point is that *the coefficients (A,B,...K) contain the masses of the bodies that have to be taken into account.* When the coefficients are estimated for two events taking place very far from any masses, the coefficient A becomes -1; B, C, and D become 1, and E to K become zero. The expression reduces to the invariant, (5), of special relativity.

The invariants of Newton and special relativity were universal—they applied to any place, any time, and any velocity. The invariant of general relativity has a universal form, but the coefficients depend on the size and location of the masses involved in the location under consideration. Thus the interval between two events in empty space is an invariant of the simple type (5), as seen from any inertial frame in empty space, but (5) is not an invariant in a region of large masses. The effect of the masses is to change what we observe as far as position and time are concerned, from what we would see if the masses were absent. However, if the details of the masses are put into the new invariant (6), it is invariant for any two frames. Two events taking place near the massive planet Jupiter will be separated by an interval s, given by (6), as measured from any frame, either on Jupiter or moving in any other locality. Thus every physical system has its own values for the coefficients.

Invariants tell us about the kind of geometry that we are using. When I fly from New York to Tel Aviv, the plane roughly follows a geodesic, following part of a great circle. Now imagine placing the Earth in a coordinate system of the kind typified in Figure 32.5. I could note the coordinates of the two cities and then attempt to find the distance between them using the Newtonian invariant (4). I would get the wrong answer. The number that I get is equal to the straight-line distance between the cities, the length of a tunnel between them. From a mathematician's point of view, the mistake I have made is to assume that the invariant of Newtonian space applied to a curved geometry, the surface of the sphere. There is an invariant that can be constructed for spherical geometry, and I need it to find the length of my flight. An invariant reveals the geometry within which it applies. If someone gives me the expressions (4) or (5), I know that he is working within a Euclidean space, a straight-line world. What can we say about the world in which (6) is an invariant?

First, it is a world that is not Euclidean. The geodesics are not straight lines, and furthermore their shape depends on the specific locality in which we are interested. What emerges from a detailed analysis is that, in general, the geodesics are more curved the greater the amount of mass in the locality, and become straight in the complete absence of mass. Einstein actually constructed an invariant that would guarantee geodesics conforming to the paths of the planets. Each planet has a different path, but that is taken care of by the flexibility of the invariant, which allows different coefficients to be constructed from different masses.

This is the origin of the concept of "warped space-time." It is apparent that almost any amount of distortion is possible, being limited only by the amount of mass in a certain locality. Some of the greatest "warpers" are black holes.

Einstein's theory does not include the force of gravity as a separate property of matter. It says that mass molds space-time in such a way that the geodesics are shaped like the observed paths of moving bodies. The space-time geodesics near the surface of the Earth ensure that bodies fall vertically. The Earth itself moves in the curved geodesics created by the Sun, as do the other planets.

Science for Pleasure

Homo supermarketus does not need the theory of relativity. But the theory represents scientific creativity at its most awesome. Einstein's incredible physical intuition, his single-minded pursuit of what he believed to be the manifestation of God through nature—Spinoza's God—his welding of mathematics to observed fact, his intellectual courage in forsaking accepted scientific dogma; all these combine to produce a theory of power and great beauty.

Einstein believed, like Pythagoras and a long line of thinkers, in the unity of nature, in the possibility of combining all the forces of nature into one grand synthesis. For about thirty years he tried to construct a unified field theory that would explain electromagnetism as he had explained gravity, in terms of space-time. He failed completely. His Achilles' heel was quantum mechanics. The man who in his youth had quantized light could never accept the probabilistic interpretation of quantum mechanics, and he left that subject untouched for decades, except for repeated attempts to convince the rest of the scientific world that they were wrong. He could not work with a theory that he believed to be seriously flawed, a theory that stretched belief in a commonsense reality "out there." Einstein was a realist: he believed in a real world, independent of the observer. But he saw it as a continuum, a kind of field in which, putting it crudely, matter manifested itself as the bunching together of the energy in the field. He didn't see quantum mechanics as having a vital role in this picture, but it is almost certain that no unified field theory can ignore quantum mechanics. It remains to be seen whether Einstein was right and nearly everyone else is wrong.

33

Cosmology

The discovery of a new dish does more for the happiness of mankind than the discovery of a star.

—Jean Anthelme Brillat-Savarin, early nineteenth-century gastronome

One can't help sympathizing with the famous chef. Most of us are chilled by the silent oceans of space. I can hear James Joyce's Stephen Dedalus, reflecting on the universe's "vast inhuman cycles of activity." Be honest: the surface of the Moon is about as intrinsically interesting to HMS as the Utah salt flats. The rest of the solar system is unfriendly; and none of us will travel to the stars. The image of space is bleak, serving as a metaphor for spiritual isolation:

> I am alone on the surface
> of a turning planet. What
> to do but, like Michelangelo's
> Adam, put my hand
> out into unknown space,
> hoping for the reciprocating touch?

R. S. Thomas, *Threshold*

There is a persistent hope that we are not alone in this colossal wasteland of hydrogen gas. Our fear of psychopathic aliens is balanced by a wish that there be someone else to keep us company, preferable a small, docile creature with big Hollywood puppy eyes. We would like to have close relatives out there, which is why the standard alien has humanoid features. Who comes out of a UFO? A little green *man.* Ask anyone.

Like the Vikings and the Polynesians, we have just begun to set sail, in primitive boats, extending this planet's thin film of life, but as yet only to our nearest neighbor. And the monsters that populate the far reaches of space are no less strange than the dragons and anthropophagi fantasized by Hakluyt and his Elizabethan colleagues. Who has not heard of pulsars and supernovas, of red giants and white dwarfs and the voracious black holes? The structure and behavior of these denizens of the cosmic deep are the clues to the history of the universe and may, in the far future, determine the fate of man, if that poor forked creature doesn't self-destruct before an asteroid obliterates us.

Cosmology is largely a product of this century. This is not a snub to Ptolemy, Kepler, Brahe, Copernicus, Newton, Halley, and generations of Babylonian, Greek, Chinese, Arab, and Maya stargazers. It was the observations and speculations of these men that culminated in Newton's grand synthesis. But it was not until the advent of giant telescopes in the twentieth century that the immense size of the universe became apparent. It was not until we began to analyze the radiation from the stars that their chemical composition was revealed. Only the advent of quantum mechanics allowed an understanding of the processes that produce energy within the stars and control the size of white dwarfs. Not until the detection of radio waves from space did we know of thousands of dark stars. Only the theory of relativity permitted us to comprehend the structure and functioning of black holes, the slowing down of the rotation of pulsars, the real meaning of the "expanding universe." And it was the detection of microwave radiation coming from space that gave us our most direct hint that the universe began with a Big Bang. Twentieth-century cosmology is one of the great breakthroughs, a breakthrough that is still proceeding.

Of the four forces known to physics, only one is important in determining the large-scale structure of the universe. Gravity is the sole major force acting between the different objects, planets, stars, galaxies, and so on that make up the coarse structure of the universe. It is not the only force. Thus light can and does exert pressure, and the magnetic field of the Sun, for example, cannot be ignored, but these are negligible factors; it is gravity that is the heavyweight.

In what follows, details will be avoided. We will look at some current ideas about the birth of the universe and will review the main types of bodies found in the heavens. We begin with our own planet, to which, as far as massive colonization is concerned, we may well be confined for at least a couple of centuries to come.

The Earth

A New Theory of the Earth, from the Original to the Consummation of All
Things, Wherein the Creation of the World in Six Days, the Universal Deluge,
and the General Conflagration, as laid down in the Holy Scriptures, are
shown to be perfectly agreeable to Reason and Philosophy.

—Title of a book by the astronomer William Whiston (1696)

A few months ago I was sitting at home in Haifa when I heard what sounded like a truckload of stones being tipped out. A small avalanche of rocks was running down the slope of the mountain on which I live. The room moved, slightly but perceptibly. The last serious earthquake in Israel was a couple of hundred years ago, but minor earth tremors are quite frequent. Haifa is situated not too far from the Jordan valley, which lies roughly in line with the Great Rift Valley of East Africa. These two valleys are the meeting place of two *tectonic plates*. The surface of the Earth is covered by about half a dozen large plates and many smaller ones. These plates are huge areas of more or less stable crust floating on a layer of partially liquid rock, the *asthenosphere*. The tectonic plates consist of the *lithosphere*, a solid layer about 70 kilometers thick, and above that the *crust*, a thin layer of earth, rock, and clay. The predominant rock on or near the surface is granite, composed primarily of oxygen, silicon, and aluminium. Tending to be lower down, but frequently appearing as outcrops, is basalt, a rock consisting mainly of oxygen, silicon, iron, and magnesium.

The plates move, at speeds of a few centimeters a year. Where they meet below oceans, their surfaces are generally nearer the asthenosphere, which sends up molten material into the joins between plates. This cools down and solidifies, and somewhere something has to give. Often the other edges of the plate are forced downward and slowly melt.

The jostling together of plates can result in the buildup of tension at their mutual boundaries, as when a drawer is jammed. The sudden release of this tension can be catastrophic. Japan lies on the boundary between the Pacific and Eurasian plates. San Francisco and Los Angeles lie on the boundary between the Pacific and American plate. A number of large and small plates come together in the area of Greece, Turkey, and the Middle East. All these areas are prone to earthquakes. Massive collisions of plates in prehistory resulted in the folding up of the Earth's surface. The Himalayas, the Andes, and the Rockies all lie along plate boundaries.

The tectonic plates are floating, and have been since they formed during the cooling of the molten Earth. The dynamic state of the Earth's surface is made dramatically evident by looking at a map of the eastern coast of the American continent and the western coast of Europe and Africa. As Sir Francis Bacon noticed, in their general shape they match each other, as though there had once been one great land mass that had split from north to south and drifted apart. The similarity of fossil remains and mineral deposits on the two sides added to the suspicion that what is two was once one. There is no doubt now, in the light of accumulated fact, that this thesis is correct.

The molten asthenosphere is about 200 kilometers wide and covers an inner layer, the *mesosphere*, which is more solid in texture and has a width of about 2000 or more kilometers. The mesosphere, lithosphere, and asthenosphere are collectively called the *mantle* of the earth. Inside this mantle is the *outer core,* which appears to be liquid metal, and which surrounds the *inner core* of the Earth, a solid ball of metal, probably mainly iron but perhaps containing nickel, about 2500 kilometers in diameter. Man has not succeeded in drilling down to depths of more than 20 kilometers. Our knowledge of the Earth's interior comes almost entirely from seismology. The behavior of shock waves generated either by natural events or by deliberately setting off explosions is, in the hands of experts, a source of information on the large-scale structure of the Earth. It's a little like the use of ultrasound in medicine.

The Earth was once entirely molten. The fact that the interior of the Earth still contains molten material, which gushes out occasionally, is the result of the continual production of heat by the decay of radioactive elements. This is what fooled those nineteenth-century scientists who tried to work out the age of the Earth from its present temperature and its estimated rate of cooling. They didn't realize that it had its own source of heat. The heat of the core, which has an estimated temperature in the range of 3000-7000° C, sets up convection currents in the less than rigid mantle, which in turn are responsible for the movement of the tectonic plates.

The Wobbling Magnet

As the Elizabethan William Gilbert guessed, the Earth is a magnet. The magnetic poles are not fixed; they meander for several miles every year. Furthermore, the North and South Poles have switched places about ten times during the past 3 to 4 million years. We know this because, long ago, rocks that were then molten were magnetized by the magnetic field of the Earth. When they cooled and solidified, they preserved the footprints of the magnetic field at the time, and from independent evidence we can date the rocks. Successive layers of rock sometimes have opposite directions of magnetization. We don't know why these switches occurred. They take about a thousand years to complete, which is fast on the geological timescale. It has been suggested that the theory of chaos may apply, but there is as yet no physical model to which the theory can be applied. Another change, which man should watch nervously, is the progressive weakening of the strength of the field, which in the last 150 years has fallen by some 7 percent.

The Earth's magnetic field dominates that of the Sun for many thousands of kilometers above the Earth's surface. The Earth's magnetic field has a vital protective role. The Sun ejects a continuous stream of protons and electrons, the "solar wind." This wind is augmented periodically by the emissions from sunspots and solar flares. The particles of the solar wind are almost completely deflected past us into space by the Earth's magnetic field. Unusually strong gusts of the solar wind can cause what are known as magnetic storms, which can disrupt electrical communication systems, and create spectacular lighting effects in the heavens—for example, the northern lights. Fluorescent lighting is based on the passage of electrons through an inert gas, such as neon. The invading electrons blast some of the atomic electrons out of their orbits, and the resultant fall of other atomic electrons into the empty orbits releases the photons which give such a ghastly color to faces in all-night cafés. This phenomenon occurs on a massive scale when a powerful gust of solar wind penetrates our atmosphere. This is the origin of the great auroras that hang in northern skies at times of enhanced solar activity.

If the Earth's magnetic field continues to weaken, we may eventually find that the solar wind is blowing uncomfortably strong. The effect of a considerable rise in the number of charged particles reaching the Earth could be an increase in radiation-induced illnesses and in the mutation rate, not to speak of the possible effects on our electronic civilization. But this is not the only threat from space.

Voltaire believed that the Earth had remained as it was when it was created except for the effects of 150 days of the Deluge, but the fact is that there have been massive global changes over the past few billion years. Life in general has managed to adapt to the slow changes that have characterized the history of the Earth, the huge changes in climate typified by the ice ages, the shifting continents, and the rise in the oxygen content of the atmosphere as plants established themselves on Earth. Occasionally, however, major catastrophes have occurred, and there may be more on the way. The main threat, apart from man's greed, is from space.

The Earth-Crossers

In 1801, in Palermo, Sicily, the monk Guiseppe Piazzi peered through his telescope and found a new heavenly body within the solar system. Ceres is about 1000 kilometers across, compared with our Moon's 3200 kilometers. This barren rock proved to be the largest of a huge number of bodies, the *asteroids,* careering around the Sun in paths lying between Mars and Jupiter. The problem is that there is a group of asteroids, the *Earth-crossers*, whose paths cross the path of the Earth.

In June 1908 a farmer sitting on his porch in the Tunguska region of Siberia was knocked off his chair by a powerful shock wave. Eight kilometers above the Earth an object, possibly an asteroid, and probably about 70 meters across, had heated up and exploded after entering the atmosphere. The sound was heard 1000 kilometers away, trees were flattened for 30 kilometers around, and the intense heat ignited the forests. And all this from an object that never reached Earth.

To penetrate our atmosphere an asteroid would have to have a minimum diameter of about 200 meters. There are plenty of those around. We can get an idea of what would happen if an asteroid with a diameter between 10 to 15 kilometers struck us. It is now generally accepted that a collision with such an asteroid occurred some 65 million years ago, leaving a crater about the size of the Netherlands, 6 miles deep and 190 miles across, off the Yucatán Peninsula in the Gulf of Mexico. The energy released was equivalent to about 200 million tons of TNT. Roughly a quarter of a million cubic kilometers of dust was thrown high into the atmosphere, forming a cloud that darkened the Sun for months and brought a mini ice age. The dinosaurs were wiped out; they couldn't accommodate to the change in climate.

There are something like 180 identified Earth-crossers, none of which is predicted to hit the Earth during the next century. But small asteroids are difficult to detect, and an asteroid measuring, say, only 3 kilometers across and traveling at 150,000 kph could raise enough dust to blot out the Sun for months. One hesitates to think what it could do if it fell on a city. In 1992 a large group of scientists, under the auspices of NASA, produced a battle plan to combat Earth-crossers. Six sophisticated telescopes at widely separated sites would search solely for asteroids. Possible paths would be predicted and the dates of collisions estimated years ahead. Is this fund-raising scaremongering? Not if you believe those astronomers who claim that there are as many as 5000 as yet undetected Earth-crossers of over half a mile diameter. Our only feasible defense against such an object would be to launch a nuclear missile and hope to blow it to pieces or divert its path. Maybe the technology that destroyed Hiroshima will one day save this planet from another ice age.

At the very edge of the solar system there is a host of comets, some of which visit our neighborhood. Some swoop in, swoop out, and are never seen again. Some swing around periodically. Ovid recounts how Venus came down invisibly into the Roman Senate to remove Caesar's soul from his body. As they ascended, his soul ignited and became a comet.

Gaia

The realization is spreading that the Earth is a collection of delicately balanced interdependent systems. The oceans and the atmosphere, the radiation of the Sun, geophysical processes, the activities of living forms—all these constitute a system of great complexity in which changes in one part may have effects that are not always easy to predict. Thus a drop in global temperature can cause significant amounts of the ocean to be added to the polar ice caps. This would result in a lowering of the level of the oceans, which in turn would lessen the pressure on the seabed, especially along the continental coasts. This could make it easier for pressurized subterranean molten rock to burst out, so that lower global temperatures could be correlated with increased volcanic activity. On the other hand, increased volcanic activity pushes more dust into the skies; if this dust contains large proportions of sulfur-containing gases, it will result in the formation of tiny droplets (aerosol) of sulfuric acid in the upper atmosphere. These are good at scattering and absorbing sunlight and thus contributing to the cooling of the Earth.

This complex interdependence was apparently first seized on by the eminent Scottish geologist James Hutton (1726–1797) to justify the supposition that the Earth behaved as a single organism. This idea, in the hands of James Lovelock, has recently been turned into a fashionable hypothesis, which sees the Earth as being an organism, with mechanisms of self-defense. The biologist Lynn Margulis has joined Lovelock in developing and promoting the theory, going under the name of Gaia (a Greek goddess), and at one time it received a great deal of publicity in the media. But, as Richard Dawkins has pointed out, the Earth can hardly be considered to be an organism, since it doesn't reproduce. It is difficult to see what is gained in scientific terms by defining an interdependent group of phenomena as one single organism, if their mutual relations are known or being researched anyway. A central point of the theory seems to be that the Earth defends itself, just as an organism does, but this is highly questionable. The Earth and its inhabitants adjust to changes, and some of those adjustments are successful, but it does not follow that they have the nature of the biological defense mechanisms produced by evolution (see Chapter 26). It is difficult to believe that the Earth as a whole has evolved in this way.

There are signs that interest in Gaia is waning. Nevertheless, if its emphasis on the unity of the natural world gets through to some of those who determine national policies, it will be doing a major service. Gaia, with its holistic aura, has inevitably been taken up by the New Age dreamers. Would they have noticed it if, instead of being named after a Greek goddess, the theory had been called coordinated interactive non-linear dynamics in the terrestrial bio- and geospheres?

The Sun

The Sun, which is 864,000 miles (1,382,000 km) in diameter, has a mass of nearly 2 x 10^{30} kilograms, equal to about 700 times the total mass of all the other planets combined. The Philosophers of Jonathan Swift's Laputa, like Helmholtz, feared that the Sun would burn out, but this will probably not happen for a billion or so years. The Sun's interior temperature has been estimated at between 8 and 15 million K, although the average temperature on the surface is only about 6000 K. The Sun has an intense magnetic field that reverses its polarity every eleven years. This field plays a major part in the Sun's activities, being responsible for the formation of sunspots and the huge archlike flares of radiant matter that appear temporarily on the Sun's surface and can reach hundreds of thousands of kilometers above the surface. For reasons that are not clear, the temperature in these *coronas* can be as hot as 2 million K, and in short bursts can even reach about 20 million K. It is electrons accelerated by the magnetic fields in these coronas that produce the radio waves present in the Sun's radiation. It is from these outbursts of matter above the Sun's surface that spurts of matter are thrown far out into space. This solar wind has been detected beyond Pluto by the spacelab *Voyager*.

The Sun is the only star that we can, at present, study at close hand, and much of what we learn can be applied to other stars, all of which, even with the best telescopes, appear to us as little more than dots. Thus, for example, the sudden, huge, short-lived rises in temperature that are observed in some stars can often probably be interpreted in terms of the solar flares that can be seen quite clearly on the Sun.

The Planets

It is convenient to define the distance of the Earth from the Sun, which is about 93 million miles, as one *astronomical unit* (1 AU). Thus if the Earth is 1 AU from the Sun, then the distances of the nearest four planets to the Sun are Mercury, 0.39 AU; Venus, 0.72 AU; the Earth; Mars, 1.52 AU. These comparatively small planets, of which the Earth is the largest, form a tight little group around the Sun and are called the *terrestrial planets* because they have an outer coat of rock over a predominantly metallic core. They consist mainly of the elements silicon, oxygen, and iron, the oxygen being primarily tied up in the silicates that form rock (see Chapter 13). Mercury (about a third of the diameter of the Earth) and Mars (about half), being considerably smaller than the Earth, cooled faster and display little volcanic activity. Mars has enough of a gravitational field at its surface (about 38% of that of the Earth) to hold on to a sparse layer of carbon dioxide. Venus, which is roughly the size of the Earth and has a comparable gravitational field, has a deep atmosphere of carbon dioxide.

Moving outward, the next four planets, the *gas giants,* are all much larger than the Earth. They contain high proportions of the two lightest elements, hydrogen and helium: Jupiter, 5.20 AU; Saturn, 9.52 AU; Uranus, 19.16 AU; Neptune, 29.99 AU. Notice how the gas giants are spread out over far greater distances than the terrestrial planets. Neptune is 2,789 million miles from the Sun.

The gas planets have solid cores but thick, partially gaseous, partially liquid wrappings, with properties appropriate to a science fiction film. Thus Jupiter's core is surrounded by an enormous ocean of liquid hydrogen, which is under the pressure resulting from Jupiter's gravitational field. This *liquid metallic hydrogen* has strange properties, being able to conduct electricity. It may well be responsible for Jupiter's powerful magnetic field. Strange auroras move over the polar regions, and great bolts of lightning fall from the atmosphere, which is about 90% hydrogen and 10% helium, with traces of other gases. The gravity of the Earth cannot hold these gases very effectively, but for its first hundred or so million years of existence, the chemical composition of the Earth's atmosphere may have resembled that of present-day Jupiter. Jupiter gives out a great deal of energy, but its source is probably the slow contraction of the planet, not the nuclear reactions that produce the Sun's radiation. (You can envisage this contraction in terms of matter falling "down" and losing potential energy, just like a boulder or the weight in Joule's experiment. The lost energy appears as increased kinetic energy.) Jupiter has sixteen moons, ranging in diameter from 6 miles to about 3300 miles, compared with our moon's 2170 miles. It was Galileo who found the four large moons of Jupiter, which he named the "Medicean planets" after the family of Grand Duke Cosimo II d'Medici of Florence. He also sent the grand duke a telescope, which was perhaps his way of applying for a research grant. He was duly appointed chief mathematician of the University of Pisa and philosopher of the grand duke.

The six innermost planets, including the Earth, have been known since prehistory. It was not until the eighteenth century that Uranus, the seventh planet, was found.

Isaac Newton made the first reflecting telescope, based on curved mirrors. Following Newton's lead, the great astronomer Sir William Herschel made a 10-foot-long reflecting telescope, 6 inches in diameter, with which, in 1781, he found the first planet discovered since prehistoric times. Herschel, who as a young man in Hanover earned a living by playing the oboe in a military band, settled in England and subsequently became the best telescope maker of his time, his crowning effort being a huge, 40-foot-long, reflecting telescope with mirrors 4 feet across. He discovered Uranus, a planet encircled by somber, coal-colored rings.

Mars is an example of seeing what you want to see. The nineteenth-century astronomers Camille Flammarion, Giovanni Schiaparelli, and Percival Lowell all looked at Mars and claimed that they saw signs of life and intelligence. Their evidence was a network of suspiciously straight lines covering great distances on the surface of the planet. Lowell published a number of convincing drawings of the planet in his book *Mars and Its Canals* (1906). On the basis of these drawings, the lines seen on Mars are almost certainly the work of intelligent beings. And then came the camera, which has one great advantage over man: it has no expectations. Émile Zola said, "We cannot claim to have really seen anything before having photographed it." One can certainly argue with this opinion,[1] but in science the camera can give an objectivity that sometimes eludes the scientist.

Photographs of Mars do not resemble Lowell's drawings at all. The canals finally faded into scientific limbo when Mars was photographed by twentieth-century space missions, showing that science fact is sometimes science fiction. Nevertheless, photographs taken from satellites in orbit around Mars have shown the presence of deep twisting scars that resemble dried up riverbeds, and it has been hypothesized that water, the sine qua non of terrestrial life, was once part of the Martian environment. There are plans to send an unmanned Russian-built robot to Mars in 2001, its mission being to drill beneath the surface to search for fossilized primitive life-forms such as algae. In the meantime, as noted earlier, what appear to be fossilized microorganisms have been detected in rocks from Mars.

All eight planets revolve in planes that are no more than 3° from that of the Earth, which suggests that they were all formed together from the same revolving disk of matter. This *nebular hypothesis* that sees the solar system as condensing out of a whirling plate of hot gas, has fallen in and out of favor over the past century. As yet there is no universally accepted theory of the origin of the solar system.

Far outside the other planets is Pluto, at an average distance from the sun of 39.37 AU. Pluto was discovered only in 1930. It is the smallest of the planets, only about the size of the Earth's Moon. Its orbit, like that of Mars, is more elliptical than those of the other planets, and is also at a marked angle to those of the rest of the planets. It may have a different origin from the other planets, which should set you UFO buffs thinking.

Or you might care to reflect on the Titius-Bode law, formulated in the eighteenth century. This says that if you take the numbers 0,1,2,4,8,16,32,64,128 . . . triple each number, add 4, and finally divide by 10, the resultant numbers give the distances of the planets (including the asteroid Ceres) from the sun in A.U. This works very well except for Neptune. Does this mean that Neptune is in some important way different from the other planets? Or is the "law" just a coincidence?

Light takes about twelve hours to cross the solar system, a negligible time by intergalactic standards. The nearest star to Earth is the Sun; the next nearest, Proxima

---

[1]As Baudelaire, who was not a realist, did. For him photography was "industry's imbecile revenge upon art." Which didn't prevent him sitting for a portrait.

Centauri, in the Centaurus constellation, is 4.8 light-years away, which is some 12 million times farther than the Moon. This comparison illustrates the magnitude of the task facing man if he wishes to reach destinations outside the solar system, which one day he might have to do.

No other planet in the solar system can support life. Our present technology gives little hope of massively colonizing another home outside the solar system. In the meantime, we are endangering our survival on Earth. Perhaps Arthur Koestler was right: *Homo sapiens* may be one of the species in the galaxy that are mentally unstable biological misfits, if there are any other species. We have had no telephone calls up to now.

34

The Cosmos and Peeping Tom

In 1682 the writer Ihara Saikaku published his *Life of an Amorous Man*. Saikaku was not one to be bound by convention. In a country where the cherry blossom is almost sacred he started one story, "I was so bored with cherry blossoms that I stayed away from the capital all spring." The fifty-four chapters of his book recount the erotic adventures of one Yonosuke, from the age of seven. By the age of fifty-four, he had had encounters with 3742 women and 725 men, which is suggestive of an unstoppable attack of hiccups rather than of promiscuity. In one episode the precocious hero, aged nine, stands on a roof watching a maidservant in her bath—through a telescope. The telescope had been put to educational use, perhaps not for the first time, but this seems to be the first such appearance in literature. Galileo—as far as we know—pointed his telescope up, not down.

Our knowledge of the universe still comes mainly via the telescope. Perhaps no branch of science depends so much on a single invention, especially when it is used in conjunction with the camera.

Objects that are too dim for the eye to see through a telescope may be recorded on a photographic plate because, unlike the eye, the plate accumulates the effect of every photon striking its surface. Thus, a photographic plate can be left exposed for a considerable time, during which it slowly records the evidence of very dim sources. The familiar photographs of galaxies are all taken with long exposure times. They are hardly visible against the night sky.

The word telescope means to "see far," which allows us to include within the term instruments that collect radiation other than visible light. The most spectacular of these is the radio telescope, two major examples of which are the huge disk at Jodrell Bank in England and the VLA (very large array), in New Mexico, which consists of twenty-seven receiver disks, each 25 meters in diameter and arranged along three arms each about 21 km long. Every range of frequencies in the electromagnetic spectrum is now used to probe the universe, but we have to use instruments carried by satellites or rockets to observe any other than visible, radio, and some IR radiation. Other wavelengths fail to penetrate the atmosphere.

Before Galileo turned his telescope to the night sky in 1610, the universe was a much smaller place. There are fewer than 5000 stars visible to the naked eye, although it must have been assumed that there were many more, in view of the common European belief that each person had his own personal star.[1] By comparison with today's telescopes, Galileo's simple lenses were toys. What he saw as single points of light, twentieth-century astronomers see as huge agglomerations of separate stars. The size of the discernible universe has expanded with the passing of the

---

[1] In the fifth century, Bishop Eusebius of Alexandria asked if "there were only two stars at the time of Adam and Eve."

centuries, and in the early years of this century, hints began to accumulate that the universe was much larger than it appeared.

On one of his journeys, Magellan recorded two smeared dots of light, now known as the Small and Large Magellanic Clouds. In 1912 the American astronomer Henrietta Leavitt (1868–1921) estimated the distance to these clouds and showed that their apparent size could be accounted for only if they were billions of miles across. They were later shown to consist of vast stretches of glowing gas and stars. By the beginning of this century, the significance of the solar system had completely faded. It was clear that we were a tiny part of a galaxy, the Milky Way, but it was not certain whether this galaxy constituted the whole universe.

On a favorable night it is possible to make out a great band of light across the sky. The Milky Way is in fact disk-shaped, its center being thicker than its edges. The reason we see a band is that we are ourselves situated in the Milky Way, about two-thirds of the way to the edge, and are looking more or less sideways through it, as was first guessed by Thomas Wright of Durham in 1750.

Only eighty years ago the American astronomer Howard Shapley thought that the Milky Way *was* the universe; it is about 120,000 light-years across, which in those days was pretty big. He was hugely wrong. He should have listened to Thomas Wright, who guessed that there were other Milky Ways, "too remote for even our Telescopes to reach."

Scattered throughout the sky are spiral objects, the *spiral nebulae*. Shapley claimed that they were comparatively near, being essentially a part of the Milky Way. Not everyone agreed with him. At a meeting of the American National Academy of Science in 1920, Heber Curtis suggested not only that the spiral nebulae were much farther away than Shapley thought but that they were in fact galaxies—"island universes" (the phrase is Kant's).[2] Was he right?

Measuring the Immeasurable

How far are the spiral nebulae? How large is the universe? We cannot begin to answer these questions unless we can measure the distance of heavenly objects. The breakthrough was made by Henrietta Leavitt, who was interested in a rather special class of stars, the Cepheids. The intensity of the light coming from Cepheids rises and falls regularly with time, with a period that varies between stars, usually being a few days but sometimes as much as three to four months. Leavitt found nearly 2500 Cepheids. Concentrating on one of the Magellanic Clouds, she found that there was a very close relationship between brightness and period for the Cepheids in the cloud. The brighter a Cepheid was, the longer its period.

The distance of the Magellanic Cloud is so great that the stars there can be regarded as all being effectively the same distance from the Earth. If you are in Los Angeles, everybody in Carnegie Hall is about the same distance from you. This means that if one Cepheid is four times brighter than another, it is not because it is nearer to you but because it is intrinsically four times brighter. Suppose that a Cepheid in the cloud has a certain brightness and a period of one week. Now look at another Cepheid in some more distant galaxy. If it has the same period, we can

---

[2]As we have seen, Kant started his academic career more as physicist than philosopher, and one of his first publications was *General History of Nature and Theory of the Heavens* (1755).

assume it has the same intrinsic brightness, and yet it is dimmer than it should be. We can deduce that the Cepheid is farther away from us than those in the Magellanic Cloud. Moreover, assuming that its real, intrinsic brightness is the same as that of a Cepheid of similar period, in the cloud, we can work out its relative distance from Earth. A star of the same intrinsic brightness that is twice as far away will be four times dimmer. A method had become available to measure the relative distances of stars, albeit a restricted class of stars. It is slightly complicated by the effects on brightness of interstellar dust clouds, but it was a huge step forward.

The next step in our story was made by the unlikely-named Vesto Melvin Slipher (1875–1969), who did to the light from spiral nebulae what Newton did to sunlight. He placed a prism in his telescope and obtained spectra. Slipher saw the kind of lines that are shown in Figure 15.6, which he readily identified as lines due to hydrogen and helium and, to a lesser extent other elements. No unrecognizable lines were found, or have been found since. The spectra of the stars is completely analyzable in terms of the spectral lines belonging to known chemical elements. What was unexpected was that the frequencies of the lines that he observed were all shifted very slightly from those obtained from the chemical elements in earthly laboratories. This strongly suggested that the sources of light were moving with respect to the observer and that frequencies were being altered by the Doppler effect (see Chapter 15). The shift varied from galaxy to galaxy. They were apparently traveling at different speeds with respect to the observer on Earth.

The strange thing about Slipher's observations was that in practically every case the frequencies were all *lower* than they should have been. There was what is called a *redshift*, so named because red light appears at the low-frequency end of the visible spectrum. Now the observed frequency of light emitted by a source moving *away* from the observer is redshifted. Slipher's finding therefore suggested that nearly all the spiral nebulae were moving away from us. Two spiral nebulae had a blueshift, indicating that they were apparently moving nearer to us. One of these was the Andromeda nebula that is approaching us at about 360,000 kph.

Why were nearly all the observed spiral nebulae moving away from us? And were they part of the Milky Way, as Shapley thought, or way out there in extra–Milky Way space, as Curtis had suggested?

The Dynamic Universe

In 1919 Edwin Hubble (1889–1953) came to the Mount Wilson Observatory. He used a new 100-inch telescope, the most powerful instrument available at that time, to look for Cepheids in galaxies. In this way, using Henrietta Leavitt's method, he hoped to measure the distances to galaxies. Were they really as far as Curtis thought they were? In December 1924 Hubble announced that he had observed a Cepheid that had a period of about one month. From its brightness he estimated that it was nearly 1 million light-years away. That figure has since been corrected to 2 million light-years. The star was in a galaxy officially named Messier 31, but it is better known as the Andromeda nebula (Figure 34.1). It is much too far away to be part of the Milky Way, and, as Hubble showed, it is probably very similar in its shape to the Milky Way, although a little larger. With one stroke, our home galaxy had been demoted to a desperately isolated whirl of stars lost in an unimaginably huge void.

Hubble showed that the universe was populated with galaxies, "island universes," that were several hundred million light-years away from us. Today the known visible universe encompasses galaxies that are thousands of millions of light-years away.

As soon as Hubble started publishing his observations, Slipher's Doppler effect suddenly became of intense interest. Were these incredibly distant galaxies all moving away from us, as Slipher's results implied? Was the universe not only much larger than people thought but also expanding?

The link between Hubble and Slipher was made by the physicist Howard Robertson. He did what all self-respecting scientists would have done—he made a graph. The magnitude of the Doppler shift can be used to calculate the speed at which a light source is moving away from an observer. Robertson plotted the estimated speed of the galaxies against the distances of the same galaxies as estimated by Hubble. He was astonished to find that *the farther the galaxy was from the Earth, the greater was the redshift.* About the same time, Hubble noticed the same correlation. In 1929 he announced that there was "a roughly linear relation between velocities and distances." The speed of recession was proportional to the distance of the galaxy. This is now known as Hubble's law, but there is a good case for calling it the Robertson-Hubble Law. Robertson didn't complain. Perhaps he was aware of the fact that Hubble was a very good boxer.

Since antiquity, astronomers had believed that the movement of the heavenly bodies was overwhelmingly cyclic. The moon circled the Earth, the planets circled the Sun. The fixed stars were fixed. Now the universe had been revealed to be not so much a carousel as a fireworks display. Some of the galaxies that Hubble observed were receding at speeds in the region of 100 million kph. But what does Hubble's law mean? If all the galaxies are moving away from us, does this mean that we are at the center of the universe? And why are they moving?

Imagine a circle of folk-dancers skipping backward, away from each other, maintaining the circle but expanding it. The first point to grasp is that *every* dancer sees *all* the others retreating from him. It is not necessary to stand in the middle of the circle to experience this distancing of each dancer from all the others. Notice also that *the farther away two dancers are from each other, the faster the distance between them increases.* This argument can immediately be generalized to three dimensions.

The interpretation of Robertson's (Hubble's) finding was that the universe, like the circle of dancers, was expanding. This was a great turning point in our efforts to write the history of the universe. Again, note carefully that the fact that all the galaxies appear to be receding from us does not mean that the Earth is at the center of the universe. Every dancer sees all the other dancers going away from him, but none is at the center of the circle. An observer on *any* galaxy will see what Hubble saw.

The next question is unavoidable: On what date did the universe start expanding, and what did it look like at that time? Regarding the time scale: on the basis of the present state of the universe, and the observed rate of expansion, cosmologists have run the video backward and estimated that the universe began its expansion

---

[3]When life first emerged on Earth, the density of matter in the expanding universe was about twice what it is today.

about 15,000 billion years ago.[3] Everything is presumed to have started with an incredible explosion, the famous Big Bang, a phrase coined by Fred Hoyle. The first man to present this startling image of the Creation was a Belgian Jesuit priest, engineer, and astronomer, Georges Lemaître (1894–1966), who proposed that the universe started as a sphere, a "primeval atom," about thirty times bigger than the Sun. The physicist George Gamow subsequently presented a related, and now defunct, model that in his and other people's hands developed into the present Big Bang theory.

How do we know that there was a Big Bang? Alternative theories have avoided the "something-out-of-nothing" conundrum by replacing it with an equally incomprehensible enigma: the eternal universe, with no beginning and hopefully no end. For many years Einstein defended this concept of a *static* universe but relinquished it in the face of the experimental evidence. At present the Big Bang theory is the favored version of the Creation. You can regard time and space as beginning with the Big Bang, but I know that whatever the cosmologists do to convince us that there was nothing before the Big Bang—because there wasn't a "before"—HMS, and the majority of scientists, will never feel really comfortable with the something-out-of-nothing scenario.

The phrase "the expanding universe" should not be taken to imply that everything in the universe is expanding. The Milky Way is not expanding, and neither are the other galaxies. Neither were the dancers in our analogy. But they are certainly moving apart from each other.

There is a common misconception about the meaning of "expansion" in this context. There is a natural tendency to see the galaxies flying through space. Here a much-used analogy helps; the universe is likened to a fruitcake baking in the oven. The currants are not moving through the expanding dough; they are moving *with* the dough. Now replace the dough with space, and you have the right picture. It is *space* that is stretching. I know it doesn't fit in with daily experience, but that's the way the cosmologists say it is. As space stretches, it stretches the light waves coming to us from the galaxies, and they drop in frequency. That is the real explanation of the redshift; it is not a genuine Doppler shift.

Another strange consequence of the expanding universe is that there are galaxies moving so fast that their light can never reach us. This implies that, compared with us, they are traveling faster than the speed of light, thus apparently breaking the condition imposed by the special theory of relativity. The difficulty can be overcome if you accept the fact that they are traveling *with* space and not *through* space. It doesn't matter if this is puzzling; the point to grasp is that there may be parts of the universe that we will never see, and those parts may be far larger than the visible universe. They may be very different, but we may never know.

We can only see the universe from the Earth or from the observatories that we send out into our immediate locality. Is this a typical view? Would the universe look completely different from another vantage point? Einstein thought about this and propounded his cosmological principle, which states that wherever you are in the universe you will see pretty much the same overall picture, rather like a man in a huge crowd. It says that the distribution of galaxies is fairly uniform throughout the universe, which conforms with what we see from here. The principle seems reasonable but cannot be proved since we cannot (yet) travel all over the universe. It certainly sweetens the life of the theoretical cosmologist.

The Galactic Host

The eternal silence of these infinite spaces terrifies me.

—Blaise Pascal, *Pensées*

They cannot scare me with their empty spaces
Between stars—on stars where no human race is.
I have it in me so much nearer home
To scare myself with my own desert places.

—Robert Frost, *Desert Places*

All the individual stars that you see with the naked eye are in the Milky Way, including our Sun and the nearest star (Proxima Centauri). There are about 100 billion other stars in our galaxy. All that you can discern of other galaxies are a few blurred smudges of light. The Milky Way is a typical galaxy, shaped like a disk with a swollen center, where most of the stars in the galaxy are found. The wispy spiral arms are characteristic of the majority of galaxies. We live in a *normal* galaxy, which just means that there is not much happening to its structure. The Milky Way turns in space, one revolution taking about 300 million years. It is this rotation that is probably responsible for the form of the arms of spiral galaxies.

Observation indicates that all stars, in our galaxy or elsewhere, have a life cycle: birth, followed by a hectic, glittering life which is ended by one of two kinds of death. The small stars go relatively quietly. The biggest stars, Gloria Swanson–like, leave the stage with a spectacular final fling. Stars are made primarily from hydrogen atoms with a minority contribution from helium. These were the two main elements left over from the Big Bang, and they were distributed thinly throughout space. The force that brought hosts of them together to form stars is gravity. The process of star formation can be pictured as a kind of gravitational snowball, although this picture is far from proven. A cluster of atoms that happened to come together by chance would tend to pull other atoms in their direction. As the mass of this cluster grows, its gravitational pull also increases, and so does its effectiveness in persuading more free atoms to fall into its lair. Huge quantities of matter eventually accumulate. The gravitational pull of the cloud not only draws in more material from its surroundings but also compresses the cloud to higher and higher densities. This process of collapse is not simple; it does not resemble a deflating balloon. Part of the gas forms a disklike structure that begins to revolve and in many cases breaks up into two lumps, forming a double star. Another possibility is that the disk condenses into a planetary system.

As the newborn star contracts, the pressure and temperature at its center rise. The speed of atoms, and therefore their kinetic energy, grows as they fall toward the center of the nascent star, and we know that the temperature of a gas is reflected in the average kinetic energy of its molecules. A typical star could have a temperature of 10 million degrees Celsius at its center. This is hot enough for the protons and other particles to react. The consequence is energy production by processes in which mass is turned into energy. In the Sun, and billions of other stars, the initial step is the fusion of two protons (hydrogen nuclei) and the subsequent loss of a positron and a neutrino to form a deuterium nucleus, that is, a proton and a neutron (see Figure 12.5):

proton + proton = deuteron + positron + neutrino

This is the hardest step. The other steps follow fairly quickly, and in the complete process, which converts protons into α-particles (*helium nuclei*), there is a net loss of mass that turns up as energy, just as it does in atomic weapons. In $E = mc^2$, the mass is multiplied by an enormous number, the speed of light squared. One kilo of matter is equivalent to 9×10^{16} joules, enough to run a 100-watt lightbulb for nearly 3 million years. The nuclear reaction is characteristic of the long adult life of a star. The tremendous kinetic energy of the particles in the hot gas prevents the gas cloud from collapsing completely under the pull of gravity. They are moving too fast to stick together. That is why the glowing star is stable only as long as it can glow, which means until the hydrogen runs out.

The great majority of stars, including our Sun, are "burning" hydrogen to give helium. The astronomers call them *main sequence* stars.

The Death of the Average Star

The stability of a star depends on a continual fight between gravitation, which tends to pull it inward, and radiation and the kinetic energy of its atoms, which produce a net outward pressure. If a star tends to shrink because of gravity, then its core would be compressed and this would increase the crowding of the gas of nuclei. These would meet and react more often, produce kinetic and radiant energy faster, and push the star outward again. The lifetime of the star, and the manner of its death, depends on its size.

Stars start to die when their central store of hydrogen begins to run out. The small ones have less fuel, but they burn it more slowly. The most massive stars, as befits their image, are big spenders and live an intense but short life. Our middle-class Sun lies in the intermediate range, its expected lifetime of 10 billion years being about 1% of that of the smallest stars but about a thousand times longer than that of the most massive stars.

The small stars die the simplest deaths. When most of the hydrogen in the core has turned to helium, there is an upsurge of energy production around the core, a kind of protest against the coming oblivion, like Beethoven on his deathbed shaking his fist at heaven. This results in a huge swelling up of the outer layers of the star, which consequently cool somewhat, so that the frequencies of the light that they emit move down into the red part of the visible spectrum. So are born the *red giants*, and in about 5 to 10 billion years our Sun will swell up in this way and approach, or envelop, the orbit of the Earth. In the early part of this process, the radiation falling on the Earth will increase thousands-fold, and life will vanish. Later we, together with Mercury and Venus, will vaporize.

Red giants can be up to a hundred times larger than the original star, but because the huge gas cloud is so far from the center of the star it starts to drift away. All that is left is a tiny, intensely hot body, usually about the size of the Earth, which is nothing to brag about on the cosmic scale. It has become a *white dwarf*, white because it is very hot. But this heat is not due to nuclear reactions; the fuel tank is empty. The star is living on past glory, glowing because it has not yet cooled down enough to disappear, a process that is long but inexorable.

The hottest object so far found in the Milky Way is a white dwarf, imaginatively named RE1502+66. This is so hot that it is not easy to see. Looking back at Figure 15.8a, we can see that the range of frequencies emitted by a heated body moves upward as the temperature rises. The surface temperature of RE1502+66 is about 170,000 K, compared with the Sun's 6000 K. It is almost too hot to emit frequencies in the visible range and indeed it emits mostly UV radiation.

A typical white dwarf is as big as the Earth but has the mass of the Sun. The compressive force of gravity is staggering. A tablespoon of white dwarf weighs about 15 tonnes, a million times more than a spoonful of sugar. The nuclei in a white dwarf are about a hundred times closer together than they are in normal matter. The electrons run free, as do the conduction electrons in a metal. We can treat them theoretically as though they were free electrons in a box. The latter problem is well known to first-year physics students. Thus does quantum mechanics become a tool for handling the structure of stars. Incidentally, the most energetic electrons in a white dwarf are moving so fast that relativity theory is needed to account for their properties.

We have seen that when main sequence stars—those that "burn" hydrogen to give helium—die, they turn into red giants and then into white dwarfs. As in the entertainment industry, some stars are "bigger" than others, and, as you know, the really big stars have somewhat different biographies from the average star. The analogy is not bad, for as we now see, the really big stars live in the fast lane; they have more intense lives and die more spectacular deaths.

The Death of the Great Stars

For much of its life, a really massive star follows the same pattern as any other star, so that when its central store of hydrogen runs out it too forms a red giant. However, because of its enormous bulk, the pressure and temperature at the center of the star rise far above that of lighter stars, so far above that "helium burning" becomes possible. Hydrogen "burning" needs a temperature of 10 to 20 million K. This is not enough to give the doubly charged helium nuclei (α-particles) enough kinetic energy to overcome the electrostatic repulsion between them. However, at 100 to 200 million degrees they have enough kinetic energy to run at each other without playing chicken. Their fusion reaction gives carbon nuclei, although the reaction requires the collision of three helium nuclei, which is far less probable than a two-body collision. Carbon was probably the first comparatively heavy element to be formed after the Big Bang. At the temperatures obtaining in giant stars, another helium nucleus can add on to the carbon nucleus to give an oxygen nucleus. These fusion reactions release even more energy, again by the conversion of mass to energy. When the temperature rises to about 500 million degrees, it is hot enough for carbon nuclei, which have a charge of six, to interact with each other and form heavier nuclei, such as neon, sodium, and magnesium. At around 1 billion degrees, oxygen nuclei can combine to give silicon, phosphorus, and sulfur. At temperatures in the region of 3 billion degrees metals, including chromium, manganese, iron, cobalt, and nickel form. More and more types of nucleus are produced and in each case fusion produces energy, but the heaviest nucleus that can be formed in this way is that of iron, because the fusion of nuclei heavier than iron does not give energy—it needs energy. Heavier nuclei are formed by the addition of neutrons to existing nuclei, a process helped by the fact that a neutron is not repelled electrostatically. The trapped neutrons can then convert to protons, thus increasing the total positive charge on the nucleus (the atomic number).

At this stage massive stars have been likened to onions, successive layers being composed predominantly of one element. The core is dominated by iron, and going outward we cross layers of silicon, neon and carbon until, in the outer regions of the red giant, the atmosphere is helium and hydrogen. This theory of *stellar nucleosynthesis*—the production of the chemical elements in the stars—which was published in 1957, is mainly due to Willy Fowler, with considerable help from Fred Hoyle. It is the stars that have generated almost all of the periodic table.

The party does not last. At a certain stage the engine stops and one of the most spectacular events in the cosmos follows. Suddenly, within a matter of seconds, the core of the star implodes. The result is a flash of energy that tears through the whole star. A huge explosion smashes the star to pieces, and for a period of hours to days it can shine brighter than the whole of the galaxy in which it resides. A *supernova* has been born. Such events are rare, but I can vouch for the excitement in the cosmologist community when they do occur, as one did in February 1987. The first recorded supernova was observed by Chinese astronomers on 4 July 1054, in the Taurus constellation. Incidentally, the earliest verifiable report of an eclipse—that which occurred in 1361 B.C.—was also recorded by Chinese astronomers.

The Fate of a Supernova

The explosion of a supernova sends out a cloud of hydrogen, helium, and heavy elements. Gravity now repeats its trick, pulling parts of the cloud inward on themselves and creating new stars, in a repetition of the process in which the first stars congealed from the matter that was created in the Big Bang. Thus the heavy stars not only live and die, they can reproduce. New stars are continually replacing old, a process that is not expected to stop for tens of trillions of years, when all the hydrogen in the universe has been "burned."

In contrast to the original star, the new star's offspring start off with heavy elements. It was through the life cycles of the massive stars and their offspring that the elements that are heavier than hydrogen and helium were synthesized and spread through the universe. The cycle can be repeated by the star's progeny. If some of the new stars are massive enough, they too will produce heavy elements, adding to the stock that they received at birth. And so it goes on. But despite the impression given by the solid nature of our planet, the heavy elements are in a complete minority in the universe, which is dominated by hydrogen and helium. Helium exists as single atoms. The next most common *molecule* in the universe, after the hydrogen molecule, is the carbon monoxide molecule, CO.

From some of the new stars, planetary systems evolve. In one of the planets of one such star, stocked with the full range of stable chemical elements, a homicidal biped evolved.

The collapse of the core of a heavy star, and the loss of the outer cloud of gas, leaves a tiny object, usually no more than about 20 miles across and consisting entirely of neutrons. Neutron stars, or *pulsars,* the first of which was discovered in 1967, are fantastically concentrated. Their diameters are about 20 million times smaller than those of the red giants that fathered them. In Chapter 5 we noted that the escape velocity from Earth is 25,000 mph. This is the minimum initial velocity of a projectile that will allow it to escape from Earth's gravitational field. For a white dwarf and a neutron star, these velocities are in the region of 10 million and 200 million mph, respectively. Pulsars are well over a hundred million times more dense than white dwarfs. The enormous density is the result of stupendous gravitational pressure. It is this pressure that appears to have forced the electrons and protons of the star's core to combine, giving neutrons. A *tablespoon* of a neutron star would weigh about 3 billion tons, about ten times the weight of the Earth's population. Densities of this magnitude confirm the emptiness of ordinary matter that was first revealed by Rutherford. The density of a neutron star is not far from the density inside a normal nucleus.

Pulsars spin around an axis up to several thousand times a second. They have tremendous magnetic fields, and electrons caught up in them accelerate toward the magnetic poles of the star. Maxwell said that accelerating charges generate electromagnetic radiation, and pulsars generate two beams of radio waves emerging from opposite sides of the star. The rotation of the star with respect to the Earth results in regular pulses of radiation being picked up down here, hence the name pulsar. Pulsars also occasionally send out gigantic amounts of visible light, equivalent to many thousands of times the total light emitted by the Milky Way. Since pulsars have huge gravitational fields, they can draw in gas from a neighboring star. This gas acquires great kinetic energy, and when its temperature reaches about 10 million K, the atoms of the gas start to emit X-rays.

Black Holes

Therefore there exists, in the immensity of space, opaque bodies as considerable in magnitude, and perhaps equally as numerous as the stars.

—Laplace

When the core of a supernova has a mass exceeding that of about three times that of the Sun, it is believed that a black hole is formed ("Black" because when light is directed at a black hole it never comes back).

Both Laplace, and before him the Reverend John Michell,[4] surmised that the attractive force of a heavenly body could be so large that the light could not flow out of it, but they were vindicated only in the twentieth century. Laplace actually calculated the condition for the escape velocity of a body to be large enough to prevent light leaving it and on the basis of questionable assumptions actually arrived at the right answer. In 1915 Einstein showed that one of the more dramatic consequences of his theory of general relativity was the formation of black holes from large masses of matter.

Black holes reached the cover of *Time*. Their notoriety is understandable; an object that swallows not only matter but also light, deserves media attention. The basic principle behind black holes is that when there is enough matter concentrated in one place it allows *nothing* to escape from it; the escape velocity (see Chapter

---

[4]John Michell (1724–1793), who was a lecturer in Greek, Hebrew, arithmetic, and geometry at Cambridge, made a telescope in 1780. He attempted to detect the pressure of light, studied magnetic force, and wrote *Essay on the Cause and Phenomena of Earthquakes* (1760), which many regard as the beginning of scientific seismology. He also designed the apparatus with which Henry Cavendish estimated the density of the Earth.

5) becomes greater than the speed of light. Of course, if no light can get out of a black hole, you can't see it, so we have to look for indications of its presence, rather like detecting the wanderings of a poltergeist.

The most common method by which black holes have been detected is through the movement of visible stars. The enormous gravitational pull of black holes has been invoked as the cause for the unexpected paths of some stars. Thus a star in the Musca constellation has been seen to be rapidly circling *nothing* at a velocity of nearly 1 million miles per hour. Obviously, it is being strongly attracted to something. Calculation, using the old-fashioned laws of motion, shows that the invisible object has a mass of about three times that of our Sun. This is about as modest as a black hole can get.

The formation of black holes is supposed to come about by the collapse of a massive star. The matter in the core of a supernova is drawn together by gravitational force. As the central mass contracts, the gravitational field at the surface of the dying star increases. The smaller a given mass is, the greater the gravitational pull at its surface.[5] At a certain stage you can imagine the formation of an imaginary surface, above the surface of the black hole. This is termed the *event horizon*, or the *Schwarzschild radius*. Any object, or radiation, that penetrates the event horizon from outside is lost forever. It is as though the black hole has isolated itself from the space-time around it. Nothing can return from this ravenous land. You might care to see the event horizon as being the consequence of the escape velocity reaching the speed of light. When no light leaves the star, we can no longer see it—a black hole has formed. This process cannot reach its final stage if there is insufficient mass to start with. Thus the mass of the Earth is far too small for its gravitational field to induce contraction or significantly slow down light. It can be shown that the minimum mass needed to form a black hole is roughly three times that of the Sun. In contrast, there are black holes that have been calculated to have masses up to about a billion times that of the Sun.

When matter is caught in the pull of a black hole and descends toward its center, it becomes compressed, heating up so much that, at around 10 million K, it starts to emit high-energy radiation. This is probably the origin of the bursts of X-rays that have been detected from the direction of suspected black holes. It is claimed that mysterious bursts of γ-rays are also caused by the energy released when black holes swallow stars, but the staggering amounts of energy associated with these bursts are not at present explicable.

Half a dozen black holes have been identified in the Milky Way, to varying degrees of certainty. There may be a black hole, having a mass about a million times that of the Sun, at the center of the galaxy. If, as is the custom of black holes, it eats anything that comes its way, it will get heavier, bigger, and more dangerous.

Strange things happen in black holes. If we follow Einstein and attribute the bending of light to the bending of the geodesics of space-time by mass, we can construct a graphic picture of a black hole in which space-time is so curved that light never escapes. Thus, on the event horizon, a beam of light directed tangentially along the horizon would circle around and around. You would be able to see the

---

[5]From Newton's law of universal gravitation, the field at the surface of a sphere of matter depends both on its mass and on the distance from the center of the sphere. If the sphere contracts by a factor of 1000, the field goes up by a factor of 1000 squared, or 1 million.

back of your head. Light attempting to escape from the black hole would slow down and become static at the event horizon because its velocity is equal to the escape velocity at that surface. An object approaching the event horizon from *outside* would accelerate because of the gravitational attraction and would approach the speed of light as it neared the horizon. Once having crossed the horizon, it would become invisible to an outside observer.

What happens to matter that falls into a black hole? First of all, it is torn to pieces. This effect is a result of the enormous increase in gravitational force between two points at different distances from the center of the hole. If you dived in headfirst you would elongate as though on a rack, until your body disintegrated. As the parts moved inward they would disintegrate further, into cells, then molecules, then atoms, nuclei, quarks, and so on. What happens at the center? We don't know. It has been suggested that matter is crushed out of existence. You don't believe it? I'm not sure that I do, using my common sense. Many physicists doubt if the presently accepted laws of physics hold at the center of a black hole. To complicate things, Stephen Hawking has suggested that in certain circumstances radiation *can* escape from a black hole. If you feel confused you have an absolute right to do so. Cosmology, especially the area of cosmology that deals with, or thinks it is dealing with, the Creation is not only often bewildering to HMS; it a rich source of (almost polemical) controversy between the leading scientists in the field. A taste of these sometimes barely gentlemanly exchanges can be savored in *The Nature of Space and Time* by Stephen Hawking and Roger Penrose (1996), in which the authors disagree on almost every important issue from the Creation to Schrödinger's cat.

Black holes apparently eat matter, but where did matter come from in the first place? Where did the universe come from? Easy: the Big Bang.[6]

---

[6]A magazine competition, intended to find a new name for the Big Bang (considered by some to be sexually suggestive), resulted in over 10,000 entries, including "The Blast from the Past," "Orgasmus Universalis" and "Hubble Bubble." The judges settled for the Big Bang.

35 | The Impossibility of Creation

When there was no heaven,
no earth, no height, no depth, no name,
when Apsu was alone,
the sweet water, the first begetter.

—The Babylonian creation myth, twelfth century B.C.

As a ten-year-old child I lay in bed and sweated out my nightly terror of death. It was not hell that worried me, but oblivion. I didn't believe in life after death. My reason refused to be humiliated by a cowardly compromise with my fears. But in the Slough of Despond I found hope: the complete inexplicability of the Creation. I asked John Donne's question: How can something appear out of nothing? For if that was possible, which it appeared to be, then anything was possible—even life after death. The fear of death faded, but I remain, as most of us do, mystified by the fact of the Creation. The Bible didn't help me. In my search for a more convincing story I discarded the Egyptians, Babylonians, and Hindus, although in my wanderings I was attracted to the Gnostics, an early Christian sect whose origins actually predate Christ and who recognized a secondary god, the Demiurge, who was responsible for the creation of evil. The Demiurge answers some very awkward questions. But not how being emerged from nonbeing.

What preceded the Creation, or has that statement no meaning? Like "How high is green?" Was time born at the Creation? And space? Aristotle said that time was created with the cosmos, and so there was no "before." Modern theories of the Creation tend to agree with him, but no one has provided a readily comprehensible solution to Donne's question: What was nothing—and how does something emerge from it? To which Locke replied: "But you will say, Is it not impossible to admit of the *making of any thing out of nothing,* since we cannot possibly conceive it? I answer, No: Because it is not reasonable to deny the power of an Infinite Being, because we cannot comprehend its operations."[1]

And once we have a universe, can we understand its physical limits? Not in everyday three-dimensional imagery. General relativity allows us to create forms of space-time that resemble the surface of an orange, thus ensuring a space-time that has no boundaries but is finite. There is no edge to this surface, no end or beginning. With a little goodwill you can feel yourself walking on the surface. It is a picture that anyone can construct, but which in the end doesn't really squash our childish(?) question: What's outside the orange, Dad?

---

[1] Saint Augustine (354–430), referring to people who ask "What was God doing before he made heaven and earth?" explains that "I keep away from the facetious reply. . . . 'He was preparing hell for people who ask awkward questions.' . . ."

The easiest way out is to accept our limitations. The explanations of the cosmologists—and there is a choice—may be convincing mathematically, but we will never be comfortable with them because they are not constructed within the kind of (separate) space and time that we experience. Our, quite unjustified, gut feeling that space and time are absolute Newtonian playing fields does not encourage friendly feelings toward professors with incomprehensible equations assuring us that we live in a warped space-time that started from a fluctuation in a vacuum.

None of the present theories pretends to give a commonsense answer to the problem of nothing and its emergence into something. The concept of fluctuations in a vacuum (see Chapter 30) is, to most people, totally unacceptable as an explanation of the Creation. I am also aware of the stories that begin, "A wedding ring has no beginning. There are objects which exist which have no beginning so why not the universe?" Once again, the mind can swallow it, but not the instincts. It has been suggested that a model for the appearance of matter is to make a video of a black hole swallowing and destroying matter, and then run the cassette in reverse. This is the so-called white hole, an entity predicted by Einstein's general relativity. It makes a good subject for an animated cartoon, but do you believe it? Where did the white hole come from?

Sir Arthur Eddington wrote, "The beginning seems to represent insuperable difficulties unless we agree to look on it as frankly supernatural. We may have to let it go at that." Not very satisfactory, but I defy any cosmologist to explain the moment of Creation in the language of HMS.

Those scientists concerned with the origins of the physical universe begin their investigations just after the Creation, not just before. The story that they have to tell will unquestionably be modified as we learn more and think deeper, but in its essentials it seems at present to give a good rough explanation of the way that *this* universe developed. As we saw, there may be others. Let's see what our tribe's creation myth says.

The Big Bang

When we speak of the Big Bang theory, we are speaking of a family of theories, differing from each other on a variety of issues but all accepting the general picture of an initial cataclysmic creation of "something," about 15 billion years ago, followed by a furious expansion. One version, which is regarded as the "standard theory," is associated with the names of Friedmann, Robertson, Walker, and Lemaître, who were, respectively, Russian, American, English, and French. This is the theory that forms the basis of our discussion. It is the most straightforward of the various theories, and is what most people are talking about when they mention the Big Bang. The majority of cosmologists feel that the outlines of the standard theory are plausible and that there is some strong supporting evidence for its general validity, but that there are some very difficult questions remaining and there may be more waiting round the corner. It should not be thought of as a stable theory, but it seems to contain a fair amount of truth. What follows is the scientific equivalent of a first draft. It is being tinkered with as I write.

The postulated early history of the universe is conveniently divided into eras, where we have to stretch that word to include minute fractions of a second as well as periods of hundreds of thousands of years. The early eras are the shortest, and the subsequent eras get longer and longer. Everything is more closely packed at the beginning, and the rate of events slows down as time progresses.

It is not wise to ask what came before the Big Bang. In some circles it is even regarded as gauche, a sign of scientific naïveté. If you wish to avoid the image of a cosmic country bumpkin, you must profess to believe the current mythology, which claims that space, time, matter, and energy were all born with the Big Bang, and that the expansion of the universe was not an expansion *into* space but an expansion *of* space. I would not blame you if you preferred the homely images of Genesis; just don't say so in smart company.

We begin not at zero but just after that, at 10^{-43} seconds. This period of time is called the Planck time. (Divide one second by ten multiplied by itself forty-three times). The theoreticians are not too united about what to say after "In the beginning there was . . .". They understandably mumble about it being difficult to apply normal concepts of space and time, or matter, to this period. As for time zero itself, the standard theory asks us to believe that the "universe" had infinite density, which in everyday terms means that it had no volume. Some suggest that the concept of time breaks down near zero. If you are uncomfortable with this, don't be afraid to say so. Most sane human beings are.

From 10^{-43} to 10^{-35} Seconds

It is supposed that at the beginning of this period, called the grand unified theory (GUT) era, all four known physical forces were united into one force. You can suppose that there was only one kind of field (not a gravitational field plus an electromagnetic field, etc.) and that this field produced a single particle that mediated a single type of force. One of the main objects of theoretical physics is to construct such a field. The drive to do so is largely based on the belief in simplicity. But as A. N. Whitehead once said: "Seek simplicity, and distrust it."

During this period, gravity separates off as a distinct force, so that there are effectively two forces in the universe: gravity and a combination of the other forces. At the end of the period, however, the strong force separates out. This separation of diverse forces out of a single unified force has been the subject of much devising of analogies. Some refer to the process as the "crystallizing out" of forces as the temperature drops, as if, at high temperatures, everything is melted together. It may be easier to see it as the separation out of the particles that mediate the forces, just as at a later stage different material particles also separate out. On the other hand, taking into account the view that all particles are packets in fields, you may prefer to think of different fields separating out of the soup. Analogies are not really that helpful. It's a case of working on the subject for thirty years, by which time you are used to it. Like quantum mechanics.

As the temperature continued to fall, small particles—the quarks and electrons—began to appear. Where did these particles come from? From the conversion of the energy of photons to mass. These particles had huge kinetic energy, which militate against their forming stable clusters. Groups of three quarks, as there are in protons and neutrons, would be unstable under frantic bombardment by violently energetic photons.

The theory assumes that, whatever it was that made up the universe in the beginning, it had uniform density throughout. In other words, wherever you were and whichever way you looked, the view was exactly the same, as if you were embedded in strawberry Jell-o. This means that the universe expanded at the same velocity in all directions.

From 10^{-35} to 10^{-32} Seconds

The frenzied activity during this instant of time makes the explosion of a hydrogen bomb look like the pop of a Christmas cracker. Whatever existed at the beginning of this period had a temperature of something like 10^{28} degrees Celsius, which is completely unimaginable. Everything was compressed into a space estimated to be about as big as a pearl—yes, all the matter/energy that exists in the present universe. It must have been hell to get to the bar. Then, according to a theory put forward by Alan Guth of MIT in 1981, came inflation; the universe expanded by a factor of about 10^{30}. To get some idea of what this means, it is as though a hydrogen atom expanded into a sphere 10 million times the diameter of the solar system.

There has been a suggestion that during this expansion gravity was a *repulsive* force, blasting the nascent universe outward at colossal velocity. It is during this first period that most of the expansion of the universe took place. The calculations have to take into account the stupendous concentration of mass, which forces the theoretician to bring in both the general theory of relativity (to give a description of space-time) and the quantum theory (to give an explanation of the behavior of matter).

At 10^{-10} Seconds

The weak force separates from the electromagnetic force about this time, so that we have all four forces that we have today.

At One-Hundredth of a Second

By this time, the pearl had expanded to roughly the size of our Sun, but this figure depends on which particular variation of the theory you are reading. Some say that the universe was already the size of the solar system at 10^{-10} seconds. It is not worth taking these differences too seriously. No one knows much about the exact, or even approximate, correlation of size and time, but all agree that the universe expanded very fast indeed at the beginning and then slowed down, keeping going until now.

The temperature had fallen to a trifling 10,000 billion degrees, and was falling rapidly, perhaps to 30 billion degrees at one-tenth of a second. As fierce as this latter temperature is, the equivalent conditions can be replicated on Earth. More accurately, it is possible to accelerate particles to the kind of speeds they should have at these temperatures. The density of matter at this stage was about a billion times that of the Earth—in other words, a teaspoon of the universe would have weighed about 16,000 tonnes (if weighed on the nonexistent Earth).

By the end of the first second, the nascent universe was probably a hectic mixture of radiation, electrons, positrons, and neutrinos, with comparatively tiny numbers of neutrons and protons formed from the rapidly disappearing isolated quarks. These protons and neutrons are essentially those present in the universe today. Regarding the electrons and positrons: our experience is that when particles are created from the destruction of photons, they come in particle-antiparticle pairs, one positively and one negatively charged. In one way this is convenient. We need roughly equal numbers, otherwise the universe would have been strongly charged—remember, the number of protons (positively charged) was relatively small. This presents a difficulty: Where have all the positrons gone? They are pretty difficult to find these days.

At One Second

The temperature had fallen to about 10 billion degrees Celsius, and the density of the universe was about 100,000 times that of the Earth. One teaspoon of universe would have weighed about 1.5 tonnes. By now the average photon did not contain enough energy (remember $E = mc^2$) to produce any of the known particles, so at this stage there was no further production of matter.

At 100 Seconds

The temperature was now down to about 1 billion degrees, similar to that at the center of the hottest known stars. This is the stage at which the nuclei of the first chemical elements were formed, not from radiation but from existing particles. Two kinds of synthesis are of interest to us: the formation of deuterium nuclei (a proton plus a neutron) and helium nuclei (two protons plus two neutrons). A delicate balance is involved here. At too high a temperature a deuterium nucleus will, if formed, immediately disintegrate into its components. At too low a temperature the repulsion between protons will not allow the helium nucleus to form. A billion degrees accommodates both conditions. After the formation of the two nuclei, an excess of neutrons was probably left over. Now, stable as the neutron is when it is inside a nucleus, it has a lifetime of only about a quarter of an hour when it is on its own, so the excess neutrons broke down to give protons (and electrons). The proton, of course, is the nucleus of the hydrogen atom. Very small amounts of other light nuclei, such as that of lithium (three protons plus three neutrons) may also have been formed. It was still too hot for electrons to become attached to these two nuclei, they would have been knocked off immediately. The average speed of a proton at 1 billion degrees is about 18 million kph.

Thus, according to theory, the net result of the activity up to 100 seconds was a sea of radiation in which there were deuterium nuclei, helium nuclei, protons, electrons, and positrons. Something like 99% of the matter in the universe was made in the first hundred seconds. None of the nuclei of elements heavier than helium existed.

A Test of the Theory to This Point

The theoreticians have calculated that at this stage the ratio of hydrogen nuclei to helium nuclei should have been ten to one, and that it is unlikely to have changed significantly since then. Furthermore, the ratio of deuterium to hydrogen at that time is calculated to have been 1 in 50,000.

We can detect the elements in the stars by examining the spectrum of their light; furthermore, we can estimate their relative abundances from the intensity of their spectral lines. The ratio of hydrogen to helium atoms in the present-day universe is about ten to one. The amount of deuterium is about 1 in 50,000. To be fair, we have to admit that some of the light arriving here from the stars has taken millions of years to get here so that we are looking at what they were like long ago. Nevertheless, the theory says that the ratios of hydrogen to helium to deuterium have not changed significantly since the first 100 seconds, so the facts are still consistent with the theory—a very impressive achievement.

Time Goes By: From 100 Seconds Onward

As the temperature continued to fall, some of the electrons managed to hang on to the nuclei, and the atoms of the chemical elements hydrogen and helium were formed. Electrons and positrons began to annihilate each other to give photons. At the end of this process, perhaps after 300,000 years, there were no positrons left, just hydrogen and helium atoms and photons, in roughly the same ratio as today, namely, about 1 billion photons for each proton or neutron. This mutual annihilation of electrons and positrons, to leave the presently observed excess of electrons, obviously implies that there must have been more electrons than positrons to begin with, otherwise both species would have disappeared completely and there would be no electrons to clothe the nuclei of the elements. This would have resulted in a positively charged universe, whereas the universe appears to be effectively neutral.

Is there any explanation of why the number of positrons and electrons differed? In those events in which radiation produces particles, it does so in pairs—particle and antiparticle, two by two. Nevertheless, there are examples of asymmetry in the particle world. For example, in 1947 an unstable particle was found in cosmic radiation. The K° meson has a lifetime of 10^{-10} seconds, which is about twenty times shorter than that of its anti-particle, thus breaking the symmetry that characterizes matter and anti-matter. (This is known to the experts as CP violation.) This doesn't directly explain the proposed inequality in the numbers of electrons and positrons, but it does show that the particle world is capable of slipping out of the straight-jacket of perfect symmetry.

After 300,000 Years

The universe is now made of atoms and radiation. The temperature is a mere 3000 degrees Celsius. Recall the shifty professor's whisky-laden breath (in Chapter 1), which had to fight its way through an astronomical number of collisions before it reached his skeptical wife. Photons in the interior of the Sun have the same problem. They are so frequently scattered by collisions that they take an average of about 1 million years to reach the surface and escape. This was what is was like up to about 300,000 years after the Big Bang, at which point the disappearance of the great majority of electrons and effectively all the positrons cleared the fog of particles so that photons could move from one end of the universe to the other without continually bumping into things. We can say that the universe became transparent. The universe has also continued to expand, albeit far less dramatically, but a highly significant change had taken place in the nature of the cosmos.

Up to about 300,000 years, radiation had dominated the universe. The photons had so much energy that if you sat down with $E = mc^2$ you would find that they could easily turn into known particles. When they did appear, particles were in far too energetic an environment to form atoms, let alone clusters of material. But the temperature fell, and from roughly 300,000 years onward atoms could form without fear of being smashed to pieces, and they could also come together in stable clumps, pulled by gravity. The stars and the galaxies were being born. And radiation was not energetic enough to give matter. Radiation and matter were divorced. The universe was still about a thousand times smaller than it is now.

The transparency of the universe, coming at a time when the temperature was about 3000 degrees, means that a flood of photons was free to roam through space. Which brings us to a very remarkable finding.

Another Test of Theory

The range of frequencies that are found in the radiation emitted by a body depend on its temperature. If the temperature of the universe was really 3000 degrees when the great flood of photons was released, then the radiation should have been mainly in the visible range and should still exist. The existence of such radiation had been predicted by the Russian-born American scientist George Gamow (1904–1968), the first person to suggest that the Big Bang had been associated with extraordinarily high temperatures. Two of Gamow's students, Ralph Alpher and Robert Herman, did some calculations and estimated that the temperature of this radiation should now be 5 K. When you see reference to the "temperature" of radiation you can think of it in terms of a heated body, preferably a blackbody. The radiation emitted by such a body is related to its temperature, so that radiation "at 300 K" is radiation with the sort of frequencies that you would measure coming from a body maintained at 300 K.

Alpher and Herman published their work in the British magazine *Nature* in 1948. It did not create a stir, although it was one of the most significant predictions ever made in the field of cosmology. In 1964, Arno Penzias (another refugee from Hitler) and Robert Wilson, working at the Bell Telephone Labs at Holmdel, were disturbed by a background buzzing that interfered with their attempts to operate a newly constructed instrument designed to pick up weak radio waves from outer space. The background noise came from all directions. Attempts to eliminate the noise failed, including sweeping out pigeon droppings from the receiver and shooting the two pigeons responsible. Finally, Penzias phoned Bernie Burke, a radio astronomer in New York. Burke suggested that the noise was radiation left over from the Big Bang. Penzias phoned Robert Dicke, the well-known Princeton physicist. Dicke had no doubts. The strange fact is that Penzias and Wilson were not firmly convinced; they were just relieved to have some kind of explanation, so that they could carry on trying to pick up radio signals from galaxies. The radiation was in the microwave and IR range, and in fact it had been observed previously, in particular by E. A. Ohm in 1961, also at Bell Labs.

This was another example of "It's not what you see, it's how you see it." Ohm, and the others who had detected inexplicable noise, regarded it as a nuisance. Penzias and Wilson happened to be doing their experiments at a time when there was considerable theoretical interest in other laboratories in the possibility of finding such radiation. When they and Dicke's group published their results and conclusions, they did so in two consecutive papers in the same journal. Gamow was furious; he had not been mentioned, and neither had Alpher or Herman. Their work had been forgotten. It is a curiosity that Penzias and Wilson were awarded a Nobel Prize for a discovery that had been predicted long before, and for observing radiation that had been observed by Ohm in 1961 but not recognized for what it was.

The irritating noise was radiation coming from the universe about 300,000 years after the Big Bang. However, the temperature of the universe at that time was probably about 3000 K, so why do we see radiation at a temperature of only 2.7 K? Because the universe was much smaller at the time and the radiation that filled it then, cooled as the universe expands. One way of looking at this cooling is to realize that, according to the theory, space itself expanded as the universe expanded, so that the wavelength of all radiation was stretched. This means that the frequency, and therefore the energy, of the radiation was lowered. The radiation, known as the cosmic background radiation (CBR), was found to be *isotropic*, which means that whichever way the receiver is turned it picks up the same intensity of radiation. This is almost as it should be, but not quite.

The universe today is not isotropic. It is *not* like a featureless bowl of strawberry Jell-o, which if you were embedded in it, would look exactly the same in all directions. You only have to see the stars to realize that. Einstein's cosmological principle says that the universe looks pretty much the same in all directions, which should be interpreted in terms of a man at the center of a Christmas pudding, not a bowl of Jell-o. The raisins are roughly evenly distributed throughout the pudding. The CBR is radiation that comes to us from the time when the universe became transparent, about 300,000 years ago. But this is exactly when matter was aggregating into the huge clumps that we call the stars and the galaxies. Thus the CBR should not be isotropic; it should not come equally from all directions but should be uneven. The gravitational field of the young galaxies should have slowed down the photons leaving their neighborhood and thus effectively reduced their frequencies, which we can interpret as a lowering of their temperature. We should see cool spots in the map of the CBR. The search for this unevenness, or *anisotropy*, was not easy. A specially commissioned satellite, the Cosmic Background Explorer (COBE), was loaded with sensitive instrumentation. In 1990 it sent back a perfect blackbody spectrum of radiation at a temperature of 2.726 K. No anisotropy was detected, the radiation looked exactly the same in all directions. Something was wrong. The negative results were threatening the Big Bang theory. COBE went on taking measurement, about 70 million in its first year in space. They paid off. As the data accumulated, tiny variations in the temperature of the radiation began to emerge. By tiny I mean about 30 millionths of a degree below the average temperature of the CBR—small, but undeniably real. The announcement of the COBE measurements was a major media event, being given prominence on the front page of many newspapers. Steven Hawking declared, "It's the greatest discovery of the century—if not of all time."

George Smoot announced the results at a meeting of the American Physical Society on 24 April 1992. He said, "Well, if you are a religious person it's like seeing the face of God." His excitement can be understood. The universe was speaking across 15 billion years. The radiation is one of the most important discoveries ever made in cosmology—it is the oldest "object" that we can see.

Eternal Expansion

Will the universe stop expanding one day? And what then? The way that this question has been approached is through the realization that the universe has an inborn tendency to collapse, just as a star does, under the influence of its own gravitation. So we can ask, as we asked of a projectile in Chapter 5, are the galaxies traveling fast enough to escape?

We saw that there was a minimum velocity below which an object could not leave Earth. Above the escape velocity, an object travels on forever; below it, it returns. Now the Big Bang resulted in a cloud of matter and radiation, which we call the universe, spreading outward and continuing to do so until today. For any object, say a planet or star, there is a net gravitational force exerted on it by all the other objects, and tending to slow it down and pull it back to the center of the universe, wherever that may be. There are two extreme possibilities concerning the future of the universe. One is that the gravitational pull of matter will eventually bring expansion to a stop. In other words, the escape velocity is too great for the expansion to keep going. Just as the stationary arrow at the top of its vertical flight starts to fall back to Earth, so the universe will slow down, and when it is stationary it will start to contract. Finally, so this plot goes, the universe will return to being a tiny, immensely compressed ball. This is the Big Crunch, and a universe that expands and then collapses is termed a *closed* universe.

An alternative scenario is based on the assumption that the mass of the universe is not enough to stop eternal expansion and it will go on forever; it will "escape." This is an *open* universe. In which universe are we living? The answer depends on how much mass there is. When a rough estimate is made, on the basis of the *visible* objects in the cosmos, it is clear that the mass is far too small to prevent eternal expansion; the universe should expand forever. But what does theory say?

The equations of general relativity, which formed the basis for the standard model of the Big Bang, do not in themselves contain a prediction of the amount of matter in the universe, any more than Newton's laws of motion tell us what the mass of the Earth is. It is Guth's addition of inflation that alters things. A theory might give a vast range of answers to the question of how much mass the universe contains. The one obtained when inflation is taken into account is extraordinary.[2] The answer is that the universe is neither open nor closed; it is *flat*, which is the formal way of saying that the mass is just enough to prevent infinite expansion but not enough to result in eventual contraction. The mass needed for this delicate balance is called the *critical mass*. The prospect for us is that the present rate of expansion will get slower and slower, approaching zero but never quite reaching it. The calculation gives a density for the present universe of about 10^{-23} grams per cubic meter, or an average of three molecules of hydrogen in every cubic meter of space. If the theory can be trusted, then we have a problem: this *calculated*, (and critical) mass is about eight to nine times larger than the *observed* mass of the universe. Should we believe theory or experiment?

Experiment says that there is *not* enough observed mass to stop the universe expanding forever. Theory says that there is. Now we can tinker with the theory in order to convince it to give us the amount of matter that we have observed in the universe, but then we lose two other accurate predictions because the CBR, the hydrogen-helium ratio, and the critical mass come as one indissoluble package that emerges conveniently from the theory. Another, Baconian, point of view is to give observation precedence and admit that the calculation could be wrong and the universe is really going to expand endlessly. The theoreticians are stubborn, and on the whole they prefer to cling to the theory and look for the "missing" matter, which has come to be known as dark matter.

---

[2]The theory, developed mainly by Alan Guth, is also extraordinary in that inflation gives a flat universe *whatever* initial conditions we start with. This seems to preclude proving anything about the beginning of the universe. The fingerprints have been rubbed off the gun.

Dark Matter

In 1993 a huge mass of glowing gas was detected, far out in space. From the temperature it is clear that the gas molecules are moving at enormous velocities, so high, in fact, that the cloud has no chance of hanging together. Yet it does, and the investigators suggest that it is held by the attraction of an invisible mass of material amounting to 20 trillion Suns. Previously there had been observations of individual bodies whose paths are inexplicable unless it is assumed that they are near to other, invisible and very massive, bodies. As far back as the 1930s, the astronomer Fritz Zwicky had used the term *missing matter* in this context.

How much of this invisible stuff is floating around? And what is it? There have been many suggestions, from new kinds of matter to our old friends the neutrinos and black holes. There are huge numbers of neutrinos in the universe, but the problem is that it is becoming more and more probable that they have no mass.[3] So they can be ruled out for the moment, but the physicists have the privilege of suggesting all kinds of new and unknown particles without risking ridicule. In order to tackle the problem of missing matter, they have invented at least three hypothetical entities, none universally accepted. And new facts are continually emerging. One type of observation, supportive of the widespread existence of dark matter, is related to the famous bending of light near the Sun. The bending of light by large masses to produce distorted images of stars is known as *lensing* (Chapter 32). The indirect evidence for the occurrence of this phenomenon is growing. Cases in which the "something" doing the lensing is invisible strongly suggest the presence of dark matter—almost by definition! Another indication of dark matter comes from theoretical calculations on the stability of certain spiral galaxies, which turn out to have masses that are unable to gravitationally hold them together but could be stable if there were large amounts of dark matter wrapped around them.

In short, it could be that the universe is at present about 90% to 95% invisible. Until we know reliably how much mass there is in the universe, we won't know whether Guth's prediction is correct—that the universe contains the critical mass that ensures that it is "flat."

Awkward Questions

I have skirted around some very complex issues and failed to mention some very weird ideas about the nature of Creation and about the past, present, and future of the universe. As one, not particularly weird, example, Einstein once postulated that there was a *repulsive* force in the universe, opposing the inward pull of gravitation. He needed such a force to maintain his model of a static universe, a model he later abandoned, stating that the repulsive force was the worst mistake he had ever made. There are some physicists who, while rejecting the idea of a static universe, suspect that such a repulsive force may in fact exist. It has never been directly detected, but who knows? It sometimes seems that in cosmology anything goes.

I emphasize again that the Big Bang theory is in fact several theories, and that they can't all be correct, even if there are signs that we may be on the right track.

---

[3]It has been proposed by some investigators that neutrinos switch around between three forms and that they *do* have a very small mass. The jury is still out.

One question to which we may never have an answer is the following: If there is a Big Crunch, will the universe be so hot and so concentrated that the conditions of the Big Bang will be recreated? And did "our" Big Bang follow such a crunch? And how many times has the cycle bang-crunch-bang been repeated? This is the phoenix with a vengeance.

Was the Universe Made for You and Me?

Newton's demonstration that the gravitational constant, G, applied to the solar system as well as to gravity on Earth, was quoted as proof that there was one supreme designer. In recent times scientists have found what some see as convincing evidence for more than chance in the structure of the universe.

It has been pointed out that if the relative strengths of the forces in the universe had not been as they are, then life would never have developed. Furthermore, there is no a priori physical reason for the universe to be as it is. If gravity had been somewhat stronger, the stars would have been compressed more strongly, and, because collisions would have been more frequent, they would have burned out far quicker. The life of the universe might not have been long enough to allow life to develop. If gravity had been somewhat weaker, the stars would never have formed—no solar system, no planets, no Eve, no Adam. If the strong force had been a bit stronger, protons would be able to hang onto each other without the necessity of neutrons being around. The single proton, the nucleus of hydrogen, would have been unstable with respect to a double proton. Hydrogen would not have existed. If the strong force had been a bit weaker, nuclei would never have formed. The only chemical element in the universe would have been hydrogen—no life! In short, it looks like something of a miracle that the fundamental forces had the strengths that they did, strengths that appear to be almost essential for the existence of a universe conducive to the emergence of life.

Several prominent physicists, including Stephen Hawking, have pushed the idea, apparently first voiced in 1957 by Robert Dicke, that we are not dealing with luck here. The so-called anthropic principle has been developed into a variety of theses, each supported by all kinds of coincidences that are shown not only to be highly improbable, but also to be conditions for the existence of life. The physicist John Wheeler has gone several steps further than most and made the extraordinary suggestion that a universe in which life could not develop would never have come into existence. One thing is certain: no one would have known if it had been created. The idea is related to those of another physicist, Robert Dicke, whose chain of reasoning starts with the question: If there is no one to see it, what point is there being a universe? Thus the condition for there being a universe is that there be a living creature capable of appreciating that the universe exists. In that case, all the laws of nature and all the forces would have to be such that this cosmic superobserver—who is literally the creator of the universe—shall have evolved. HMS may be unaware that sober scientists have this sudden urge to walk on the wild side. Their imagination has been stimulated in part by some intriguing (Pythagorean) relationships between the numbers that characterize the universe. A simple example revolves around the number 10^{39}, which is roughly equal to the ratio of the electrostatic force between an electron and a proton to the gravitational force between them. But it is also roughly the ratio of the radius of the known universe to the radius of the electron and, lo and behold, when squared it gives 10^{78}, which is about the number of particles estimated to be in the universe. These ideas have been expanded, but I will not follow that trail here.

The anthropic principle and particularly the cosmic number game have provoked either the awe that comes from seeing the light or the ridicule engendered by Scientology. My own feeling, for what it is worth, is that the anthropic principle is fascinating, unprovable, and unlikely. The principle always seems to me to be a substitute for the religious belief in the special creation of man. It is a theory that allows the hard-bitten materialist with the soft center to lead God back on the stage, and to use the value of physical constants to hint that our advent had been predicted in handwritten (Hebrew, of course) holy scripts, slightly before the Big Bang.

As for cosmic numbers, they are an easy target for ridicule, but I would say that, in one sense at least, that ridicule is misplaced. The question of why the universe is like it is is far from trivial. Thus all electrons have the same mass. Why? And why that particular mass? And what determined the ratio of the gravitational to the electromagnetic force? Again the simple answer is: that's how it happened, but in this case that is not satisfactory. I am not suggesting that there was a guiding hand, but that we don't understand the mechanism by which the hosts of elementary particles turned out to be like the standardized products of a production line rather like the idiosyncratic pebbles on the beach. We have no answer to these questions, and finding a solution will tell us something fundamental about the structure of the cosmos. It must be admitted, even by all those with whom I share disbelief, that at present by far and away the most likely hypothesis is that Zeus rules.

Cosmology Comes of Age

The twentieth century has seen our knowledge of the present and past of the universe grow into a self-confident discipline with some very significant achievements to its credit. We have no reason to suppose that we have seen as far as we can, nor to conclude that we have detected more than a small portion of the dark matter which we believe is out there. Nevertheless, an impressive start has been made on rationalizing the cosmos. The Hubble telescope, circuiting 300 or so miles above Earth, has extended by a factor of about a thousand the volume of space open to our view. And we are exploring not only space but also time, for the farther away the objects we observe, the longer the light has taken to arrive here. When we gaze at distant galaxies, or at the birth of a far-off supernova, we are watching objects and events as they were billions of years ago. And all this is being been done from the Earth and some satellites mainly confined to the solar system—a ludicrously limited platform on which to put our instruments.

Cosmologists have ventured to construct theories explaining the history of the universe. One of the tests of a theory is whether it can assimilate new facts naturally. The Big Bang theory was capable of doing this with the CBR and with the ratios of hydrogen to helium, and hydrogen to deuterium. It should be remembered that we are dealing with events that happened some 15 billion years ago under conditions far removed from anything man can create or observe today. The miracle is that the theory is moderately successful, but the theoreticians would be the first to admit that it is not successful enough.

If the theoretician had one wish, it would probably be to start off with a set of equations, incorporating within them the axioms of quantum mechanics and general relativity, and to show that the birth and development of the universe was an inevitable consequence of these equations, which would provide him with the wave function of the universe. From this wave function he would derive the masses and other properties of all the particles. There are physicists who believe that this dream of an overall theory, or something very like it, will become reality. In his inaugural lecture, on taking the chair of Lucasian Professor of Mathematics at the University of Cambridge in 1982, Stephen Hawking expressed the opinion that "there are some grounds for cautious optimism that we may see a complete theory within the lifetime of some of those present here." Other eminent physicists hold similar views. But Hawking's optimism was based on a theoretical development that looked very promising at the time but has been slowly evaporating. A history of repeatedly dashed optimism, is, as Popper have would explained, no guarantee that the next chamber will not reveal the gold-encrusted sarcophagus of the pharaoh. It might be argued in general that since science can explain more and more phenomena it must, by definition, be approaching an explanation of everything. Don't hold your breath.

Decision Time for Planet Earth

No other human activity rivals science in its ability to erect cities of the imagination on one hand and mold our material existence on the other. This century began with quantum mechanics and draws to its close with the sequencing of the human genome. We have been handed a set of tools with which we can relieve man's estate or create misery. Whether we build a golden city or a gallows depends on us, and the minimum we can ask of ourselves is to understand not only the political environment in which we live but also the awesome potential that science holds. *Man's attitude toward science matters,* but that attitude is varied in the extreme and, unfortunately, often dangerously unbalanced. Let us take a walk through the sociological garden, starting with the naysayers.

36

The Tree of Death

When man ate from the tree of knowledge he elected to find a short-cut to the Godhead. He attempted to rob the Creator of the divine secret, which to him spelled power. What has been the result? Sin, disease, death.

—Henry Miller

Ironically—considering that it was Miller who wrote these sentences—sin, disease, and death are today more associated with sex than with Miller's target, science. But he is far from alone; antagonism to science is widespread, and at the end of our journey it is worth pausing to see why the search for knowledge is not universally applauded. I came up against the opposition early on.

My father's table talk installed in me a number of archetypal heroes of whom the elite, for me, were Louis Pasteur, Jesse Owens, and Ted "Kid" Lewis, the last being a ferocious Jewish boxer who lived in the same slums in the East End of London where I spent my childhood. My father gave me boxing lessons, and I developed into an aggressive middleweight, but Pasteur won out. The emotion-laden story of the boy bitten by a rabid dog, and made whole by Pasteur's primitive vaccine, turned me toward science. At twelve years of age I lay in bed and saw myself performing a series of incisive experiments with a few test tubes and a microscope, culminating in a dramatic evening at the Albert Hall. "This is it, ladies and gentlemen, the cure for cancer." I raised high the vial of clear violet liquid. Tier after tier of luxuriously bearded nineteenth-century scientists and luxuriously bosomed beauties rose to create a torrent of applause. My father encouraged my choice; like Bacon, he believed that science was intended for the relief of man's estate.

It was thus with complete bewilderment that, at the age of sixteen, I came across the words of William Blake: "Art is the Tree of Life, Science is the Tree of Death." No ambiguity, no mincing of words. As I grew up, I came to realize that my "rabid dog" view of science was far from universal. What was particularly riling, in adolescence, was that the opposition seemed to be drawn heavily from the ranks of my favorite poets. Only occasionally was there a note of grudging awe for science, as in W. H. Auden's well-known confession: "When I find myself in the company of scientists I feel like a shabby curate who has strayed by mistake into a drawing room full of dukes."

The nonscientist's attitudes towards science range from admiring puzzlement to virulent antagonism. To the scientist, some of these attitudes are understandable, some are infuriatingly irrational. All have their importance, because science affects our daily lives and is also heavily dependent on public funding. Public funding can be influenced by public opinion. What forms that opinion?

Is Science Safe? The Mad Scientist Syndrome

> When I was a child the mad scientist was a figure of derision and contempt, a fictional joke. How times have changed. The world is now full of these mad scientists—mad doctors tampering with embryos, mad physicists and chemists working out more refined methods of termination. . . . What can be done about these loonies?
>
> —Reader's letter, London *Evening Standard*, 31 January 1994

Diagonally across the road from one of the houses I lived in as a child, was the Odeon "picture palace," where for a penny I spent my Sunday mornings, at the kid's show. The program was preceded by a recital on the Wurlitzer organ, which mysteriously and garishly rose out of the ground and sunk back into the depths about ten minutes later. I was allowed a packet of Smith's Crisps. Horses starred in most of the films. Occasionally a scientist was featured. At the Odeon, scientists were not normal: they had spade beards, pince-nez spectacles, bushy hair, and crude German accents. Their laughter had two modes—sinister or maniacal. They were given to periodical fits of eye rolling during which they were likely to throw switches or pour bubbling liquid from one smoking flask to another. Sometimes I struck gold and was treated to the spine-chilling announcement: "With this I shall be Master of the World!!!" "This" was either the bubbling liquid or a metal bread-box to which were glued a few clock dials and from which leaped an occasional anemic spark. The convention was never broken—scientists were kooks, dangerous foreign kooks.

The scientist as a threat to civilization is more than a "B" movie stereotype. The satanic mills of the Industrial Revolution, the cloud above Hiroshima, the DDT in animal fat, carcinogenic food additives, thalidomide babies, and a series of other horror stories have, for many an HMS, rightly or wrongly, formed the image of science. The scientist himself may be seen not as literally mad but as too concerned with his research or his position to act responsibly with the box of matches in his hands. How, the public asks, can a man get up in the morning, eat his cornflakes, hug his kids, kiss his wife, and drive to his Defense Department laboratory where he is working on a virus that causes paralysis of the respiratory system? Is this not a kind of madness?

The question of the dangers of irresponsible science[1] recur persistently in the media. Unfortunately, emotionalism, often fueled by misleading and inaccurate sensationalism and preconceived attitudes, has usually prevented meaningful debate. Add to this the disagreements between experts, and the occasional dishonesty of big companies and local or government authorities, and it is often not apparent who to believe, or what attitude to adopt to certain areas of scientific research. Suspicion is certainly understandable when it comes to projects that threaten our health, or even our existence. But before you form an opinion, first get the facts. Second, don't simplify. And don't blame science if business-driven technology is really to blame.

---

[1] I use the term *irresponsible science* as a shorthand for the catastrophic scenarios of the previous paragraph. In that science cannot in itself be irresponsible, it is a misleading phrase, but it is convenient.

Take pollution. It is an easy word to write on a placard or spray on a wall (with ozone-friendly spray), but it is a complex subject that should not be mindlessly simplified. Some sources of pollution, for example, the centuries-old custom of burning wood to obtain heat, can hardly be laid at science's door. The smoke-laden skies of industrial England were not created by science. Pollution is not an invention of modern science, and the danger of burning fossil fuels would not even be apparent if the scientist had not researched the atmosphere and the physical properties of carbon dioxide. A significant part of the damage to the atmosphere is caused by volcanic gases and the gaseous emissions of cattle. Man is not the only culprit.

Of course, there are pollution problems that can certainly be laid at science's door, including the chemicals partially responsible for the ozone hole, the creation and indiscriminate use of DDT, and the effects of hormone-disrupting chemicals in the food chain. In no case was the scientific project initiated with the knowledge that there was an environmental threat. The worst aspect of the DDT story was the money-powered effort of big business to downplay the dangers of the insecticide. The same thing is happening with the tobacco companies and smoking. This is where HMS, acting within a democratic society, has a pivotal role to play.

Much of the blame directed at irresponsible science has been posted to the wrong address. Science is not synonymous with technology. The stereotypical scientist is interested in how nature works; the stereotypical technologist is interested in making more profitable soap powders. As a scientist, I am offended when science is falsely accused and the real criminal, irresponsible technology, roams the streets free. The tank was not invented by a scientific "loony," neither was the sword, the musket, the bow, the bayonet, or gunpowder. And neither science nor high-technology is responsible for alcohol, heroin, cholesterol, saturated animal fats, cholera, the common cold, poliomyelitis, leprosy, influenza, the population explosion, or AIDS. Where is a sense of balance? Plague, hunger, gunfire, religious or political fanaticism, and man's inhumanity to man have killed many more people than have died from atomic weapons or faulty vaccination. Which is not to say that atomic weapons are a blessing or that vaccination techniques cannot be improved, but it is hardly an advertisement for human reason when the antivaccination lobby highlights the handful of fatal vaccinations and glosses over the millions who have been protected from disease and death.

And yet Miller's indictment cannot be waved away.

The Tree of Knowledge

Miller is not alone in claiming that the scientist has ignored the lesson of Genesis in stealing from the tree of knowledge; that nature has its secrets and it was not intended that they should be revealed. Man has transgressed and brought down catastrophe upon himself. Such statements, you might say, are made by those who know little or nothing about science and are happy to have found a mystical excuse for remaining ignorant. Having said which, it is undeniable that man's curiosity has not always led him straight toward the earthly paradise.

The Judeo-Christian myth defines knowledge as something that was stolen, for which act Adam and Eve were expelled from Eden. (Not surprisingly for a male chauvinist society, it was the woman who was blamed.) In the crueler Greek variation, Prometheus paid the price of eternal torment for stealing the secret of fire from the gods.

The myth of Prometheus speaks to us because it condenses, in the hideous sufferings of the protagonist, the danger and the occasional sense of transgression that accompany our probing of the natural world. And it symbolizes, as myths often do, a very real problem, a problem that requires a rational answer, not one based on sloppy mysticism: *Is the scientist to be permitted to investigate everything in nature?* The question worries HMS, and he has reason to worry.

The spontaneous reply of the Baconian observer is "yes, all nature is there for us to study." Bacon himself was not so sure. In *New Atlantis* he puts this opinion into the mouth of his natural philosopher: "We have consultations, which of the inventions and experiences which we have discovered shall be published, and which not: and take all an oath of secrecy, for the concealing of those which we think fit to keep secret: though some of those we do reveal sometimes to the state, and some not." In contrast, Descartes held that the secrets of nature were in fact the mathematical laws that lay beneath the visible reality, and he had no doubt that, by using his powers of observation and his reason, man could and should reveal this knowledge. But in those days science had a negligible practical effect on society. Today, HMS may not be so liberal, and some scientists are deeply troubled by the question. Newton held that a scientist should not reveal that part of his work which could conceivably lead to dangerous practical developments. Robert Oppenheimer, after Hiroshima and Nagasaki, expressed his doubts in religious terms: "The physicists have known sin." That is a powerful statement of the fact that, like some political and religious doctrines, some scientific research can lead to suffering and death. It is this that should worry us and that demands rational debate.

Obviously the scientists must have a part in such debate; the scientist is both the Stealer of the Secrets, since that is his vocation, and the Keeper of the Secrets by virtue of his expertise. But he cannot thereby assume the role of an Elder of the Tribe, and it is certainly debatable whether he should be allowed an absolutely free hand. Inquiry is one of our most basic instincts, and there will be those who will protest against restrictions, in the name of academic freedom.[2] But science is no longer, as it was in the seventeenth century, a harmless activity of a few curious minds, in modest private or university laboratories. Very large sums of money are put into government-sponsored research and into the research laboratories of private companies. The type of work being done in some laboratories can have an enormous impact on our lives, just as the obscure experiments of a handful of European scientists in the 1930s unlocked the atomic era. And now the double helix is here.

There should be public input into the question of academic freedom in science. Would you like to live near a laboratory that was putting a defective human gene into a virus or bacteria? Would you like your son to work on biological or chemical weapons? Should experiments of this kind be forbidden? I refuse to hide behind the excuse that it is not I who will make the decision to randomly wipe out civilians. A scientist cannot, or should not, be unaware of the possible uses of his work, although this is a very difficult area. Rutherford's experiments led to the atomic bomb, but he couldn't have known it.

At present the tree is available to everyone. Whether the scientist should be free

---

[2]Defined as the right to speak before you think.

to pick what fruit he wants is a question that should concern all of us. HMS has a role in that discussion, and he is also needed as a watchdog.

Controlling Technology

HMS must be politically active, and scientifically aware, if he is to control technology. It was political action that began the elimination of the more obvious curses of the Industrial Revolution—which was a technological, not a scientific, revolution—and allowed us to guiltlessly enjoy its benefits. (To pretend that there were no benefits is just bloody-mindedness.) It was political action in the 1960s, sparked by Rachel Carson's *Silent Spring*, that triumphed over the U.S. Department of Agriculture and the big chemical companies that profited from the wild overuse of insecticides. *Informed* political action is essential, and yet too often we hear anti-science opinions that have not got past the mad scientist stage. This is not amusing. It is dangerous, because it directs anger at the wrong targets. It is essential that the public have a rational, informed, and unprejudiced voice in discussion of the uses and misuses of scientific research. These matters are, as they say, too important to be left to the scientists.

Scientific research is here to stay. Our modest aim here is to draw attention to the fact that HMS has a central role to play in the future of that research. In matters that affect our daily lives, HMS should certainly have his say, but the easy demonization of science or scientists is not a worthy starting point. The consequences of science and technology are too important to be judged from a seat in the Sunday morning kid's show.

Down with Science?

The genuine drawbacks associated with science and technology have engendered dreams of turning back the clock on scientific discovery so as to return to a better, more natural world, forsaking the crazed Dr. Hackenschmidt for the noble savage. This is sentimental naïveté.

Without the benefits of science we would return not to a Technicolor, William Morris land of pink-cheeked folk dancing farmers but to a world in which the majority of men (as in the Third World today) would live "short brutish lives," hounded by disease, killed off by infections that are now controllable, living in bug-infested habitations and clothes and unsanitary cities, and lucky to have a life span of four decades—if childbirth was survived. One of my children underwent open-heart surgery at the age of six; the other two suffered from influenza in their early years. My guess is that none of them would have survived had they been born a century ago. Primitivism has its pleasures, but, in terms of health, and social and educational deprivation, it is too costly for most of us. Nature is red in tooth and claw. Nevertheless, there are those who claim that too high a price has been paid for the improved standard of living of HMS. It is fashionable in hand-wringing circles to detect a spiritual void at the center of the consumer society, and to implicate science in the creation of that void. Which brings us to the science trashers.

The Science Trashers

Science is an overhyped facade, a self-supporting club that covers up its errors and is continually falling over itself. This has been the theme of a number of authors who have made a profession out of trashing science. Their books, because of their provocative content, sell well and get reviewed and quoted in the media. For this reason alone they need to be confronted. The trashers must not be allowed to mislead HMS.

Science is a huge, remarkably varied activity. At present many hundreds of thousands of research projects are going on, in dozens of countries. The projects range from the most abstruse aspects of quantum mechanics to the effect of food composition on human health. The quantity of observational data is staggering. A standard ploy of the trashers is to pick out conflicting data, or data that are suspect because of faulty procedures. Does any rational person believe that all the data collected by scientists in all the studies ever made would be faultless? In almost all cases the opposing camps eventually converge. There are always scientists ready to detect the mistakes of other laboratories! If there is any human activity that is self-critical it is science, but the trashers dig up conflicting studies, question the objectivity of the scientist, and declare the death of science. In fact, their whole approach is heavily dependent on the showcasing of the (expected) failures of science and on the maliciously blind underemphasis on its successes. Failed theories fuel their critical fires. They revel in tales of shaky concepts, multiple theories designed to explain the same phenomena, fraudulent results, laws that are not exact, antibiotics that induce resistance, backbiting scientific rivals, grant-grabbing professors, imperialistic heads of laboratories. You name it, and the trashers will dig it up for you. For them anything less than perfection is failure. *With this sort of approach it is possible to discredit the totality of human culture.*What about those weak (and stolen!) plots in Shakespeare? Or Tchaikovsky's sugary sentimentalism? And, good God, didn't the Egyptians know *anything* about perspective?

Every human activity, *because it is human,* is imperfect. There have been scientific frauds—sometimes euphemistically termed hoaxes—and there have been incompetent scientists. This no more discredits science than a plagiarizing musician discredits music.

The trashers harp on the inertia of the scientific establishment, an inertia that I have documented in this book. There is inertia, but for every case that I have quoted, the truth finally came out. And unlike politics or religion, there are no immovable facts in science, there is no principle that cannot be questioned. Even Newton's laws were modified after 250 years. The trashers see this as a sign of weakness. A normal man would see it as a sign that the cosmos is immensely difficult to understand, but that we are understanding more and more, as evidenced by our exponentially increasing ability to predict the behavior of the material world. That ability does not always grow in the way suggested by the schoolbook's idealized account of science: the miraculous succession of observation, theory, and experiment. In this, at least, the trashers are right: scientific progress can sometimes be messy, shortsighted, and the result of accident rather than planning. But look at the incredible structure that has evolved over the centuries, and compare that to the petty-minded and basically dishonest niggling of the naysayers.

The trashers sit opposite their word processors (the fruit of Basic Scientific research, for short, BSc), turn on the light (BSc), wear artificial fabrics (BSc), take their blood pressure medication (BSc), phone their friends (BSc), look at the TV news (BSc) through their plastic glasses (BSc), listen to their radios (BSc), get their injections against influenza (BSc), and so on and on. And what are they typing? About the failure of science! If they had any self-respect, they would live in a virgin forest, writing their manuscripts on parchment with a quill, forgoing any modern medicine or comforts, bereft of electronic communications. To take one of the greatest and most successful human adventures and pick trivial holes in it, like a thin-lipped aunt running her finger over a dusty shelf, requires a special kind of mean-mindedness. Science, by its very nature, *demands* criticism, but not by someone who would condemn the Taj Mahal for demolition because some of the stones are cracked.

The Ice-Cold Clock?

In my final year at high school an intense and talented girl taking "Arts," as we called them in those days, told me that I was "too bright to take science." I quoted the mathematician G. H. Hardy, "Archimedes will be remembered when Aeschylus is forgotten." (Frankly, I would have given my vote to Aeschylus, but I was on the defensive.) The argument left one phrase permanently in my memory: "Your universe," she said, putting a pale hand on my forearm, "your universe is an ice-cold clock."

Two years later I was studying chemistry at University College London. One evening, after watching the young Tom Courtney in the College Drama Club's performance of *Ring Round the Moon,* we ended up in the "Orange Tree." Towards closing time Jill, who was reading English, leaned across the sticky table and, in Chelsea tones slurred by gin, asked, "And what's the Moon to you, Silver? A lump of sodium chloride or something?" She raised her near-empty glass, "Buy me another one. We'll drink to your mechanical sodding universe." I bought her another two.

"Whatever the Sun may be," said D. H. Lawrence, "it is certainly not a ball of flaming gas." Helios, the sun god, has more sex appeal than a cloud of gas, however hot. Lawrence spoke for my friend Jill, and for many others who see science systematically chipping away at the mysterious, but generally benign, *unknown* and arrogantly replacing it with the dull, prosaic, down-to-earth *known.* The mechanical universe, the "ice-cold clock," is not something you want to curl up with on a winter's night. Genesis 9:13 reads: "I do set my bow in the cloud, and it shall be for a token of a covenant between me and the earth." The scientist's rainbow is the result of the different refractive indices of the various frequencies of light that make up solar radiation. But man evidently prefers mystery to math, and the intrusion of science into the movements of the planets and the stars, into the living cell and into that final sanctuary of the spirit, the mind, has undoubtedly cast a chill over that warm, blurred garden, the theocentric universe. The scientist, ruthlessly buying up desirable property, appears to many people to be building an automated factory in the middle of the garden. For this reason, science has, for some, become an unwanted neighbor. And indeed, most scientists would make very hard work of explaining how the concept of the soul fits into the material universe, where there is nothing but "atoms and the void." Was this what Blake meant when he said that science was the tree of death? The death of religion? Of imagination? Both have been frequently suggested.

Science, since the fifteenth century, has been seen as a threat to the dogmatic Judeo-Christian account of the Creation and the universe, and even to religion itself:

> That vast moth-eaten musical brocade,
> Created to pretend we never die.

Do these lines from Philip Larkin's "Aubade" encapsulate the real fear? That the mechanical universe dispenses with the immortal soul? Many distinguished scientists think not.[3] And even if science has genuinely discredited revelation, does that discredit science?

As to imagination, it is true that the first recorded stoned writer, De Quincey, declared that science was the greatest threat to the facility to dream, but he was wrong. Consider the following improbable scenario: An astronomer observes events that took place 5 billion years ago (it takes that long for light from the star he is observing to reach the Earth, light generated by explosions on an unbelievable scale, at incomprehensible distances). The astronomer lives on a tiny planet, in a universe that is mostly gas or empty space, Eliot's "vacant interstellar spaces, the vacant into the vacant." And yet on this planet, out of immense fiery continents of molten rock, over hundreds of millions of years, a solid crust solidified, the wild clouds rained down to form tempest-swept primeval seas, and in those seas an incredible process began. Matter, the cloth of the universe, tentatively, tenuously developed, by a slow improbable transformation, into a form so complex and so organized that it can comprehend its own history and, however pitifully, plan its own future. This is the strangest dream of all. It is a story that has the power of primitive myth, of Gilgamesh, of Adam. *It is a dream created only by science.* There is enough wonder in the physical world.

It was a poet, Wordsworth, not a scientist, who wrote: "the beauty in form of a plant or an animal is not made less but more apparent as a whole by more accurate insight into its constituent properties and powers." And Voltaire wrote to Newton: "I do not see why the study of physics should crush the flowers of poetry. Is truth such a poor thing that it is unable to tolerate beauty?" Science is certainly our prime weapon against superstition and irrationalism, but in a world in which science flourishes—with or without a God—love and fear still remain, as do pleasure and regret, poetry and humor, art and music. The arts are not lessened by the sciences. Blake was mistaken: man's ineradicable gift, his questing curiosity, the divine discontent, is the common source of the arts *and* the sciences.

The Obscurity of Science?

The public aspects of science are important enough for HMS to make the effort to understand the issues involved and not to fall for the shallow antagonism of the naysayers. The effort is vital. The expert may know more about his subject than the layman, but his political and social judgment are no better than that of his lay friends. The nauseating prostitution of Nazi and Soviet scientists to the immoral aims of their states reinforces the theory that DQ, the decency quotient, is not positively correlated with IQ. Science policy needs the input of HMS. This does not mean that business executives and factory foremen should brush up on their quantum mechanics, but you don't need a degree in science to understand the need for,

---

[3]Immortality has been subjected to political attack: "the idea of the soul's immortality is a superstition encouraged by rulers to keep their subjects docile." So wrote the sixteenth-century Paduan radical Pietro Pomponazzi.

and to participate in, say, open discussion on the way that new drugs are marketed and tested, whether genetic analysis of employees should be allowed, or whether huge sums should be devoted to areas of science that are almost certain to be of no practical use. HMS might protest that science is too obscure for him to attain a meaningful understanding of a specific subject, within the constraints of his working day. It is true that very few laymen can sight-read and understand the Watson-Crick paper on the double helix structure of DNA. So should HMS give up? No!

Bertrand Russell opined that if a scientist could not explain what he was doing in layman's terms then he didn't really understand what he was doing himself. This is not quite accurate, but all the same, most of the important topics in science can be made intelligible to the layman, and this is particularly true of those areas of science relevant to the welfare of this planet. The obscurity of science is an obstacle but definitely not an insurmountable obstacle, otherwise there wouldn't be popular science magazines. HMS must not use this obscurity as an excuse to mindlessly slam science and avoid rational debate on scientific problems.

Know Your Enemy

If you choose to criticize science or scientists, because you feel that in a specific instance they are damaging the community, materially or spiritually, do it from a position of strength. Know what you are talking about. Don't let the "science-is-an-unmitigated-curse" extremists or the "tree of knowledge" mystics get at you, any more than you would blindly accept every statement made by the drug company scientists or their lawyers. If someone tells you that "those mad scientists" are responsible for modern man's exposure to dangerous levels of radiation, go to the library. You'll find that since the dawn of life about 55% of the radiation to which living things are exposed comes from the *naturally occurring* gas radon, about 8% from cosmic rays, about 11% from *naturally occurring* radioactive isotopes in our food, and so on. The major man-made source of radiation encountered by HMS are the X-rays used for medical diagnosis. And, by the way, I have met only one scientist who was certifiably mad, and he was harmless.

Those who waffle about forbidden fruit may be excused for using the language of myth, but not for facing some of mankind's most urgent problems by opening the doors of emotion and closing the gates of reason. Iconoclasts are essential, but only if they aim at being as informed, as rational and as passionate as a Rachel Carson.

If you still have a gut feeling that we are being punished for having stolen fire from the Gods and apples from the tree, you might care to reflect that, apparently just for kicks, benevolent nature has bestowed on us earthquake, hurricane, drought, disease, famine, overpopulation, and the threat of annihilation by collision with asteroids. To conclude, as Henry Miller did, that science is responsible for the major ills of mankind is lazy thinking. Far from damning us, the fruits of the tree of knowledge could be our salvation. Prometheus and Adam should demand a retrial.

37

"What the Devil Does It All Mean?"

The popular image of science is formed primarily by science-based technology. As evidenced by what we buy, what we use, and what we covet, we are a gadget-based civilization. To suggest that the tidal wave of consumer goods is responsible for what has been seen by some as the emptiness at the heart of modern life is facile in the extreme. No other era has had so readily available to it great music, great art, and great literature. Cheap travel, the mass electronic media, and mass publishing have spread the products of the creative arts deeper and wider than at any time in history. However, it is true that in general those who have benefited from technology couldn't care less about its scientific basis. Rare is the dancing teenager or madrigal buff who can tell you that the canned music he or she is listening to comes by courtesy of the basic science that eventually led to the transistor, the semiconductor laser, and magnetic films. Most of the main *concepts* of modern science are as familiar to the man in the street as the language of the Aztecs. Sometimes, as in the case of theoretical particle physics, it doesn't really matter. Sometimes, as in the case of genetic engineering, it really does, and unfortunately scientific ignorance is typical of a society whose attitudes toward science can hardly be said to be balanced or informed.

The naysayers have been given their say. Now we turn the microphone over to HMS, to the humanists, to the philosophers, and to the scientists themselves.

HMS Looks at Science

Georges Braque and Pablo Picasso called each other Wilbur and Orville, after the Wright brothers. They were not alone in their aeromania. On 28 September 1909, a Prague newspaper published an article by Franz Kafka called "The Aeroplanes at Brescia," an account of an airplane race. Kafka was clearly hooked. In the same year the lure of the machine was spelled out explicitly by the Italian poet Emilio Marinetti. The *Futurist Manifesto* totally repudiated the past and declared that art should create an authentically modern sensibility founded on the beauty and vitality of the machine. But HMS did not need a manifesto to arouse his lust for gadgets, and it is the plethora of technological products that issue from Sony, General Motors, and other industrial labs that dominates HMS's image of science.

In the early 1920s the automated production line turned America into an object of mass envy and admiration. Technology, not art or basic science, was confirmed as civilization's status symbol. Few heeded the words of Dean Inge, in the 1920 Romanes Lecture: "The European talks of progress because by the aid of a few scientific discoveries he has established a society which has mistaken comfort for civilization."

The avalanche gathered force. The radio and later the television set, initially objects of curiosity and envy, became prosaic and indispensable items in households that often lacked a single book except for the two holy texts of the era: the Bible and the Sears catalogue. The washing machine, the spin drier, air-conditioning, the long-playing record, the magnetic tape cassette, the microwave oven, the compact disc, the video recorder, the fax, electronic mail, and the cellular phone have added to the domestic, and hence immediately obvious, benefits of technology. Add medical instrumentation, antibiotics, synthetic hormones and vitamins, synthetic fabrics and colors, the automobile, the airplane, and, for a very select few, the space shuttle. We live—and breathe—in an environment that everywhere exhibits the work of *Homo technologicus*.

As judged by its *practical* achievements, science has a strong claim to be a worldwide success story. The amount of government and private-sector funding for scientific research far outstrips anything spent in previous centuries. There are science correspondents on all the "quality" newspapers, and science slots are standard in the program scheduling of most major TV networks and radio stations. Scientists are regularly interviewed, on the news or on talk shows, primarily on medical or ecological matters. Food products contain analyses of sodium, cholesterol, polyunsaturates, fiber, and calories. There is a flood of books (of which this is another), and a variety of magazines, devoted to the popularization of science and technology. Science museums have interactive, pedagogic exhibits. In England and Israel it is possible to take a degree in science by following the courses of the Open University on TV.

Nevertheless, HMS is aware that technology has its drawbacks, and at the end of the century there is, as compared to its beginning, a vastly increased public awareness of the dangers of applied science.[1] The responsibility of the scientist toward society is a topic that occupies scientists and nonscientists alike.

As early as 1925 Bertrand Russell wrote, "At present (science) is teaching our children to kill each other, because many men of science are willing to sacrifice the future of mankind to their own momentary prosperity." He was obviously unaware of the scale of scientific salaries, but all the same it cannot be denied that there have been many scientists who have conveniently closed their eyes and ears to the concerned voices around them:

> No tyrant ever fears
> his geologists or his engineers.
>
> W. H. Auden, *Marginalia*

As a scientist I regret that the welter of science-based technology is too often the window through which HMS sees science. In the twentieth century the wonderfully unifying, if often strange, *ideas* of science have not become common currency.

HMS and the Ideas of Science

Consider the following quartet of notables, each representing a different aspect of

---

[1]When I graduated from University College London, I, together with the others who had been awarded first-class honors, received a letter from the Biological and Chemical Warfare Establishment. It was superficially tempting in that it offered a relatively assured future, a reasonable salary, and a promise that I could, if allowed to, publish my research results—in the *Journal of Pathology*.

the creative life of the seventeenth century: John Donne, John Dowland, Sir Peter Lely, and Sir Isaac Newton. Admittedly, James I said of Donne that his poetry resembled the peace of God, for "it passeth all human understanding,"[2] but most of Donne's writings are immediately understandable. Dowland, the great lutenist, was ahead of his time with his occasional deliberate discords, but his compositions were immensely popular, his *Lachrimae* being a best-seller in Europe. Newton's mechanics was not overly obscure, and an educated man could recognize its contribution to our understanding of the cosmos. Finally, anyone can enjoy the Playmate of the Month portraits of Sir Peter Lely. All in all, the scientific and cultural activities of his day were accessible to seventeenth-century HMS.

Now take a prominent mid-nineteenth-century quartet: Dickens, Mendelssohn, Courbet, and Darwin. Courbet initially came under criticism for his depiction of his subjects, but he presented no visual puzzles to the beholder. The others were all a part of middle-class culture. I could have chosen Maxwell as a difficult-to-understand scientist, but an educated layman could grasp the meaning of his laws. The educated mid-Victorian may not have agreed with, or comprehended, everything happening on the scientific and cultural front, but he broadly understood what was being written, painted, and played.

Move to the early years of the twentieth century and weigh this quartet, chosen from the major figures of the time: James Joyce (after 1922), Alban Berg, Picasso (after 1907), and Einstein. The connection with HMS has collapsed completely. Language failed to communicate, music was (to the ears of the time) cacophonous, painting (to the eyes of the time) was incomprehensible, a description that also fitted the theory of relativity. Alternative quartets do little to help: Ezra Pound, Schoenberg, Braque, and Schrödinger; or T. S. Eliot, Stravinsky, Kandinsky, and Heisenberg. I could have chosen a more accessible artist, say Matisse, but most works of his that we enjoy today were regarded as difficult in those days. Of course there was an ongoing popular culture of traditional literary forms, crooners, and paintings of puppies, but the intellectual mountain climbers were completely out of sight of the base camp. Contemporary comment, from layman and critic alike, leaves no doubt whatsoever that the creative world, both scientific and humanist, appeared to HMS to have forsaken its senses. Obscurity seemed to be obligatory.

As the century progressed, the twentieth-century eye acclimatized to much of modern art, and the twentieth-century ear accepted Stravinsky and Berg—even if one doesn't often hear people whistling Schoenberg in the bath. The era of willfully obscure poetry slid quietly into history and although dadaism in its purest form became passé, Beckett and Ionesco played to full houses. Only science, especially physics, continued to drift away from HMS. The tribe does not understand the language of the priests.

The Obscurity of Physics

> "... We get a trifle weary,
> at Mr. Einstein's theory."
>
> —Herman Hupfeld, "As Time Goes By"

---

[2]The king obviously liked this comment, for he also used it in respect of Francis Bacon's *Novum Organum*.

The outstanding scientific concepts born in the twentieth century, apart from the genetic code, have been in the field of physics. For HMS, most of these developments remain profoundly obscure. Despite the shelves of "popular" books on quantum mechanics, the Big Bang, chaos, black holes, and so on, the average reader finds himself in a state akin to that of a medieval peasant attending mass in a great Gothic cathedral: he is awed, he knows that something terribly important is being said, but he understands very little because it's all in Latin. He takes the wafer and remains mystified.

In one way the failure to communicate doesn't really matter. No one will be asked to vote on whether or not he wants Big Bangs and chaos in his neighborhood and, as the charismatic Richard Feynman said, some science is just not understandable. If Feynman didn't understand, then we are all excused.

After ninety years, relativity and quantum mechanics still remain enigmas to HMS, and, along with chaos and cosmology, have done almost nothing to form his thought processes. The conceptual bases of modern physics remain the preserve of a limited section of the scientific community. Newton's achievements and methods were well enough understood to be a real source of fuel for the fires of the Enlightenment. Darwin and Faraday were within comparatively easy grasp of the layman. That can hardly be said of Einstein and Heisenberg, Schrödinger and Poincaré.

In October 1847, Maria Mitchell built an observatory on desolate Nantucket Island. She gazed out and discovered a previously unknown comet. By 1900 it was clear that few averagely educated ladies, or gentlemen, could set up their own scientific laboratory and hope to compete with the professionals. Science had become a profession. HMS was puzzled. So was the world of the arts.

The Humanities Look at the Sciences

> Language and science are abbreviations of reality; art is an intensification of reality.
>
> —Ernst Cassirer

In the nineteenth century the cracks between science and the humanities were obvious, and by the mid-twentieth century the once happy couple were barely speaking, although there are some prominent exceptions. Berthold Brecht envied the duty of the scientist to look at any theory, however successful, and challenge its validity: "A technique for getting irritated by familiar, 'self-understood,' accepted facts was built up by science with great care and there is no reason why art should not adopt that immensely useful attitude." In *The Sacred Wood*, T. S. Eliot revealed a knowledge of chemistry. In a striking image he compares the poet to a chemical catalyst: "Just as the platinum catalyst used to induce the combination of oxygen and sulphur dioxide to give sulfuric acid,[3] is left unchanged at the end of the reaction, so the poet's mind would, after creating the poem, emerge inert, neutral, and unchanged." But despite similar examples, it has to be said that the interest, and understanding, of most creative artists in the twentieth century has been superficial or nonexistent.

---

[3] An irritatingly pedantic note: The product of the reaction is sulfur trioxide. This has to be dissolved in water to give sulfuric acid.

Undigested scientific concepts have been used to embroider texts or to create interest as titles of paintings, sculptures, or television series. The double helix makes occasional guest appearances in poetry. Time is frequently warped in space epics. One of the few writers who has made more than trivial use of modern science is the playwright Tom Stoppard, but he enjoys, and is highly successful at, juggling abstract ideas (He has confessed that he is expecting a professional scientist to get up in the middle of a performance of one of his plays and shout "Bollocks!") It is the suggestive terminology of science, the occult connotations of the terms *uncertainty*, *warped space*, *cloning*, and *black holes*, that have generally been the means through which science has made an appearance. In spite of Apollinaire's conclusion that relativity was the spiritual precursor of cubism, it is certainly not, except in rare cases, the *understood* concepts of science that have fertilized art or literature. This is not to say that the humanities have suffered. The problem is that when a complete and influential section of the community is ignorant of the intellectual content of science, its attitude toward important scientific and technological issues becomes unbalanced.

A Painful Divorce?

World War I did nothing to beautify the image of science in the eye of the humanist. The deep disillusionment with the direction in which Europe had been driven by its accepted values led to a wave of romanticism, antithetical, among other things, to rationalism in general and deterministic science in particular. The flagship of the reaction to science was Oswald Spengler's widely read *Decline of the West* (1918), in which nonrational, mystical ideas were touted as the antidote to the scientific subversion of civilization and nature. Spengler's demotion of reason and his singling out of deterministic causality as the serpent in the garden came at a time when advances in the quantum theory appeared to be burrowing under the deterministic foundations of classical science from within. But few humanists were aware of the full meaning of the collapse of Newton's mechanical universe; the more popularly sensational aspects of relativity took the limelight: the bending of light, the effect of space travel on aging. In the meantime, Spengler and pseudoscientific mystics such as J. W. Dunne spoke to the antirationalistic leanings of many intellectuals in the humanities. The gap widened.

In 1959, C. P. Snow, the physicist and novelist, gave a now famous lecture entitled "The Two Cultures and the Scientific Revolution," describing, as he saw it, the profound schism between the humanities and the sciences. Snow suggested, among other things, that a fully educated man should have a knowledge of the sciences, and he made it clear that most humanists didn't meet this criteria. He underlined the antagonism of the British literary establishment to science and technology, and its ignorance of the achievements and conceptual richness of science. There was a furious response from Dr. F. R. Leavis, who ruled the English literature roost at Cambridge for a good part of the mid-twentieth century.[4] Leavis let loose: "The intellectual nullity is what constitutes any difficulty there may be in dealing with Snow's panoptic pseudo-cogencies, his parade of a thesis: a mind to be argued with—that is not there." Evidently Snow had touched a sensitive spot, but Leavis

---

[4]The axiom of the French farce-writer Feydeau could be applied to Snow and Leavis: "Whenever two characters must under no circumstances meet, I immediately bring them together."

had a point. The rather cardboard characters in Snow's novels and his stodgy reference to the period 1914–1950 as a "misguided period" in the history of literature hardly made him the ideal emissary between the two camps. The two-cultures row spread rapidly into the intellectual magazines and the media. A decade later, Leavis was still on the defensive. He let it be known that, in his opinion, the English department was the heart of his university, the center about which all else revolved. The statement is symptomatic of the gap.

It is pitiful to claim that the English department at Cambridge was the heart of the university during an era when Watson and Crick were determining the structure of DNA and, in the same university, Frederick Sanger was performing the first aminoacid sequencing of a protein (insulin), for which he received the Nobel Prize. And if that isn't enough, the astronomer Martin Ryle was reaching into deep space, detecting scores of radio stars, work for which he was knighted and also awarded the Nobel Prize. These people were changing the future of humanity, and altering the way we see the cosmos. During that period there was no one doing any significant *original* writing in the English department. There was, of course, criticism, criticism of criticism, and so on—infinite reflections in parallel mirrors.

Was the professor of Latin equally convinced of the centrality of his subject when Isaac Newton was occupying himself with trivia such as the law of universal gravitation? And by the way, who *was* the professor of Rhetoric at Padua when Galileo first turned a telescope on the night sky? One often despairs of academics. Consider another professor, Michael Oakshotte, a distinguished writer on philosophy who, when referring to science, declared that there was no room for "vocational training in a university." Science, said Oakeshotte, dealt with "objects and observations," not "ideas and thoughts"! One can only throw up one's hands; perhaps Charles Snow was right about the two cultures, but surely the time has come to expect a little more awareness on *both* sides.[5]

It may be that one of the roots of the fairly prevalent antagonism of academic humanists to science is an unconscious resentment of the emergence of a high-profile, worldwide community that sometimes implicitly or explicitly claims to be involved in a pursuit that is more important than the humanities, not only in respect to its applications but also in a wider sense. It is all right (but not in my opinion accurate) if a distinguished historian like David Thomson writes that in the twentieth century, "the scientist . . . came to dominate the whole sphere of creative endeavour." For scientists to express similar sentiments smacks of cultural imperialism. Unfortunately, science is seen as somewhat larger than life by some who practice it.

Science in the Mirror

> I like my face in the mirror,
> I like my voice when I sing.
> My girl says it's infatuation—
> I know it's the real thing.
>
> —Kit Wright, "Every Day in Every Way"

---

[5]Newton set a bad example, branding poetry as "a kind of ingenious nonsense" but he could have pleaded that he was only paraphrasing Saint Bernard, who in the twelfth century spoke of the "lies of poetry."

The increasing obscurity of much of twentieth-century science leaves ample room for the scientist to present science to HMS in, often justifiably, glowing terms. But HMS may lack the knowledge to meaningfully question the justification. The publicity given to science, the increasing intrusion of science into the lives of HMS, the marvelous consistency of the scientist's cosmos—all these have had their effect on the *scientist's* image of science. Scientists occasionally grossly overestimate the value of their research. Lately this seems to be an occupational risk that is particularly severe among particle physicists. Hyperbole often has budgetary as well as philosophical implications.

False Gods

"My name is Ozymandias, king of kings:
Look on my works, ye Mighty, and despair!"
Nothing beside remains. Round the decay
Of that colossal wreck, boundless and bare.
The lone and level sands stretch far away.

—Percy Bysshe Shelley

HMS should be on his guard against Ozymandiism. In the old days I might have said that there were patent medicine salesmen in town. Consider the interesting example of the superconducting supercollider (SCSC) mentioned in Chapter 12, a project that illustrates not only the clash of scientific and budgetary interests but also the selling of science to the public. The hope was that this mighty machine would induce the "final particle," the Higgs boson, to appear. Professor Leon Lederman, a Nobel prize-winning theoretical physicist, has dubbed the Higgs boson "the God particle." No less! With such PR, one can understand why the thought of this shy but tempting morsel flicking aside its fan and revealing itself for a billionth of a second, is inducing ecstasy in the audience of particle physicists. But the striptease could be a trifle pricey, perhaps in excess of $11 billion. And the lady may not deliver everything expected of her. Proponents of the SCSC spent much effort in lobbying the powers-that-be in order to be assured of sufficient funding for the giant accelerator. They cannot be accused of underestimating the importance of the apparatus. A prominent physicist has used the phrase "propaganda war" in referring to the statements of the SCSC lobby. What is HMS to make of all this? Is the deification of the Higgs boson warranted? Is the expense justified?

I hope that the Higgs boson will one day be found. It will be a magnificent technical and theoretical triumph, like a great Bobby Fisher game. But let's get it in perspective: tax-paying HMS should be aware that the discovery will not help to solve any outstanding problem in medicine, chemistry, biology, or any other field of human endeavor, except theoretical particle physics and possibly cosmology. There is almost no doubt whatsoever that the discovery will be of no practical value to HMS. This is not the discovery of atomic structure, which allowed an explanation of the periodic table and thus created modern chemistry, materials science, molecular physics, and molecular biology.

Admittedly, science has other justifications besides utility, although I have acquaintances who see the concept of "useless" basic science as an effete, elitist avoidance of the real problems with which scientists should be occupied. I do not agree with them, partly because useless science has often turned out to be useful, and partly because I believe in the disinterested search for knowledge. But the "elitist" objection to the search for the Higgs boson has nothing to do with the feeling among many scientists that the significance of the search has been greatly overplayed by some very prominent physicists.

Maybe subconsciously aware of the probable practical irrelevance of the Higgs boson, and seeking to justify the preposterous expense involved in looking for the God particle, another Nobel Prize–winning physicist has gone on record as predicting that when the Higgs boson is discovered, "the news that nature is governed by impersonal laws will percolate through society, making it increasingly difficult to take seriously astrology or creationism or other superstitions." One can see the scene, the day after the Higgs boson is discovered and the *National Enquirer* announces: "Bouncy Boson Bared!" The bookshops will be emptied of occult books, the horoscopes will vanish from the daily papers, the Creationists will crumple, the streets of San Francisco will be littered with the discarded paraphernalia of loony sects. Not the day after, and not a century after. The Age of Enlightenment, and of the reason-worshiping French philosophes, was followed by the irrational Reign of Terror. Many of the heroes of the sixteenth- and seventeenth-century scientific revolution were deeply interested in the occult, in the so-called Hermetic writings, and in magic in general; one only has to look at the lives of John Dee, Boyle, Bruno, Paracelsus, Kepler, and many others. Boyle had an ongoing correspondence with magicians and alchemists and only avoided an attempt to contact the other world by his fear that he would get through to evil spirits. Newton, the heralder of the Age of Reason himself, believed firmly in the mystic aspects of alchemy and of Pythagorean thought. Of the 1752 books in his library, 170 were on occult subjects, including the Kabbala, Rosicrucianism, and plain old-fashioned magic. In 1890, a president of the Royal Society, Sir William Crookes, was inducted into the Hermetic Order of the Golden Dawn. The physicist Sir Oliver Lodge was taken in by pretty girls wearing skimpy costumes and pretending to be psychics, and I know two respectable practicing physicists who believe that Uri Geller can psychically bend spoons. If anything was a publicly trumpeted triumph of reason in this century, it was the experimental confirmation of Einstein's prediction of the bending of light going past the Sun, and yet (amazingly enough), it was followed by the rise of the Nazi party, based on irrational racial myths. The Rosicrucians flourished, millions read their daily horoscope, and Scientology went on its brainless way.

One only has to go through the "astrology" or "occult" shelves of Barnes and Noble or Doubleday to see that, in the day of the genetic code, antibiotics, and nuclear power, the general public still provides a sure market for the irrational. The cold truth is that the Higgs boson will hardly stir HMS. It is far too late in the history of science and society to expect the discovery of a useless (to HMS) particle to create anything more than a weeklong flurry of, probably inaccurate, newspaper articles and a few TV interviews. This is not, for HMS, Newton setting the heavens and terrestrial mechanics in order, or Darwin putting man in his place. This is not the gift of the genetic code. Reason will not advance one millimeter. HMS beware! This is science looking in the mirror and seeing itself very much larger than life. Not infatuation, but true love. Several physicists have doubts about whether the *scientific* gains following the discovery of the Higgs boson will be staggeringly significant. The particle is important, but not that important. I would suggest that titles as grandiose as "The God particle" be reserved for a general method, if there is one, of defending man against viral disease. If anything is guaranteed to bolster the image of reason and counterbalance the negative connotations that science aroused after the atomic bomb, it would be the victory over AIDS and other major diseases. And it is arguable that far more could be done for the mental health of man by diverting some of the billions devoted to accelerators, to plain old-fashioned improvements in the elementary school system. Reason is built block by block in the kindergarten, not proton by proton in a supercollider.

Particle physics is a major part of the scientific endeavor and a towering example of the ability of the mind of man to seek out order in the cosmos; but it is evidently seen as even more than that by at least one of its very distinguished practitioners. In his fascinating book *The First Three Minutes*, Steven Weinberg writes:

> Men and women are not content to comfort themselves with tales of the gods and giants, or to confine their thoughts to the daily affairs of life; they also build telescopes and satellites and accelerators, and sit at their desks for endless hours working out the meaning of the data they gather. The effort to understand the universe is one of the few things that lifts human life a little above the level of farce, and gives it some of the grace of tragedy.

Which puts the human race in its place! Accelerator builders on one side and on the other those who, when not occupied with the daily affairs of life, comfort themselves with myths.

Could we perhaps have a list of the "few" other occupations in life that are worthwhile? I know teachers and nurses who have done more to give life meaning than a lab full of data analysts. And what is to become of my wife, who is an actress, and my acquaintances, among them businessmen, doctors, workmen, writers, painters, housewives, all of whom desperately want to throw down their color-by-number copies of Jack the Giant-killer and build accelerators and sit at their desks for endless hours analyzing data, to give some meaning to their lives? And what of Verdran Smilovitz, the last remaining member of a string quartet, the other members of which had been killed, who played his cello in the ruined streets of Sarajevo to "show that civilization was not dead"? It may be that there is no purpose in the long run, that we are, in a few million or billion years, doomed to extinction, but it is not the accelerator builders who will have lifted us above the level of farce. This example of "Mirror, mirror on the wall . . ." reminds me of J. D. Bernal, the Marxist physicist who predicted, and anticipated with approval, government by scientists, who would "emerge as a new species and leave humanity behind."[6] Well, at least we know one of those who has contributed to the farce.

Overevaluation is not confined to some particle physicists. (I say "some": I have colleagues in the field.) A number of gene sequencers have also started to walk on water. Again: HMS, beware!

Earlier on I mentioned the Human Genome Project, the attempt to determine the complete base sequence of human DNA. It has been claimed, by some of those involved in the sequencing, that this knowledge is the key to what it is to be a human being.

---

[6]Bernal would have felt at home in the kingdom of Bensalem in Bacon's *New Atlantis,* where the only statues raised were to inventors.

The human genome controls most, but not all, of our physical characteristics. But the genome is not the man. A genetic factor is probably involved in our thought mechanisms, but not in a simple fashion. Identical twins have identical genes, yet when brought up in different environments they can exhibit very different behavior.[7] How would a complete knowledge of their genes explain who they are, or who any of us are, in any meaningful *human* sense?

Human beings are more than their genetic components, even if some of those components control most of their physical and some of their mental characteristics. This is *not* to say that there is an additional unidentifiable component, call it the soul if you will. It is just to stress that we are the most complex systems in the universe and that that complexity is partially built-in and partially, but *very substantially*, the result of our interactions with the inanimate world and with other human "systems." To claim that such a system can be completely understood in terms of the series of bases on DNA is not even a scientific statement. Do the sequencers really believe that the workings of the human brain can be completely described once the gene sequence is known? Ask those working on the brain. There is evidence that the development of the central nervous system in the embryo is partly a random process, and that the "wiring" of the brain is affected by the nature of the experiences that feed into it. The fallacy of those who believe that "the genome is the man" is their implicit acceptance that genes are conceptually equivalent to elementary particles. The particles are the *only* basis of all matter; their properties completely control the properties of individual molecules. Genes are not the only determinants of man's nature—and, incidentally, science is not the only jewel in his crown.

Big Fleas Have Little Fleas . . .

The canonization of the genome is consistent with the old reductionist hope that it will be possible to explain everything in Creation in terms of the properties of the simplest components of matter. Reductionism is a creed that tends to be popular with particle physicists.

Part of the accepted system of beliefs of medieval man, at least in Europe, was the existence of the Chain of Being. This was a hierarchical ordering of the Creation from the lowest level, the inanimate world, up through plants and animals to man and the angels. There was, of course, no implication of evolution here. Every member of the chain had been created individually. It is tempting to revive this idea in the light of modern science, with the additional premise that at each level of complexity, systems can be explained on the basis of the properties of the level below them.

With the possible exception of the angels, everything on Earth, and probably in the Heavens, is made of quarks and leptons. Now just as we can explain the structure and behavior of atoms in terms of the elementary particles, and the behavior and structure of molecules in terms of those of atoms, the reductionist's dream is to carry on climbing upward along our ladder and explain the structure and behavior of living cells and living organisms in terms of their component molecules. The top of the ladder now beckons; perhaps we can explain man and his mind in terms of

---

[7]This was one good reason given by Aristotle for doubting the validity of astrology.

the nature of cells—and ultimately quarks? And then science will have explained everything.

Is this a meaningful program? Niels Bohr thought not, stating that biology would never be entirely explained in terms of Chemistry. Schroedinger disagreed, but we don't have to go up to the level of mind to run into difficulties. We saw when dealing with entropy that the macroscopic behavior of a large ensemble of particles cannot be deduced from the properties of the individual particles themselves. At least from a practical point of view, reductionism has to be put on hold for the moment. Some might say forever.

The belief in reductionism has something of the Cartesian desire to start with a simple, verifiable basis and systematically construct from it the whole Creation. We don't yet know if this can be done, but the seemingly rational supposition that complex systems can always be *completely* explained in terms of the properties of their simpler components is at present not proved. In any case, the statement that King's College Chapel was built by J. Quark and Company, even if true, adds nothing whatever to our real understanding of the plan, purpose, and wonder of that glorious structure.

Which does not imply that it is a fruitless exercise to climb up and down the ladder of the sciences. Most levels of science have benefited from looking down to the level below, and the cross-fertilization of different disciplines, as exemplified by biophysics, is a distinctive aspect of twentieth-century science. Standing on the ladder of the sciences, molecular biology, for example, has profited immeasurably from looking down to chemistry and molecular physics.

Science and Philosophy Weigh Each Other Up

In this century, the attempt to understand the nature of the physical world has pushed us toward the borders of philosophy. Can science talk to philosophy? In the age of quantum mechanics this is not a pointless question.

Scientists on the whole have little time for philosophy. Put bluntly, philosophy seems irrelevant to science. And yet the scientist's feeling that the nature of the Big Bang and the subsequent evolution of the universe is approaching a solution, the geneticist's successful probing of the human genome, the doctor's manipulation of embryos, the claims that we are beginning to understand consciousness—all these have maneuvered science into confrontations with the ethics or the epistemology of the philosopher and the morality of religion. Science and philosophy persist in flirting with each other, even if science has rarely profited.

The twentieth century has seen a growing stream of books purporting to link science with God, with meaning, and with our place in the cosmos. At the really woolly end of the spectrum is the science-reveals-God literature. The classic modern statement was made by a geologist, the Jesuit Pierre Teilhard de Chardin, whose book *Man's Place in Nature* (1966) attained temporary cult status. Here we are in the land of unprovable statements, abuse of language, and everything that has given mysticism a bad name. That there is poetry, awe, and mystery in science is part of its appeal, but it is intellectually sloppy to take, for example, a concept like energy, use it in a loose metaphorical sense, and pretend that the word *energy* that appears in your text is the same as, or has the remotest resemblance to, the word *energy* that appears in a scientific paper. To do so is merely a highbrow equivalent of the TV psychic, emoting about "focusing cosmic energy."

More incisive thinkers have considered the relationship between science and philosophy.

"What the Devil Does It All Mean?"

A person may be supremely able as a mathematician, engineer, parliamentary tactician or racing bookmaker; but if that person has contemplated the universe all through life without ever asking "What the devil does it all mean?" he (or she) is one of those people for whom Calvin accounted by placing them in his category of the predestinately damned.

—George Bernard Shaw, *The Adventures of the Black Girl in Her Search for God*

You can almost hear Shaw's Irish accent as he says "divil." Can science help in a search for meaning? Has science any relevance to the "big questions" that the philosophers used to ask? For many, the answer to Shaw's question lies not in science or philosophy but in religion (once defined as man's attempt to speak to the weather). But most of us, looking at history, can see no meaning issuing from religion's diverse gods.

Science can certainly engender awe for the workings of the universe, and this is one step along the path to the rose-tinted sentimentalism of the "How-marvelous-is-thy-handiwork" school of thought, with its supposition of a benevolent Designer. It is a pleasant, warm, harmless trap into which to fall. But Science has revealed nothing that is unambiguously indicative of purpose. If we are honest with ourselves, we have to accept that all evidences of design seem to be due to evolution, coincidence, or an order inherent in nature itself. The laws of motion, Maxwell's laws, the second law of thermodynamics, the double helix—none of them give the slightest support to the hypothesis of a Creator, except insofar as we throw up our hands and say, "Well, it must have been created by someone." And this returns us to the deist fold.

Nevertheless, although faith appears to have almost no interface with science, scientists in this century are being forced to ask questions that are normally regarded as the territory of the philosophers. For example, is science a valid means of examining reality? In this century quantum mechanics has led us a very long way from commonsense reality. One has to say of certain experiments, "If this is reality, it is beyond our comprehension." Are Kant's *phenomena*, the appearance of things, perhaps wildly different from his *noumena*, the true nature of things? If so, is science, along with other human activities, a kind of dream, albeit a very useful dream? In the world of elementary particles, what does "reality" mean? Sense-data? Readings on instrument panels?

> Ah, solving that question
> Brings the priest and the doctor
> In their long coats
> Running over the fields.[8]

---

[8]From "Days," by Philip Larkin, from *The Whitsun Weddings*, Farber & Farber, London (1969).

And, one might add, the philosophers and the scientists. And what do the philosophers make of science's picture of reality?

Through a Glass Darkly

> Philosophers consistently see the method of science before their eyes, and are irresistibly tempted to ask and answer questions in the way that science does. This tendency . . . leads the philosopher into complete darkness.
>
> —Ludwig Wittgenstein, *The Blue Book*

> Philosophy is virtually empty without science.
>
> —A. J. Ayer

For at least 300 years, philosophers have flirted with the scientific method. Wittgenstein recognized this and was not pleased, but his dire warning remains largely unheeded, and it is typical of twentieth-century philosophy that many of its major practitioners, including Wittgenstein himself, had mathematical or scientific roots, and some have taken science very seriously indeed.

In contrast, most contemporary scientists regard philosophy as completely irrelevant to science, or indeed to life in general. Things started differently. Partially spurred on by the rise of physics, three major seventeenth-century philosophers set out to construct universal systems of thought that would provide all-inclusive accounts of the eternal truths of religion, the findings of science, and the problems of metaphysics. They all tried to incorporate ethics within their schemes. Descartes, Leibniz, and Spinoza all failed. And none of them succeeded in linking science with philosophy in a practical sense, although all of them tried.

The nature of knowledge might be thought to be a basic concern not only of philosophers but also of scientists. Indeed, the primary objective of the British empiricists who followed Locke was to find a method of differentiating between, on the one hand, reliable and meaningful propositions concerning reality and, on the other, propositions that were either meaningless or beyond our ability to verify. Most scientists ignored this epistemological swamp until this century, when quantum mechanics seeped over the border into philosophy.

Three central questions that have resisted philosophy since Socrates have become the respectable concern of science as much as of philosophy: What is matter? What is mind? What criteria are there for truth? The scientist has had little help from the philosophers. Modern philosophy is either irrelevant to science or antiscience, in spite of the fact that many twentieth-century philosophers were concerned with the nature of science or mathematics.

The first major book on philosophy published in this century was *Fundamental Laws of Arithmetic* (1903) by the German logician Gottlob Frege (1848–1925). Frege attacked two old problems: What is the nature of mathematical truth? and What are numbers? His work on the bases of mathematics was fatally wounded by Bertrand Russell, and subsequently Frege became concerned with questions of language, a central concern of many philosophers in this century and a turning away from the traditional problems of knowledge (How and what do I know?) to questions of meaning (What am I saying?). Frege is regarded as the founder of "analytical" philosophy, but he has had little direct effect on science.

Bertrand Russell (1872–1970), although not primarily concerned with science, believed in the unity of science and philosophy; they were merely two related aspects of our knowledge of the world. Like Frege, he made his first professional mark with work on the bases of mathematics in his great tome, *Principia Mathematica* (1910–1913), written with A. N. Whitehead. Russell's empirical philosophy was based on *observed qualities* (sense-data), not on the supposition of an unobservable substance or substances that "have" those qualities. He wanted to construct entities out of sense-data, not out of objects that were only inferred, in his eyes unjustifiably, from those sense-data. This approach is reminiscent of Heisenberg and Bohr: only observables mean anything. Unfortunately, Russell then complicated matters by saying that sense-data are different for different people; everyone has his own personal sense-data. Most people, especially scientists, get impatient at this kind of approach to reality. The vast majority of scientists believe in a real, external world that is analyzable through our senses. Russell in the end decided that, to save the foundations of science, one had to accept the existence of permanent objects and substance. Contrary to Wittgenstein, he recommends the scientific method (in *On Scientific Method in Philosophy* [1914]) to those attacking philosophical problems. Philosophers should, he said, be "impregnated with the scientific outlook." But his philosophy has had no practical effect on science.

Moving back to the Continent, one of the most influential philosophers during the past decades was yet another mathematician, the Moravian Edmund Husserl (1859–1938). Husserl's plea to make "all-embracing self-investigation" the basis of science, if acceded to, would probably put an end to most natural science as we know it. Husserl closes his best-known book, *Cartesian Meditations* (1931), with a quotation from Saint Augustine: "Do not wish to go out; go back into yourself. Truth dwells in the inner man." A catastrophic working philosophy for a scientist.

Husserl's student Martin Heidegger (1889–1976) was an opponent of science and the technological society. His temperament led him along the dangerous path of German romanticism, which ended with his flirting with Nazism. Some have excused this aberration by supposing that he looked to the Nazis to combat the anti-cultural, technological society of the time. If this is so it only confirms the suspicion that many brainy men are incredibly naïve.[9] It is disconcerting to think of a philosopher telling his students, as Heidegger did in 1933, "Do not let doctrines and ideas be the rules of your being. The Führer himself, and he alone, is the present and future German reality and its rule." His students were required to salute him. In a 1966 interview for *Der Spiegel*, Heidegger dissociated himself from his statements of the 1930s—absent-mindedly forgetting to mention that he was a Gestapo informer.[10]

Heidegger's writings fall into the highest German tradition of obscurity. Thus his description of the self is "a being such that in its being its being is in question." Which is OK if you like that kind of thing.[11]

---

[9]I have a pamphlet entitled "What Are You Going to Do about It?" written in 1936 by the undoubtedly bright Aldous Huxley, suggesting that the way to defeat Hitler and prevent war was to organize small discussion groups of people dedicated to the idea of peace. Doubtless, Panzer General Rommel trembled in his boots.

[10]I cannot refrain from noting another example of the discrepancy between the professional and political IQs of apparently rational men. Russell, at a public meeting, declared that Kennedy and Macmillan were ". . .wicked and abominable. They are the wickedest people that ever lived in the history of man." Which lets Pol Pot and Himmler off the hook.

[11]Those who have a weakness for debunking may appreciate Paul Edward's opinion that Heidegger has given us "huge masses of hideous gibberish which must be unique in the history of philosophy."

Heidegger saw the great metaphysical question as being why there is anything rather than nothing. He believed that man had lost the sense of being, of just existing. The fault, as he saw it, lay mainly with science and technology. The central question on which he focussed—the meaning of being—is an ancient and haunting one and he faced it straight on, accepting that we come from oblivion and return to oblivion: "these things are not otherwise but thus." It is a stance, reminiscent of Camus and Sartre, that speaks to those among us who see no evidence for the supernatural.

A major twentieth-century philosophical movement which was largely created by scientists and has had a strong, but sometimes questionable, effect on scientists in this century is Logical Positivism. It is far more understandable than most continental philosophies. Ernst Mach, whom we have already met, is often regarded as the spiritual father of the movement, but d'Alembert, the joint instigator of the *Encyclopédie*, and Auguste Comte were not blameless. The school was initially centered in Vienna under the leadership of Moritz Schlick (1882–1936).

Logical positivism is empiricism taken to extremes. It places strong emphasis on observation and measurement. Scientists, especially physicists, often tend to sympathize with the basic principle of logical positivism: that *the meaning of any proposition is its method of verification.* As the Vienna Circle's English disciple, A. J. Ayer, put it: " . . . a sentence is factually significant to a given person if, and only if, he knows how to verify the proposition which it purports to express." This has clear echoes of Heisenberg's approach to quantum mechanics, in which the "meaning" of an observation is taken to be simply the result of an experiment done on a known system with a known piece of apparatus and a defined procedure. The Vienna Circle considered that two kinds of verification were acceptable: experience and logical necessity. Propositions which are true in the limited sense that they are condemned to be true by the meaning of the symbols contained within the proposition, would be termed *analytic* propositions, in Kant's vocabulary. On the other hand, if I observe that any two of a given set of bodies fall with the same acceleration, then I am using experience to verify the proposition that all bodies in that set fall with the same acceleration. Propositions of this type, depending for their truth on the observed facts of nature, are examples of Kant's *synthetic* propositions. This reduction of possible knowledge to either tautologies or propositions which are empirically verifiable is directly traceable to David Hume who asked of a proposition: "*Does it contain any abstract reasoning concerning quantity or number?* No. *Does it contain any experimental reasoning concerning matter of fact and existence?* No. Commit it then to the flames: for it can contain nothing but sophistry and illusion." These criteria cut the ground from under metaphysics. Questions such as the meaning of existence are reduced to nonquestions, since they are not susceptible to either logical analysis or experimental verification. More down-to-Earth questions also lose all meaning, such as, "What is right and wrong?" As Schlick said, such questions have no answers, not because they are difficult but "simply because they are not questions"! Schlick considered that a practical system of ethical behavior could be built up from man's experience, but it had no *philosophical* validity. This is a difficult attitude to challenge.

Some outstanding physicists accepted logical positivism, either implicitly or explicitly. They were in practice phenomenologists—dealing solely in observables and the observed relationships between them; they had given up on the *noumena* of Kant, the supposed hidden reality. The same "black box" attitude to reality shows up in the behaviorist school of psychology, associated particularly with J. B. Watson and B. F. Skinner, in which it was not done to speak of mental states, since they could not be directly observed.

A difficulty that the logical positivists never solved was how to accommodate science in their system. Since most of them had been scientists, they wanted to preserve science as meaningful, but it was not easy. Many of the basic concepts of Science cannot be verified either logically or by observation. As we saw, Mach rejected the concept of atoms because they were neither observable nor, in his opinion, logically necessary. Kaufmann famously missed out on a Nobel Prize because he was a logical positivist in outlook and refused to interpret his experiments in terms of a particle (the electron) that he could not see (see Chapter 12).

If you want to annoy a logical positivist, ask him if the verifiability principle stands up to its own criteria for verifiability. You could also point out that unless you *first* understand a proposition, you have no idea how to verify it. This means that a proposition may really have meaning, but because you don't understand it you don't know how to verify it, and thus assume that it has no meaning.

The logical positivists belong to the school of philosophers who have tended to concentrate less on problems than on how to solve them. The tradition goes back at least as far as Hume. In this context the so-called "scientific method," which has the reputation of being a successful means of problem solving, has had an attraction for some modern philosophers. Thus the "logic of scientific discovery" was long been a primary concern of the philosopher Karl Popper. We met Popper in Chapter 2, where we discussed the best-known aspect of this work, the question of the verifiability of scientific theorems. In the scientific community, Popper is probably by far and away the best-known and respected modern philosopher, partly because he is a blessed example of the fact that writing impenetrable prose is not a precondition for being a German-speaking philosopher.

As might be expected, modern American philosophy contains a strong streak of practicality, often bringing it closer to science than modern European philosophy. The recognized leader of the pragmatic school of philosophy was William James (1842–1910), the older brother of the novelist Henry James. To sum it up in its most concise form: *A proposition is true if the results that follow from it are useful.* This is an ideal philosophy for a practicing scientist—or for an automobile manufacturer. Since scientists usually continue using a theory as long as it is useful, they are implicitly using the criterion of pragmatism to determine whether the theory is true. Once the theory does not give correct (useful) results, it is discarded. John Dewey (1859–1952) was even nearer to the scientific frame of mind, in that he stressed the need to accept that any idea could be proved to be mistaken; one might approach, but possibly never reach, truth.

A concern with the method of finding the truth often goes hand in hand with an interest in the way that we use language. In fact, some would see modern Anglo-Saxon philosophy as far too concerned with the meaning of words and far too little concerned with the classic moral questions that still remain unanswered. But the dilemmas of quantum mechanics have sharpened our need to define our terms.

The outstanding name in linguistics in recent years has been that of Noam Chomsky, who has revolutionized our attitude toward language. His theory of the deep structure of languages suggests that we have inborn, inherited grammatical patterns which are common to all languages. Chomsky addresses the nature of knowledge: " . . . the general character of knowledge, the categories in which it is expressed or internally represented, and the basic principles that underlie it, are determined by the nature of the mind." This is a direct negation of Locke's belief that there is no innate knowledge of either principles or ideas. Locke famously insisted that when we come into the world our minds are "white paper, void of all characters, without any ideas," a view of man echoed by Sartre's insistence that there is no built-in human nature, that we are completely free. Anti-Lockean views were expressed by the great physicist Wolfgang Pauli (1900–1958) in an essay on Johannes Kepler: "The process of understanding in nature, together with the joy that man feels in understanding, . . . seems therefore to rest upon a correspondence, a coming into congruence of preexistent internal images of the human psyche with external objects and their behavior." The suspicion that man imposes a structure on the universe is related to the controversies, considered in Chapter 9, surrounding the ideas of philosophers such as Kuhn and Feyerabend.

If anyone delved profoundly into the meaning of language—obsessively and unnecessarily so, in Karl Popper's view—it was Wittgenstein. He had been deeply impressed by the work of two of the greatest scientists of the late nineteenth century, Heinrich Hertz and Ludwig Boltzmann. It was not so much their science that concerned him as their attitude toward Science. Boltzmann in particular felt that all that we were really doing when we stated physical laws was using a series of linguistic representations of reality. To relate force and mass, as Newton had done in his laws of motion, was to relate labels in such a way that we could use the relations for predictive purposes. To read anything more into the terms *force* and *mass* was to presume more than we can know. One is reminded once again of the logical positivists although it is a mistake to put Wittgenstein in their camp. Wittgenstein put it this way: "Philosophy is a battle against the bewitchment of our intelligence by means of language."

Wittgenstein was basically antiscientific. He knew that science was partly driven by a desire to generalize, and he rejected generalization. Scientific questions were of no great interest to him; they merely addressed the working of the natural world.

Wittgenstein spent much of his later years examining the way in which language may shape our reality. This is not a subject that is irrelevant to science. The use of words like *matter, space, mass* or *life*, is charged with connotations that arise from our education and the way those words have been used in the past. It is worth remembering that Einstein's questioning of the meaning of the word *simultaneity* was more than idle philosophizing.

Is Philosophy Relevant to Science?

> There have been times when science and philosophy were alien, if not actually antagonistic to each other. These times have passed.
> —Max Planck (1936)

I have done a gross disservice to some of the outstanding intellectuals of this century. I have taken video clips of their life's work or summarized tiny pieces of complex philosophies in a few glib phrases, but, as explained earlier, in certain areas this book is not a map, but a sign-post.

Ask the next scientist you meet if philosophy has affected his life or work. Get ready for a blank stare. This complete avoidance of philosophy may have to change. We must be prepared to face a growing number of philosophical problems arising from the ethical aspects of modern science. Less controversially, we saw the difficulties which quantum mechanics has introduced into our ideas of reality, and, for some of us, the nature of reality is, perhaps, *the* philosophical question. Shimony, in *The Reality of the Quantum World* (1988), writes, "We live in a remarkable era in which experimental results are beginning to elucidate philosophical questions." He was referring to the challenging fact that the questions raised by the fundamentals of quantum mechanics are now impinging very directly on the nature of the relationship of the observer's perceptions to the external "real" world. This is an old question, but we have hardly advanced since, on the one hand, Locke wrote that the "bulk, figure, number, situation, and motion" of bodies are real properties "whether we perceive them or not," and, on the other, Berkeley avowed that " . . . as to what is said of the absolute existence of non-thinking things without any relation to their being perceived, that seems perfectly unintelligible . . . nor is it possible they should have any existence, out of the minds or thinking things which perceive them." There is no knowing where, if anywhere, these two divergent roads to the understanding of matter will lead. But, because of quantum mechanics, they are becoming scientific as well as philosophical questions.

An Endless Road or a Dead End?

My feeling is that the attempt to understand the basic nature of reality may well be a losing game. It could be that we have reached, or are fast approaching, the limits of human comprehension. Dare I suggest that we may never be capable of forming a "commonsense," easily visualized picture of what we choose to call reality? Perhaps because we are part of the system or because we haven't got the right hardware. Thomas Hobbes wrote, in *De homine* (1658), that language "is the connexion of names constituted by the will of men to stand for the series of *conceptions* of the things about which we think" (my italics). The conceptions are, perhaps, as far as we can get. It is possible that, as in the case of quantum mechanics, we will just have to be satisfied with theories that give the right answers even if we don't really understand why.

As a working scientist I sympathize with the approach of the American philosopher, V. O. Quine (1908–), who sees " . . . the conceptual scheme of science as a tool, ultimately, for predicting future experience in the light of past experience." Up to now that tool has been enormously effective.

A Note on Complete Uncertainty

The foundations of science are uncertain, but mathematics has often been looked to as a system which was potentially foolproof. By which I mean that if a suitable set of axioms could be found, and ways of manipulating them agreed upon, then it would be possible to state unambiguously whether a statement involving the subjects covered by the axioms, was true or false. Thus it was taken for granted that once the integers, and the rules for handling them (addition, multiplication, etc.), had been defined, any statement about the integers could be proved to be true or false, the proof only involving the axioms. Mathematics thus had a certainty denied to any other human construction of the mind; it was a completely self-sufficient system, requiring no extraneous factors to be taken into account in proving *anything* that lay within the scope of its basic axioms. In particular it was superior to science in this respect—especially after Popper.

The man who upset the applecart, in 1931, was the Austrian mathematician and logician Kurt Gödel (1906–1978). There are a number of equivalent ways of stating his result, perhaps the clearest of which is that no *finite* set of axioms is sufficient to form the basis for all true statements concerning integers. No matter how many axioms form the basis of mathematics, there will be statements about integers which cannot be proved to be either true or false. This is an amazing finding and for mathematicians at the time, a severe psychological jolt.[12]

In addition to the suspicion that we may never fully understand the physical universe, we are now faced with the fact that we are forever limited in our ability to construct a noncontradictory system of mathematics. The twentieth century has been maliciously unkind to Man's intellectual pretensions.

Toward the end of his life, Gödel feared that he was being poisoned, and he starved himself to death. His theorem is one of the most extraordinary results in mathematics, or in any intellectual field in this century. If ever potential mental instability is detectable by genetic analysis, an embryo of someone with Kurt Gödel's gifts might be aborted.

---

[12]Incidentally, Gödel does not say that there cannot be a limited system of axioms, within the whole set of axioms, which cannot be used to prove any statement *involving only those axioms*, provided that we can hope to use the remaining axioms, if we are in trouble proving that something is true or false *within the framework of the smaller set.* The problem is that, if we take all the axioms, we have none left over to help us.

X

Cross My Hand with Silver

In which some thoroughly unreliable predictions are hazarded.

38 | The Future

> We all believe that it isn't possible to get to the moon; but there may be people who believe that it is possible and that it sometimes happens. We say: These people do not know a lot that we know. And, let them never be so sure of their belief—they are wrong and we know it.
>
> If we compare our system of knowledge with theirs then theirs is evidently the poorer one by far.
>
> —Ludwig Wittgenstein, "On Certainty"

This will be a very short chapter—not because the apocalypse is upon us but because the success rate of futurology is probably on a par with that of astrology. There was at one time, especially in the 1960s, a fashion for committees of savants to issue documents purporting to predict the broad lines of the future. The authors usually considered it advisable to use extensive data processing, as if shaky assumptions could somehow be corrected by expensive hardware.

I am prepared to bet a modest sum that there is not *one* document issued before the fall of the Berlin Wall that predicts even the rough timing of that event, or the subsequent disintegration of Eastern Europe. On the other hand, it is quite possible that an astrologer did. After all, the sixteenth-century French court futurologist, Nostradamus, in his last prophecy, foresaw that England would be the dominating world power for 300 years, which, if taken as the period from Elizabeth to Victoria, works out fairly well—plus or minus, as the scientists say. (He added that the "Portuguese will not be content," which I'm quite prepared to believe.) He certainly did much better than Wittgenstein. Really successful prophecy went out with the Old Testament, partially because human history is chaotic in the scientific sense, continually demonstrating oversensitivity to initial conditions, a fact appreciated by Blaise Pascal when he declared that the whole course of history would have been changed if Cleopatra's nose had been a different shape. Vive le nez.

Predicting the future of basic science is a losing game. The twenty-first century may bring a successful unified field theory, and perhaps hidden variables will save quantum mechanics. We might really understand the Big Bang, dark matter, and the mind. My guess is that there will be multiple theories of everything, several for each universe, and that foreigners will be blamed for everything, in all universes. But I really don't know, and this is typical of the marvelous unpredictability of basic science. Any day a stranger may knock on the door and irrevocably change your life.

As to applied science, there are a few safe bets about developments in the early years of the coming century: the scientifically based development of new materials; the understanding and control of the immune system; the spread of gene therapy; ecology-related research, and so on. A fairly obvious list. The crystal ball also re-

veals that computers will get faster, paint will become more durable, and detergents will wash even whiter. Inessential gadgets will proliferate and remain socially essential.

Many of the new directions in applied science will have to be carefully monitored by HMS. We have spoken of some of these areas, particularly those based on
the human genome, but the information explosion is also a phenomenon that needs
watching, and one that has already begun to change our lives.

A combination of basic science and technology has created what has been
termed the third industrial revolution. We watch distant wars in realtime in our
living rooms. I can sit down opposite a computer terminal in Haifa and tap into
huge databases, assembled in distant countries. My computer and I are part of a
vast pseudoneural network that is spreading over the globe with almost threatening
speed. The information explosion is changing the way we live. The availability of
information can be a blessing. The ability to access libraries and data that are in another town or another continent is not only a gift to academics, doctors, media
workers, and writers; it also expands dramatically the educational means available
to nine-to-five man. Without moving from your armchair, you can make use of a
huge library, with recorded plays, concerts, documentaries, and sports events.

Information is flowing into terminals everywhere, but this is not always a good
thing. Difficult questions of privacy and of censorship inevitably arise when an
easily accessible worldwide network exists. How much about my life and health,
recorded in a government computer, should be available to anyone who is curious?
Should pedophiles and rabid racists be allowed to channel their messages into
every home?

A subtler question is that of creativity. I'm not at all sure that I know what terminal watching is doing to children's minds. Observe a child, sitting for an hour or so
in front of a video game—usually consisting of some brutal, strutting figure beating
the living daylights out of an equally punklike automaton. My own children have
avoided this plague, but will theirs? Will they lack the creativity that goes into
building a sledge, or a primitive boat, out of whatever materials are available? Are
we receiving too much and creating too little? A youngster reading this might say,
"I don't believe it. He wants to ban computers and go back to 'the good old times.'"
I don't. As I sit typing this on a word processor, I thank Faraday for technology, as I
do when I am ordering airline tickets or getting into my car, but I do not believe that
the new times are necessarily better in every respect. I do believe that the monitor-
ing of information technology is a legitimate concern of HMS, and that, where children are concerned, we must strive to maximize the educational and recreational
advantages of the "information highway" and fight to minimize its deleterious effects. Making sadistic pornography available to children does not come under my
definition of free speech.

A final thought. The two most self-confident activities of mankind are religion
and politics. No one is surer of himself than a believer. This self-confidence is
based on a fundamental rigidity, a stubborn refusal to really hear the other side, to
admit for one moment that there might be something *basically* wrong with the accepted dogma. A believer may be prepared to say that we are all the children of one
God, but he doesn't usually switch from Islam to Catholicism. Science, on the other
hand, is completely open-minded—despite the history of inertia. Any monument
can be demolished, any belief forsaken. It is exactly this liberating acceptance of

the possibility that our minds can mislead us that underlies the magnificent successes of science. Scientists are not invariably ecstatic when their scientific beliefs are undercut by better theories or new facts. But in the end, the scientific community gives in to change because, on the average, we refuse to be irrational—or to be seen to be irrational by our colleagues. It is the (reluctant!) willingness to be shown to be wrong that has so often led us in the direction of being partially right. Science, like art, is continually seeing the world anew. This is part of the joy of science.

Annotated Bibliography

I have gone for readability and accessibility, in general avoiding referring to sources that are only available in specialized libraries or archives. I have kept the list short, since almost all the books mentioned serve as jump-off points to other sources.

At the university level there are a plethora of undergraduate textbooks dealing with the scientific ideas presented in this book but they are not usually the kind of books that appeal to the layman.

The lives of hundreds of scientists and short summaries of their work are contained in the *Dictionary of Scientific Biography,* the sixteen volumes of which were published between 1970 and 1980, by Scribner. Scientists who died after 1980 are not included.

Part One

Hollis, Martin. *Invitation to Philosophy.* Basil Blackwell, 1991. A short, lucid, jargon-shunning introduction. A pleasure to read.

Kuhn, T. S. *The Structure of Scientific Revolutions.* University of Chicago Press, 1970. An easily digestible classic, although there is much to argue with in his approach.

O'Hear, Anthony. *An Introduction to the Philosophy of Science.* Oxford University Press, 1989. A good starting point for those interested in the subject. Contains a useful bibliography.

Popper, Karl. *Conjectures and Refutations.* Routledge and Kegan Paul, 1962. A basic course in Popper's view of things.

Popper, Karl. *The Logic of Scientific Discovery.* Hutchison, 1959. A modern classic. Helpful on the theory of probability.

Quinton, Anthony. *Bacon.* Oxford University Press, 1990. A short, excellent summary of Bacon's life and thought.

Williams, Bernard. *Descartes, The Project of Pure Enquiry.* Penguin Books, 1985. A book that cuts no philosophical corners. Not always easy, but stimulating.

Part Two

Brecht, Berthold. *The Life of Galileo.* Methuen, volume 1, 1960. This play, which presents Galileo in an unflattering light, might get you arguing about whether he should have recanted or not. "He who does not know the truth is merely an idiot. But he who knows it and calls it a lie, is a criminal".

Cohen, I. Bernard. *The Birth of the New Physics.* Penguin Books, 1987. A readable and scholarly account of the transition from the physics of Aristotle to the era of Galileo and Newton.

De Santillana, Giorgio. *The Crime of Galileo.* Heinemann, 1958. A good account of one of the critical intellectual confrontations in scientific history.

Diderot, Denis. *Lettres à Sophie Volland.* Gallimard, 1984. Nothing to do with science, but a unique window on the man behind the *Encyclopédie,* and on his times. The English translation appears to be out of print.

Fauvel, J., R. Flood, M. Shortland and R. Wilson, eds. *Let Newton be!* Oxford University Press, 1990. An entertaining collection of essays on various aspects of Newton's work. The accounts of his non-scientific or pseudo-scientific interests are particularly interesting.

Gay, Peter. *The Enlightenment, An Interpretation.* W. W. Norton & Co., 1977. A fascinating, authoritative text which includes a detailed discussion of Newton's role, as seen by a historian.

Merton, Robert K. *Science, Technology and Society in Seventeenth-Century England.* Harper and Row, 1970. This is a classic study by one of the leading experts in the field.

Popper, Karl. *Objective Knowledge.* Clarendon Press, Oxford, 1979. Read the section showing how Kepler generalized on the basis of sketchy data, and how Newton built on Kepler's laws knowing them to be approximations to the truth. So much for "the scientific method."

Turnbull, H. W., ed. *The Correspondence of Isaac Newton.* Cambridge, 1960. The letters bring the reader a little closer to the introverted human being behind the stereotype of the "great scientist."

Part Three

Friedel, Robert, and Paul Israel. *Edison's Electric Light: Biography of an Invention.* Rutgers University Press, 1985. An excellent book showing how a brain wave was turned into a commercially viable product.

Latour, Bruno, and Steve Woolgar. *Laboratory Life.* Princeton University Press, 1986. This highly unusual book is a sociological study of a major scientific laboratory by a French philosopher and an English sociologist. Skeptical in tone, but almost always objective, it should be read by all scientists (to embarrass them), and all interested laymen (to amuse them).

Merton, Robert K. *Sociology of Science.* Chicago University Press, 1973. An absorbing book that has a solid, well researched feel about it.

Williams, L. Pearce. *Michael Faraday.* Basic Books, 1965. An interesting, solid biography that places emphasis (perhaps too much) on the influence of *Naturphilosophie* on Faraday's thought.

Part Four

Gordon, J. E. *Structures.* Penguin Books,1978. A dated but very readable, informative and completely non-technical tour of part of the world of materials.

Harrison, J. F. C. *Late Victorian Britain.* Fontana Press, 1990. An interesting account of the social circumstances and shifts at the end of the 19th century.

Knight, David. *The Age of Science.* Blackwell, 1986. A good summary of the advance of science in the nineteenth century.

Lucretius. *On the Nature of the Universe.* Trans. by Ronald Latham. Penguin Books, 1951. Of no scientific value, but a landmark in man's attempt to understand reality without invoking the supernatural.

Wilson, David. *Rutherford, Simple Genius.* MIT Press, 1983. Ambiguously titled biography of the great no-nonsense experimentalist.

Part Five

Atkins, Peter. *The Second Law.* Scientific American Library, 1984. The author is acknowledged as an outstanding science writer. This book, on entropy and energy, will require the nonscientist to concentrate, but it's worth it.

Gleick, James. *Chaos: Making a New Science.* Abacus, 1987. An excellent introduction to chaos, spiced with anecdotes about those who have contributed to the theory.

Honderich, Ted. *How Free Are You?* Oxford University Press, 1993. A really clear discussion of the problem of determinism.

Mandelbrot, Benoit. *The Fractal Geometry of Nature.* W. H. Freeman, 1982. The inventor of the term *fractal* reveals the wonders of the fractal world.

Ruelle, David. *Chance and Chaos.* Penguin Books, 1993. A first class work of popularization by a distinguished theoretical physicist.

Wilson, S. S. "Sadi Carnot," *Scientific American,* vol. 254 (1981): 134. Carnot, who more or less invented the second law, could have been a TV "personality" had he lived today.

Part Six

Bannister, Robert C. *Social Darwinism: Science and Myth in American Social Thought.* Temple University Press, 1979. Chronicles the rise of the phenomenon in the United States, its origins in the work of Herbert Spencer, and its connections to eugenics and racism.

Darwin, Charles. *On the Origin of Species.* Harvard University Press, 1975. Facsimile of the first edition. This book hardly needs an introduction from me. Unless you are a naturalist, best taken in small doses.

Dawkins, Richard. *The Blind Watchmaker.* Penguin Books, 1991. A superbly readable defense of Darwinism by a confirmed evolutionist and atheist. A model of scientific popularization.

Gosse, Edmund. *Father and Son.* Penguin Books, 1989. A classic Victorian autobiography, which includes a touching account of one man's unsuccessful struggle to accommodate both his religious faith and the new doctrine of Darwinism.

Kamin, Leon. *The Science and Politics of IQ.* Wiley, 1974. A sober, reasoned book that has had much influence on sober, reasonable people.

Kitcher, Philip. *Abusing Science: The Case Against Creationism.* MIT Press, 1982. If you need ammunition against the creationists, you will find plenty here.

Lucas, J. R. *Wilberforce and Huxley, A Legendary Encounter. Historical Journal,* 22(1979): 313. Against my principles I include an article in a learned journal. My reason is that this account of the championship fight suggests that it ended nearer a draw, rather than the knockout usually attributed to Huxley.

Miller, Jonathan. *Darwin for Beginners.* Random House, 1982. The doctor who became a theater director occasionally returns to his roots, in this case producing an amusing popular book illustrated with cartoons.

Olby, Robert. *The Path to the Double Helix.* University of Washington Press, 1975. A very thorough analysis of the different trails that lead to the great discovery.

Tennyson, Alfred, Lord. *In Memorium.* Norton, 1973. Apart from the famous poem this edition includes a wide range of critical essays, including discussions of the effect of mid-Victorian science on the poem and on the poet, who was a friend of Darwin.

Watson, James. *The Double Helix.* Weidenfeld and Nicolson, 1968. The story of one of the major advances in the history of science as told by one of the men responsible. Fascinating science, but not always completely fair when it comes to personal matters. If you want the story as Crick might have told it see the book by Olby.

Part Seven

Badash, Lawrence. "Werner Heisenberg and the German Atomic Bomb." *Physics and Society,* 16(1987): 10. Some claim that he deliberately slowed down the project, others say not. Perhaps we will never know.

Boscovich, R. *A Theory of Natural Philosophy.* Translation of the 1763 edition. MIT Press, 1966. Plenty of interesting raisins in this imaginative text.

Einstein, Albert, and Leopold Infeld. *The Evolution of Physics.* Simon and Schuster, 1967. Originally published in 1938, this is a lively and comprehensible history of the main theories in physics, with a touch of philosophy thrown in.

Feynman, Richard. *The Character of Physical Law.* Penguin Books, 1992. Feynman was a genius—and he knew it. He was one of the most gifted physicists and explainers of physics of the past decades. This book discusses general issues in physics and also shows us how Feynman's brilliantly intuitive mind worked.

Hey, Tony, and Patrick Walters. *The Quantum Universe.* Cambridge University Press, 1987. A visually attractive and absorbing account of the history of the quantum theory, with emphasis on its role in particle physics.

Weinberg, Steven. *The Discovery of the Subatomic Particles.* W. H. Freeman & Co., 1990. Written for nonscientists. Primarily the story (well-told) of the proton, neutron and electron. Quarks barely get a mention.

Part Eight

Barrow, J. D. and F. J. Tipler. *The Anthropic Cosmological Principle.* Oxford University Press, 1986. Try reading this and see whether it convinces you that the universe was designed for man.

Einstein, Albert, and Peter Smith. *Relativity: The Special and General Theory.* 1917. If you can get hold of it, this is relativity straight from the man himself. The special theory in particular is lucidly explained.

Gribbin, John. *In Search of the Big Bang.* Corgi, 1987. Gribbin is a well-known and effective popularizer.

Lovelock, James. *Gaia.* Oxford University Press, 1982. A book by the originator of the theory. A rich source of information on planet Earth, although personally I'm not convinced by the basic idea behind Gaia.

Miller, Ron, and William K. Hartmann. *The Grand Tour: A Traveller's Guide to the Solar System*. Workman Publishing, 1981. Lavishly illustrated with photographs and spectacular graphics.

Pais, Abraham. *Subtle Is the Lord*. Oxford University Press, 1982. This is an authoritative description of Einstein's work, with considerable reference to his personal life. A layman will find some parts obscure and may get more out of the book by Clifford Will.

Weinberg, Stephen. *The First Three Minutes*. Fontana, 1983. Weinberg is a Nobel Prize-winning theoretical physicist with an easy writing style. This account of the Big Bang has been overtaken by subsequent theory but remains a good overall picture of the conventional picture of the Big Bang after the first one-hundredth of a second.

Will, Clifford. *Was Einstein Right? Putting General Relativity to the Test*. Basic Books, 1986.

Part Nine

Broad, William, and Nicholas Wade. *Betrayers of the Truth*. Simon and Schuster, 1982. The authors revel in cases of scientific fraud and claim that the scientific establishment is not equipped to detect fraud. They tend to wildly overstate their case, but the incidents they discuss make uncomfortable reading for a professional scientist.

Carson, Rachel. *Silent Spring*. Houghton Miflin, 1963. A monument to her fight for ecological sanity in a country where big business didn't want to know about anything but profits.

Hoffmann, Roald. *The Same and Not the Same*. Columbia University Press, 1995. The author is a Nobel Prize-winning chemist, with a poet's eye. This delightful book of short essays linking the scientific and humanistic worlds is a good book to curl up with.

Holton, Geraldo. *Science and Antiscience*. Harvard University Press, 1994. A distinguished author skillfully undermines those antiscientists whose case is built on unreasoning prejudice.

Johnson-Laird, Philip. *The Computer and the Mind*. Fontana Press, 1993. Despite my avoidance of the nature of the mind, I cannot resist recommending this readable, but not always easy, book on mentality because it is an important illustration of the effect of science (in the guise of the electronic computer) on man's image of himself.

Maxwell, Nicholas. *From Knowledge to Wisdom*. Oxford University Press, 1984. The author contends that the fact that science is divorced from ethical values is not a point in its favor but a guarantee of disaster. I disagree, but you may think otherwise.

Morgan, Michael, Joseph Moran, and James Wiersma. *Environmental Science*. William C. Brown, 1993. A wide-ranging, well-illustrated discussion of the environment and the dangers it faces.

Olsen, Richard G., ed. *Science as Metaphor: The Historical Role of Scientific Theo-*

ries in Forming Western Culture. Wadsworth, 1971. A collection of generally absorbing essays that covers a wider range than this book since it includes psychology.

Russell, Bertrand. *Why I Am Not a Christian.* Routledge, 1992. Not too relevant to this book, apart from a discussion of determinism, but how lucky the secular are to have this sharp-witted iconoclast on their side.

Russell, Bertrand. *A History of Western Philosophy.* Allen and Unwin, 1961. Lively, holds the reader, but hasn't much to say about modern philosophers. My professional philosopher friends tell me that Russell was sometimes too sure of himself, but don't let them put you off.

Ziman, John. *An Introduction to Science Studies: The Philosophical and Social Aspects of Science and Technology.* Cambridge University Press, 1985. An impressive survey.

Index